D1270626

OXFORD ENGINEERING SCIENCE SERIES

THE OXFORD ENGINEERING SCIENCE SERIES

Engineering Rheology

Second Edition

ROGER I. TANNER

Professor of Mechanical Engineering
University of Sydney

OXFORD
UNIVERSITY PRESS

OXFORD

UNIVERSITY PRESS

Great Clarendon Street, Oxford OX2 6DP

Oxford University Press is a department of the University of Oxford.
It furthers the University's objective of excellence in research, scholarship,
and education by publishing worldwide in

Oxford New York

Athens Auckland Bangkok Bogotá Buenos Aires Calcutta
Cape Town Chennai Dar es Salaam Delhi Florence Hong Kong Istanbul
Karachi Kuala Lumpur Madrid Melbourne Mexico City Mumbai
Nairobi Paris São Paulo Singapore Taipei Tokyo Toronto Warsaw

with associated companies in Berlin   Ibadan

Oxford is a registered trade mark of Oxford University Press
in the UK and in certain other countries

Published in the United States
by Oxford University Press Inc., New York

© Roger I. Tanner, 1985, 1988, 2000

First edition 1985
Revised edition 1988
Second edition 2000

A catalogue record for this book is available from the British Library

Library of Congress Cataloging in Publication Data

Tanner, Roger I.
Engineering rheology/Roger I. Tanner.—2nd ed.
p. cm.—(Oxford engineering science series)
Includes bibliographical references and indexes.
1. Rheology. 2. Kinematics. 3. Non-Newtonian fluids. I. Title. II. Series.
TP156.R45.T36 2000   620.1′064 21—dc21   99-045852

ISBN (Hbk)
0 19 856 473 2

Typeset by Newgen Imaging Systems (P) Ltd., Chennai, India

Printed in Great Britain on acid-free paper by
T.J. International Ltd.

PREFACE TO SECOND EDITION

This edition again set out to provide a guide for those wishing to make predictions about the behaviour of non-Newtonian materials in engineering; it contains much new material following the rapid growth and change in the subject of rheology over the past 15 years.

The basic orientation has not been changed, so the original preface is still relevant, but the scope has been widened, and some important topics not discussed in earlier editions are now covered. In particular:

(a) The question of fluid slip at solid walls is discussed in Chapter 3. It is clearly of paramount importance to know whether or not slip occurs when solving boundary-value problems relating to practice.

(b) Much more discussion of materials models with a yield stress is included in Chapter 4. These models are frequently useful in practice and also there are some interesting connections with plasticity theory.

(c) A section on suspensions of particles has been included in Chapter 5; this is a very active field of research and application.

(d) Chapter 8 has been completely recast in view of the extremely rapid growth of the use of computing in rheology. While I believe the pioneering days of computing are now over, the intelligent use of these methods is still not routine, and they need to be applied carefully, which is one theme of this chapter. For example, appropriate boundary conditions for viscoelastic flows are now included. Regretfully, it has not been possible in a single volume to do justice to the rapidly developing field of computation that deals with direct 'molecular' simulation—for example, molecular and Brownian dynamics—and this section is therefore brief.

(e) Considerable additions to the stability section in Chapter 10 have been made in order to reflect recent progress in this difficult area.

Many other additions and deletions have been made, many new and up-to-date references and some new problems have been included, but no attempt to substantially cover areas such as liquid crystals and electrorheology was possible. In the field of thermal rheology much research remains to be done, and so Chapter 9 remains fairly conservative in tone.

What I believe has emerged over the 20 years since I began to write this book is a reasonable consensus on polymer fluid description, and an ability to solve applications via computing. Happily, many basic puzzles also remain to be solved, especially in the melt–solid transition area and in application to non-polymeric systems.

In closing this new Preface I wish to thank many helpers and collaborators, but especially my colleague Professor Nhan Phan-Thien FAA who has done so much to further rheological research on so many fronts. The sudden death in 1994 of my former co-worker, Jack Pipkin, is also noted with my deepest regrets. The Acknowledgements section also records my thanks to those who assisted in reading and criticizing parts of the work; Paul Briozzo did the new drawings and Lyn Kennedy did a lot of the word processing—to them, my gratitude.

Finally, my heartfelt thanks go to Ena, who saw virtue in educating me, and to Elizabeth, who has helped in so many ways while research and book were being produced. To both I dedicate this edition, which is being launched into the second century and the new millenium of Rheology.

Sydney, N.S.W.   R.I.T.
September 1999

PREFACE TO FIRST EDITION

This book sets out to provide a guide, with some illustrations, for those who wish to make predictions about the behaviour of non-Newtonian fluids in engineering. It presupposes some previous contact with fluid mechanics and heat transfer, and so is intended for use at the graduate level by engineering students and others. Later chapters go beyond what it is reasonably possible to teach in a typical one-semester course and lead on to currrent topics of research, including the devising of effective numerical methods for flow prediction. Problems are given at the end of each chapter, and a booklet of solutions may be obtained from the author for a nominal charge.

The plan of the book is centred around kinematics. From the coarse-scale point of view fluid flow often looks much the same whether it is Newtonian or non-Newtonian, but there are very significant differences in detailed flow behaviour and in the levels of stress that occur in the two cases. After surveys of some typical non-Newtonian behaviour and of classical continuum mechanics, the latter using mainly Cartesian tensor and matrix notation for simplicity, the main kinematic division into shearing and stretching flows is introduced in Chapter 3. In the simplest cases no specific constitutive equation or rheological model is needed for application to typical experimental situations. Many practical cases are not simple, so two chapters (4 and 5) are devoted to the development of constitutive equations which are expected to be useful in predictions. Both continuum and molecular methods of devising constitutive laws are discussed because the combination of the two is believed to be more powerful than either used alone. The need to have some idea of the basic kinematics before electing to use one or other constitutive equation is emphasized. Two chapters on applications then follow; one on nearly-viscometric or shearing flows, exemplified by lubrication and calendaring, and one on stretching flows, exemplified by spinning and film-blowing. The choice of examples here was difficult and many interesting cases have had to be left out; the main lines of attack for other cases follow once the principal kinematic features have been discovered. When a change of flow type takes place, for example, from one viscometric flow to another, or from a viscometric flow to a stretching flow, then one is practically forced to use numerical methods and this is recognized in Chapter 8. Again, only a few examples are analysed in detail. Up to this point the effect of temperature on flows has been (mainly) ignored, but it is of such great importance that Chapter 9 is devoted to thermal effects. The introduction of this factor at this point is convenient since it is extremely difficult to solve any practical non-isothermal problem without numerical help. Finally, Chapter 10 deals with stability and turbulence, both of which are important topics now being researched.

The overall emphasis on kinematics owes much to my former mentor and collaborator at Brown University, Jack (A. C.) Pipkin. Similarly, it was at that institution that Robert Nickell introduced me to the use of finite elements for solving problems in non-Newtonian fluid mechanics. The idea of combining continuum and molecular ideas in the development of constitutive models is clear in the pioneering works of A. S. Lodge and R. B. Bird and their influence can be seen. I am also indebted to the many other colleagues and students, especially Bruce Caswell, who have assisted in general ways. R. A. Antonia (Chapter 10), M. M. Denn (Chapter 10), C. J. S. Petrie (Chapter 7) and Nhan Phan-Thien (Chapters 6, 9, and 10 especially) were helpful in reading and commenting on the chapters indicated. Useful suggestions were also made by Matti Keentok, Bob (R. C.) Warren, Mark Bush, and others. I am grateful to all the above for their advice, either explicit or implicit, and to Lyn and Stephanie for their typing skills.

The present document is, then, not a complete atlas but a guide to some fairly rugged territory. There are probably many uncharted rocks and some crevasses which will no doubt be found the hard way. It is these unexpected problems that makes rheology such a fascinating subject for research.

Mittagong, N.S.W.   R.I.T.
March 1984

Preface to revised edition.
The present edition is largely unchanged from that of 1985 except that errors have been corrected and some of Chapter 8 has been revised in view of the rapid progress in numerical simulation.

Mittagong, N.S.W.   R.I.T.
January 1988

ACKNOWLEDGEMENTS

We record here with gratitude the assistance of the following organizations and individuals.

Academic Press Inc.
We thank them for permission to reproduce
Fig. 1.6 from F. R. Eirich (ed.) *Rheology*, Vol. 5, Copyright 1969 (article by J. P. Tordella).
Fig. 8.21 from C. D. Han, *Rheology in polymer processing*, Copyright 1976.

American Chemical Society
Fig. 7.7 is reprinted with permission from *Industrial and Engineering Chemistry Fundamentals*, Vol. 19, p. 195, Copyright 1980, American Chemical Society.

American Institute of Chemical Engineers
Figs 3.10–3.13 are reproduced with permission from the *A.I.Ch.E. Journal*, Vol. 18, p. 540, Copyright 1972.
Fig. 5.8 is reproduced with permission from the *A.I.Ch.E. Journal*, Vol. 22, p. 910, Copyright 1976.
Fig. 7.8 is reproduced with permision from the *A.I.Ch.E. Journal*, Vol. 21, p. 796, Copyright 1975.
Fig. 9.8 is reproduced with permission from the *A.I.Ch.E. Journal*, Vol. 12, p. 1196, Copyright 1966.
Fig. 10.4 is reproduced with permission from the *Chemical engineering progress* Symposium Series, *Drag reduction* (ed. J. G. Savins and P. S. Virk), Copyright 1971 (article by S. K. F. Karlsson, M. Sokolov, and R. I. Tanner).

American Institute of Physics
Fig. 8.28 is reproduced with permission from the article by R. L. Gear, M. Keentok, J. F. Milthorpe, and R. I. Tanner, *Physics of Fluids*, Vol. 26, p. 8, Copyright 1983.

American Society of Mechanical Engineers
We acknowledge with thanks reproduction of Fig. 6.17 from the article by J. A. Tichy and W. O. Winer from the ASME *J. Lubrication Technology*, Vol. 100, p. 56, 1978 and Fig. 9.3 from the article by S. Bair and W. O. Winer, ASME *Journal of Lubrication Technology*, Vol. 114, p. 8, Copyright 1992.

Annual Reviews Inc.
Figs 1.5, 3.14 and 4.20 are reproduced, with permission, from the *Annual Review of Fluid Mechanics*, Vol. 9, Copyright 1977 by Annual Reviews Inc. Fig. 7.5 is reproduced, with permission, from the *Annual Review of Fluid Mechanics*, Vol. 12, Copyright 1980 by Annual Reviews Inc.

Applied Science Publishers

Fig. 9.2 is reproduced with permission from *The rheology of lubricants* (ed. T. C. Davenport), Copyright 1973 (article by R. F. Pywell). Fig. 9.7 is reproduced with permission from *Computational analysis of polymer processing* (ed. J. R. A. Pearson and S. M. Richardson), Copyright 1983.

Cambridge University Press

We thank them for permission to reproduce Fig. 9.12 from the article by H. B. Phuoc, and R. I. Tanner, Thermally-induced extrudate swell, *J. Fluid Mechanics*, Vol. 98, Copyright 1980, Cambridge University Press.

Carl Hanser Verlag, Munich and Vienna

We thank them for permission to reproduce Fig. 9.1 from P. Kennedy, *Flow analysis of injection molds*, p. 23, Copyright 1995.

Elsevier Science Publishers B.V.

We thank them for permission to reproduce

Figs 3.6 and 3.7 from the *Journal of Non-Newtonian Fluid Mechanics*, Vol. 6, p. 303, Copyright 1980.

Fig. 3.20(a–d) from the *Journal of Non-Newtonian Fluid Mechanics*, Vol. 52, p. 153, Copyright 1994; article by Y. W. Ooi and T. Sridhar.

Fig. 3.21 from the *Journal of Non-Newtonian Fluid Mechanics*, Vol. 52, p. 137, Copyright 1994; article by D. M. Binding, J. Maia, and K. Walters.

Figs 5.13, 5.14, and 5.15 from the *Journal of Non-Newtonian Fluid Mechanics*, Vol. 76, p. 249, Copyright 1998; article by G. Lielens, P. Halin, I. Jaumain, R. Keunings, and V. Legat.

Fig. 5.23 from the *Journal of Non-Newtonian Fluid Mechanics*, Vol. 27, pp. 225–6, Copyright 1997; article by J. A. Byars, R. J. Binnington, and D. V. Boger.

Fig. 6.4 from the *Journal of Non-Newtonian Fluid Mechanics*, Vol. 73, p. 120, Copyright 1997; article by B. P. Williamson, K. Walters, T. W. Bates, R. C. Coy, and A. L. Milton.

Fig. 8.8 from the *Journal of Non-Newtonian Fluid Mechanics*, Vol. 31, p. 264, Copyright 1989; article by S. Pilitsis, and A. N. Beris.

Fig. 10.5 from the *Journal of Non-Newtonian Fluid Mechanics*, Vol. 14, p. 182, Copyright 1984; article by K. P. Jackson, K. Walters, and R.W. Williams.

Table 9.6 from D. W. van Krevelen, and P. J. Hoftyzer, *Properties of polymers*, 2nd edn, Copyright 1976.

International Union for Pure and Applied Chemistry

Figs 3.15–3.19 inclusive and Tables 3.1–3.7 inclusive are reprinted with permission from the article by J. Meissner, *Pure and Applied Chemistry*, Vol. 42, Copyright 1975, IUPAC.

McGraw-Hill Book Co.
Figs 6.10 and 6.11 are reproduced with permission from S. Middleman, *Fundamentals of polymer processing*, Copyright 1977, McGraw-Hill Book Co.

Pergamon Press Ltd
The following items are reprinted with permission:
Fig. 3.1 from *Mechanics Today*, Vol. 1, article by A. C. Pipkin and R. I. Tanner, p. 265, Copyright 1972, Pergamon Press Ltd.
Table 1.5 and Fig. 6.21 from W. R. Schowalter, *Mechanics of non-Newtonian fluids*, Copyright 1978, Pergamon Press Ltd.
Figs 8.26 and 8.27 from *Computers and Fluids*, Vol. 6, article by K. R. Reddy and R. I. Tanner, p. 86, Copyright 1978, Pergamon Press Ltd.

Plenum Press
We thank them for permission to reproduce
Fig. 10.8 from the article by J.-C. Chang and M. M. Denn, in G. Astarita, G. Marrucci, and L. Nicolais (ed.) *Rheology*, Vol. 3, *Applications*, p. 11, Copyright 1980 by Plenum Publishing Corporation.

Society of Polymer Science (Japan)
Fig. 4.8 is adapted with permission from material in the article by Y. Einaga, K. Osaki, M. Kurata, S. Kimura, and M. Tamura, *Polymer J. (Japan)*, Vol. 2, Copyright 1971.

Dr Dietrich Steinkopff Verlag
We thank them for permission to reproduce
Table 2.1 from *Rheologica Acta*, Vol. 18, p. 681, Copyright 1979.
Fig. 9.5 from *Rheologica Acta*, Vol. 9, p. 155, Copyright 1970.
Fig. 9.15 from *Rheologica Acta*, Vol. 19, p. 168, Copyright 1980.
Fig. 4.2 (a–d) and Fig. 4.3 from *Rheologica Acta*, Vol. 24, p. 325, Copyright 1985; article by H. A. Barnes and K. Walters.
Fig. 4.4 from *Rheologica Acta*, Vol. 25, p. 554, Copyright 1986; article by D. C. H. Cheng.
Fig. 4.12 from *Rheologica Acta*, Vol. 18, p. 688, Copyright 1979; article by M. H. Wagner.
Fig. 5.26(a, b) from *Rheologica Acta*, Vol. 32, pp. 460–1, Copyright 1993; article by A. I. Johma and P. A. Reynolds.

Society of Rheology
We thank the Editor for permission to reproduce
Fig. 3.23 from the *Journal of Rheology*, Vol. 35, p. 503, Copyright 1991; article by S. Hatzikiriakos and J. M. Dealy.
Fig. 5.25 from the *Journal of Rheology*, Vol. 24, p. 804, Copyright 1980; article by F. Gadala-Maria and A. Acrivos.

Springer-Verlag GmbH

We thank them for permission to reproduce Fig. 5.19 from *Advances in Polymer Science*, Vol. 5, p. 267, Copyright 1968; article by G.C. Berry and T.G. Fox.

John Wiley and Sons, Inc.

We thank them for permission to reproduce the following, all of which are the copyright of John Wiley Sons, Inc.

Fig. 3.5 from the *Transactions of the Society of Rheology*, Vol. 14, p. 483, Copyright 1970.

Figs 4.17 and 4.18 from the *Transactions of the Society of Rheology*, Vol. 13, p. 471, Copyright 1969.

Figs 5.9–5.11 from the *Transactions of the Society of Rheology*, Vol. 19, p. 37, Copyright 1975.

Fig. 9.4 from *J. Polymer Science*, Vol. 19, p. 111, Copyright 1956.

Fig. 7.6 from *J. Applied Polymer Science*, Vol. 20, p. 181, Copyright 1976.

Fig. 7.9 from *J. Rheology*, Vol. 22, p. 280, Copyright 1978.

Fig. 5.1 reprinted from R. B. Bird, R. C. Armstrong, and O. Hassager, *Dynamics of Polymeric Systems*, Vol. 1, Copyright 1977.

Figs 9.13 and 9.14 reprinted from Z. Tadmor and C. Gogos, *Principles of polymer processing*, Copyright 1979.

Table 9.3 (part) from J. D. Ferry, *Viscoelastic properties of polymers*, 3rd edn, Copyright 1981.

Fig. 3.22 from the *Journal of Polymer Science*, Vol. 10, p. 1067, Copyright 1972; article by G.V. Vinogradov *et al.*

Thanks are also due to R. B. Bird, D. V. Boger (Figs 1.5 and 8.19), B. Caswell (Table 8.6), M. J. Crochet (Fig. 8.22), M. M. Denn, C. D. Han, K. H. Hunt (Fig. 3.2), A. S. Lodge, S. Middleman, N. Phan-Thien (Figs 8.39–8.40), R. Keunings (Figs 5.13–5.15), E.Mitsoulis, G. McKinley, S. Hatzikiriakos (Fig. 3.23), W.R. Schowalter, S.-C. Xue (Figs 8.15–8.18, 8.41), P. Kennedy and R. Zheng (Fig. 9.16), for help with sundry material and/or criticisms of draft text.

CONTENTS

NOTATION

Vectors, tensors, and matrices are in bold-face type. An overbar denotes a mean value. A tilde (˜) over a symbol denotes a Laplace transform. The trace of a tensor $\mathbf{A}$ is denoted by $\text{tr}\mathbf{A}$, and the transpose of $\mathbf{A}$ by $\mathbf{A}^{\mathrm{T}}$. Vectors and tensors are given both in direct (for example, $\mathbf{a}$) and component (for example, a_i) notation. Symbols which have more than one meaning are listed with a semi-colon dividing the meanings.

Roman symbols

a	length of link or submolecule; bubble radius
$\mathbf{a}(a_i)$	acceleration vector; also unit vector tangential to slip surface
a_T	time–temperature shift factor
A	amplitude parameter; area
AR	aspect ratio
$\mathbf{A}^{(n)}(A_{ij}^{(n)})$	nth Rivlin–Ericksen tensor
b	magnitude of $\mathbf{b}$
$\mathbf{b}(b_i)$	unit vector normal to slip surface; vector
B	dimensionless pressure difference
$\mathbf{B}(B_{ij})$	Finger strain defined as $\mathbf{C}^{-1}$; second Rivlin–Ericksen tensor [eqn (4.84)]
c	distance between cone tip and plate; mass fraction; bearing clearance
$\mathbf{c}(c_i)$	unit vector normal to $\mathbf{a}$ and $\mathbf{b}$ in slip plane
C	creep response function
$\mathbf{C}(C_{ij})$	strain tensor $\mathbf{F}^{\mathrm{T}}\mathbf{F}$
C_D	drag coefficient
$C_{p,v}$	specific heat at constant pressure, volume
C_{ijkm}	elasticity tensor
d	diameter
$\mathbf{d}(d_{ij})$	rate-of-deformation tensor
D	diameter; depth; diffusion coefficient [eqn (5.24)]
D/Dt	material derivative $\partial/\partial t + v_j \frac{\partial}{\partial x_j}$
(De)	Deborah number
D_R	draw ratio
e	internal energy per unit mass
$\mathbf{e}$	strain, eqn(5.181)
$E(t)$	tensile relaxation function
E, E_0	activation energy; Young's modulus.
E_{ij}	classical linear strain tensor $\frac{1}{2}\left(\frac{\partial u_i}{\partial x_j}+\frac{\partial u_j}{\partial x_i}\right)$
E_1	see eqn (7.9)
Ei	see eqn (7.12)
f	non-linear factor in step strain responses [eqn (5.166)]; friction factor [eqn (10.110)]
$\mathbf{f}(f_i)$	body force vector per unit mass
F	total force; front factor [eqn (9.30)]
$\mathbf{F}$	force vector
$\mathbf{F}(F_{ij})$	deformation gradient $\partial r_i/\partial x_j$
g	gravitational acceleration

g_i	stability sensitivity coefficients; elastic moduli (eqn 2.105)
G	pressure gradient; shear modulus
$G(t)$	relaxation modulus
$G'(\omega)$	storage modulus
$G''(\omega)$	loss modulus
G^*	complex modulus $G' + iG''$
G_g	glass modulus
G_e	equilibrium modulus
(Gz)	Graetz number
$\mathbf{G}$	Strain tensor $\mathbf{C} - \mathbf{I}$ [eqn (4.42)]
h	distance; film thickness; memory factor [eqn (4.76)]; enthalpy; heat transfer coefficient
$\mathbf{h}$	network vector (Fig. 5.20)
h_t	heat transfer coefficient
H	distance; height; memory factor [eqn (4.74)]
$H(t)$	unit step function
$H(\lambda)$	distribution function for relaxation times
$\mathbf{H}$	strain measure [eqn (4.63)]
$\mathbf{i}_\alpha$	unit vector in α-direction
$\mathbf{I}(\delta_{ij})$	unit tensor
$I_{1,2}$	integrals [eqn (3.28)]; modified Bessel functions
$\mathbf{J}$	flux vector
J	creep compliance
J_g	glass creep compliance
J_e	equilibrium creep compliance
J^*	complex compliance $J' - iJ''$
k	power-law consistency parameter; Boltzmann constant; thermal conductivity
K	dimensionless spring constant; drag coefficient; bulk modulus
l	cell spacing; length
L	length
$L(\lambda)$	retardation spectrum function
$\mathbf{L}(L_{ij})$	velocity gradient tensor $\partial v_i/\partial x_j$
m	mass; $M/2\pi r_0^3$ [eqn (3.34)]; kinematic classifier [eqn (3.95)]
M	moment; molecular weight
n	power-law exponent
n_0	number of molecules per unit volume
$\mathbf{n}$	outward-pointing unit normal vector
N	number of 'molecules'; Bagley correction [eqn (3.62a)]
N_1	first normal stress difference [eqn (3.17)]
N_2	second normal stress difference [eqn (3.17)]
N_i	shape function
(Nu)	Nusselt number
$p, p^+, \bar{p}$	pressure
P	pressure; probability [eqn (5.8)]
(Pe)	Peclet number
(Pr)	Prandtl number
q	$\equiv Q/\pi R^3$; flux per unit width [eqn (6.8)]; source strength
$\mathbf{q}(q_i)$	heat flux vector
Q	volumetric flow rate; potential [eqn (4.83)]

Q^*	potential
r	radial coordinate; radius; $z\sqrt{\rho/\eta\lambda}$
r_f	radius of flat on tip of cone
$\mathbf{r}$	position of particle at time t'
R	gas constant (8.3143 J/mole K); radius; relaxation function; initial position; Rayleigh number
$\mathbf{R}(R_{ij})$	orthogonal matrix
$\mathbf{R}$	dumbbell end-to-end vector
(Re)	Reynolds number
s	time difference $t - t'$; heating per unit mass; entropy; amount of shear [eqn (5.65)]; constant
$\mathbf{s}$	unit vector tangential to streamline
S	surface; Sommerfeld number; dimensionless stress
$\mathbf{S}$	dimensionless stress [eqn (5.95)]
Sg	stretching dimensionless number
S_R	recoverable shear [eqn (10.93)]
t, t'	times; thickness (Fig. 8.38)
$\mathbf{t}(t_i)$	traction vector
T	temperature; as superscript denotes transpose of a tensor; $\Psi_1/2\eta$ in eqn (4.85); tension
T^*	$\dfrac{\Psi_1}{2\eta} + \dfrac{\Psi_2}{\eta}$ [eqn (4.85)]; dimensionless tension [eqn (7.61)]
(Ta)	Taylor number
u	x-component of velocity or displacement
$\mathbf{u}(u_i)$	displacement vector; unit vector
u_*	friction velocity
U	velocity; potential function
U_+	dimensionless mean velocity
v	y-component of velocity or displacement
$\mathbf{v}(v_i)$	velocity vector
V	volume; velocity
$\mathbf{V}$	solution vector
w	z-component of displacement or velocity
w_{ij}	component of vorticity tensor $\dfrac{1}{2}\left(\dfrac{\partial v_i}{\partial x_j} - \dfrac{\partial v_j}{\partial x_i}\right)$
W	strain–energy function; load on bearing
(Wi)	Weissenberg number
$\mathbf{x}(x_i)$	position vector
x_i	Acierno structural parameter
X	power-law drag coefficient
y	coordinate
Y	coordinate; function
z	coordinate
z_+	dimensionless distance [eqn (10.103b)]
Z	initial position of particle

Greek symbols

α	angle; viscosity–temperature coefficient; coefficient of volumetric expansion; Giesekus parameter [eqn(5.111)]
β	pressure–viscosity exponent; slip coefficient
$\beta(h)$	network destruction function

β_i	stability coefficients
γ	shear strain
$\dot{\gamma}$	shear strain-rate
$\hat{\gamma}$	amplitude of shear strain in sinusoidal shearing
Γ	gamma function; time const. for MPTT model
$\delta(\omega)$	phase angle
$\delta(t)$	delta or impulse function
δA	element of area
δ_{ij}	Kronecker delta or unit tensor
$\delta \mathbf{F}$	element of force
Δ	dilatation
Δp	pressure difference
$\Delta \gamma$	strain increment
ε	tensile strain; dimensionless force [eqn(7.33)]; PTT parameter
$\dot{\varepsilon}$	elongational strain-rate
$\dot{\varepsilon}_i$	principal extension rate in i-direction
$\varepsilon_{\mathbf{R}}$	recoverable portion of tensile strain
ε_H	Hencky (logarithmic) strain
ζ	friction coefficient
η	shear viscosity
η'	dynamic viscosity G''/ω
η''	$\equiv G'/\omega$
η_{E}	Trouton or elongational viscosity
η^*	complex viscosity $\eta' - \mathrm{i}\eta''$
$[\eta]$	intrinsic viscosity
θ	angular position; dimensionless temperature [eqn (7.73)]
Θ	initial angular position of particle
κ	thermal diffusivity; Kármán constant
λ	time constant; stretch ratio
$\bar{\lambda}$	Lamé modulus
Λ	second viscosity coefficient [eqn (2.68)]; length at time t/initial length
μ	friction coefficient; viscosity ratio [eqn (7.71)]
$\mu(t)$	memory function
ν	kinematic viscosity η/ρ
ξ	Phan-Thien slip parameter [eqn (5.160)]; z/h
$\xi(t)$	pseudo-time [eqn (9.25)]
ρ	density; v' [eqn (7.36)]; radius of curvature
σ	surface tension coefficient; tensile stress; sinusoidal shearing stress
σ_i	principal stresses
$\boldsymbol{\sigma}(\sigma_{ij})$	stress tensor
$\hat{\sigma}$	amplitude of stress in sinusoidal shearing
τ	shear stress
τ_w	shear stress at wall
$\boldsymbol{\tau}(\tau_{ij})$	extra stress tensor $\boldsymbol{\sigma} + p\mathbf{I}$
ϕ	angular co-ordinate; velocity potential; probability function [eqn (5.134)]; volume concentration
$\phi(\tau)$	fluidity function
Φ	solution vector [eqn (8.18)]
χ	Stokes stream function; swelling ratio; spring potential [eqn (5.42)]
ψ	stream function; probability density [eqn (5.24)]; function [eqn (9.56)]

Ψ_1, Ψ_2 first and second normal stress coefficients
ω rotation rate; frequency
$\boldsymbol{\omega}$ vorticity vector $\nabla \times \mathbf{v}$
Ω rotation rate

Special symbols
I_c, II_c, III_c First, second, and third invariants of tensor $\mathbf{c}$ respectively.
∇ gradient operator
$\langle \ \rangle$ denotes average [eqn (5.10)]
$\Delta/\Delta t$ convected derivative [eqn (5.54)]
$\mathcal{L}$ $\lambda\mathbf{L} - \frac{1}{2}\mathbf{I}$ dimensionless modified velocity gradient; also $\mathbf{L} - \xi\mathbf{d}$ [eqn (5.161)]
$\mathcal{D}/\mathcal{D}t$ derivative operator [eqn (5.177)]

1
INTRODUCTION TO RHEOLOGY

1.1 What is rheology?

RHEOLOGY is the science of deformation and flow of materials. The principal theoretical concepts are *kinematics*, dealing with geometrical aspects of deformation and flow; *conservation* laws, dealing with forces, stresses, and energy interchanges, and *constitutive* relations special to classes of bodies (for example, elastic bodies). The constitutive relations serve to link motion and forces to complete the description of the flow processes, which may then be applied to solve engineering problems arising in polymer processing, lubrication, food technology, printing, and many other technologies.

All real materials possess a microstructure at the molecular, crystal or higher level. In mechanics we are often not interested in specifying the material in such detail, as the large-scale phenomena of interest usually involve the average behaviour of a very large number of units of the microstructure. Thus it is convenient to work with an idealized *continuum* model of the material, whose microstructure is not specified. Some systems (for example, multiphase flows, reinforced concrete) have several regions which have very different material properties, and each of these may need to be considered separately. While these more complex systems need to be borne in mind, we will concentrate on homogeneous continua to begin with.

Continuum mechanics is the mathematical study of the response of such ideal bodies to applied forces and deformations. The early success of the theory of (small-strain) elasticity, which is entirely a mathematical construction, raised considerable expectations from the methods of continuum mechanics. The only physical inputs to small-strain elasticity theory are the ideas in Hooke's law and the observation that most metals undergo only a fraction of a percentile strain before yielding. In fluid mechanics the success was almost as complete. In the isothermal case, only the Newtonian viscous law and a connection between density and pressure are needed as experimental inputs to the highly precise Navier–Stokes equations. These two examples are atypical as we shall find that most other continuum mechanics descriptions cannot match the simplicity and elegance of these linear models. Thus the pure continuum approach becomes less and less useful as the complexity of response increases. The response of polymer solutions and polymer melts is very complex (see Section 1.3) and recent methods have tended to combine continuum mechanics with ideas obtained by thinking about the microstructure of the bodies under study. We shall term this study field, *rheology*.

1.2 An historical note

The term rheology was invented by E. C. Bingham and his Lafayette College (Easton, PA, USA) associates in 1929 (see Table 1.1); the Greek root 'rheo' implies flow. Clearly, the study of rheology preceded the invention of the name, and despite their exclusion from Table 1.1, one needs to look back to the beginnings of solid and fluid mechanics in the seventeenth and eighteenth centuries to trace the origins of rheology. Hooke (1676) and Newton (1687) were the first to set down quantitative connections between forces and deformations, and so it is appropriate to use the terms Hookean and Newtonian for elastic and viscous bodies respectively. The disciplines of elasticity and fluid dynamics developed via the work of Leonhard Euler (1755), Navier (1821, 1823), and Cauchy (1827). Navier developed theories of fluid flow (1821) and elasticity (1823) beginning from 'molecular' ideas, and Cauchy set down the essential mathematical ideas for handling shear and tension as twin aspects of the stress tensor. Stokes (1845) essentially completed Navier's work on fluids. The book by Dugas (1988) gives a concise history of mechanics which can be consulted for the origins of these classical theories.

Before the verification of the classical theories of elasticity and fluid dynamics was completed, marked deviations from Hooke's law began to be observed. (The ability to deform metals permanently was long a matter of common observation; such behaviour, which is clearly not elastic, is a part of the general study of 'plasticity').

We begin by noting the work of Wilhelm Weber (1835) on silk threads. At that time, there was a general interest in improving galvanometer construction and the use of silk fibres for instrument suspensions was common. Weber noticed that the elasticity of silk fibres in tension was not perfect. He applied a tensile load to a silk fibre and noticed an immediate (elastic) extension; this was followed by a continued slow extension with time. Removal of the load led to an immediate contraction. This kind of behaviour was already well known in metals, but, to Weber's surprise, he found that the silk fibre eventually recovered its original length. This differentiates the elastic after-effect from previously observed elastic or plastic behaviour in metals and marks the point of departure for our *viscoelastic* studies.

Table 1.1 Definition of rheology

'Rheology is the study of the deformation and flow of matter.'

E. C. Bingham (1929)

Flow of electrons and heat were excluded by agreement and it was also agreed that because rheologists were interested in the properties and constitution of the 'matter' under investigation, hydro- and aerodynamics and the classical theory of elasticity would be excluded.

While Weber only conducted what we now call creep and recovery tests, he also stated what would happen in a relaxation test, where a sudden strain was imposed and maintained. Weber also discovered the need to 'condition' the silk fibre by imposing a few load/unload cycles, before repeatable behaviour was observed. The mechanical conditioning eliminated some non-recoverable creep observed in the first few loading cycles.

Looking at the state of *theoretical* work at these times, we note Kelvin's (1865) concept of 'viscosity of solids'. In 1867, James Clerk Maxwell put forward the idea that the 'viscosity in all bodies may be described independently of hypothesis' by the equation

$$\frac{d\sigma}{dt} = E\frac{d\varepsilon}{dt} - \frac{\sigma}{\lambda}, \tag{1.1}$$

where σ is the stress, ε is the strain, E is Young's modulus, and λ is the time-constant. No real explanation was given for this equation and Maxwell used it to calculate gas viscosity, given by the product $E\lambda$. He was aware of the experiments of Weber and others and noted that the simple exponential decay of stress implied by (1.1) did not agree with their data; he proposed that λ should depend on the stress, an idea which has since been frequently revisited. It is perhaps ironic that the concepts of the rivals Hooke and Newton were united forever by Maxwell in his equation (1.1).

O. E. Meyer (1874) assumed that the shear stress τ and strain γ could be written in the form:

$$\tau = G\gamma + \eta\frac{d\gamma}{dt}, \tag{1.2}$$

where G and η are material constants. This expression actually describes the so-called Kelvin–Voigt body; Kelvin did experiments on the damping of metals and applied the concept implied by (1.2) without writing down any formulae and Voigt later generalized the Meyer idea to anisotropic media. In justice, one ought to refer to a Kelvin–Meyer body, or simply a Meyer material.

A very important contribution to the subject was made by Ludwig Boltzmann, who is best known for his work on the kinetic theory of matter and the concept of entropy. Boltzmann's early work in 1874, written as he attained his thirtieth year, was apparently motivated by the lack of generality in Meyer's (1874) formulation, and the paper contains a long criticism of Meyer's work.

Considering the isotropic viscoelastic case, Boltzmann assumed that the stress at time t not only depended on the strains at that time, but also on those in previous times; it was explicit assumed that the longer the interval from the present to the past time, the smaller would be the contribution to the current stress resulting from a given (past) strain. This is an expression of the now familiar principle of 'fading memory'. The assumption of (linear) superposition was also made, with a footnote that stated that the principle of superposition would not hold for large deformations. This was then followed immediately by the now accepted general theory of linear viscoelasticity, in the form (for the shear stress τ)

$$\tau(t) = G\gamma(t) - \int_0^\infty \mathrm{d}\omega \phi(\omega)\gamma(t - \omega). \tag{1.3}$$

Here, ϕ is a memory function and $\omega = t - t'$, t' being a past time; γ is the shear strain. Boltzmann deduced several ways of finding $\phi(\omega)$ from torsional experiments using various strain patterns; relaxation, free vibration and short steps of strain were among the patterns considered.

Some experiments on glass fibres in torsional vibration were shown to agree well with Boltzmann's calculations, and the closing remarks in the paper concern the way the general theory collapses to the viscous case for certain forms of the memory function ϕ. We conclude that Boltzmann's is the first successful rheological theory.

Soon afterwards, the viscosity of gelatin solutions was shown by Schwedoff (1890) to depend on shear rate, and curious phenomena, some described in Section (1.3), began to be noticed. Weissenberg (1949) drew attention to these phenomena, but even earlier, in the 1930s, the need for truly three-dimensional non-linear theories began to be apparent.

Early work by Fromm (1933, 1947) developed constitutive equations for non-linear viscoelastic bodies, and Oldroyd (1950) set up the now-accepted conceptual framework for developing non-linear constitutive equations. A more abstract approach to this framework was later given by Noll (1955). The book by Lodge entitled *Elastic liquids* (1964) has been very influential in that it recognized the value of incorporating information from the microstructural level into the formulation of constitutive equations, especially for complex bodies. Lodge also made many advances in experimental methods and pointed out the need to work with both shear and elongational flows when exploring material response.

A more complete sketch of the history of the subject is given by Tanner and Walters (1998).

The emergence of rheology as a separate field in 1929 occurred when the mechanical behaviour of important industrial materials like rubber, plastics, clays, paints, and many biological fluids began to attract increasing attention in the fields of physics, mechanics, and mathematics, in addition to the traditional interest from colloid chemists.

None of the scientific societies existing at that time provided a forum which brought together all those concerned with these problems, and by 1928 a number of prominent workers in both Europe and the United States agreed on the need for a new interdisciplinary organization and journal. The formation of the Society of Rheology in 1929 marks the start of international organization in this newly-recognized field.

The *Journal of Rheology* (1929–32) provided an important medium of international communication between rheologists for three years, with articles, communications, and abstracts covering activities throughout the world. After

1932, the Society was unable to continue a separate publication for several years, and its papers appeared in special issues of other journals until the *Transactions of the Society of Rheology* was established in 1957. This journal assumed in 1978 the old title of *Journal of Rheology*.

Two other specialist journals (*Rheologica Acta* and *Journal of Non-Newtonian Fluid Mechanics*) are now available. In addition, some of the national groups also publish bulletins and literature surveys; much useful rheological information also appears in the polymer science, colloid science, and engineering literature.

1.3 Some rheological phenomena

It is well known that liquids with complex structure, such a macro-molecular solutions and melts, soap solutions, and suspensions behave in unexpected ways and are not described by the constitutive equations (the Navier–Stokes equations) appropriate for Newtonian liquids. Some of the 'non-Newtonian' or rheological effects which have been observed are:

(*a*) *Shear-rate-dependent viscosity*. Most of these fluids display 'shear thinning'; that is, the viscosity decreases with increasing shear rate, sometimes reaching 10^{-2} or 10^{-4} of the zero-shear-rate viscosity. The viscosity is often the most important property for engineering calculations. Figure 1.1 shows a definition sketch of simple shearing; the shear stress τ (or σ_{yx}) is linked to the shear rate ($\dot{\gamma}$) by the (shear) viscosity η, where

$$\tau = \eta U/d = \eta\dot{\gamma}. \tag{1.4}$$

Figure 1.2 shows a large variation (about 5000-fold) of viscosity in a polymer solution.

(*b*) *Normal-stress effects in steady flows*. Polymeric fluids usually exhibit a number of second-order effects associated with the inequality of normal stresses in steady shear flow and related simple flows. These include: the 'Weissenberg effect', in which a macromolecular fluid climbs up a rotating rod (Fig. 1.3); the reversal of direction of the secondary-flow pattern when a disc rotates liquid in a beaker (Fig. 1.4) and the slight bulging of the surface of a liquid as it flows down a trough (Fig. 3.6).

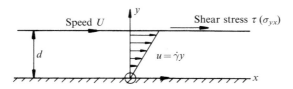

Fig. 1.1 Simple shearing. Fluid is contained between two infinite paralled plates separated by a distance *d*. The top plate moves with a constant velocity *U* in the *x* direction. The shear rate $\dot{\gamma} = U/d$.

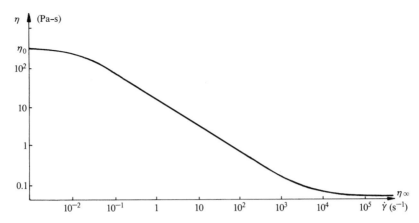

Fig. 1.2   Shear-thinning in a typical non-Newtonian fluid, Separan AP-30 in glycerol/water.

Fig. 1.3   Weissenberg or rod-climbing effect. Beaker on right is Newtonian (Glycerol) showing a surface depression caused by inertia. Non-Newtonian fluid on the left (Separan AP-30) climbs the rotating rod.

(c) Transient responses in unsteady shear flows. A wide variety of small-strain experiments, such as 'dynamic testing' (that is, oscillatory flows), stress relaxation, creep, and recoil have been used to obtain information on mechanical properties (see Section 1.5.2). In addition some large-amplitude time-dependent

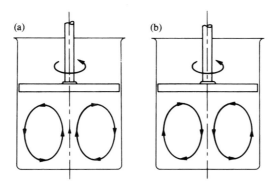

Fig. 1.4 Reversal of flow pattern in beaker. A disc at the top rotating in its plane drives the flow. The secondary flows are shown. On the left the Newtonian fluid moves up in the centre; on the right a non-Newtonian fluid moves down in the centre (After C. T. Hill, *J. Rheology*, **16**, 213, 1972).

experiments have also been performed: these include recoil, shear flow with superposed oscillatory motion and the overshoot in stress at the starting up of a shear flow.

(d) Other phenomena. The above examples involve mainly shearing flows. In other flows there are additional phenomena, such as: swelling of extrudates issuing from a die (Fig. 8.23); development of a toroidal vortex in the inlet flow to a die; the 'Uebler effect', which is the abrupt stopping of large bubbles in the accelerating velocity field where the fluid flows from a reservoir into a tube; the tubeless-siphon effect, in which a siphon continues to operate even though the upstream end has been withdrawn from the fluid (Fig. 1.5); the unsteady-state behaviour in elongational flow, in which apparently a steady-state elongational flow cannot be attained; 'melt fracture' in which very irregular extrudates are produced (Fig. 1.6); and the solid-like behaviour of polymeric liquids subjected to high-speed impact tests.

The reader is recommended to view these and other phenomena in the films by H. Markovitz[†] and K. Walters[‡] and his group, and in the book by Boger and Walters (1993).

1.4   Purposes and aim of rheological theory

Newtonian fluids do not exhibit the curious effects described above, yet parts of the mathematical descriptions of both Newtonian and non-Newtonian fluids are similar; they obey the same rules for the kinematic descriptions and conservation principles. Generally, the velocity field (v_i) is described with respect

[†] 'Rheological behaviour of fluids'. A film produced by Educational Services Inc. (Watertown, Mass., USA) for the US National Committee on Fluid Mechanics.
[‡] 'Non-Newtonian fluids'. A film produced at the Department of Applied Mathematics, University of Wales, Aberystwyth, UK (1980).

Fig. 1.5 Ascending free siphon. The black band marks the end of the glass tube. The fluid is a solution of Separan AP-30 in a mixture of 75 per cent glycerol and 25 per cent water.

to fixed, Eulerian axes (x_i) and time t, so $v_i = v_i(x_i, t)$, in both cases. Often the liquid density changes very little, so the conservation of mass and the conservation of momentum and energy have identical forms. The principles of continuum mechanics are set out in Table 1.2, and they apply to all materials considered here, since we generally assume symmetry of the stress tensor ($\sigma_{ij} = \sigma_{ji}$). At each point in the material one needs to solve for six unknown

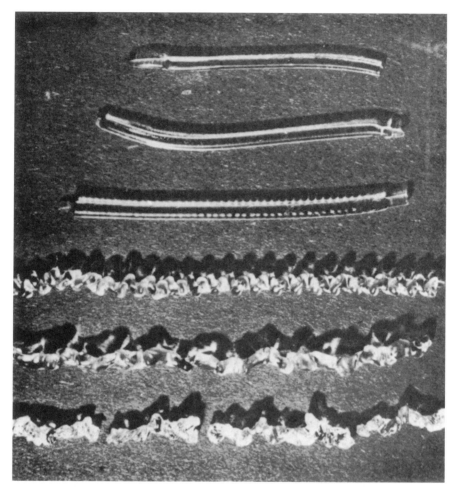

Fig. 1.6 Melt fracture. Specimens of polymethyl-methacrylate extruded at 170 °C through a short circular capillary 1.2 mm diameter × 0.2 mm long. Instability sets in the second specimen from the top; flow rate increases from top to bottom.

Table 1.2 Principles of continuum mechanics

1. Conservation of mass;
2. Stress concept;
3. Symmetry of stress tensor (from balance of angular momentum);
4. Stress-equations of motion (from balance of linear momentum);
5. Deformation and strain analyses;
6. Energy conservation.

stress components plus three displacements (compressible elasticity) or, for incompressible materials where a pressure p appears, we have p plus three velocity (v_i) or displacement components to find. We will mainly consider incompressible materials in this book, and then there are ten unknowns, while the general principles supply only four equations connecting the σ_{ij}, p, and v_i unknowns. We need six *constitutive* equations connecting σ_{ij} and the motion, and also *boundary conditions*. The latter will usually be given in terms of either the forces per unit area on the boundary (tractions) or the velocities, or a combination of the two. (For static elasticity problems, we substitute displacement for velocity in these and other remarks.) Table 1.3 shows the complete set of equations for an incompressible *Newtonian* fluid, for any *flow problem*. For other incompressible

Table 1.3　A typical continuum mechanics problem

Let x_i be the components of the particle position vector **x**. We choose an incompressible Newtonian fluid as an example.

Equations
　1. Conservation of mass. (Incompressibility)
Compactly, $\partial v_i/\partial x_i = 0$; writing this out,[†]

$$\frac{\partial v_1}{\partial x_1} + \frac{\partial v_2}{\partial x_2} + \frac{\partial v_3}{\partial x_3} = 0 \qquad (1.5)$$

This is one equation connecting the three components (v_i) of the velocity vector **v** $(i = 1, 2, 3)$.
　2. Angular momentum conservation $\sigma_{ij} = \sigma_{ji}$ (equality of cross-shears; this reduces the number of independent components of the stress tensor σ_{ij} from 9 to 6).
　3. Linear momentum conservation.

$$\frac{\partial \sigma_{ij}}{\partial x_j} + \rho(f_i - a_i) = 0. \qquad (1.6)$$

This yields three equations connecting six independent stress components (σ_{ij}); the density ρ and the body forces per unit mass, f_i, (usually gravity) are known; the acceleration components a_i depend on the velocities and are not new unknowns.
　4. For an incompressible *Newtonian* fluid we have, by definition, the six constitutive equations

$$\sigma_{ij} = -p\delta_{ij} + \eta\left(\frac{\partial v_i}{\partial x_j} + \frac{\partial v_j}{\partial x_i}\right), \qquad (1.7)$$

where p is the pressure and the symbol δ_{ij} (the unit tensor) is such that

$$\delta_{ij}\begin{cases} \text{is zero if } i \neq j \\ \text{is unity if } i = j. \end{cases}$$

Here η is the viscosity.
Clearly, $\sigma_{ij} = \sigma_{ji}$ automatically here, so the stress tensor is symmetric.

　　[†]Here and elsewhere, a repeated index is to be summed over unless stated otherwise.

fluid models, we retain (1.5) and (1.6) but (1.7) will change. In contrast, Table 1.4 shows a set of boundary conditions; they are generally *problem*—but not *fluid-dependent*. Note that p does not occur as a boundary condition; in general p does not occur in boundary conditions even in a Newtonian viscous fluid, although attempts to use it are common; physical problems usually give rise to either traction or velocity boundary conditions. (Exceptionally, it may be necessary to fix the pressure at a single point to prevent pressure indeterminancy in cases where velocity boundary conditions only are applied.)

In summary, for isothermal problems, the above shows the constituents of continuum mechanics: given a constitutive equation and boundary conditions, the problem is reduced to mathematics. Thermal and other effects add complications, but the path is similar, and we see that the development of suitable constitutive relations is the key problem in rheology and continuum mechanics as it is these relations which differentiate one material from another.

Notice that we have spoken of models in much of the above; Table 1.3 sets out the incompressible Newtonian fluid *model*. For any *real* fluid, which is

Table 1.4   Typical boundary conditions for an example

All the equations in Table 1.3 are to be used in all incompressible Newtonian flows; it is the *boundary conditions* which define specific problems. Figure 1.7 shows a complex problem (extrusion). The boundary conditions here are:
(i) on AB we can set in a parabolic (or any other) velocity profile. Hence labelling

$$x_1 = z \quad v_1 = v_z$$
$$x_2 = r \quad v_2 = v_r$$
$$x_3 = \theta \quad v_3 = v_\theta$$

then (for example) on AB

$$\begin{cases} v_z = 2\bar{w}\left(1 - \frac{r^2}{R^2}\right) & \text{(where } \bar{w} \text{ is the mean} \\ v_r = v_\theta = 0 & \text{velocity in the tube).} \end{cases}$$

The motion is axisymmetric, since $v_\theta = 0$.
(ii) on BC we assume no slip at the solid wall, so that the boundary conditions are

$$v_r = v_z = v_\theta = 0.$$

(iii) On CD we have a free surface and we assume that air drag and surface tension will produce some known stresses (often zero). Thus the tractions (force/unit area) are given on the free surface. Ignoring circumferential shear stresses because of axisymmetry, this is equivalent to giving σ_{nn} and σ_{ns}, the normal stress and the tangential shear stress. $\sigma_{n\theta}$ and $\sigma_{\theta s}$ are zero by symmetry.
(iv) On DE we may be given a force/unit area σ_{zz} (often zero). We can assume $\sigma_{rz} = 0$, or assume no radial velocity here (a mixed condition).
(v) On the centreline AE there is no radial velocity ($v_r = 0$) and there also is no shear stress ($\sigma_{rz} = 0$).

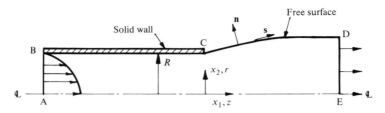

Fig. 1.7 Boundary conditions for extrusion.

more or less compressible and has a more or less constant viscosity, the model behaviour is an approximation to the real behaviour, which may or may not be useful. It is the aim of rheological theory to devise relevant models for use in various flows; the same fluid may be represented by different models in different circumstances.

Various levels of sophistication in the models are possible, ranging from one with variable viscosity up to models that can describe and interrelate, at least qualitatively, a wide variety of phenomena. Often the use of the constitutive relations in hydrodynamic problem solving has been (roughly) inversely proportional to the complexity of the relation. The result is that in the past for complex hydrodynamic flows one usually used a simple variable viscosity model or some very simple non-linear viscoelastic model, while elaborate equations were reserved for steady shearing and other simple flow patterns. Fortunately, effective computer methods to solve complex problems have now appeared, so that more complex descriptions are now more widely used. In this book some constitutive equations which are known to be useful are presented; the applications include the following broad categories of problems:

(i) *Dimensional analysis and experimental data.* Mapping flow regimes using parameters from a constitutive relation in order to classify flows.

(ii) *Relations among rheological measurements*

(a) Rigorous relations for cross-checking results of rheological experiments; for example, relations between steady shear flow quantities and various time-dependent shear flow quantities.

(b) Rough estimates of one rheological property from measurements of another; for example, prediction of normal stresses from shear stresses; use of the Cox and Merz (1958) relation to estimate shearing viscosity from the complex viscosity.

(iii) *Solving fluid dynamics problems*

(a) Calculations for analysing rheological experiments; for example, secondary flows in a non-circular tube.

(b) Computations intending to explain the observed phenomena; for example, extrudate swell.

(c) Engineering calculations; for example, mould filling, wire coating, viscous heating in screw extruders, flow patterns in mixers.

Considerable intuition has been, and will continue to be, used to select a constitutive equation for these applications, and the choice may be somewhat subjective. In many areas, there is such a dearth of trustworthy data that discrimination among constitutive equations is difficult.

1.5 Description of material behaviour in shear

Although it is vital to consider both shear and non-shearing motions to fully understand material behaviour (see Fig. 1.5) the present chapter will deal only with shearing motion; more general motions will be discussed later.

Shearing is easily visualized (Fig. 1.1) and for Newtonian fluid behaviour the shear stress τ and shear rate $(\dot{\gamma})$ are linearly related, as in eqn (1.4). The viscosity η may vary with temperature and pressure (Chapter 9), but not with time or shear rate. Many materials are non-Newtonian and display much more complex responses. For example, a meat product sample subjected to a set of steady shear rates gave the result shown in Fig. 1.8. The arrows show the sequence of testing—from low to high shear rate, then back down to zero. The hysteresis loop is fairly repeatable from one sample to another and this sample clearly shows that a one–one relation between shear stress and shear rate does not exist here. This fluid is not appreciably elastic, and structural changes on shearing cause it to show a time-dependence (diminution of stress, at a constant $\dot{\gamma}$). If left at

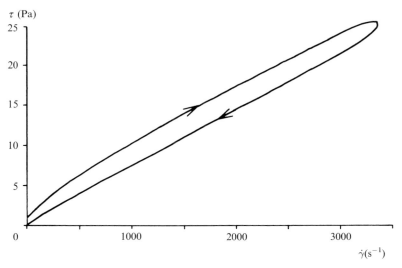

Fig. 1.8 Meat product flow curves, showing hysteresis. The arrows show the sequence of testing.

rest, the sample recovers and the results can be repeated. This behaviour is termed *thixotropy*; see Barnes (1997) for a survey of work in this field. Rarely, fluids can *build* structure instead of destroying it.

In this case, there is a gradual formation of structure by shear, whereas so far the properties of structured materials have been discussed on the basis that shearing tends to destroy structure.

In these fluids, there is often a critical amount of shear beyond which reformation of structure is not induced and breakdown occurs. This behaviour is observed with dilute aqueous solutions of vanadium pentoxide and bentonite.

There are other materials, however, in which structure only forms under shear and gradually disintegrates when at rest. This is usually termed 'rheopexy'. Even so, this behaviour is only noted at moderate rates of shear, for if shearing is rapid the structure does not form. A suspension of ammonium oleate behaves in this way. Consider the flow of this material through a capillary tube. At a moderate pressure difference the flow is rapid at first and then decreases as the structure builds up. At a high pressure difference the flow is always rapid and does not fall off because the structure does not build up at high rates of shear.

The sample of Fig. 1.8 was tested in a viscometer which automatically advanced $\dot{\gamma}$ with time. If a thixotropic material is subject to a constant shear rate for a very long time, then one can often generate a unique $\tau - \dot{\gamma}$ curve, but it will often differ from values found from tests like that of Fig. 1.8 which was created from fluid which had been 'resting'.

Figure 1.9 shows some time-dependent data for a wheat flour dough. In this case no steady state exists, and the sample eventually ruptures in a solid-like manner at a critical strain. Sometimes chemical reactions lead to permanent structural changes, such as one finds in thermosetting plastics (Broyer and Macosko 1976), which is another class of time-dependent phenomena.

The structural changes discussed above take place over a certain time span which is often much longer than the viscoelastic response times which we shall discuss below, but the division between material models which undergo quasi-permanent structural changes and those with very long relaxation times is somewhat arbitrary, and the relevant model will depend on the time-scale of the processes of interest.

1.5.1   Time-independent non-Newtonian fluids

In many cases one finds that the concerns of the previous section are minimal, and that structural changes with time can be ignored. For inelastic, non-Newtonian fluids a possible model is given by the form (1.8) (or its inverse)

$$\frac{du}{dy} = \dot{\gamma} = f(\tau). \tag{1.8}$$

This equation implies that the rate of shear ($\dot{\gamma}$) at any point in the fluid is a function of the shear stress (τ) at that point. Such fluid models may be termed non-Newtonian viscous fluids, or generalized Newtonian fluids.

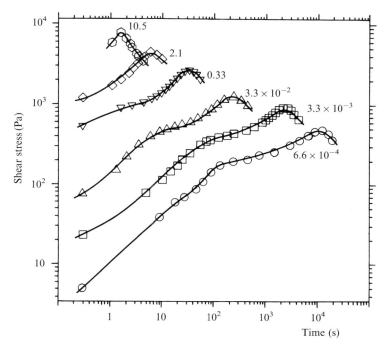

Fig. 1.9 Shear stress as a function of time, at different shear rates, for a wheat flour dough. The shear rates $\dot{\gamma}(s^{-1})$ are noted next to the curves.

These fluids may conveniently be subdivided into three distinct types depending on the nature of the function in eqn (1.8). These types are

(a) Bingham plastics (and related bodies);

(b) 'pseudo-plastic', or shear-thinning fluids

(c) 'dilatant' or shear-thickening fluids

and typical flow curves for these three fluids are shown in Fig. 1.10 (a)–(c); the linear relation typical of Newtonian fluids is also shown.

Note that linearity of the $\tau - \dot{\gamma}$ relationship in shearing is not a complete definition of a Newtonian fluid model; a non-Newtonian fluid model is one not obeying the eqns (1.7). For example, a second-order fluid, as we shall see (Section 4.5), does not obey these equations but does have a constant shear viscosity. In the simple situation discussed here, the second-order fluid and the Newtonian fluid are not distinguishable.

Bingham plastics. A Bingham plastic is characterized by a flow curve which is a straight line having an intercept τ_y, on the shear-stress axis. The yield stress, τ_y, is the magnitude of the stress which must be exceeded before flow starts.

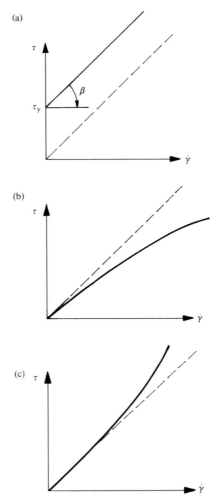

Fig. 1.10 Shear stress (τ) vs. shearing rate ($\dot{\gamma}$) for (a) Bingham body. τ_y is the yield stress and the tangent of the slope β gives the incremental viscosity. (b) Pseudoplastic or shear-thinning fluid. (c) Dilatant or shear-thickening fluid. Dashed lines show Newtonian behaviour.

Hence this model substance is really a solid, not a fluid. The rheological equation for a Bingham plastic may be written ($\dot{\gamma} \geqslant 0$)

$$\begin{aligned} \tau - \tau_y &= \eta_p \dot{\gamma}; &\quad \tau > \tau_y \\ \dot{\gamma} &= 0; &\quad |\tau| \leqslant \tau_y, \end{aligned} \tag{1.9}$$

where η_p, the plastic viscosity, is the slope of the flow curve. For $\dot{\gamma} < 0$, the curves in Fig. 1.10 are to be continued as odd functions of $\dot{\gamma}$.

The concept of an idealized Bingham plastic is convenient in practice because some fluids approximate this type of behaviour more or less closely. Common examples are slurries, drilling muds, greases, oil paints, toothpaste, and sludges. The explanation of Bingham plastic behaviour is that the fluid at rest contains a three-dimensional structure of sufficient rigidity to resist any stress less than the yield stress, τ_y. If this stress is exceeded the structure disintegrates and the system behaves as a Newtonian fluid under a shear stress $\tau - \tau_y$; when the shear stress falls below τ_y the structure is reformed. Further discussion of this model is given in Section 4.2.

Pseudo-plastic (shear-thinning) fluids. Pseudo-plastic fluids show no yield value and the typical flow curve for these materials indicates that the ratio of shear stress to the rate of shear, which is termed the viscosity,[†] falls progressively with shear rate and the flow curve becomes linear only at very high rates of shear. This limiting slope is known as the viscosity at infinite shear and is designated η_∞ (Fig. 1.2). The logarithmic plot of shear stress and rate of shear for these materials is often found to be linear over many decades of shear rate, with a slope between zero and unity. As a result, an empirical functional relation known as the 'power law' is widely used to characterize fluids of this type. This relation, which was originally proposed by de Waele (1923) and Ostwald (1925), may be written as

$$\tau = k|\dot{\gamma}|^{n-1}\dot{\gamma}, \tag{1.10}$$

where k and n are constants $(n < 1)$ for the particular fluid: k is a measure of the consistency of the fluid, the higher k the more 'viscous' the fluid; n is a measure of the degree of non-Newtonian behaviour, and the greater the departure from unity the more pronounced are the non-Newtonian properties of the fluid. The index n may often be regarded as constant over several decades of shear rate. Table 1.5 gives some representative values of k and n.

The viscosity for a power-law fluid may be expressed in terms of k and n since

$$\eta = \tau/\dot{\gamma} = k|\dot{\gamma}|^{n-1}, \tag{1.11}$$

and since $n < 1$ for pseudo-plastics the viscosity function decreases as the rate of shear increases. This type of behaviour is characteristic of high polymers, polymer solutions and many suspensions. One physical interpretation of this phenomenon is that with increasing rates of shear the molecules (or the structure) are progressively aligned. Instead of the random intermingled state which exists when the fluid is at rest the major axes are brought into line with the direction of flow and the viscosity decreases. This concept is further discussed in Chapter 5.

The Herschel–Bulkley equation combines the Bingham and power-law bodies. Other empirical equations which have been used to describe pseudo-plastic

---

[†] In many older books this ratio is often termed the 'apparent viscosity'. We shall not use this term.

Table 1.5 Power-law parameters

	Range of $\dot{\gamma}(\text{s}^{-1})$	$k(\text{Pa-s}^n)$	n
54.3% cement rock in water	10–200	2.51	0.153
23.3% Illinois yellow clay in water	1800–6000	5.55	0.229
Polystyrene at 422 K	0.03–3	1.6×10^5	0.4
1.5% carboxymethyl cellulose (CMC) in water	10^2–10^4	9.7	0.4
0.7% CMC in water	2×10^3–3×10^4	1.5	0.5
3% polyisobutylene in decalin	25–200	0.94	0.77
0.5% hydroxyethylcellulose in water	293 K –	0.84	0.509
	313 K –	0.30	0.595
	333 K –	0.136	0.645
1% poly(ethylene-oxide) in water	293 K –	0.994	0.532
	313 K –	0.706	0.544
	333 K –	0.486	0.599

All data are at room temperature (300 K) except where indicated. Data adapted from W. R. Schowalter (1978), *Mechanics of non-Newtonian fluids*, p. 139, Pergamon Press, Oxford. See also Table 1.6.

behaviour are the following with the names of their inventors:

$$
\begin{aligned}
\text{Prandtl} \qquad & \tau = A \sin^{-1}(\dot{\gamma}/C) \\[4pt]
\text{Eyring} \qquad & \tau = \dot{\gamma}/B + C \sin(\tau/A) \\[4pt]
\text{Powell–Eyring} \qquad & \tau = A\dot{\gamma} + B \sinh^{-1}(C\dot{\gamma}) \\[4pt]
\text{Williamson} \qquad & \tau = A\dot{\gamma}/(B + |\dot{\gamma}|) + \eta_\infty \dot{\gamma} \\[4pt]
\text{Ellis} \qquad & 1/\eta = 1/\eta_0 + m^{-1/n}(\tau^2)^{(1-n)/2n} \\[4pt]
\text{Casson (Solid)} \qquad & \sqrt{\tau} = \sqrt{A} + \sqrt{\eta_0 \dot{\gamma}} \quad (\tau > A).
\end{aligned}
\tag{1.12}
$$

There are many others (Wilkinson 1960).

In these equations, A, B, and C, η_0, η_∞, m, and n are constants which are typical of a particular fluid. An especially useful form has been described by Carreau (see Bird *et al.* 1977, pp. 210–11). He sets

$$
\frac{\eta - \eta_\infty}{\eta_0 - \eta_\infty} = \left[1 + (\lambda\dot{\gamma})^2 \right]^{(n-1)/2}
\tag{1.13}
$$

which combines all the power-law region and the two Newtonian regions of the complete curve (Fig. 1.2). Table 1.6 is derived from his data; values are for room temperature conditions (300 K) unless otherwise stated. A slightly better fit can be obtained by replacing 2 by a, where $a \sim 1.3$, in (1.13), giving rise to the Yasuda model.

Table 1.6 Parameters for Carreau fluid behaviour

Fluid	η_0 (Pa-s)	η_∞ (Pa-s)	λ (s)	n
2% Polyisobutylene in Primol 355	923	0.15	191	0.358
5% Polystyrene in Aroclor 1242	101	0.059	0.84	0.380
0.75% Separan-30 in 95/5 mixture by weight of water-glycerol	10.6	0.010	8.04	0.364
7% Al soap in decalin and m-cresol	89.6	0.010	1.41	0.200
Polystyrene at 453 K	1.48×10^4	0	1.04	0.398
High-density polyethylene at 443 K	8920	0	1.58	0.496
Phenoxy-A at 485 K	1.24×10^4	0	7.44	0.728

In fitting these forms of curve η_∞ can be taken as being of the order of the solvent viscosity; if η_0 is known then λ and n will be the parameters that must be found; n will usually be clearly defined from a log–log plot.

These equations are considerably more difficult to use than the power-law but if one needs a close fit for numerical work, they can be useful. For illustrative analytical work most of the eqns (1.12) are too complex, and it is difficult to find any physical connection between structure and the values of most of the constants.

Dilatant (shear thickening) fluids. Dilatant fluids are similar to pseudo-plastics in that they show no yield stress but the viscosity for these materials increases with increasing rates of shear. The power-law equation is again often applicable but in this case the index n is greater than unity. (Compare Table 1.5.)

This type of behaviour was originally discussed in connection with concentrated suspensions of solids by Osborne Reynolds (1885). He suggested that when these concentrated suspensions are at rest, the voidage is at a minimum and the liquid is only sufficient to fill the voids. When these materials are sheared at low rates, the liquid lubricates the motion of one particle past another and the stresses are consequently small. At higher rates of shear the dense packing of the particles is broken up and the material expands or 'dilates' slightly and the voidage increases. There is now insufficient liquid in the new structure to lubricate the flow of the particles moving past each other and the applied stresses have to be much greater. The formation of this structure causes the viscosity to increase rapidly with increasing rates of shear.

The term 'dilatant' has since come to be used for fluids which exhibit the property of increasing viscosity with increasing rates of shear. Many of these, such as starch pastes, are not true suspensions and do not dilate on shearing. The above explanation therefore does not apply and the term 'shear thickening' is a more accurate one.

In the process industries dilatant fluids are much less common than pseudo-plastic fluids but when the power-law is applicable the treatment of both types is similar.

1.5.2   Viscoelastic materials

So far we have mainly been considering materials in steady shearing, where they can appear to be viscous and inelastic. In many cases of interest this description is insufficient, as materials are often both elastic and viscous.

In the classical linearized theory of elasticity, the stress in a sheared body is taken to be proportional to the amount of shear. In a Newtonian fluid the shearing stress is proportional to the *rate* of shear. In most materials, under appropriate circumstances, effects of both elasticity and viscosity are noticeable. If these effects are not further complicated by the time-dependent behaviour mentioned above (Section 1.5), we call the material *viscoelastic*. The account below follows the method of Pipkin (1972). From the broader and more unified point of view that the theory of viscoelasticity affords, we will be able to see that perfectly elastic deformation and perfectly viscous flow are idealizations that are approximately realized in some limiting conditions.

For some materials it is these limiting conditions that are most easily observed. The elasticity of water and the viscosity of ice may pass unnoticed. In describing the behaviour of materials mathematically, we use idealizations that depend strongly on the *circumstances* to be described, and not only on the nature of the material; thus we will find that the distinctions between 'solid' and 'fluid' and between 'elastic' and 'viscous', are not absolute distinctions between types of materials.

(a) Stress relaxation. We will consider the behaviour of material in inertia-less simple shearing motion. The sample is assumed to be homogeneously deformed, with the amount of shear $\gamma(t)$ variable in time. Let $\tau(t)$ be the shearing stress on the material, as before. We first consider the single-step *shear* history $\gamma(t) = \gamma_0 H(t)$ (Fig. 1.11). Here $H(t)$ is the Heaviside unit step function, zero for negative t and unity for t zero or positive. If the material were perfectly elastic, the corresponding stress history would be of the form $\tau(t) = \tau_0 H(t)$, constant for t positive. If the material were an ideal viscous fluid, the stress would be instantaneously infinite during the step, and then zero for all positive t. Because we have $\tau = \eta\dot{\gamma}$ for the Newtonian viscous fluid, we have, for $t > 0$:

$$\gamma(t) = \frac{1}{\eta}\int_{-\infty}^{t} \tau\,\mathrm{d}t = \gamma_0, \tag{1.14}$$

and $\gamma = 0$ for $t < 0$. Thus τ has the form of a delta-function of strength $\eta\gamma_0$ (Fig. 1.11).

Observation of real materials shows that neither of these idealized limiting cases is accurate; the stress usually decreases from its initial value rapidly at first, then more gradually, and finally approaches or 'relaxes' to some limiting value τ_∞ (Fig. 1.12).

The limiting value is to some extent a subjective matter, due to measurement problems, but it is a convenient idea. If the limiting value is not zero, we call the

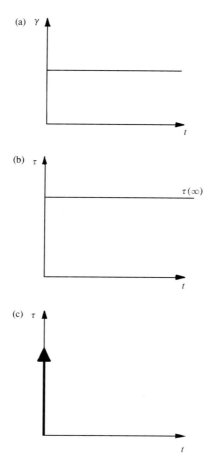

Fig. 1.11   (a) Step shear strain applied at $t=0$. (b) Elastic response to step shear strain. (c) Viscous response to step shear strain. Heavy arrow represents a delta or impulse function.

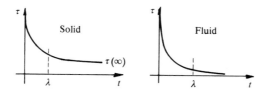

Fig. 1.12   Solid and fluid response to step strain.

material a solid, and if it is zero, and the approach to zero is sufficiently rapid, we call the material a fluid. Evidently the classification depends on an interaction between the nature of the material and the nature of the observation. Let λ be the relaxation time. Here, we provisionally take λ to be some order-of-magnitude

Table 1.7   Viscoelastic fluid properties

Fluid type	Temperature (K) T	Relaxation time (s) λ	Zero-shear viscosity (Pa-s) η_0	Rigidity η_0/λ (Pa)
Water	293	$\sim 10^{-12}$	0.001	10^9
Mineral oil	303	7×10^{-10}	0.5	7×10^8
Poly-dimethysiloxane	303	10^{-6}	0.3	3×10^5
	398	1.7×10^{-4}	100	6×10^5
Low-density polyethylene	388	10	2×10^5	2×10^4
	513	0.1	3000	3×10^4
High-density polyethylene	453	0.07	2000	3×10^4
	493	0.05	1000	2×10^4
High-impact polystyrene	443	7	2×10^5	3×10^4
	483	3	1×10^5	3×10^4
0.5% Hydroxyethyl-cellulose in water	300	0.1	1.3	13
2% Polyisobutylene solution in Primol oil	300	100	1000	10
Glass	300	$> 10^5$	$> 10^{18}$	$\sim 5 \times 10^{10}$

estimate of the time required for stress relaxation to approach completion. Some idea of the spread of relaxation times for various materials can be found from Table 1.7. The distinction between solid and fluid is usually made on the basis of a subjective comparison of the relaxation time and the time of observation. For example, at room temperatures glass is a supercooled liquid, but unless one is interested in times of the order of centuries, it is best regarded as a solid.

Silicone (silly) putty seems odd because its relaxation time is commensurable with our attention span. It will bounce, like an 'elastic solid', the process being complete before there is time for much stress relaxation. It will also flow, like a 'viscous fluid', before human patience runs out.

In the stress relaxation experiment on a sheared sample, the relaxation time λ may be so short that it escapes observation, and the experimenter may then conclude that he or she is dealing with a perfectly elastic solid or a fluid, as the case may be. If λ is so long that no stress relaxation is observed during the period of the experiment, again the observer may conclude that the material is perfectly elastic. We call materials viscoelastic, and use appropriate mathematical models, when the relaxation time and the period of observation are not greatly different.

(b) Creep. Now suppose that the material is subjected to a one-step stress history $\tau(t) = \tau_0 H(t)$ (Fig. 1.13). The response of an elastic solid would be $\gamma(t) = \gamma_0 H(t)$, constant shear for t positive. In a viscous fluid, the shear would increase at a constant rate, $\gamma(t) = \tau_0 t/\eta, \eta$ being the viscosity. Viscoelasticity theory recognizes more refined observations which show departures from these idealizations (Fig. 1.14). The shear may at first jump; in this case the instantaneous

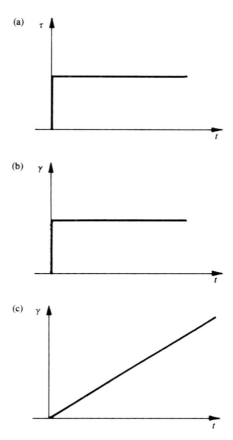

Fig. 1.13 Creep test response. (a) Shear stress applied. (b) Elastic shear strain response (c) Viscous shear strain response.

response is elastic; if it does not jump, it is like a viscous fluid initially. The shear then continues to increase but does so more and more slowly. If it appears to approach a limiting value $\gamma(\infty)$, the material is said to be solid. If it appears to increase linearly after a long time, the material is called fluid. In some cases it is impossible to tell which type of behaviour is occurring, if either. For example, if the amount of shear were increasing in proportion to $t^{1/2}$, one might become convinced by looking at limited data that a limiting value had been reached, or that the shear were increasing linearly, depending on one's preconceived ideas. In such cases a knowledge of the material microstructure is useful.

(c) Response functions. Let $R(\gamma, t)$ be the *stress relaxation* function, the stress t units of time after application of a shear step of size γ. Let $C(\tau, t)$ be the *creep*

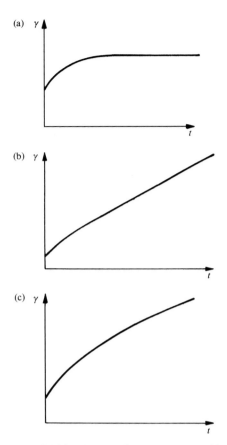

Fig. 1.14 Creep curves. (a) Solid-like shear strain response to suddenly applied shear stress. (b) Fluid-like response. (c) This curve cannot be classified as fluid or solid with the data given.

function, the shear t units of time after application of a stress τ. R and C are supposed to be zero for t negative.

If the material is isotropic (meaning that it has no distinguishable directions) it is evident by symmetry that R is an odd function of γ and C is odd in τ. Hence, assuming smooth dependence and supposing that γ and τ are small, we have

$$R(\gamma, t) = G(t)\gamma + \mathrm{O}(\gamma^3), \tag{1.15}$$

and

$$C(\tau, t) = J(t)\tau + \mathrm{O}(\tau^3). \tag{1.16}$$

The coefficients of the linear terms are the *linear stress relaxation modulus* $G(t)$ and the *linear creep compliance* $J(t)$ (Fig. 1.15). The values of these functions at $t = 0+$ are denoted G_g and J_g (g for glass), and the values at $t = \infty$ are G_e and J_e (e for equilibrium), provided that these values exist. If $J(t)$ tends to increase like $(t + \lambda)/\eta_0$ for large t, η_0 is the *steady-shearing viscosity* and λ is the *mean relaxation time*.

Immediately after application of a step in stress or strain, the response is independent of whether it is the stress or the strain which is to be held constant in the future. Hence, at time $0+$, $\tau = G_g\gamma$ and $\gamma = J_g\tau$. Thus, we find that the initial values of G and J are reciprocal:

$$J_g G_g = 1. \tag{1.17}$$

If the stress and strain approach limiting values, for viscoelastic materials it is irrelevant which one was held absolutely constant for all positive time. Thus, after an infinite time, both $\tau = G_e\gamma$ and $\gamma = J_e\tau$ are true.

Hence,

$$J_e G_e = 1. \tag{1.18}$$

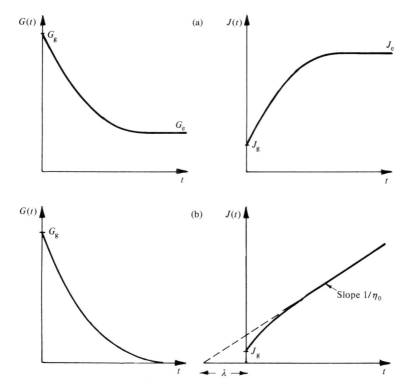

Fig. 1.15   Relaxation modulus G and creep compliance J for (a) solids, (b) fluids.

In fact, $J(t)$ and $G(t)$ are roughly reciprocal at all times; their exact relationship is discussed in Chapter 2, Section 2.6.4, eqn (2.97). They are exactly reciprocal only in the two limiting cases mentioned, and in the trivial case of perfectly elastic response. The reciprocal relations for the two limits can be viewed as consequences of the *assumptions* that the instantaneous response and the equilibrium response are elastic.

(d) Spring-dashpot models. To get some feeling for linear viscoelastic behaviour, it is useful to consider the simpler behaviour of analogue models constructed from linear springs and dashpots. As analogues for stress and strain, we use the total extending force and the total extension. It is, of course, not necessary to assume that the material is constructed of literal springs and dashpots.

The spring is an ideal elastic element obeying the linear force-extension relation $\tau = G\gamma$. Its relaxation modulus is $G(t) = GH(t)$, and its creep compliance is $J(t) = JH(t)$. Here J is $1/G$. [Fig. 1.16(a).]

The dashpot is an ideal viscous element that extends at a rate proportional to the applied force, $\dot{\gamma} = \tau/\eta$. Hence, $J(t) = tH(t)/\eta$ and $G(t) = \eta\delta(t)$, where δ is the impulse or δ-function discussed above. [Fig. 1.16(b).]

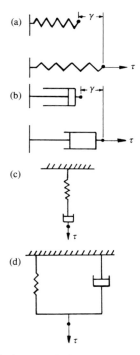

Fig. 1.16 Mechanical models. Displacement is the analogue of strain, force the analogue of stress. (a) Elastic body. (b) Viscous body. (c) Maxwell fluid. (d) Kelvin–Meyer solid.

When two elements are combined in series, their compliances are additive. Thus, for example, the Maxwell model [cf. eqn (1.1)] consisting of a spring and a dashpot in series [Fig. 1.16(c)] has the creep compliance

$$J(t) = (J_g + t/\eta)H(t). \tag{1.19}$$

When two elements are combined in parallel, their moduli are additive. The Kelvin–Meyer model [Fig. 1.16(d)] consisting of a spring and dashpot in parallel has the modulus

$$G(t) = G_e H(t) + \eta \delta(t). \tag{1.20}$$

(e) Superposition. Knowledge of either one of the single-step response functions $G(t)$ or $J(t)$ is sufficient to allow us to predict the output corresponding to any input, within the linear range in which stresses proportional to γ^3 and strains proportional to τ^3 can be neglected.

First, we note that in $G(t)$ and $J(t)$, t is the time *lag* since the application of strain or stress. An input $\gamma(t) = \gamma_0 H(t - t_0)$ would be accompanied by an output $\tau(t) = \gamma_0 G(t - t_0)$; we say that the response is time-translation-invariant. It is here that we insist that only $t - t_0$ appears in the response; if t alone appears then the material is time-dependent in the sense discussed in Section 1.5.

Next, consider the stress response to a two-step history (Fig. 1.17):

$$\gamma(t) = H(t - t_1)\Delta\gamma_1 + H(t - t_2)\Delta\gamma_2. \tag{1.21}$$

The stress can depend on $t, t_1, t_2, \Delta\gamma_1$ and $\Delta\gamma_2$. We assume that it is a smooth function of the step sizes:

$$\tau(t) = G_1(t, t_1, t_2)\Delta\gamma_1 + G_2(t, t_1, t_2)\Delta\gamma_2 + O(\Delta\gamma^3). \tag{1.22}$$

We will neglect the higher-order terms, although it is not essential to the argument to do so.

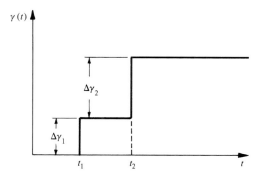

Fig. 1.17   Double-step strain input.

Since this expression is assumed to be valid for all small-enough step sizes, it holds in particular when $\Delta\gamma_2 = 0$. But in that case, the history is a single-step history, for which the stress response must be $G(t - t_1)\Delta\gamma_1$. Hence $G_1(t, t_1, t_2)$ is identified as $G(t - t_1)$. Similarly, $G_2(t, t_1, t_2)$ must be identical with $G(t - t_2)$. Thus, if

$$\gamma(t) = H(t - t_1)\Delta\gamma_1 + H(t - t_2)\Delta\gamma_2$$

then

$$\tau(t) = G(t - t_1)\Delta\gamma_1 + G(t - t_2)\Delta\gamma_2,$$

or, more compactly,

$$\gamma(t) = \sum_1^2 H(t - t_\alpha)\Delta\gamma_\alpha \rightarrow \tau(t) = \sum_1^2 G(t - t_\alpha)\Delta\gamma_\alpha. \tag{1.23}$$

In the linear approximation, the stress is just the sum of the stresses corresponding to each strain step taken separately. Coupling effects depending on both $\Delta\gamma_1$ and $\Delta\gamma_2$ jointly may occur but only in higher-order, non-linear approximations.

The same arguments apply equally well to histories with an arbitrary number of steps. The sums over $\alpha = 1, 2$, above become sums over $\alpha = 1, 2, \ldots, N$.

We can approximate any physically realizable strain history arbitrarily well by a step history involving an arbitrarily large number of arbitrarily small steps. We assume that the response of the material is such that when two strain histories are close together, so are the corresponding stress histories. Then the stress corresponding to any given strain history is nearly the same as the stress for a close-by step history. By passing to the limit in the sums above, we obtain the integral form

$$\gamma(t) = \int_0^t H(t - t')\,d\gamma(t') \rightarrow \tau(t) = \int_{-\infty}^t G(t - t')\,d\gamma(t'). \tag{1.24}$$

Since $G(t)$ is zero for negative t, we can set the upper limit of integration equal to t or ∞ or anything in between, as convenience dictates. The lower limit will usually be written as $-\infty$ in order to be non-committal about when the shearing began, but we implicitly restrict attention to histories with $\gamma = 0$ for all times prior to some starting time, unless and until we find that it makes any sense to do otherwise.

If $\gamma(t)$ has a jump discontinuity $\Delta\gamma_0$ at time t_0, the corresponding contribution to the integral is, of course, $G(t - t_0)\Delta\gamma_0$. Wherever $\gamma(t)$ is differentiable, by $d\gamma(t)$ we mean $\dot{\gamma}(t)dt$.

The relation above is the *stress-relaxation integral* form of the stress–strain relation. All of the preceding remarks hold just as well if we regard the stress history as the input and the strain as the output. We then obtain the *creep integral* form of the relation:

$$\gamma(t) = \int_{-\alpha}^{t} J(t - t') \, d\tau(t'). \tag{1.25}$$

The two forms of the relation must be equivalent, as they describe the same material behaviour. One can also integrate (1.24) by parts and recover the Boltzmann form (1.3). We shall further develop linear viscoelastic theory in Chapter 2 (Sections 2.6.3–6).

1.6 Summary

The present chapter has sought to present some of the main problems in rheology to the reader, to survey some of the unusual flow phenomena that occur, and to discuss typical responses in one-dimensional shearing flows.

The latter is traditional rheology and constituted the main avenue of enquiry until about 1945. Since then the field has developed further as attempts have been made to explain the phenomena sketched here. This book continues to explore these more modern aspects. However, the older approach is still useful and provides some basis for analyses of practical problems—in some cases material is so poorly characterized that nothing else is warranted in any case. Thus no apology is needed for spending time on this phase of rheology. The connection of one-dimensional rheology with real materials is given by Houwink (1953, 1971). We have tried, finally, to present the dilemma always posed by experimental data acquired over limited time domains (long or short). It is not always possible to deduce immediately if one is seeing, for instance, stress relaxation or material degradation, and in this respect it is always useful to make use of whatever knowledge is available of the material's microstructure when setting up mathematical models. This we shall attempt to do in the subsequent discussion, but before proceeding, we briefly review the classical areas of continuum mechanics in Chapter 2.

References

Barnes, H. A. (1997). *J. Non-Newtonian Fluid Mech.*, **70**, 1.
Bingham, E. C. (1929). *J. Rheology*, **1**, 93.
Bird, R. B., Armstrong, R. C., and Hassager, O. (1977). *Dynamics of polymeric liquids,* Vol. 1, *Fluid mechanics*, Wiley, New York.
Boger, D. V. and Walters, K. (1993). *Rheological phenomena in focus*, Elsevier, Amsterdam.
Boltzmann, L. (1874). *Sitzber. Kgl. Akad. Wiss. Wien, Math-Naturwisss. Classe*, **70**, 275.
Broyer, E. and Macosko, C. W. (1976). *Am. Inst. Chem. Engrs J.*, **22**, 268.
Cauchy, A. L. (1827). *Ex. de Math.*, **2**, 42, 60, 108.
Cox, W. P. and Merz, E. H. (1958). *J. Polymer Sci.*, **28**, 619.
De Waele, A. (1923). *Oil Color Chem. Ass. J.*, **6**, 23.
Dugas, R. (1988). *A history of mechanics* (trans. J. R. Maddox). Dover, New York.

Euler, L. (1755). *Mem. de l'Acad. de Sci., Berlin*, **11**, 217.
Fromm, H. (1933). *Ing. Archiv.*, **4**, 432.
Fromm, H. (1947). *ZaMM*, **26**, 146.
Hooke, R. (1676). *A description of helioscopes, and some other instruments*. London.
Houwink, R. (1953). *Elasticity, plasticity and structure of matter*. Harper Press, Washington, DC.
Houwink, R. and De Decker, H. K. (1971). *Elasticity, plasticity and structure of matter*. Cambridge University Press.
Kelvin, Lord (W. Thomson) (1865). *Proc. Roy. Soc. London*, **14**, 289.
Lodge, A. S. (1964). *Elastic liquids*. Academic Press, London.
Maxwell, J. C. (1867). *Phil. Trans. Roy. Soc. Lond.*, **157**, 49.
Meyer, O. E. (1874). *Pogg. Ann. Physik*, (6), **151**, 108.
Navier, C. L. M. H. (1821). *Ann. de Chemie*, **19**, 244.
Navier, C. L. M. H. (1823). *Bull. Soc. Philomath.*, 177.
Newton, I. (1687). *Philosophiae naturlis principia mathematica* (translated A. Motte). Cambridge University Press 1934.
Noll, W. (1955). *J. Rat. Mech. Anal.*, **4**, 3.
Oldroyd, J. G. (1950). *Proc. Roy. Soc. Lond.*, **A 200**, 523.
Ostwald, W. (1925). *Kolloid-Z.*, **36**, 99.
Pipkin, A. C. (1972). *Lectures on viscoelasticity theory*. Springer-Verlag, New York.
Reynolds, O. (1885). *Phil. Mag.*, [5] **20**, 469.
Schwedoff, T. (1890). *J. Physique* [2] **9**, 34.
Stokes, G. G. (1845). *Trans. Cambr. Phil. Soc.*, **8**, 287.
Tanner, R. I. and Walters, K. (1988). *Rheology: an historical perspective*. Elsevier, Amsterdam.
Weber, W. (1835). *Ann. Phys Chem.*, **34**, 247.
Weissenberg, K. (1949). *Proc. 1st Intl. Cong. on Rheology*, **2**, 114.
Wilkinson, W. L. (1960). *Non-Newtonian fluids*. Pergamon Press, Oxford.

Problems

1. Consider Maxwell's equation (1.1). Suppose the input strain ε has the form

(i) $\varepsilon = 0 \quad t \leq 0$
(ii) $\varepsilon = \varepsilon_0 \sin \omega t, \quad t > 0$

(a) If the material is in a stress-free state for $t \leq 0$, find the stress for $t > 0$.

(b) As $t \to \infty$, show that the steady-state response becomes sinusoidal. As $\lambda\omega$ becomes very large, show that the response becomes elastic, and that as $\lambda\omega$ becomes very small, it becomes viscous.

(c) Consider the Kelvin–Meyer model (1.2) with the same excitation. What do you conclude about its behaviour?

2. Write out the equations of motion [(1.6), see Table 1.3] in full. Substitute (1.7) into these equations to eliminate the stresses and find the three Navier–Stokes equations for the velocity components and the pressure.

3. A film-coating arrangement is shown in Fig. 1.18.
Set up boundary conditions on the coating fluid, including the effect of surface tension on the free surface. What role does the atmospheric pressure play if the fluid is incompressible?

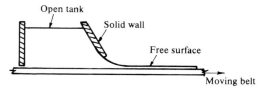

Fig. 1.18 Problem 3.

4. You are asked to devise a simple model of the kind $\tau = f(\dot\gamma)$ for the material shown in Fig. 1.8 which is to be pumped at a rate of 0.2 litre through a 10-mm diameter circular tube. The object of the exercise is to estimate the pressure drop which will be involved. What model would you advise? Assume the flow is laminar, and check this assumption after the solution is obtained.

5. An incompressible isothermal Newtonian fluid is located between two infinite parallel planes at $z = 0$ and $z = h$. The upper plane moves at a speed U in the x-direction and the lower one is stationary. In addition, there is a pressure drop of magnitude ΔP per unit length in the x-direction. Find the velocity profile and the volume rate of flow.

6. In a steady *elongational* flow the velocity field is given by (cylindrical coordinates z, r)

$$w = v_z = \dot\varepsilon z$$
$$u = v_r = -\dot\varepsilon r/2$$

and $v_\theta = 0$; the elongation rate is $\dot\varepsilon$.

Using a Newtonian model, compute the stresses σ_{zz} and σ_{rr}. Find the pressure by equating σ_{rr} to zero (that is, neglecting atmospheric pressure). Hence show that the elongational viscosity $\eta_E \equiv \sigma_{zz}/\dot\varepsilon = 3\eta$, where η is the shear viscosity.

7. Use the data in Fig. 1.2 and fit to it

   (i)   a power-law model

   (ii) a Carreau model

   (iii) a Yasuda model

8. A fluid is to be modelled as a Bingham body. Suppose it flows down a wide, long inclined plane in a parallel sheet (Fig. 1.19). Find the velocity field and the amount of fluid passing the plane $x = 0$ per unit time and per unit length in the z-direction.

9. Consider the flow of (a) a Bingham body, (b) a power-law fluid. Ignoring gravity and surface tension, but not the acceleration terms in the equation of motion, do a dimensional analysis of the flow around a fixed sphere to find the relevant dimensionless variables on which the drag coefficient $C_D (\equiv F/\frac{1}{2}\rho U^2 R^2)$ depends.

Here F is the drag on the sphere, ρ is the fluid density (constant) and U is the speed of the fluid approaching the sphere, radius R.

10. Find the stress relaxation modulus for a Maxwell model [Fig. 1.16(c)].

11. By integrating by parts the result (1.24) show that you can recover the Boltzmann equation (1.3). What is the relation between $G(t)$ and Boltzmann's ϕ-function? For the case when $G = \text{constant} \times \exp(-t/\lambda)$ show that the response is the same as a Maxwell model.

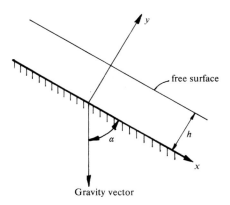

Fig. 1.19   Problem 8.

12. Suppose that the shearing stress $\tau(t)$ on a viscoelastic sample is varied arbitrarily from time $t = 0$ until $t = t_1$, and then is kept at zero. Show that

(a) if $\left. \dfrac{\mathrm{d}J}{\mathrm{d}t} \right|_{t=\infty} = J'(\infty) = 0$.

then the material returns to its initial state after a long time, so $\gamma(\infty) = 0$;

(b) if J behaves otherwise, so that $J'(\infty) = \frac{1}{\eta} \neq 0$

$$\gamma(\infty) = \frac{1}{\eta} \int_0^{t_1} \tau(t)\,\mathrm{d}t.$$

2
REVIEW OF CONTINUUM MECHANICS

2.1 Introduction

THE present chapter is intended as a reference source for the remaining chapters. We use vectors and dyadics (orthogonal base vectors only), Cartesian tensors, and matrix representation of tensors. All of these ideas are easily accessible (for example, Jeffreys and Jeffreys 1956) and will not be commented upon extensively. Besides the balance laws for mass, momentum, and energy we also need constitutive equations and boundary conditions. The latter have already been mentioned (Section 1.4) and will be discussed as they arise in problems. Constitutive relations for non-linear materials are discussed later; in this chapter we consider the classical constitutive laws for the Hookean solid, the Newtonian fluid, and the linear viscoelastic body of Boltzmann. General principles to be obeyed by non-linear constitutive relations are also discussed prior to considering these relations in later chapters.

2.2 Stress

The concept of stress is the way continuum mechanics specifies the mechanical interaction between one part of a material body and another. Here we will develop the classical theory of stress in a simple way. From the microstructural approaches discussed elsewhere in this book one finds directly that the stress tensor is symmetric; furthermore, the author is not aware of any successful experimental detection of lack of symmetry in polymers. Thus, this symmetry is a constitutive assumption which limits the class of materials we shall study; however, there is every reason to expect that it is a good assumption unless some specific knowledge about the microstructure suggests otherwise. Examples of such unusual microstructures can be found in discussions of liquid crystal mechanics (Ericksen 1961).

2.2.1 The stress tensor

In mechanics, we recognize two types of interaction between particles: between touching particles and by action at a distance. In considering a system of particles we must specify the manner in which one particle is influenced by all the others. In a similar way, in continuum mechanics we have to consider the interaction between one part of the body and another.

However, since in a continuum even the smallest volume is deemed to contain a very large number of particles, it would be difficult to approach the interaction problem through the particle concept. Instead, consider a body occupying a

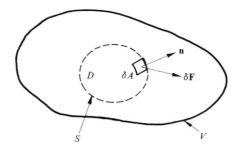

Fig. 2.1 Stress vector definition.

spatial region V at some time (Fig. 2.1). Imagine a closed surface S within the body. Consider the interaction between the material outside S and that in the interior. This interaction can be divided into two kinds: one, due to the action-at-a-distance type of forces such as gravitation and electromagnetic forces, can be expressed as force per unit mass, and is called the *body* force; the other, due to the direct action across the boundary surface S is called the *surface* force (or contact force). To describe the surface force let us consider a small surface element of area δA on the imaginary surface S. From a point on δA, draw a unit vector **n** normal to δA, with its direction *outward* from the interior of S. Then we can distinguish the two sides of δA according to the direction of **n**. Let the side to which this normal vector points be called the positive side. Consider the portion of material lying on the positive side. This part exerts a force $\delta \mathbf{F}$ on the other part which is situated on the negative side of the normal. The force $\delta \mathbf{F}$ depends on the location and size of the area and the orientation of the normal. We make the assumption that as δA tends to zero, the ratio $\delta \mathbf{F}/\delta A$ tends to a definite limit $d\mathbf{F}/dA$, and that the moment of the force acting on the surface δA about any point within the area vanishes in the limit. The limiting vector will be written as

$$\mathbf{t} = \frac{d\mathbf{F}}{dA}. \tag{2.1}$$

The limiting vector **t** is called the *traction* vector or the stress vector and represents the force per unit area acting on the surface. There is no a priori reason why the interaction of the material on the two sides of the surface element δA must be momentless. Clearly, the most general action on δA from the outside would consist of a force, as we have assumed, plus a moment ($\delta \mathbf{M}$). Here we will suppose that as δA tends to zero the effect of any residual moment vanishes, so that $\delta \mathbf{F}$ describes the mutual action across δA completely. Similarly, considering actions-at-a-distance (gravity, for example) on a small volume of material, we will suppose that no body moments are present, and that a body force vector per unit mass is sufficient to describe actions at a distance completely. With these assumptions one finds, from an angular momentum balance, that the stress

tensor is symmetric. While theories involving moments and couple stresses can
be constructed (Truesdell and Noll 1965) we will not use them here.

2.2.2 Definition of stress: Notation

Consider a point O in a body and an element of area δA (Fig. 2.2), which is normal
to the direction x. The surface or contact force $\mathbf{t}\,\delta A$ can be resolved along the x, y,
z directions respectively into components; then the components of force per unit
area in these directions can be found. These components per unit area will be
called σ_{xx} (direct stress), σ_{xz} and σ_{xy} (shear stresses) respectively; in the case of
the shear stresses the first suffix denotes the direction of the normal to the
area, and the second the direction of resolution. We shall take a tensile stress
as positive (pressure negative); the reader should be aware of the opposite convention that is used sometimes. Figure 2.3 shows positive stress components
on the faces of a block. By considering the moments due to the stresses on
the block (or any other shaped piece) we can easily see that, in the absence of

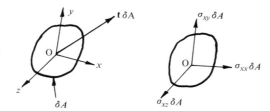

Fig. 2.2 Resolution of traction (or stress) vector.

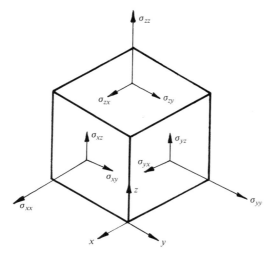

Fig. 2.3 Notation used for stress.

couples on the faces, as assumed in Section 2.2.1, that the stress tensor is symmetrical, so that

$$\sigma_{xy} = \sigma_{yx} \text{ etc.} \tag{2.2}$$

A more formal derivation of this result is given by Huilgol and Phan-Thien (1997, pp. 95–6). We now find it convenient to replace x by x_1, y by x_2 and z by x_3, so that σ_{yz} for example, is written as σ_{23}.

If the stress components on the face normal to **n** are σ_{ij}, $(i, j = 1, 2, 3)$ then the components t_i of **t** are connected to the stress components by the decomposition (or resolution of forces)

$$t_i = \sigma_{ij} n_j, \tag{2.3}$$

where n_j are the components of the outward normal unit vector **n**. (We always assume that a sum is taken over repeated indices unless the contrary is stated.)

2.2.3   Properties of the stress tensor

We can represent the stress tensor $\boldsymbol{\sigma}$, with components σ_{ij}, by regarding $\boldsymbol{\sigma}$ as a symmetric matrix of which σ_{ij} are the components, or by writing it in dyadic form, using unit vectors along three orthogonal axes.

All the properties of symmetric matrices are enjoyed by the stress tensor, and we know that (Jeffreys and Jeffreys 1956):

(1) The stress tensor possesses a transformation law from axes **x** to **x'** so that if we write

$$\mathbf{x}' = \mathbf{R}\mathbf{x} \quad (x_i' = R_{ij} x_j), \tag{2.4}$$

where **R** is an orthogonal matrix whose elements are given by

$$R_{ij} = \cos(x_i', x_j)$$

(one should read this definition as "R_{ij} is the cosine of the angle between x_i' and x_j"), then

$$\boldsymbol{\sigma}' = \mathbf{R}\boldsymbol{\sigma}\mathbf{R}^T \quad (\sigma_{ij}' = R_{im} R_{jn} \sigma_{mn}), \tag{2.5}$$

where $\mathbf{R}^T$ is the transpose of **R**. Since $\boldsymbol{\sigma}$ is symmetric $\boldsymbol{\sigma}^T = \boldsymbol{\sigma}$.

The reader is reminded that the transpose $\mathbf{R}^T$ of an orthogonal matrix **R** is also the inverse $\mathbf{R}^{-1}$. Hence $\mathbf{R}^T\mathbf{R} = \mathbf{R}^{-1}\mathbf{R} = \mathbf{I}$, the unit matrix. Taking the determinant of this expression, we find $\det \mathbf{R} = \pm 1$. The negative sign occurs in transformations that change from a right- to a left-handed coordinate system or vice versa. We shall not normally permit such transformations, and hence $\det \mathbf{R} = 1$, so that **R** is a proper orthogonal matrix.

The transformation (2.5) in the plane case is conveniently represented in terms of a single angle θ giving the rotation of the $\mathbf{x}'$ system relative to the $\mathbf{x}$ system, and the Mohr circle is a convenient graphical representation of this transformation; see Problem 2.2 and Timoshenko and Goodier (1951).

(2) At each point in the field of σ_{ij} there is a traction vector $\mathbf{t}$ (components t_i) equal to $\sigma_{ij}n_j$, where the n_j define a unit vector and a plane normal to this vector. If we try to find a plane so that its normal $\mathbf{n}$ is a vector parallel to $\mathbf{t}$, then there is no shear on this plane (see Fig. 2.2), and we can write

$$\mathbf{t} = \boldsymbol{\sigma}\mathbf{n} = \sigma\mathbf{n}, \tag{2.6}$$

where σ is a constant of proportionality. In order that there be a non-trivial solution for $\mathbf{n}$ it is necessary, from (2.6), that

$$\det(\boldsymbol{\sigma} - \sigma\mathbf{I}) = 0, \tag{2.7}$$

where $\mathbf{I}$ is the unit matrix.

This is the characteristic equation for $\boldsymbol{\sigma}$. Expanding the determinant in (2.7) gives, for our 3×3 matrices, a cubic equation for σ. The three values of σ ($\sigma = \sigma_1$, σ_2, σ_3) are the eigenvalues or principal values of $\boldsymbol{\sigma}$. In the present context they will be called the principal stresses. It can be shown (see, for example, Jeffreys and Jeffreys 1956) that σ_1, σ_2, and σ_3 are *real* numbers. Each principal stress has a direction $\mathbf{n}$ associated with it as a solution of (2.6), and it may be proved that these vectors are mutually perpendicular. On planes normal to these directions the shear stresses are zero; the direct stresses will be ordered so that $\sigma_1 > \sigma_2 > \sigma_3$.

(3) Expansion of eqn (2.7) gives the cubic characteristic equation for σ

$$\sigma^3 - I_\sigma\sigma^2 + II_\sigma\sigma - III_\sigma = 0, \tag{2.8}$$

where the coefficients are given by

$$I_\sigma = \sigma_{ii} = \operatorname{tr}\boldsymbol{\sigma} \tag{2.9}$$

$$II_\sigma = \tfrac{1}{2}(\sigma_{ii}\sigma_{jj} - \sigma_{ij}\sigma_{ij}) \tag{2.10}$$

$$III_\sigma = |\sigma_{ij}| = \det\boldsymbol{\sigma}. \tag{2.11}$$

The quantities I_σ, II_σ, III_σ are called the first, second, and third invariants of the stress tensor respectively. These and similar invariants are important in the study of isotropic materials.

If the stresses are principal, so that $\boldsymbol{\sigma}$ is diagonal, with values σ_1, σ_2, σ_3, then, following the summation convention, eqns (2.9)–(2.11) take the simple form

$$I_\sigma = \sigma_1 + \sigma_2 + \sigma_3 \tag{2.9a}$$

$$II_\sigma = \tfrac{1}{2}\left[(\sigma_1 + \sigma_2 + \sigma_3)^2 - \sigma_1^2 - \sigma_2^2 - \sigma_3^2\right] \tag{2.10a}$$

$$= (\sigma_1\sigma_2 + \sigma_2\sigma_3 + \sigma_3\sigma_1).$$

$$III_\sigma = \sigma_1\sigma_2\sigma_3. \tag{2.11a}$$

We can write out (2.9)–(2.11) in full for a set of Cartesian axes (x, y, z):

$$I_\sigma = \sigma_{xx} + \sigma_{yy} + \sigma_{zz} \tag{2.9b}$$

$$II_\sigma = (\sigma_{yy}\sigma_{zz} + \sigma_{zz}\sigma_{xx} + \sigma_{xx}\sigma_{yy})$$

$$- (\sigma_{yz}^2 + \sigma_{xz}^2 + \sigma_{xy}^2) \tag{2.10b}$$

$$III_\sigma = \det \boldsymbol{\sigma} = \sigma_{xx}\sigma_{yy}\sigma_{zz} + 2\sigma_{yz}\sigma_{zx}\sigma_{xy}$$

$$- \sigma_{xx}\sigma_{yz}^2 - \sigma_{yy}\sigma_{xz}^2 - \sigma_{zz}\sigma_{yx}^2. \tag{2.11b}$$

These stress invariants do not change as we rotate axes (say from $\mathbf{x}$ to $\mathbf{x}'$). To see this note that the roots of eqn (2.8) (σ_1, σ_2, σ_3) are fixed numbers independent of whatever axes we measure from; the principal stresses in fact *describe* the stress state in magnitude, but not orientation. Hence, because the principal stresses are fixed, so must the coefficients of the cubic (2.8) also be fixed. I_σ, II_σ, III_σ are useful for characterizing a given stress state when the material is isotropic. It is, of course, possible to prove that (2.9)–(2.11) are invariant by other methods which are more laborious.

(4) In the above it has been assumed that all the principal values are different. If any two (or all three) are equal, then some symmetry occurs and the principal directions are not unique. Usually this is of little consequence in applications, and we shall not spend time on these exceptions; they are fully described in many texts on matrix algebra.

(5) It is of interest to note that the principal stresses are extremal values of the direct or normal stresses. The extremal values of *shear* stress occur on three planes whose normals are orthogonal to one another and inclined at 45 degrees to the principal axes. It can be shown that these extremal shear stresses are (see Timoshenko and Goodier (1951, p. 218)).

$$\tfrac{1}{2}|\sigma_1 - \sigma_2|, \quad \tfrac{1}{2}|\sigma_2 - \sigma_3| \quad \text{and} \quad \tfrac{1}{2}|\sigma_3 - \sigma_1|,$$

and hence that the largest is $\tfrac{1}{2}|\sigma_1 - \sigma_3|$.

2.2.4 Isotropic functions of tensors

We shall frequently wish to express a scalar function of a tensor or matrix so that the value of the scalar function does not depend on any special co-ordinate system. The invariants I_σ, II_σ, and III_σ (or combinations of these quantities) are used to perform this operation. If, for example, the viscosity η is a function of the stress state, we write

$$\eta(\boldsymbol{\sigma}) = \eta(I_\sigma, II_\sigma, \ III_\sigma). \tag{2.12}$$

The scalar viscosity is independent of coordinate rotations as is appropriate for an isotropic body.

We shall now consider how to address the problem of writing down a matrix or tensor expression that depends on the stress state (or other state described by a second order tensor or matrix) in a direction-independent way. To begin, we note that (2.8) is satisfied when $\sigma = \sigma_1$, σ_2 or σ_3 (supposed unequal). These three equations can be rewritten in matrix form.

$$
\begin{bmatrix} \sigma_1 & & \\ & \sigma_2 & \\ & & \sigma_3 \end{bmatrix}^3 - I_\sigma \begin{bmatrix} \sigma_1 & & \\ & \sigma_2 & \\ & & \sigma_3 \end{bmatrix}^2
$$

$$
+ II_\sigma \begin{bmatrix} \sigma_1 & & \\ & \sigma_2 & \\ & & \sigma_3 \end{bmatrix} - III_\sigma \begin{bmatrix} 1 & & \\ & 1 & \\ & & 1 \end{bmatrix} = \mathbf{O}. \tag{2.13}
$$

Since the nth-power of a diagonal matrix $\{\sigma_1, \sigma_2, \sigma_3\}$ is just another diagonal matrix $\{\sigma_1^n, \sigma_2^n, \sigma_3^n\}$, we can write $\{\sigma_1^n, \sigma_2^n, \sigma_3^n\}$ as $\boldsymbol{\sigma}^n$. Hence (2.13) may be expressed compactly as

$$\boldsymbol{\sigma}^3 - I_\sigma \boldsymbol{\sigma}^2 + II_\sigma \boldsymbol{\sigma} - III_\sigma \mathbf{I} = \mathbf{O}. \tag{2.13a}$$

Thus $\boldsymbol{\sigma}$ satisfies its own characteristic equation, and this result is known as the Cayley–Hamilton theorem. Equation (2.13a) enables one to express $\boldsymbol{\sigma}^3$ as a function of lower powers of $\boldsymbol{\sigma}$:

$$\boldsymbol{\sigma}^3 = I_\sigma \boldsymbol{\sigma}^2 - II_\sigma \boldsymbol{\sigma} + III_\sigma \mathbf{I}, \tag{2.13b}$$

and by multiplying (2.13b) by $\boldsymbol{\sigma}$ we can extend this rule to $\boldsymbol{\sigma}^4$:

$$\boldsymbol{\sigma}^4 = I_\sigma \boldsymbol{\sigma}^3 - II_\sigma \boldsymbol{\sigma}^2 + III_\sigma \boldsymbol{\sigma}, \tag{2.13c}$$

by reapplying (2.13b) we find

$$\boldsymbol{\sigma}^4 = (I_\sigma^2 - II_\sigma)\boldsymbol{\sigma}^2 + (III_\sigma - I_\sigma II_\sigma)\boldsymbol{\sigma} + I_\sigma III_\sigma \mathbf{I}. \tag{2.13d}$$

The net result of such reductions enables us to replace the integer powers of $\sigma(n \gtrless 3)$ by a function of the form, say

$$\sigma^n = \alpha_n \sigma^2 + \beta_n \sigma + \gamma_n \mathbf{I}, \tag{2.14}$$

where α_n, β_n, and γ_n are functions of the invariants of σ. This result is a special case of the general rule for isotropic tensor functions of a single tensor (Truesdell and Noll 1965, p. 27 *et seq.*):

$$\mathbf{F}(\sigma) = f_2 \sigma^2 + f_1 \sigma + f_0 \mathbf{I}, \tag{2.15}$$

where f_0, f_1 and f_2 are functions of the invariants of σ. The isotropic property of the representation (2.15) can be stated as follows. Under a rotation $\mathbf{R}$ of axes the left-hand side takes up a value $\mathbf{F}(\mathbf{R}\sigma\mathbf{R}^T)$. In an isotropic function we demand that this be equal to $\mathbf{R}\mathbf{F}(\sigma)\mathbf{R}^T$. The form (2.15) satisfies this criterion, and gives a useful rule for expanding tensor functions of a single variable.

2.2.5   Remarks about other tensors and coordinate systems

All of the above properties of the stress tensor are mathematical consequences that follow from the symmetry of the tensor and the transformation rule (2.5). Hence any symmetric tensor quantity that transforms according to (2.5) will also possess these qualities.

While the discussion above has used Cartesian coordinates, any orthogonal set of axes (for example, cylindrical or spherical polar coordinates) will behave in the same way, since only a point in space is dealt with in the transformation.

2.3   Motion and deformation

We shall usually find it convenient to consider a fixed Eulerian frame of reference as is normally employed in fluid mechanics. Thus we can consider a particle P in a body (Fig. 2.4) which is presently (time t) located at a position (or place) $\mathbf{x}$ with

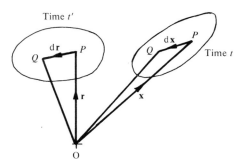

Fig. 2.4   Motion of a body from time t' to time t.

respect to an origin O. At some other time t' it will be at a place **r** with respect to O. Thus we regard **x**, the Eulerian place, as fixed in space; it is occupied by different particles (in general) at different times. The particle traces out a *path line*. The actual trajectory of P is given by $\mathbf{r}(t')$, and

$$\mathbf{r}(t) = \mathbf{x}. \tag{2.16}$$

2.3.1 Velocity and acceleration: Streamlines

The velocity **v** of particle P is given by

$$\mathbf{v}(t') = \frac{d\mathbf{r}(t')}{dt'}. \tag{2.17}$$

Similarly, we can define the particle velocity in the usual Eulerian way, so that for each **x**, at time t, there is a velocity

$$\mathbf{v} = \mathbf{v}(\mathbf{x}, t). \tag{2.18}$$

Clearly, when (2.17) is evaluated at time t and place **x**, it must give the same vector as (2.18), so that the two are equivalent. In fact, the argument of **v** in (2.17) could be written $(\mathbf{x}, t')$ so as to remind one that the particle P is one which passes through **x** at t.

The velocity vector is tangent to a streamline at a point. Thus if $d\xi$ is a portion of a streamline, we have the defining equation

$$d\xi = k\mathbf{v}, \tag{2.19}$$

where k is a constant.

In steady Eulerian flows the streamlines and the particle pathlines coincide. The acceleration of the particle P at point **x** is given by the usual hydrodynamic derivative (in Cartesian coordinates).

$$a_i(\mathbf{x}, t) = \dot{v}_i(\mathbf{x}, t) = \frac{\partial v_i}{\partial t} + v_j \frac{\partial v_i}{\partial x_j} \equiv \frac{D v_i}{D t}. \tag{2.20}$$

The D/Dt derivative will thus be used as a notation for the particle-following (or material) derivative as is usual in fluid mechanics. For other orthogonal coordinate systems, we can write

$$\mathbf{a} = \frac{\partial \mathbf{v}}{\partial t} + (\mathbf{v} \cdot \mathbf{\nabla})\mathbf{v} = \frac{D\mathbf{v}}{Dt}, \tag{2.20a}$$

where the grad operator is interpreted in the relevant coordinate system.

The Appendix at the end of this book tabulates results in several coordinate systems.

2.3.2   Velocity gradient, rate of deformation, and vorticity

In fluid dynamics we compute the velocity gradient as the rate of change of velocity with respect to the (fixed) Eulerian coordinate system. Formally we have $v_i = v_i(x_i, t)$.

The rate of change of the velocity component v_i with respect to the coordinate x_j is given by $\partial v_i / \partial x_j$, or, for the total change $\mathrm{d}v_i$ (at a fixed time):

$$\mathrm{d}v_i = L_{ij}\,\mathrm{d}x_j = \frac{\partial v_i}{\partial x_j}\,\mathrm{d}x_j, \tag{2.21}$$

or, in direct notation

$$\mathrm{d}\mathbf{v} = \mathbf{L}\,\mathrm{d}\mathbf{x},$$

where $\mathbf{L}$ is the velocity gradient tensor.

It is tempting to write $\mathbf{L} = \boldsymbol{\nabla}\mathbf{v}$, but this is not logical since $(\boldsymbol{\nabla}\mathbf{v})_{ij} = \partial v_j / \partial x_i$, which is the tensor L_{ji}, the transpose of L_{ij}. Since $\mathbf{L}$ is not symmetric, we shall write:

$$\mathbf{L}^T = \boldsymbol{\nabla}\mathbf{v} \quad \left(\text{or } L_{ji} = \frac{\partial v_j}{\partial x_i}\right). \tag{2.22}$$

Then (2.22) can be cast into forms appropriate for non-cartesian systems; see Appendix.

We split L_{ij} into symmetric and antisymmetric parts

$$L_{ij} = \frac{1}{2}\left(\frac{\partial v_i}{\partial x_j} + \frac{\partial v_j}{\partial x_i}\right) + \frac{1}{2}\left(\frac{\partial v_i}{\partial x_j} - \frac{\partial v_j}{\partial x_i}\right), \tag{2.23}$$

or

$$L_{ij} = d_{ij} + w_{ij}, \tag{2.24}$$

which defines the rate of deformation tensor (d_{ij}) and the vorticity tensor (w_{ij}) respectively. The principal values of the symmetric d_{ij} describe the rates of stretching of filaments in the principal directions (see eqn 2.35a), while the three non-zero components of the antisymmetric w_{ij} are alternatively computed from half of the vorticity $\frac{1}{2}\boldsymbol{\nabla}\times\mathbf{v}$. Precisely, $w_{12} = -\frac{1}{2}\omega_3$, $w_{13} = \frac{1}{2}\omega_2$, $w_{23} = -\frac{1}{2}\omega_1$ where $\boldsymbol{\omega} = \boldsymbol{\nabla}\times\mathbf{v}$. The vorticity gives the solid-body rate of spin at a point in the fluid; if $\boldsymbol{\omega} = 0$, then the flow is *irrotational*.

2.3.3   Deformation gradient

The quantities d_{ij} and w_{ij} are sufficient for Newtonian fluid mechanics but additional quantities are required to describe solid materials or materials with memory. Looking at Fig. 2.4, if P and Q are neighbouring points separated by a small distance d**x** at time t, and by a distance d**r** at time t', then several ways of defining strain tensors exist.

We observe that a particle P which is at a place **x** at time t is at **r** at time t', so

$$\mathbf{r}(t') = \mathbf{r}(\mathbf{x}, t) \quad [\text{or } r_i = r_i(x_i, t)] \tag{2.25}$$

describes the deformation. As we move to a neighbouring particle Q (Fig. 2.4) the mapping (2.25) means that the relation between d**r** and d**x** is:

$$dr_i = \frac{\partial r_i}{\partial x_j} \, dx_j \tag{2.26}$$

We define the deformation gradient tensor **F** relating d**r** and d**x** as

$$d\mathbf{r} = \mathbf{F} \, d\mathbf{x} \quad (dr_i = F_{ij} \, dx_j), \tag{2.27}$$

or

$$F_{ij} = \frac{\partial r_i}{\partial x_j}. \tag{2.28}$$

Note that **F** is not necessarily symmetric.

Alternatively we have, for convenience in non-Cartesian orthogonal systems

$$\mathbf{F}^T = \mathbf{\nabla} \mathbf{r}. \tag{2.29}$$

Often $\mathbf{F}(t')$ is called the relative deformation gradient, and is given a subscript to denote what time it is relative to. Since we mainly use $\mathbf{r}(t')$ and the gradient with respect to **x**, we shall not use an elaborate notation. Clearly $\mathbf{F}(t) = \mathbf{I}$, since d**x** and d**r** then coincide.

In an incompressible medium the volume of an element of material $(dx_1 \, dx_2 \, dx_3)$ transforms under **F** into an element of equal volume in **r**-space. Thus the Jacobian of the transformation **F** must be unity, or

$$\det \mathbf{F} = 1. \tag{2.30}$$

2.3.4 Strain

To relate the magnitudes of d**x** and d**r** we can form dr^2 and find

$$dr^2 = d\mathbf{r} \cdot d\mathbf{r} = d\mathbf{r}^T \, d\mathbf{r} = d\mathbf{x}^T \mathbf{F}^T \mathbf{F} \, d\mathbf{x}. \tag{2.31}$$

The quantity $\mathbf{F}^T \mathbf{F}$ will be called the strain tensor; often it is termed the right relative Cauchy–Green strain tensor (Truesdell and Noll 1965), but we shall mainly use it (and its inverse) and hence the simplified terminology will be preferred. We define the strain tensor $\mathbf{C}$ as

$$\mathbf{C} = \mathbf{F}^T \mathbf{F}, \tag{2.32}$$

and the inverse strain tensor as

$$\mathbf{C}^{-1} = \mathbf{F}^{-1}\mathbf{F}^{-T}. \tag{2.33}$$

Clearly, $\mathbf{C}$ and $\mathbf{C}^{-1}$ are both symmetric tensors. In any medium the principal values of $\mathbf{C}$ represent the squares of the relative elongation ratios of filaments of material. Similarly, it may be shown (Truesdell and Noll 1965) that the principal values of $\mathbf{C}^{-1}$ measure relative area changes in an incompressible medium, in which, from (2.30), we must have $\det \mathbf{C} = \det \mathbf{C}^{-1} = 1$.

2.3.5   Rate of change of strain. The Rivlin–Ericksen tensors

Suppose the pathlines of the motion are known and that the strain history $\mathbf{C}(t')$ is known for all time t'. One can then differentiate $\mathbf{C}$ with respect to t', assuming we have a sufficiently smooth motion. We have, from (2.31) and (2.28),

$$C_{ij}(t') = F_{ki}F_{kj} \equiv \frac{\partial r_k}{\partial x_i}\frac{\partial r_k}{\partial x_j}. \tag{2.34}$$

Differentiating with respect to t' we obtain, since r_k is a function of t', (but x_j is not)

$$\frac{dC_{ij}}{dt'} = \frac{\partial r_k}{\partial x_i}\frac{\partial v_k}{\partial x_j} + \frac{\partial v_k}{\partial x_i}\frac{\partial r_k}{\partial x_j}. \tag{2.35}$$

Hence, at $t' = t$, when $r_k = x_k$, we have

$$\left.\frac{dC_{ij}}{dt'}\right|_{t'=t} = \delta_{ki}\frac{\partial v_k}{\partial x_j} + \frac{\partial v_k}{\partial x_i}\delta_{ki} = \frac{\partial v_i}{\partial x_j} + \frac{\partial v_j}{\partial x_i} = 2d_{ij} = A_{ij}^{(1)}, \tag{2.35a}$$

where $\mathbf{A}^{(1)}$ is the first Rivlin–Ericksen tensor, and δ_{ki} is the unit tensor; $\delta_{ki} = 1$ if $k = i$ and is zero otherwise.

Differentiating (2.35) again, we obtain

$$\frac{d^2C_{ij}}{dt'^2} = \frac{\partial v_k}{\partial x_i}\frac{\partial v_k}{\partial x_j} + \frac{\partial r_k}{\partial x_i}\frac{d}{dt'}\left(\frac{\partial v_k}{\partial x_j}\right) + \frac{d}{dt'}\left(\frac{\partial v_k}{\partial x_i}\right)\frac{\partial r_k}{\partial x_j} + \frac{\partial v_k}{\partial x_i}\frac{\partial v_k}{\partial x_j}. \tag{2.36}$$

Now using the chain rule of differentiation, we have:

$$\frac{d}{dt'}\left(\frac{\partial v_k}{\partial x_j}\right) = \frac{d}{dt'}\left(\frac{\partial v_k}{\partial r_m}\frac{\partial r_m}{\partial x_j}\right) = \frac{d}{dt'}\{L_{km}(t')\}\frac{\partial r_m}{\partial x_j} + \frac{\partial v_k}{\partial r_m}\frac{\partial v_m}{\partial x_j}. \tag{2.37}$$

Hence, we find, as $t' \to t$,

$$\left.\frac{\mathrm{d}^2 C_{ij}}{\mathrm{d}t'^2}\right|_{t' \to t} = 2L_{mi}L_{mj} + \frac{\mathrm{d}}{\mathrm{d}t'}\{L_{ij} + L_{ji}\} + L_{im}L_{mj} + L_{jm}L_{mi}$$

$$= \frac{\mathrm{d}}{\mathrm{d}t'}A_{ij}^{(1)} + L_{mi}A_{mj}^{(1)} + L_{mj}A_{mi}^{(1)}. \tag{2.38}$$

We term this quantity the second Rivlin–Ericksen tensor $A_{ij}^{(2)}$.

Because of the fact that we have followed a particle by working in the t' time framework, the time derivative $\mathrm{d}\mathbf{A}^{(1)}/\mathrm{d}t'|_{t' \to t}$ in eqn (2.38) must be replaced by $\mathrm{D}A_{ij}^{(1)}/\mathrm{D}t$ when the result is referred to fixed Eulerian co-ordinates. By continuing this differentiation process one finds a recursion formula for $A_{ij}^{(n+1)}$ in terms of L_{ij} and the $A_{ij}^{(n)}$ (Rivlin and Ericksen 1955). In Eulerian terms (referred to $\mathbf{x}$ and t) we find

$$A_{ij}^{(n+1)} = \frac{\mathrm{D}A_{ij}^{(n)}}{\mathrm{D}t} + L_{mi}A_{mj}^{(n)} + L_{mj}A_{mi}^{(n)}, \tag{2.39}$$

or

$$\mathbf{A}^{(n+1)} = \frac{\mathrm{D}\mathbf{A}^{(n)}}{\mathrm{D}t} + \mathbf{L}^{\mathrm{T}}\mathbf{A}^{(n)} + \mathbf{A}^{(n)}\mathbf{L}. \tag{2.39a}$$

Clearly, the Rivlin–Ericksen tensors are symmetric.

We can thus replace C_{ij} in sufficiently smooth motions by a Taylor series in the $\mathbf{A}_{ij}^{(n)}$. Since $C_{ij}(t)$ is the unit tensor δ_{ij}, then we find

$$C_{ij}(t') = \delta_{ij} - A_{ij}^{(1)}(t - t') + \frac{1}{2!}A_{ij}^{(2)}(t - t')^2 + \cdots \tag{2.40}$$

which is sometimes useful.

2.3.6 *Some kinematic examples—shear and planar elongation*

(a) Consider a shearing deformation in the $x_1 - x_2$ plane where the velocity field is given (Fig. 1.1) in Eulerian coordinates by $\mathbf{v} = (\dot{\gamma}x_2, 0, 0)$; that is, flow is in the x_1-direction, with a shear rate $\dot{\gamma}$. The particles move along straight lines and their positions are described by

$$r_1(t') = x_1 + \dot{\gamma}x_2(t' - t)$$
$$r_2(t') = x_2; \quad r_3(t') = x_3. \tag{2.41}$$

Hence, $\mathbf{r}(t) = \mathbf{x}$, and $v_1(t') = \mathrm{d}r_1/\mathrm{d}t' = \dot{\gamma}x_2 = \dot{\gamma}r_2$. The streamlines are clearly the lines of constant x_2, x_3, and the acceleration of a particle is zero. The velocity gradient tensor is, in matrix form,

$$
\mathbf{L}(t) = \begin{bmatrix} \dfrac{\partial v_1}{\partial x_1} & \dfrac{\partial v_1}{\partial x_2} & \dfrac{\partial v_1}{\partial x_3} \\[2mm] \dfrac{\partial v_2}{\partial x_1} & \dfrac{\partial v_2}{\partial x_2} & \dfrac{\partial v_2}{\partial x_3} \\[2mm] \dfrac{\partial v_3}{\partial x_1} & \dfrac{\partial v_3}{\partial x_2} & \dfrac{\partial v_3}{\partial x_3} \end{bmatrix} \tag{2.42}
$$

and all terms except $\partial v_1/\partial x_2$, equal to $\dot{\gamma}$, are zero in shearing. The rate of deformation tensor then has $d_{12} = d_{21} = \tfrac{1}{2}\dot{\gamma}$, all other terms zero, and $w_{12} = -w_{21} = \tfrac{1}{2}\dot{\gamma}$, all other terms being zero.

The deformation gradient $\mathbf{F}$ can be computed from (2.41) to be, in matrix form,

$$
\mathbf{F}(t') = \begin{bmatrix} 1 & \dot{\gamma}(t'-t) & 0 \\ 0 & 1 & 0 \\ 0 & 0 & 1 \end{bmatrix} \tag{2.43}
$$

and det $\mathbf{F} = 1$, as required.

The strain tensor $\mathbf{C}(t')$ can be computed from (2.43) to be

$$
\mathbf{C}(t') = \mathbf{F}^T\mathbf{F} = \begin{bmatrix} 1 & \dot{\gamma}(t'-t) & 0 \\ \dot{\gamma}(t'-t) & 1 + \dot{\gamma}^2(t'-t)^2 & 0 \\ 0 & 0 & 1 \end{bmatrix}. \tag{2.44}
$$

Note, $\mathbf{C}$ and $\mathbf{d}$ are symmetric; while $\mathbf{F}$ and $\mathbf{L}$ are not; $\mathbf{w}$ is antisymmetric. $\mathbf{C}^{-1}$ is also symmetric and has the same pattern of entries as $\mathbf{C}$:

$$
\mathbf{C}^{-1}(t') = \begin{bmatrix} 1 + \dot{\gamma}^2(t-t')^2 & -\dot{\gamma}(t'-t) & 0 \\ -\dot{\gamma}(t'-t) & 1 & 0 \\ 0 & 0 & 1 \end{bmatrix}. \tag{2.44a}
$$

One can compute $\mathbf{A}^{(1)} = 2\mathbf{d}$, and $\mathbf{A}^{(2)}$ from (2.39a). $\mathbf{A}^{(2)}$ is of the form, since $D\mathbf{A}^{(1)}/dt = \mathbf{0}$, where all elements are zero except $(\mathbf{A}^{(2)})_{22}$, equal to $2\dot{\gamma}^2$.

In this case, one can show that all $\mathbf{A}^{(n)}$ for $n \geq 3$ are zero.

(b) In a contrasting case, we consider planar elongation where $v_1 = \dot{\varepsilon}x_1$, $v_2 = -\dot{\varepsilon}x_2$, $v_3 = 0$; the elongation rate is $\partial v_1/\partial x_1 = \dot{\varepsilon}$. Particles move according to eqn (2.17) so that

$$
\frac{dr_1}{dt'} = \dot{\varepsilon}r_1; \quad \frac{dr_2}{dt'} = -\dot{\varepsilon}r_2; \quad \frac{dr_3}{dt'} = 0. \tag{2.45}
$$

The solutions of (2.45) satisfying $\mathbf{r}(t) = \mathbf{x}$ are

$r_1 = x_1 \exp[\dot{\varepsilon}(t' - t)];$

$r_2 = x_2 \exp[-\dot{\varepsilon}(t' - t)],$

$r_3 = x_3.$

Note the exponential separation of particles, in contrast to the algebraic separation in the shear case. The streamlines are hyperbolic, $x_1 x_2 = \text{constant}$. The $\mathbf{L}$ tensor is diagonal, $[\dot{\varepsilon}, -\dot{\varepsilon}, 0]$; the acceleration $\mathbf{a}$ is $\dot{\varepsilon}^2 (\mathbf{i} x_1 + \mathbf{j} x_2)$.

The vorticity is zero, and all matrices are diagonal: $\mathbf{d} = \text{diag} \{\dot{\varepsilon}, -\dot{\varepsilon}, 0\}$, so $\mathbf{d} = \mathbf{L}$, $\mathbf{F} = \text{diag} \ (X, X^{-1}, 1)$, where $X = \exp[\dot{\varepsilon}(t' - t)]$, and $\mathbf{C}^{-1}$ is diag $(X^{-2}, X^2, 1)$. The second Rivlin–Ericksen tensor $\mathbf{A}^{(2)} = \text{diag} \ (2\dot{\varepsilon}^2, 2\dot{\varepsilon}^2, 0)$.

2.4   Conservation of mass, momentum, and energy

We have already considered, in effect, the conservation of angular momentum by insisting on the symmetry of the stress tensor (Sections 2.2, 2.2.1, and 2.2.2). We now consider the conservation of mass, linear momentum, and energy. For most of this book, up to Chapter 9, the energy conservation principle will not be needed.

There are many derivations of the consequences of the conservation principles. The simplest methods usually involve a brick-shaped small element and consideration of fluxes across the surfaces and accumulations of quantities within the brick. For example, see Timoshenko and Goodier (1951) for a derivation of the equations of motion following from momentum conservation. Here we will take a more general approach based on a control volume (Fig. 2.5). In order to proceed, we first need a purely mathematical result, the Reynolds transport theorem, which enables us to compute the rate of change of certain volume integrals.

2.4.1   Reynolds transport theorem

The physical laws to be considered are formulated in terms of a constant collection of particles[†] and we need to be able to transform these into a fixed Eulerian formulation. This may be done by using the Reynolds transport theorem:

$$\frac{\mathrm{d}}{\mathrm{d}t} \int_{\text{particles}} \rho \psi \, \mathrm{d}\nu = \int_V \frac{\partial}{\partial t} (\rho \psi) \mathrm{d}\nu + \int_s \rho \psi \mathbf{v} \cdot \mathbf{n} \, \mathrm{d}A. \tag{2.46}$$

Referring to Fig. 2.5, we see that the volume V with surface S is fixed in space and the material cloud of particles under consideration drifts through the control volume. At time t we choose that the material particles occupy the control volume. The quantity ψ in (2.46) may be any specific (that is, per unit mass)

[†] In this context a 'particle' is shorthand for 'a small quantity of matter' and does not specify a microstructure.

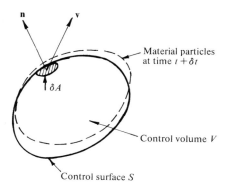

Fig. 2.5 Control volume showing its relation to a material cloud of particles.

quantity of a scalar, vector, or tensor nature. Then $\rho\,\mathrm{d}\nu$ is a mass element, and multiplication by ψ gives an element of the quantity which is to be integrated over the control volume. An elementary proof of (2.46) is given by Streeter and Wylie (1979).

To demonstrate the truth of (2.46) let $\int_{\text{particles}} \rho\psi\,\mathrm{d}\nu = P$. Then

$$\frac{\mathrm{d}P}{\mathrm{d}t} = \int_{\text{particles}} \frac{\mathrm{d}}{\mathrm{d}t}\{\rho\psi\,\mathrm{d}\nu\}. \tag{2.47}$$

The inversion of integration and differentiation is permitted because we are integrating over a fixed set of particles. Now $\rho\psi$ is supposed a function of $\mathbf{x}$, t, and hence (2.47) becomes

$$\frac{\mathrm{d}P}{\mathrm{d}t} = \int_V \frac{\mathrm{D}(\rho\psi)}{\mathrm{D}t}\,\mathrm{d}\nu + \int_V \rho\psi\frac{\mathrm{D}}{\mathrm{D}t}(\mathrm{d}\nu). \tag{2.48}$$

Now the rate of change of the material element $(\frac{\dot{}}{\mathrm{d}\nu})$ needs to be computed.

Let the volume element $\mathrm{d}\nu$ belong to the fixed $(\mathbf{x})$ coordinate system. After deformation let the element be $\mathrm{d}\nu^*$ in the $\mathbf{r}$-system, say. The ratio of the two volumes is the determinant of the transformation matrix (see Jeffreys and Jeffreys 1956, for example). In the present case, (2.27), Section 2.3.4, gives $\mathbf{F}$ as the transformation matrix and so

$$\mathrm{d}\nu^* = \det\mathbf{F}\,\mathrm{d}\nu.$$

Noting that $\mathrm{d}\nu$ is fixed in space, we can differentiate this with respect to time (over a fixed set of particles) to find (Mase 1970, p. 115)

$$\overline{\dot{\mathrm{d}\nu^*}} = \mathbf{\nabla}\cdot\mathbf{v}\,\mathrm{d}\nu^*.$$

This formula must also apply at the instant t, when $\mathbf{r} = \mathbf{x}$, and so one finds

$$\overline{\mathrm{d}\nu} = \boldsymbol{\nabla} \cdot \mathbf{v} \, \mathrm{d}\nu.$$

If (2.48) is rewritten in terms of the fixed control volume (V) we find

$$\frac{\mathrm{d}P}{\mathrm{d}t} = \int_V \left\{ \frac{\partial}{\partial t}(\rho\psi) + (\mathbf{v} \cdot \boldsymbol{\nabla} + \boldsymbol{\nabla} \cdot \mathbf{v})\rho\psi \right\} \mathrm{d}\nu$$

$$= \int_V \left\{ \frac{\partial}{\partial t}(\rho\psi) + \boldsymbol{\nabla} \cdot (\mathbf{v}\rho\psi) \right\} \mathrm{d}\nu, \tag{2.49}$$

or, using the divergence theorem on the last member of (2.49) we recover (2.46).

Equation (2.46) states that the rate of increase of a material quantity is equal to the rate of increase of that quantity in those particles inside the fixed control volume plus the net flux of the quantity through the boundaries of the control volume. This formula may be applied to the angular momentum of a system to deduce that $\sigma_{ij} = \sigma_{ji}$ under the conditions we assumed in Section 2.2.1 (Huilgol and Phan-Thien 1997).

2.4.2 Conservation of mass

We can apply (2.46) by setting $\psi = 1$. Then we have $\int_{\text{particles}} \rho \, \mathrm{d}\nu = m$, the total mass of the material system.

We assume $\mathrm{d}m/\mathrm{d}t = 0$; that is, the total mass of the particles is constant and hence,

$$0 = \int_V \frac{\partial}{\partial t}\rho \, \mathrm{d}\nu + \int_s \rho\mathbf{v} \cdot \mathbf{n} \, \mathrm{d}s. \tag{2.50}$$

Using the divergence theorem the second term in (2.50) can be turned into a volume integral over V

$$0 = \int_V \left(\frac{\partial\rho}{\partial t} + \boldsymbol{\nabla} \cdot \rho\mathbf{v} \right) \mathrm{d}\nu. \tag{2.51}$$

Since V is arbitrary,[†] we must have at all places $\mathbf{x}$

$$\frac{\partial\rho}{\partial t} + \boldsymbol{\nabla} \cdot \rho\mathbf{v} = 0. \tag{2.52}$$

[†] In this and similar arguments one must always bear in mind that the scale of the microstructure is small enough so that the 'infinitesimal' element $\mathrm{d}\nu$ still contains enough material so that the continuum hypothesis is valid. For an elaboration of this argument see L. C. Woods, *Math. Chron.* **9** (1980) 75.

We shall almost exclusively be concerned with materials of constant density, hence (2.52) reduces to

$$\mathbf{\nabla} \cdot \mathbf{v} = 0 \quad \left(\frac{\partial v_k}{\partial x_k} = 0 \right). \tag{2.53}$$

2.4.3 Conservation of momentum

The linear momentum of an element $d\nu$ is $\rho \mathbf{v} \, d\nu$, and hence in this case we use $\psi = \mathbf{v}$ in (2.46). The rate of change of linear momentum of the material particles is the net force on the control volume due to surface and body forces.

$$\text{Total force} = \frac{d}{dt} \int_{\text{particles}} \rho \mathbf{v} \, d\nu \tag{2.54}$$

To obtain an expression for the total force we note that the total body force on the particles in the control volume is

$$\int_V \rho \mathbf{f} \, d\nu, \tag{2.55}$$

where $\mathbf{f}$ is the body force vector per unit mass.

The total surface force is the integral of $\mathbf{t} \, dA$ [see eqn (2.1)] over the surface S:

$$\int_s \mathbf{t} \, dA. \tag{2.56}$$

Adding together these forces and equating to the rate of change of momentum in the Eulerian frame we find

$$\int_s \mathbf{t} \, dA + \int_V \rho \mathbf{f} \, d\nu = \int_V \frac{\partial}{\partial t} (\rho \mathbf{v}) \, d\nu + \int_s \rho \mathbf{v} \mathbf{v} \cdot \mathbf{n} \, dA. \tag{2.57}$$

Now we can express $\mathbf{t}$ (in components) by eqns (2.3) and (2.57) becomes, in coordinate form

$$\int_s \sigma_{ij} n_j \, dA + \int_V \rho f_i \, d\nu = \int_V \frac{\partial}{\partial t} (\rho v_i) d\nu + \int_s \rho v_i v_j n_j \, dA. \tag{2.58}$$

Use of the divergence theorem yields, from (2.58),

$$\int_V \left\{ \frac{\partial \sigma_{ij}}{\partial x_j} + \rho f_i \right\} d\nu = \int_V \left\{ \frac{\partial}{\partial t} (\rho v_i) + \frac{\partial}{\partial x_j} (\rho v_i v_j) \right\} d\nu. \tag{2.59}$$

If account is taken of the equation (2.52) for mass conservation, and noting that the integrand must vanish everywhere, one finds the equations of motion

$$\frac{\partial \sigma_{ij}}{\partial x_j} + \rho f_i = \rho \left\{ \frac{\partial v_i}{\partial t} + v_j \frac{\partial v_i}{\partial x_j} \right\} = \rho a_i. \qquad (2.60)$$

Forms of this equation for various coordinate systems are given in the Appendix at the end of this book.

2.4.4 Conservation of energy

If e is the specific internal energy of a particle and $\frac{1}{2}v^2$ is the specific kinetic energy of that particle, then the first law of thermodynamics states that the rate of increase of energy of the material is equal to the rate of working of the exterior forces on the material minus the net rate of heat loss from the body. Now the rate of working can be expressed as the sum of a surface integral and a volume integral, and hence we write

$$\int_s t_i v_i \, dA + \int_V \rho f_i v_i \, dv - \int_s q_i n_i \, dA + \int_V \rho s \, dv$$
$$= \frac{d}{dt} \int_{\text{particles}} \rho \left(e + \frac{1}{2} v^2 \right) dv. \qquad (2.61)$$

Here, q_i is the heat flux vector (positive outwards), and we have included the term of the form $\int_V \rho s \, dv$, the heating term. In the present studies this form of heat transfer is not important and we will be able to ignore it in applications; a standard example (Carslaw and Jaeger 1959) of its use is in a wire heated by an electric current.

Following the paths outlined in Sections 2.4.2 and 2.4.3 we arrive at the energy equation

$$\rho \frac{De}{Dt} = \sigma_{ij} d_{ij} - \frac{\partial q_i}{\partial x_i} + \rho s. \qquad (2.62)$$

Forms of the terms in this equation in various coordinate systems are given in the Appendix. It should be noted that it is usually necessary to express e and $\mathbf{q}$ in terms of the temperature and/or its gradient.

The first law does not exhaust the subject of thermodynamics. However, we shall not need to appeal directly to the restrictions on processes or constitutive equations imposed by the second law of thermodynamics and the reader is referred to the literature for further discussion. A source which uses fluid mechanics as a context is Woods (1975).

2.5   General rules for constitutive equations

Here we discuss some basic principles which govern formulation of constitutive equations as a preliminary to the treatment of materials with responses more complicated than that of the classical Newtonian fluid. By the term constitutive equation we refer to an equation relating the stress on some material to the motion of that material. Only mechanical interactions are considered, and although in some cases the effects of electric and thermal fields may need to be included, we shall not do so here. In this section, the major requirements which we expect constitutive equations to satisfy are outlined.

2.5.1   Coordinate invariance, determinism, and local action

The requirement of coordinate invariance merely states explicitly what is always assumed for physical laws: their validity does not depend upon expression in a unique coordinate system. When this principle is applied to constitutive equations, the nature of the physical law requires that the principles of coordinate invariance apply to transformations between inertial coordinate systems (that is, the coordinate systems are not undergoing relative acceleration) at any instant of time. The requirement ensures that one will not obtain a new 'law' every time a different coordinate system is used. The principle also establishes the obvious value of using coordinate-free (for example, vector and tensor) notation to describe constitutive equations. The idea of determinism states that, though fluids may exhibit memory, they cannot possess foresight. Consequently, the most general constitutive equation relates the stress in a material at, say, the present time, to the present and previous experience of the material. We say that the stress is determined by the history of the material.

A third idea, introduced by Oldroyd (1950), is that only the neighbouring particles in a material should be involved in determining the stress at a point. This local action principle is consistent with the idea of short-range forces between molecules in a real material, while excluding long-range forces, such as those of electrostatic origins, which are included as body forces.

2.5.2   Principle of material objectivity

This principle is an expression of the belief that response of a material to a given experience or history of motion is independent of any motion of the person observing the response and the history. The principle itself has a long, confused history (Tanner and Walters 1998). It appears to have been enunciated near the turn of the century by Zaremba (1903). However, it was first generally made known to rheologists through the important paper of Oldroyd (1950).

In order to discuss the principle of material objectivity we must first define what is to be meant by a change of reference frame (or observer). By this phrase we mean a time-dependent but spatially homogeneous transformation of space and time; that is, a transformation of an event $(\mathbf{x}, t)$ into a corresponding event $(\mathbf{x}', t')$ according to the transformation rule

$$\begin{aligned}\mathbf{x}' &= \mathbf{c}(t) + \mathbf{R}(t)\mathbf{x} \\ t' &= t - \alpha,\end{aligned} \tag{2.63}$$

where $\mathbf{c}(t)$ is an arbitrary vector-valued function of time, $\mathbf{R}(t)$ is an arbitrary time-dependent orthogonal transformation, and α is an arbitrary constant.

Entities which are invariant under the change of frame (2.63) are said to be frame-indifferent or objective. By invariant we mean that a scalar is unchanged, that a vector preserves its same physical meaning; that is, if

$$\mathbf{v} = \mathbf{y} - \mathbf{x},$$

then

$$\mathbf{v}' = \mathbf{y}' - \mathbf{x}',$$

and that a (second-order) tensor is equivalent to a linear transformation which, when operating on an objective vector, yields an objective vector. These requirements lead one to conclude that under the change of frame (2.63), objective scalars, vectors, and tensors transform, respectively, according to

$$\begin{aligned}b' &= b \\ \mathbf{v}' &= \mathbf{R}(t')\mathbf{v} \\ \mathbf{A}' &= \mathbf{R}(t')\mathbf{A}\mathbf{R}^T(t').\end{aligned} \tag{2.64}$$

Not all entities possessing physical significance transform objectively. For example, we can consider the deformation gradient tensor $\mathbf{F}$ defined in (2.28). We have

$$F_{ij} = \frac{\partial r_i(t')}{\partial x_j}, \quad F'_{ij} = \frac{\partial r'_i(t')}{\partial x'_j}.$$

Now if

$$x'_i = R_{ij}(t)x_j + c_i(t),$$

then

$$r'_i(t') = R_{ij}(t')r_j(t') + c_i(t').$$

Hence, by the chain rule

$$F'_{ij} = \frac{\partial r'_i}{\partial x'_j} = \frac{\partial r'_i}{\partial r_m}\frac{\partial x_k}{\partial x'_j}\frac{\partial r_m}{\partial x_k} = R_{im}(t')R_{jk}F_{mk},$$

or

$$\mathbf{F}'(t') = \mathbf{R}(t')\mathbf{F}(t')\mathbf{R}^T(t). \tag{2.65}$$

This is *not* the rule (2.64) because of the different times involved, and hence $\mathbf{F}$ is not objective.

The principle of material objectivity is an assertion that constitutive equations are to be frame-indifferent or, equivalently, must transform objectively under the change of frame (2.63). Physically, this principle embodies the belief that a constitutive equation expresses material behaviour, and that this behaviour must be indifferent to the motion of an observer. As noted by Truesdell and Noll (1965), one can also state what is essentially the same result by requiring that the response of a material be indifferent to rigid body translations and rotations of the material. It may be shown that the stress tensor $\boldsymbol{\sigma}$ is objective and that the strain tensor $\mathbf{C}$ and the Rivlin–Ericksen tensors $\mathbf{A}$ are also objective.

It is easy to construct physical systems where this principle does not hold. As an example, Ryskin and Rallison (1980) have shown that the principle does not hold for a dilute suspension of spheres when the microscale Reynolds number is not negligible. In most cases we shall merely use the above ideas to check constitutive relations produced by other methods; it is generally necessary to have some microstructural knowledge to apply this principle with confidence.

2.6   The classical constitutive relations

The classical constitutive equations are those for an inviscid fluid, a linear viscous fluid, the linear elastic body and the linear viscoelastic body. One might also consider the theory of plasticity as classical but we shall not examine it here. These cases will now be treated to introduce some ideas involved in formulating constitutive models and in the solution of problems.

In Chapter 1 we discussed material response in one-dimensional shear motions. Now it is required to connect these simple cases to fully three-dimensional flows. Generally, for (isotropic) compressible materials, one needs to consider both the shear and the bulk (or change of volume) response. The shear part of the total stress tensor will be denoted by τ_{ij}, and the bulk portion by $-p\delta_{ij}$. In the compressible case the pressure p is a function of the volumetric strain (and other variables possibly) while τ_{ij} is independent of volume change.

We write

$$\sigma_{ij} = -p\delta_{ij} + \tau_{ij}. \tag{2.66}$$

Each one-dimensional shearing motion yields $\sigma_{ij} = \tau_{ij}(i \neq j)$ and τ_{ij} may be described by a suitable response as in Chapter 1. The bulk notion needs extra material response information which is not deducible from shearing motion alone. By contrast, in the incompressible case, which is our main consideration here, the pressure p is not connected to the motion via a constitutive equation, and must be found using the incompressibility condition (2.53) and the momentum balance (2.60).

2.6.1 Linear viscous fluids and inviscid fluids

In any fluid at rest, there is always a pressure (p) determined by equilibrium considerations. By definition, no shear stresses occur in a fluid at rest, and the pressure acts equally in all directions. Therefore the constitutive model for any fluid at rest is simply

$$\boldsymbol{\sigma} = -p\mathbf{I} \quad (\sigma_{ij} = -p\delta_{ij}). \tag{2.67}$$

Euler took this over in his theory of inviscid flow and it is the simplest constitutive model possible. It contains no shear stresses and hence is not a good model of real fluid behaviour. A more complex model is the linear viscous (Newtonian) fluid (Stokes 1845):

$$\sigma_{ij} = -p\delta_{ij} + \Lambda d_{kk}\delta_{ij} + 2\eta d_{ij}, \tag{2.68}$$

where Λ is a second viscosity coefficient which in general is not related to η, the usual viscosity coefficient measured in shearing motion. For incompressible fluids $d_{kk} = 0$ and so the value of Λ is irrelevant and we find

$$\sigma_{ij} = -p\delta_{ij} + 2\eta d_{ij} \equiv -p\delta_{ij} + \eta A_{ij}^{(1)}. \tag{2.69}$$

Note that for any incompressible material we have the incompressibility constraint ($\nabla \cdot \mathbf{v} = 0$) and p must be found by using the equation of motion (2.60) and boundary conditions; it cannot be found from the constitutive equation alone. It is often convenient with incompressible fluids to write

$$\sigma_{ij} + p\delta_{ij} = \tau_{ij} \quad (\boldsymbol{\sigma} + p\mathbf{I} = \boldsymbol{\tau}), \tag{2.70}$$

$\boldsymbol{\tau}$ is then the extra stress (or deviatoric stress) and is the part of the stress tensor which can be computed from the constitutive equation when the motion is known. Substitution of (2.69) into the equations of motion gives the Navier–Stokes equations. Thus we get three equations plus the incompressibility condition to find the three velocity components and the pressure. All of the above constitutive forms can be seen to satisfy all of the general principles for constitutive laws laid down in Section 2.5; the relation (2.69) is *isotropic* because it has been made so; we expect isotropic behaviour from fluids with a random isotropic microstructure. Before leaving these fluids we shall examine their complete responses in simple shearing flow and in elongational flow. In the former we shall consider the velocity field $\mathbf{v} = \mathbf{i}\dot{\gamma}y$ (Fig. 2.6), where $\mathbf{i}$ is a unit vector parallel to the x-axis, and consider only the incompressible case. Then the proposed velocity field satisfies the incompressibility condition (2.53) and

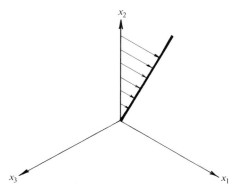

Fig. 2.6 Simple shearing kinematics; x_1 is the flow direction, x_2 is the velocity gradient direction, and x_3 is the neutral direction. The speed $v_1 = \dot{\gamma} x_2$; the other velocity components v_2 and v_3 are zero.

from (2.69) we find

$$\sigma_{zy} = \sigma_{xz} = 0$$
$$\sigma_{xx} = \sigma_{yy} = \sigma_{zz} = -p$$
$$\sigma_{xy} = \eta\dot{\gamma} = \tau(\text{constant}). \tag{2.71}$$

We now check that (2.71) and the assumed velocity field satisfy the equations of motion, under the assumption that no variation of quantities in the x and z directions occurs. Some assumption about the body force must be made; it will be ignored here for simplicity. (If required a gravity body force can be easily added; it simply produces a hydrostatic pressure distribution.) Then we find $p = $ constant, the level being set externally by some boundary condition and not influenced by the flow. We emphasize that since the hydrostatic pressure p is not set by the flow, only the *differences* of normal stresses

$$N_1 = \sigma_{11} - \sigma_{22},$$
$$N_2 = \sigma_{22} - \sigma_{33}, \tag{2.72}$$

and the shear stress τ are of rheological interest. The quantity N_1, the first normal stress difference, is defined by the direct stresses along two axes, one along the flow direction ($x_1 = x$) and the second along the flow gradient direction ($x_2 = y$). The second normal stress difference, N_2, is defined using $x_2 (= y)$ and $x_3 (= z)$. For the Newtonian fluid, N_1 and N_2 are both zero, and, by assumption, η is constant.

When $\dot{\gamma}$ is a function of time, and conditions are such that inertia can be ignored, then the same solution will hold; the shear stress τ merely follows the shear rate exactly, with no delay or change of waveform. Thus the Newtonian fluid has instantaneous response and no memory. A second basic flow field is steady elongation. Here, the velocity field is assumed to be

$$\mathbf{v} = \tfrac{1}{2}\dot{\varepsilon}\{2\mathbf{i}x - \mathbf{j}y - \mathbf{k}z\}. \tag{2.73}$$

(This is not the same field as that described in Section 2.3.6). Thus the elongational rate $\partial u/\partial x$ is equal to $\dot{\varepsilon}$, and $\partial v/\partial y$ and $\partial w/\partial z$ are both $-\tfrac{1}{2}\dot{\varepsilon}$. The flow is symmetric around the x-axis and satisfies the incompressibility conditions (2.53). A possible way of producing the irrotational velocity field (2.73) is to elongate a viscous rod, so that it is a tensile test, rather than a shear test. We shall ignore body forces and accelerations. It is then clear that the only quantity of rheological interest is $\sigma_{xx} - \sigma_{yy}$, since all shear stresses are zero and $\sigma_{yy} = \sigma_{zz}$. Again, the absolute values of the direct stresses will be set by some boundary condition, which in turn determines p. We find

$$\sigma_{xx} - \sigma_{yy} = 3\eta\dot{\varepsilon} \equiv \eta_E\dot{\varepsilon}. \tag{2.74}$$

The quantity η_E is the Trouton (1906) or elongational viscosity; in a Newtonian fluid it is exactly 3η. Unsteady elongations, in which $\dot{\varepsilon}$ is a function of t, simply make the left-hand side of (2.74) time-dependent also, following $\dot{\varepsilon}$.

2.6.2   Linear elastic behaviour

Here we expect to relate the stress tensor to a *strain*. Suppose a body is at rest in the configuration $\mathbf{x}$ at time t. If it is then strained ($t' > t$ in this case) so that the particle which was at the place $\mathbf{x}$ is now at $\mathbf{r}$ then we can define the displacement vector $\mathbf{u}$ as

$$\mathbf{u} = \mathbf{r} - \mathbf{x} \quad (u_i = r_i - x_i). \tag{2.75}$$

The strain tensor C_{ij} can then be written in terms of $\partial u_i/\partial x_j = F_{ij} - \delta_{ij}$, giving, from (2.32) and (2.34)

$$C_{ij} = \delta_{ij} + \frac{\partial u_i}{\partial x_j} + \frac{\partial u_j}{\partial x_i} + \frac{\partial u_k}{\partial x_i}\frac{\partial u_k}{\partial x_j}. \tag{2.76}$$

This strain measure is non-linear. In classical elasticity theory it is supposed that all displacement gradients are small, so that we can rewrite (2.76) as

$$2E_{ij} \equiv C_{ij} - \delta_{ij} = \frac{\partial u_i}{\partial x_j} + \frac{\partial u_j}{\partial x_i} + \mathrm{O}\left\{\left(\frac{\partial u_i}{\partial x_j}\right)^2\right\}. \tag{2.77}$$

E_{ij} is the classical small-strain tensor. There is no pressure term in a *compressible* solid, hence the most general linear relation (Hooke's law) between stress and strain is

$$\sigma_{ij} = C_{ijkm}E_{km}. \tag{2.78}$$

This reflects the physical assumption that σ_{ij} vanishes when the strain is zero. Note a further approximation here: σ_{ij} now rightly belongs to the configuration after displacement ($\mathbf{r}$ – coordinates) but in classical elasticity theory no account is taken of the difference between $\mathbf{r}$ and $\mathbf{x}$. There is no requirement (as there is in a fluid) that the material should be isotropic, but if microstructural considerations indicate that it is (for example, a polycrystalline metal) then it can be shown (see Mase (1970)) that (2.78) reduces to

$$\sigma_{ij} = \overline{\lambda} E_{kk} \delta_{ij} + 2G E_{ij}, \tag{2.79}$$

where $\overline{\lambda}$ and G are moduli (the Lamé constants).

If the material is incompressible, then the stress is not completely determined by the displacements, and (2.79) must be rewritten as

$$\sigma_{ij} + p\delta_{ij} = 2G E_{ij}. \tag{2.80}$$

Here, because of the approximation to C_{ij}, the constitutive model may not be objective; this is of no consequence when the theory is used correctly within its limitations on small strains and displacements.

If one subjects the linear elastic material to the steady flow fields mentioned in the previous section, then there are no steady solutions, and the stresses rise linearly in time. The stress solutions for constant *strains* are identical to the stress fields in a Newtonian fluid with constant *strain-rates*.

2.6.3   Linear viscoelastic behaviour

The main features of linear viscoelastic behaviour have been discussed in Section 1.5.2 for one-dimensional shearing motions. The result obtained for the shear stress, eqn (1.24), was

$$\tau(t) = \int_{-\infty}^{t} G(t - t') \, d\gamma(t'). \tag{1.24}$$

By appropriately defining an elongational relaxation function $E(t)$ (Pipkin 1986) in place of $G(t)$ we can also explore the extensional behaviour in a form similar to (1.24). This should be borne in mind, although we shall not write down explicit relations for extensional motions.

A suitable three-dimensional form for incompressible materials may be found by replacing $\dot{\gamma}(t')$ by the first Rivlin–Ericksen tensor $A_{ij}^{(1)}$ and the shear stress τ by the extra stress tensor τ_{ij}:

$$\sigma_{ij} + p\delta_{ij} = \tau_{ij} = \int_{-\infty}^{t} G(t - t') A_{ij}^{(1)}(t') \, dt'. \tag{2.81}$$

Then $A_{ij}^{(1)}$ can be interpreted as the rate of strain; (2.81) gives the same result as (1.24) for the shear component of σ_{ij}. (It will be necessary to regard $A_{ij}^{(1)}$ as a generalized function in order to deal with step jumps in strain. In this case $A_{ij}^{(1)}$ will formally be written but one has to bear in mind that delta functions are now implied in its definition.) Furthermore, because tr $\mathbf{A}^{(1)}$ vanishes, so does the trace of the extra stress or stress deviator (τ_{ij}), and the sum of the direct stresses is just $-3p$, as in a Newtonian fluid; similarly, after motion has ceased, σ_{ij} approaches a hydrostatic state, as expected. Equation (2.81) is in fact the proper generalization of (1.24) to three dimensions for an isotropic incompressible material. We shall derive (2.81) by an alternative method later. When the fluid is compressible, or for viscoelastic solids, the matter is not as simple as one then has two memory kernels corresponding to $\overline{\lambda}$ and G in eqn (2.79) (Bland 1960; Pipkin 1986); in eqn (2.81) $G(t)$ corresponds to the constant shear modulus G only. The form of the linear viscoelasticity relation for compressible materials can be written (Pipkin 1986)

$$\sigma_{ij} = \delta_{ij} \int_{-\infty}^{t} \overline{\lambda}(t - t')\dot{\theta}(t')\,\mathrm{d}t' + 2 \int_{-\infty}^{t} G(t - t')\,\mathrm{d}_{ij}(t')\,\mathrm{d}t', \tag{2.81a}$$

where $\dot{\theta} = d_{kk}$, the rate of volume strain, and $\overline{\lambda}$ is an appropriate memory function. In addition, one could also have a purely elastic term in (2.81a).

Equations (2.81) and (2.81a) hold only in the linear regime; in this case doubling the strain doubles the stresses, for example. The range of strain which may be employed before leaving the linear range varies quite widely; a 10 per cent strain may well be linear for some polymeric materials, but Phan-Thien and Safari-Ardi (1998) have shown that bread doughs need to be subjected to less than about 0.1 per cent; (0.001) shear strain amplitude to be in the linear range.

To complete this section we shall investigate some simple cases to parallel the sections above. If we consider a suddenly started simple shearing motion then the results are the same as eqn (2.71) for a viscous fluid, except that the shear stress τ is not reached immediately. All the direct stresses are the same, equal to the externally imposed pressure, and the normal stress differences N_1 and N_2 are zero.

Similarly, except for the delay in reaching the final state, the response in steady elongational flow is the same as (2.74). We have imposed the fluid-like behaviour here, of course; in case we have solid-like behaviour then the results correspond to those in Section 2.6.2.

Thus the interest in incompressible linear viscoelastic fluid behaviour resides in the shear response. With a step strain input of magnitude γ the shear component of $\mathbf{A}^{(1)}$ is a delta function $\gamma\delta(t)$. Hence the relaxation function $G(t)$ is the impulse response of the fluid, in terms familiar in linear systems theory. For the Newtonian fluid we have the special form

$$G(t - t') = \eta_0 \delta(t - t'), \tag{2.82}$$

where relaxation is immediate.

Use of this relation in (2.81) gives the Newtonian fluid case

$$\sigma_{ij} + p\delta_{ij} = \eta_0 \int_{-\infty}^{t} \delta(t - t')A_{ij}^{(1)}(t')\,\mathrm{d}t = \eta_0 A_{ij}^{(1)}(t). \tag{2.83}$$

2.6.4   Sinusoidal linear viscoelastic response

We have mainly concentrated so far on step-strain signals, but a very important test procedure applies a time-sinusoidal signal to a sample in the linear region and measures its response. Useful sources of information are Ferry (1970), Pipkin (1986) and Lodge (1964).

Let us suppose that the input signal to the specimen is a sinusoidal shear strain described by the complex expression ($i = \sqrt{-1}$)

$$\gamma = \hat{\gamma}\exp(i\omega t), \tag{2.84}$$

(where the real part of this and later expressions is understood to be relevant).

The (complex) shear stress response τ is then, because the system is linear

$$\tau = \hat{\sigma}\exp(i\omega t), \tag{2.85}$$

where ω is the oscillation frequency of input and output. Then from (2.84)

$$\dot{\gamma} = i\omega\hat{\gamma}e^{i\omega t}. \tag{2.86}$$

Substituting in (2.81) we get

$$\hat{\sigma}e^{i\omega t} = i\omega\hat{\gamma}\int_{-\infty}^{t} G(t - t')\exp i\omega t'\,\mathrm{d}t'. \tag{2.87}$$

We define the *complex modulus* G^* as $\hat{\sigma}/\hat{\gamma}$; hence from (2.87), with $t - t' = s$,

$$G^* = i\omega \int_{0}^{\infty} G(s)e^{-i\omega s}\,\mathrm{d}s. \tag{2.88}$$

Separating into real and imaginary components

$$G^* = G' + iG'' = \int_{0}^{\infty} \omega G(s)\sin \omega s\,\mathrm{d}s + i\int_{0}^{\infty} \omega G(s)\cos \omega s\,\mathrm{d}s. \tag{2.89}$$

We will term $G'(\omega)$ the *storage modulus* and $G''(\omega)$ the *loss modulus*. We can also define the *complex viscosity* $\eta^* = \hat{\sigma}/\hat{\dot{\gamma}}$, hence

$$\eta^* = \eta' - i\eta'' = \frac{G^*}{i\omega} = \frac{G''}{\omega} - i\frac{G'}{\omega}. \tag{2.90}$$

Fig. 2.7 Sinusoidal shearing response. The shear signal (γ) lags behind the stress signal (σ) by an angle δ; $0 < \delta < \pi/2$.

We call $G''/\omega = \eta'$ the dynamic viscosity; it is equal to the steady-flow Newtonian viscosity for a Newtonian fluid. As $\omega \to 0$, from (2.89) and (2.90) we find the zero-shear rate viscosity η_0:

$$\eta_0 = \int_0^\infty G(s)\,\mathrm{d}s. \tag{2.91}$$

Using the inverse Fourier transform theorem, we find, from (2.89)

$$G(s) = \frac{2}{\pi} \int_0^\infty \frac{G''}{\omega}\cos\omega s\,\mathrm{d}\omega \tag{2.92}$$

$$= \frac{2}{\pi} \int_0^\infty \frac{G'}{\omega}\sin\omega s\,\mathrm{d}\omega. \tag{2.93}$$

If we record the results of experiments we have a *phase shift* δ (Fig. 2.7) and clearly if $\delta = 0$ $G'' = 0$. Hence $\tan\delta = G''/G'$.

Example
For the Maxwell model $G(t) = G_g \mathrm{e}^{-t/\lambda}$.
Hence

$$G^*(\omega) = i\omega \int_0^\infty G_g \mathrm{e}^{-s/\lambda}\mathrm{e}^{-i\omega s}\,\mathrm{d}s = i\omega\lambda G_g/(1 + i\lambda\omega).$$

Rationalizing the denominator,

$$G^*(\omega) \equiv G'(\omega) + iG''(\omega) = G_g[\lambda^2\omega^2 + i\lambda\omega]/(1 + \lambda^2\omega^2). \tag{2.94}$$

When $\lambda\omega \to \infty$, the behaviour is elastic, and $G' \to G_g$, $G'' \to 0$. As $\omega \to 0$, the behaviour is viscous, $G' \to 0$, $G'' \to \eta_0\omega$. The phase angle is $\tan^{-1}(1/\omega\lambda)$, going from $90°$ as $\omega \to 0$ to zero as $\lambda\omega \to \infty$.

One can regard $|G^*|$ as a measure of 'signal' attenuation.

When we have an *elastic solid*, $\delta = 0$ (lossless); for a *viscous fluid* $\delta = \pi/2$. We can argue that δ cannot exceed certain limits on physical grounds. The rate of dissipation of energy/unit volume $= \tau\dot{\gamma}$ instantaneously; that is, if $\dot{\gamma} = \hat{\gamma}\cos\omega t$ and $\gamma = (\hat{\gamma}/\omega)\sin\omega t$ then the average dissipation per cycle is $(\hat{\sigma}\hat{\gamma}/2)\sin\delta$. In order for the dissipation to be positive we must have $\pi \gtrless \delta \gtrless 0$. The stored energy is $\tau\gamma$; when averaged over a cycle, it is $\frac{1}{2}(\hat{\sigma}\hat{\gamma}/\omega)\cos\delta$. If this is also to be positive, $\pi/2 \gtrless \delta \gtrless 0$, which is what is observed in experiments.

We have shown that (γ_0 is real here)

$\tau(t) = G^*(\omega)\gamma_0 e^{i\omega t}$ in this case.

If we had started from the creep integral description (1.25) and performed the same kind of analysis we would obtain $\gamma(t) = J^*(\omega)\tau$. Eliminating τ from these equations shows that

$$J^*(\omega)G^*(\omega) = 1. \tag{2.95}$$

While this result is good for solid materials where J_e exists (Fig. 1.15) there are convergence problems for fluids because no finite J_e exists (Pipkin 1986).

In Chapter 1 we stated that $J(t)G(t) \approx 1$, but in fact (Pipkin 1986) $JG \leq 1$, and the exact relation between J and G can be found by taking Laplace transforms of (1.24), and (1.25) and then eliminating the transforms of the stress (τ) and strain (γ). If $\tilde{J}(s)$ is the Laplace transform of $J(t)$ and $\tilde{G}$ is the Laplace transform of $G(t)$, then we have

$$\begin{aligned} \tilde{\tau}(s) &= s\tilde{G}(s)\tilde{\gamma} \\ \tilde{\gamma}(s) &= s\tilde{J}(s)\tilde{\tau}, \end{aligned} \tag{2.96}$$

and eliminating $\tilde{\tau}$ and $\tilde{\gamma}$ we find the result

$$s^2 \tilde{J}\tilde{G} = 1. \tag{2.97}$$

Example: Maxwell fluid
Here $G(t) = G_g e^{-t/\lambda}$ and if we note that the Laplace transform of a quantity $y(t)$ is given by

$$\tilde{y}(s) \equiv \int_0^\infty \exp(-st)y(t)\,\mathrm{d}t$$

we find

$$\tilde{G}(s) = G_g/(s + 1/\lambda).$$

From (2.97), we find $\tilde{J}(s) = 1/s^2\tilde{G}$ or

$$\tilde{J}(s) = (1 + 1/\lambda s)/G_g s.$$

Inverting the transform we find $J(t) = H(t)G_g^{-1}(1 + t/\lambda)$ which agrees with (1.19).

Another useful concept is the *relaxation spectrum* defined by

$$G(t) = \int_{\lambda=0}^{\infty} H(\lambda) \exp(-t/\lambda) \, \mathrm{d} \ln \lambda. \tag{2.98}$$

Note the (traditional) logarithmic basis for λ.
Thus, if we set $\lambda = \frac{1}{p}$, $\mathrm{d}\lambda = \frac{-1}{p^2} \mathrm{d}p$

$$G(t) = \int_{0}^{\infty} H\left(\frac{1}{p}\right) \exp(-pt) \frac{\mathrm{d}p}{p}. \tag{2.99}$$

Thus, $G(t)$ is the Laplace transform of

$$\frac{H\left(\frac{1}{p}\right)}{p} \qquad \text{or} \qquad \lambda H(\lambda).$$

We can regard $H(\lambda)$ as the density (on a log basis) of elements of the form $e^{-t/\lambda}$ making up $G(t)$. We can define analogously the retardation time spectrum $L(\lambda)$ using the creep compliance J:

$$J(t) = J_0 + \int_{0}^{\infty} L(\lambda)(1 - e^{-t/\lambda}) \frac{\mathrm{d}\lambda}{\lambda}. \tag{2.100}$$

Now substitute (2.98) in (2.89)

$$\left.\begin{matrix} G' \\ G'' \end{matrix}\right\} = \int_{0}^{\infty} \omega \left\{ \begin{matrix} \sin \omega s \\ \cos \omega s \end{matrix} \right\} \left| \int_{0}^{\infty} H(\lambda) e^{-s/\lambda} \frac{\mathrm{d}\lambda}{\lambda} \right|.$$

Interchanging the order of integration

$$\left.\begin{matrix} G' \\ G'' \end{matrix}\right\} = \int_{0}^{\infty} \omega \frac{\mathrm{d}\lambda}{\lambda} H(\lambda) \int_{0}^{\infty} e^{-s/\lambda} \left\{ \begin{matrix} \sin \omega s \\ \cos \omega s \end{matrix} \right\} \mathrm{d}s. \tag{2.101}$$

We find

$$G' = \int_{0}^{\infty} \frac{\omega^2 \lambda^2 H(\lambda)}{1 + \omega^2 \lambda^2} \, \mathrm{d} \ln \lambda \tag{2.102}$$

$$\eta' = \frac{G''}{\omega} = \int_{0}^{\infty} \frac{H(\lambda) \, \mathrm{d}\lambda}{1 + \omega^2 \lambda^2}. \tag{2.103}$$

These results can be reduced to Stieltjes or Fourier transforms by a change of variables. By numerical inversion—'exact' or by analytical approximations—we can find $H(\lambda)$. In practice, it is usually more accurate to use G'' in lossy materials.

Ferry (1970) gives many useful rules for finding spectra. For consideration of temperature effects on G^* see Chapter 9.

Special case—Maxwell liquid
If we take $H(\lambda)$ to be a set of N delta functions, *a discrete spectrum*, that is,

$$H(\lambda) = \sum_{n=1}^{N} \lambda_n g_n \delta(\lambda - \lambda_n) \tag{2.104}$$

then from (2.98)

$$G(t) = \sum_{n=1}^{N} g_n \exp(-t/\lambda_n). \tag{2.105}$$

Differentiating, we get dG/dt as

$$\mu(t) = \frac{dG}{dt} = -\sum_{n=1}^{N} \frac{g_n}{\lambda_n} \exp(-t/\lambda_n). \tag{2.106}$$

When $n = 1$ we have a *Maxwell liquid*, one of the two simplest linear viscoelastic materials. For the Maxwell fluid, if $\lambda_1 = \lambda$, $g_1 = \eta_0\lambda$, from (2.102) we again find (2.94):

$$G' = \frac{\eta_0\omega^2\lambda}{1 + \omega^2\lambda^2} \tag{2.94a}$$

$$\frac{G''}{\omega} = \eta' = \frac{\eta_0}{1 + \omega^2\lambda^2} \tag{2.94b}$$

$$\tan\delta = \frac{1}{\lambda\omega}, \tag{2.94c}$$

giving the frequency response.

2.6.5   *Fitting data for linear viscoelastic materials*

So far we have discussed representing $G(t)$ by a sum of discrete exponentials (2.105). Roscoe (1950) showed that all spring-dashpot models can be reduced to two canonical forms—either a series of Maxwell elements in parallel or a set of Kelvin–Meyer elements in series. Thus there is considerable support for the use of discrete spectra models.

    If a discrete spectrum is chosen to fit a set of experimental data, then the number of relaxation times (N) in (2.104) needs to be specified, together with the values of the relaxation times (λ_n) and their strengths g_n in (2.105). Then the discrete line forms of (2.102) and (2.103) are

$$G'(\omega) = \sum_{i=1}^{N} g_i \frac{\omega^2 \lambda_i^2}{1 + \omega^2 \lambda_i^2} \tag{2.102a}$$

$$G''(\omega) = \sum_{i=1}^{N} g_i \frac{\omega \lambda_i}{1 + \omega^2 \lambda_i^2} \tag{2.103a}$$

Unfortunately, only a finite range of frequencies is available, and the problem of finding a suitable fit is ill-posed; many spectra will reproduce G' and G'' fairly accurately. There is a vast literature on computing the inversion problem, and we only cite a paper by Brabec *et al.* (1997) which compares several methods of computation, the papers by Baumgaertel and Winter (1989, 1992) which seek to fit a minimum number of spectral lines, and the use of a variant of Weese's NLREG method by Phan-Thien and Safari-Ardi (1998). Often 1–2 spectral lines per decade of frequency have been found adequate. Various regularization methods of fitting (regression, least squares) can often produce spectra that reproduce the G', G'' data within a few per cent over the given range, which is often within experimental accuracy. If the range of experimental data is from ω_{min} to ω_{max} then the relaxation times needed are often considered to lie in the range ω_{max}^{-1} to ω_{min}^{-1}. However, this is only a crude approximation, and the data can only be reliably determined on a shorter interval. According to Davies and Anderssen (1997, 1998), the interval of accurate spectra lies in the range $e^{\pi/2}\omega_{max}^{-1}$ to $e^{-\pi/2}\omega_{min}^{-1}$, more than a decade shorter than the simpler estimate. The lowest frequency limit is usually set by the inability of the instrument's force transducer to respond to the very small forces produced at very low frequencies, so that noise becomes a problem. At the high-frequency end, the instrument mass dominates the response, and the material impedance often cannot be measured accurately. Hence near the limits of ω_{min} and ω_{max} data may not be accurate, thereby making the deduction of spectra even more difficult.

Alternatively, one might try and fit a relaxation curve $G(t)$ directly, using (2.105), and compute G' and G'' from this fitting, or sometimes one can use $G(t)$ over a suitable interval, without resorting to discrete spectra. We now give some examples.

Example 1: Polyethylene melt (Laun 1978)

Table 2.1 shows a one-line per decade least-squares fit over eight decades of the data shown in Fig. 2.8. The curves show slight ripples not present in the experimental data.

Example 2: Polyvinyl chloride

A second set of data is shown in Fig. 2.9 for a polyvinyl chloride (PVC) sample at various temperatures (°C). PVC is not truly molten at the temperatures at which it is formed, and at higher temperatures than those shown it decomposes, so Fig. 2.9 shows data for a soft solid. It is noticeable that for temperatures

Table 2.1 Linear relaxation data for low-density polyethylene at 150 °C

$$G(t) = \sum_{i=1}^{8} g_i \exp(-t/\lambda_i)$$

i	$\lambda_i(s)$	$g_i(Pa)$
1	10^3	1.00
2	10^2	1.80×10^2
3	10	1.89×10^3
4	1	9.80×10^3
5	0.1	2.67×10^4
6	10^{-2}	5.86×10^4
7	10^{-3}	9.48×10^4
8	10^{-4}	1.29×10^5

Data of Laun (1978)

above 100 °C the relaxation data is very well fitted by a power-law rule

$$G(t) = G(1)t^{-B} \tag{2.107}$$

Taking the 140 °C curve, we find $G(1)$ is 0.749 MPa and B is 0.09, where t is in seconds. An alternative exponential fit to the $G(t)$ curve using a standard least-squares fitting package is shown in Table 2.2. It uses seven relaxation times.

The curves agree closely with the experimental data for $G'(\omega)$ (Fig. 2.9). The measured data are shown by crosses and the lines are computed from (2.102a) for the line spectrum. Alternatively, one can find the $G'(\omega)$ corresponding to the power-law fit (2.107) directly. From (2.89) we find

$$G' = G(1)\omega^{0.09}\Gamma(0.91)\cos(.09\pi/2)$$

where Γ is the gamma function, or $G' = 0.787\,\omega^{0.09}$ MPa, which is plotted as the upper curve in Fig. 2.9(b). The agreement is a self-consistency test of the data sets.

2.6.6 Solving linear viscoelastic boundary-value problems

For incompressible linear viscoelastic materials, $G(t)$ contains all the information needed to describe the flow. For compressible materials a second, volumetric response kernel is also needed, corresponding to the $\overline{\lambda}$-modulus in equation (2.79). We shall refer to Bland (1960) for a discussion of compressible solid materials, and here we discuss only incompressible materials, which are completely described when $G(t)$ is known.

We list the principal types of linear viscoelastic boundary-value problems. These are classified as

(a) Quasi-static

(b) Sinusoidal oscillation

(c) Wave propagation.

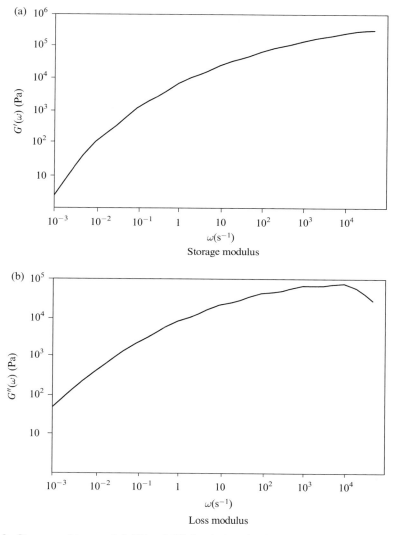

Fig. 2.8 Storage and loss moduli (G' and G'') for the low-density polyethylene melt I, Table 2.1. The slight ripples in the curves could be removed by using more relaxation times. G' and G'' vanish, as $\omega \to 0$, like ω^2 and ω respectively.

In class (a) we can take such problems as a pressure (or displacement) suddenly applied to the inside of a cylinder of material. Flow starts at $t = 0$ and the hole gets bigger, but with many 'plastic' materials the process is so slow that one can ignore the inertia terms $[\rho(\mathrm{D}\mathbf{v}/\mathrm{D}t)]$ in the equations of motion. This is the quasi-static approximation.

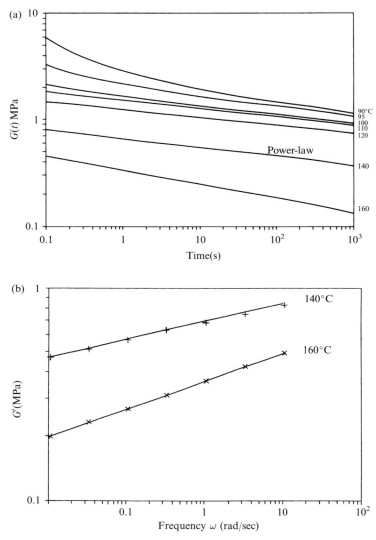

Fig. 2.9 (a) $G(t)$ for a polyvinyl chloride sample at various temperatures (°C). At higher temperatures relaxation follows a power-law. (b) $G'(\omega)$ for polyvinyl-chloride sample of Fig. 2.9(a). —— Computed from $G(t)$, where $G'(\omega) = \int_0^\infty \omega G(s) \sin \omega s \, ds$. × Direct measurement of G'.

Suppose we consider the specific case when the boundary velocities are given as $\mathbf{v}_0(t)$, then we have

$$\boldsymbol{\sigma} = -p\,\mathbf{I} + \int_0^t G(t - t')\mathbf{A}^{(1)}\mathrm{d}t', \tag{2.81}$$

Table 2.2 Relaxation modes for PVC AT 140 °C

i	$\lambda_i(s)$	$g_i(MPa)$
1	10^{-2}	2×10^{-5}
2	0.1	0.261
3	1	0.170
4	10	0.117
5	10^2	0.105
6	10^3	0.068
7	10^4	0.406

and the equations of motion and mass conservation:

$$\mathbf{V} \cdot \boldsymbol{\sigma} = 0; \quad \mathbf{V} \cdot \mathbf{v} = 0.$$

We can take the Laplace transform of all these equations obtaining

$$\tilde{\boldsymbol{\sigma}} = -\tilde{p}\mathbf{I} + \tilde{G}\tilde{\mathbf{A}}$$
$$\mathbf{V} \cdot \tilde{\boldsymbol{\sigma}} = 0; \quad \mathbf{V} \cdot \tilde{\mathbf{v}} = 0;$$

taking the Laplace transform of the boundary conditions we have $\tilde{\mathbf{v}}|_{\text{boundary}} = \tilde{\mathbf{v}}_0$, and so

$$\nabla\tilde{p} = \tilde{G}\mathbf{V} \cdot \tilde{\mathbf{A}} = \tilde{G}\mathbf{V}^2\tilde{\mathbf{v}}. \tag{2.108}$$

Thus, with the exception of changing $\tilde{G}$ for η in (2.108) the results are *identical* to those for the Newtonian problem. Hence, one may solve (2.108) and $\mathbf{V} \cdot \tilde{\mathbf{v}} = 0$ by the usual methods for viscous cases and at the end replace η by $\tilde{G}$, then invert the transforms.

As a specific example we consider a Couette flow start-up. Here, in the absence of inertia, the viscous solution is independent of material properties and hence so is the viscoelastic velocity field. However, the stresses are different. We assume r, θ, z coordinates:

$$\mathbf{v} = \{0, r\Omega(r, t), 0\}.$$

Thus mass is conserved and we may, from symmetry, assume $\partial p/\partial\theta = 0$. Also

$$\mathbf{A}^{(1)} = \begin{bmatrix} 0 & r\frac{\partial\Omega}{\partial r} & 0 \\ r\frac{\partial\Omega}{\partial r} & 0 & 0 \\ 0 & 0 & 0 \end{bmatrix}.$$

Thus all stress deviators except $\sigma_{\theta r}$ are zero, and the only equation of equilibrium not satisfied reduces to

$$\frac{\partial}{\partial r}(r^2 \sigma_{r\theta}) = 0,$$

or, integrating,

$$\sigma_{r\theta} = \frac{M}{2\pi r^2} \quad \text{where} \quad M = M(t).$$

Laplace transforming, we find $\tilde{M} = 2\pi r^2 \tilde{\sigma}_{r\theta}$.
    Also, from the constitutive equation

$$\tilde{\sigma}_{r\theta} = \tilde{G} r \frac{d\tilde{\Omega}}{dr}.$$

Thus

$$\tilde{M} = \tilde{G} 2\pi r^3 \frac{d\tilde{\Omega}}{dr}, \quad \text{or} \quad \frac{\tilde{M}}{2\pi \tilde{G}} \frac{dr}{r^3} = d\tilde{\Omega},$$

and

$$\tilde{\Omega} = -\frac{r^{-2}}{2} \frac{\tilde{M}}{2\pi \tilde{G}} + c(s),$$

where s is the Laplace transform variable. Suppose that at

$$\begin{cases} r = R_i & \Omega = \Omega_i(t)H(t), \quad \text{then} \quad \tilde{\Omega} = \tilde{\Omega}_i \\ r = R_0 & \Omega = 0, \qquad\qquad\qquad \tilde{\Omega} = 0. \end{cases}$$

Hence

$$\tilde{\Omega} = \frac{\tilde{M}}{4\pi R_0^2 \tilde{G}}\left(1 - \frac{R_0^2}{r^2}\right)$$

and

$$\tilde{\Omega}_i = \frac{\tilde{M}}{4\pi R_0^2 \tilde{G}}\left(1 - \frac{R_0^2}{R_i^2}\right). \text{ Then}$$

$$\frac{\tilde{\Omega}}{\tilde{\Omega}_i} = \frac{1 - R_0^2/r^2}{(1 - R_0^2/R_i^2)} \text{ which is the same as the Newtonian solution.}$$

The stresses and moments are different; in the Newtonian case

$$\tilde{G} = \eta_0, \quad \text{because} \quad G = \eta_0 \, \delta(t),$$

and hence

$$M_N = \frac{4\pi R_0^2 \eta_0}{(1 - R_0^2/R_i^2)} \Omega_i.$$

For the viscoelastic case

$$\tilde{M} = \frac{4\pi R_0^2}{(1 - R_0^2/R_i^2)} \tilde{G}\tilde{\Omega}_i \quad \text{or} \quad \tilde{M} = \text{constant} \times \tilde{G}\tilde{\Omega}_i.$$

Suppose we have a Maxwell material. Then

$$G = \frac{\eta_0}{\lambda} e^{-t/\lambda},$$

and, Laplace transforming,

$$\tilde{G} = \frac{\eta_0}{\lambda} \int_0^\infty e^{-st} e^{-t/\lambda} \, dt = \frac{\eta_0}{\lambda} \frac{1}{\left(s + \frac{1}{\lambda}\right)}.$$

Suppose we have a sudden start-up

$$\Omega_i = \Omega_0 H(t) \quad \text{or} \quad \tilde{\Omega}_i = \frac{\Omega_0}{s},$$

where s is the Laplace transform variable and

$$\tilde{M} = \frac{C\eta_0\Omega_0}{\lambda} \frac{1}{s\left(s + \frac{1}{\lambda}\right)},$$

where C is a constant.

Inverting the transforms, we find

$$M = C\eta_0\Omega_0[H(t)][1 - \exp-(t/\lambda)] \quad \text{or} \quad \frac{M}{M_N} = 1 - e^{-t/\lambda}.$$

There are a good many such problems in linear viscoelasticity where a *correspondence principle* can be established between the viscous (or elastic) problem and the viscoelastic problem in the Laplace transform domain. For further examples see Bland (1960) and Pipkin (1986).

In type (b) problems we must use a slightly different treatment from quasi-static problems, with Fourier transforms instead of Laplace transforms. Usually inertia is not to be neglected and many of the difficulties are common to types (b) and (c).

In type (c) problems inertia cannot be neglected. As a simple example we will consider a sudden step velocity applied to a half-space (Rayleigh problem) in which the velocity u of the boundary plane $y=0$ is $u_0 H(t)$ in the x-direction. We assume the velocity field is of the form $\mathbf{v} = \{u(y,t),0,0\}$.

Then we have, where τ is the shear stress σ_{xy},

$$\tau(t) = \int_{-\infty}^{t} G(t-t')\dot{\gamma}(t')\,\mathrm{d}t'$$

where

$$\dot{\gamma}(t) = \frac{\partial u(y,t)}{\partial y}.$$

Continuity is satisfied. The only component of the equations of motion (2.60) that is not identically satisfied is

$$\rho\frac{\partial u}{\partial t} = \frac{\partial \tau}{\partial y}.$$

Taking Laplace transforms and substituting for $\tilde{\tau}$,

$$\rho s \tilde{u} = \frac{\partial \tilde{\tau}}{\partial y} = \tilde{G}\frac{\mathrm{d}^2 \tilde{u}}{\mathrm{d}y^2}.$$

This may be readily solved to give

$$\tilde{u} = A(s)\exp{-y}\sqrt{\left(\frac{\rho s}{\tilde{G}(s)}\right)} + B(s)\exp{y}\sqrt{\left(\frac{\rho s}{\tilde{G}(s)}\right)}.$$

Clearly, $B(s)=0$ as we have $\tilde{u}(\infty,s) = 0$ as a boundary condition. To evaluate $A(s)$ we Laplace transform the boundary conditions at $y=0$

$$\tilde{u}(s,0) = \frac{u_0}{s} = A(s).$$

Thus

$$\frac{\tilde{u}}{u_0} = \frac{1}{s}\exp{-y}\sqrt{\left(\frac{\rho s}{\tilde{G}(s)}\right)}.$$

The transform inversion is often difficult; in case we have a *viscous fluid* we have

$$G(t) = \eta_0\delta(t) \qquad \tilde{G} = \eta_0$$

and we get the well-known result

$$\frac{\tilde{u}}{u_0} = \frac{1}{s} \exp - y \sqrt{\left(\frac{s}{\nu_0}\right)},$$

where $\nu_0 = \eta_0/\rho$.

Inversion gives

$$u = u_0 \, \mathrm{erfc} \frac{y}{2\sqrt{(\nu_0 t)}} \equiv u_0 \left(1 - \frac{2}{\sqrt{(\pi)}} \int_0^z e^{-q^2} dq \right),$$

where $z = y/2\sqrt{(\nu_0 t)}$.

This is equivalent to a diffusion of momentum.

In case we have an *elastic solid*

$$\begin{cases} G(t) = GH(t) \\ \tilde{G} = G/s \end{cases}$$

$$\frac{\tilde{u}}{u_0} = \frac{1}{s} \exp - y \sqrt{\left(\frac{\rho s^2}{G}\right)} = \frac{1}{s} \exp - ys \sqrt{\left(\frac{\rho}{G}\right)}. \tag{2.109}$$

We know that the inverse transform of $F(t - a)$, denoted as $L\{[F(t - a)]\} = e^{-as} LF(t)$, and thus the inverse transform of (2.109) gives

$$u = u_0 H \left[t - y\sqrt{\left(\frac{\rho}{G}\right)}\right]. \tag{2.110}$$

Equation (2.110) describes a shear wave propagating at speed $\sqrt{(G/\rho)}$. Now for a Maxwell liquid,

$$G(t) = \frac{\eta_0}{\lambda} e^{-t/\lambda} \quad \therefore$$

$$\tilde{G}(s) = \frac{\eta_0}{\lambda} \frac{1}{(s + \frac{1}{\lambda})} = \frac{\eta_0}{\lambda s + 1} \tag{2.111}$$

$$\frac{\tilde{u}}{u_0} = \frac{1}{s} \exp - y \sqrt{\left(\frac{s(1 + \lambda s)}{\eta_0} \rho\right)}.$$

The exact inversion of eqn (2.111) (Tanner 1962) is

$$\frac{u}{u_0} = \exp(-t/\lambda) H \left[t - y\sqrt{\left(\frac{\rho \lambda}{\eta_0}\right)}\right]$$

$$+ \frac{r}{2} \int_0^{t/\lambda} \frac{e^{-\xi/2}}{\sqrt{(\xi^2 - r^2)}} I_1 \left[\frac{1}{2} \sqrt{(\xi^2 - r^2)}\right] \times H(\xi - r) \, d\xi, \tag{2.112}$$

where

$$r = y\sqrt{\left(\frac{\rho}{\eta_0 \lambda}\right)},$$

and I_1 is a modified Bessel function. Equation (2.112) describes a damped wave. These and other results for more complex cases were given by Tanner (1962). More complex values of $\tilde{G}$ usually only allow asymptotic or numerical inversions for u. See also Section 6.8.3 for a graphical description of (2.112).

In summary, we see that many difficult problems in linear viscoelasticity can be solved by transform methods. Numerical plus transform methods have also been used (Booker 1973). In many cases it is useful to find the linear viscoelastic response to obtain the qualitative changes with viscoelasticity; to obtain quantitative changes in non-Newtonian fluid mechanics we may often need to move outside the linear régime.

2.7   Non-linear viscoelastic behaviour

There are several possible paths for studying the response of non-linear fluids. In the earliest phase, (Weissenberg 1949, Oldroyd 1950) ideas of material symmetry were discussed for simple flows. Later (1945–56), many equations were written down on a fairly *ad hoc* basis using generalizations of the Newtonian fluid and linear viscoelasticity.

In most cases these were intended as a complete description of fluid behaviour, but they often failed to agree in detail with experimental results. Several writers (Green and Rivlin 1957; Noll 1958) then postulated that the stress was a *functional* of the history of the strain tensor $\mathbf{C}$ [eqn (2.32)].

Although this general statement must be correct for many fluids, unfortunately one can predict little from this approach that cannot be found by Weissenberg's (1949) symmetry arguments; see, for example, the book on viscometric flows by Coleman, Markovitz, and Noll (1966). More recently guidance as to the form of the response functional has been sought in molecular or structural arguments and we shall consider this approach in Chapter 5.

Many useful results for *restricted flow classes* can be found without resort to the formal functional approach. We discuss these in Chapter 3, together with relevant experimental work. Then we show (Chapter 4) how the older approaches approximate to these basic experimental results. Finally, attempts to seek the form of the response for general flows using microstructural arguments are outlined in Chapter 5.

References

Baumgaertel, M. and Winter, H. H. (1992). *J. Non-Newtonian Fluid Mech.*, **44**, 15.
Baumgaertel, M. and Winter, H. H. (1989). *Rheol. Acta*, **28**, 511.
Bland, D. R. (1960). *The theory of linear viscoelasticity*. Pergamon, Oxford.
Booker, J. R. (1973). *J. Engn. Maths*, **7**, 101.

Brabec, C. J., Rogl, H., and Schausberger, A. (1997). *Rheol. Acta*, **36**, 667.

Carslaw, H.S. and Jaeger, J.C. (1956). *Conduction of heat in solids*. Oxford University Press.

Coleman, B. D., Markovitz, H., and Noll, W. (1966). *Viscometric flows of non-Newtonian fluids*. Springer-Verlag, New York.

Davies, A. R. and Anderssen, R. S. (1997). *J. Non-Newtonian Fluid Mech.*, **73**, 263.

Davies, A. R. and Anderssen, R. S. (1998). *J. Non-Newtonian Fluid Mech.*, **79**, 235.

Ericksen, J. (1961). *Trans. Soc. Rheol.*, **5**, 23.

Ferry, J. D. (1970). *Viscoelastic properties of polymers*. Wiley, New York.

Green, A. E. and Rivlin, R. S. (1957). *Arch. ration Mech. Anal.*, **1**, 1.

Huilgol, R.R. and Phan-Thien, N. (1997). *Fluid mechanics of viscoelasticity*. Elsevier, Amsterdam.

Jeffreys, H. and Jeffreys, B. S. (1956). *Methods of mathematical physics*. Cambridge University Press.

Laun, H. M. (1978). *Rheol. Acta*, **17**, 1.

Lodge, A. S. (1964). *Elastic liquids*. Academic Press, London.

Mase, G. E. (1970). *Continuum mechanics*. McGraw-Hill, New York.

Noll, W. (1958). *Arch. ration. Mech. Anal.*, **2**, 197.

Oldroyd, J. G. (1950). *Proc. R. Soc.*, **A 200**, 523.

Phan-Thien, N. and Safari-Ardi, M. (1998). *J. Non-Newtonian Fluid Mech.*, **74**, 137.

Pipkin, A. C. (1986). *Lectures on viscoelasticity theory*, 2nd edn. Springer-Verlag, New York.

Rivlin, R. S. and Ericksen, J. L. (1955). *J. ration. Mech. Anal.*, **4**, 323.

Roscoe, R. (1950). *Br. J. Applied Phys.*, **1**, 171.

Ryskin, G. and Rallison, J. M. (1980). *J. Fluid Mech.*, **99**, 513.

Stokes, G. G. (1845). *Trans. Camb. phil. Soc.*, **8**, 287.

Streeter, V. L. and Wylie, E. B. (1979). *Fluid mechanics*, 7th edn, p. 89. McGraw-Hill, New York.

Tanner, R. I. (1962). *Z. angew. Math. Phys.*, **13**, 573.

Tanner, R. I. and Walters, K. (1998). *Rheology: An historical perspective*. Elsevier, Amsterdam.

Timoshenko, S. and Goodier, J. N. (1951). *Theory of elasticity*. McGraw-Hill, New York.

Trouton, F. T. (1906). *Proc. R. Soc.*, **A 77**, 426.

Truesdell, C. and Noll, W. (1965). *The nonlinear field theories of mechanics*. Springer-Verlag, Berlin.

Weissenberg, K. (1949). *Proc. int. Congr. Rheol.* (1948). North Holland, Amsterdam.

Woods, L. C. (1975). *The thermodynamics of fluid Systems*. Oxford University Press.

Zaremba, S. K. (1903). *Bull. Acad. Sci. Cracovie*, **85**, 380–594.

Problems

1. Use the summation convention on repeated indices and the definition of the unit tensor δ_{ij} to show that

(i) $\delta_{ii} = \delta_{ij}\delta_{ij} = \delta_{ij}\delta_{ik}\delta_{jk} = 3$

(ii) $\delta_{ij}\delta_{jk} = \delta_{ik}$

(iii) $\delta_{ij}A_{ik} = A_{jk}$.

2. A primed set of Cartesian axes ($\mathbf{x}'$) is obtained by rotation through an angle θ about the x_3 axis. Show that the components of the rotation matrix ($\mathbf{R}$) relating the two

sets of axes ($\mathbf{x}' = \mathbf{R}\mathbf{x}$) are

$$\mathbf{R} = \begin{bmatrix} \cos\theta & \sin\theta & 0 \\ -\sin\theta & \cos\theta & 0 \\ 0 & 0 & 1 \end{bmatrix}.$$

Show that $\mathbf{R}$ is an orthogonal matrix.

3. For the vectors $\mathbf{a} = 3\mathbf{i} + 4\mathbf{k}$, $\mathbf{b} = 2\mathbf{j} - 6\mathbf{k}$ and the dyadic $\mathbf{D} = 3\mathbf{ii} + 2\mathbf{ik} - 4\mathbf{jj} - 5\mathbf{kj}$, find the products $\mathbf{a} \cdot \mathbf{D}$, $\mathbf{D} \cdot \mathbf{b}$, and $\mathbf{a} \cdot \mathbf{D} \cdot \mathbf{b}$.

4. Determine the principal values and principal directions of the symmetric tensor $\mathbf{T}$ whose matrix representation is

$$\mathbf{T} = \begin{bmatrix} 3 & -1 & 0 \\ -1 & 3 & 0 \\ 0 & 0 & 1 \end{bmatrix}.$$

Show that the principal axes form a right-handed set of orthogonal axes, and that when transformed to these axes, the tensor ($\mathbf{T}'$) is diagonal, with entries (1, 2, 4). Calculate the invariants of $\mathbf{T}$.

5. Using the results of Problem 4 find $\sqrt[3]{\mathbf{T}}$.

6. The Stokes solution for the drag on a stationary sphere (radius a) in a Newtonian creeping flow of undisturbed velocity $U\mathbf{i}$ shows that at a point on the surface of the sphere the three components of the traction vector $\mathbf{t}$ are

$$t_x = -\frac{x}{a}p_0 + \frac{3}{2}\eta\frac{U}{a}; \quad t_y = -\frac{y}{a}p_0; \quad t_z = -\frac{z}{a}p_0.$$

Show that the total drag on the sphere is $6\pi\eta aU$.

7. A rectangular block of material is 20 mm long in the x direction, 6 mm high in the y direction, and 2 mm thick. A compressive load of 800 N and a shearing force are applied uniformly on the two surfaces normal to the x-axis ($x = \pm 10$ mm). On the top surface (plane $y = 3$ mm) is a shear force of 300 N directed along the negative x-direction. If the block is in equilibrium, find the maximum normal and shearing stresses. Also find the normal and shearing stress components on a diagonal plane which contains the z-axis and the block edge formed by the intersection of the planes $x = 10$ mm, $y = 3$ mm.

8. Does equilibrium exist for the following stress distribution in the absence of body force?

$$\sigma_{xx} = 3x^2 + 4xy - 8y^2, \quad \sigma_{xy} = -\tfrac{1}{2}x^2 - 6xy - 2y^2$$
$$\sigma_{yy} = 2x^2 + xy + 3y^2, \quad \sigma_{zz} = \sigma_{xz} = \sigma_{yz} = 0.$$

9. Consider an incompressible viscous (Newtonian) fluid. Show that the equation for the vorticity $\boldsymbol{\omega}$ is, in three dimensions,

$$\frac{D\boldsymbol{\omega}}{Dt} = (\boldsymbol{\omega} \cdot \mathbf{V})\mathbf{v} + \nu\nabla^2\boldsymbol{\omega}$$

where **v** is the velocity vector, ν is the kinematic viscosity and D/Dt is the material derivative. Show that in a two-dimensional motion the term containing **v** vanishes.

10. Show that in a plane flow the introduction of the stream function $\psi(x, y)$, where the velocity components are $u = \partial\psi/\partial y$, $v = -\partial\psi/\partial x$, satisfies the conservation of mass equation, Show that in a Newtonian creeping plane flow the stream function ψ satisfies the biharmonic equation. $\mathbf{V}^4\psi = 0$. Draw streamlines for the plane Poiseuille flow field $\mathbf{v} = (3/2)\mathbf{i}(1 - y^2)$ $(1 \geqslant y \geqslant 0)$.

11. In axisymmetric incompressible flows we can introduce the Stokes stream function, $\chi(r, z)$, so that

$$v_r = \frac{1}{r}\frac{\partial\chi}{\partial z}, \quad v_z = -\frac{1}{r}\frac{\partial\chi}{\partial r}.$$

Draw accurately the streamline pattern for the Poiseuille flow in a tube of radius 1, where $\mathbf{v} = 2\mathbf{k}(1 - r^2)$. Compare with the plane case.

12(a). A bar 100 mm long with a square cross-section of 400 mm^2 is deformed homogeneously so that the 100 mm side is extended to 120 mm and the cross-section remains square and is $333\frac{1}{3}$ mm^2. Define convenient co-ordinates and compute components of $\mathbf{F}(t')$ and $\mathbf{C}(t')$. Let t' refer to the time corresponding to the initial state and t the time for the present (deformed) state.

(b). The bar is deformed from its initial state as shown in Fig. 2.10. The cross-section is not changed. Compute $\mathbf{F}(t')$ and $\mathbf{C}(t')$ for this deformation.

13. Suppose one postulates a material such that the stress at a material point X is a linear function of the velocity gradient at that material point. Thus,

$$\boldsymbol{\tau}(X, t) = K\mathbf{L}(X, t),$$

where $\mathbf{L}^T = \mathbf{V}\,\mathbf{v}$. Show that the principle of material objectivity can only be obeyed if

$$\tau = K\mathbf{d},$$

where $\mathbf{d} = \frac{1}{2}(\mathbf{L} + \mathbf{L}^T)$.
Hint: Write the equation as $\boldsymbol{\tau} = K(\mathbf{d} + \mathbf{w})$, where $\mathbf{w} = \frac{1}{2}(\mathbf{L} - \mathbf{L}^T)$, and show that this must be reduced to $\boldsymbol{\tau} = K\mathbf{d}$ because, for two equivalent motions, $\mathbf{w}' = \mathbf{R}\mathbf{w}\mathbf{R}^T + \dot{\mathbf{R}}\mathbf{R}^T$.

14. Prove that the principal axes of the stress and strain tensors coincide for a homogeneous isotropic elastic body (small strain theory).

15. Show that the tensor $\mathbf{DA}/Dt$ is not objective, where $\mathbf{A}$ is a symmetric objective tensor.

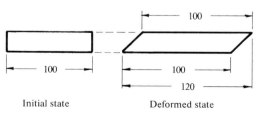

Initial state   Deformed state

Fig. 2.10 Problem 12.

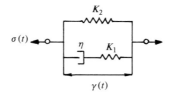

Fig. 2.11 Problem 17. Standard solid.

16. Consider a compressible Newtonian fluid with a constitutive equation of the form $\sigma_{ij} = -p\delta_{ij} + \Lambda\delta_{ij}d_{kk} + 2\eta d_{ij}$ where Λ and η are viscosities. Find the condition that the mean normal pressure $-\frac{1}{3}\sigma_{ii}$ is equal to the thermodynamic pressure p.

17. The standard linear solid is a mechanical model relating components of stress and strain through the form shown in Fig. 2.11 where K_1 and K_2 are spring constants and η is a viscosity. Write a differential operator relationship between stress and strain for this model and obtain the corresponding relaxation function, creep function and complex modulus. Sketch the real and imaginary parts of the complex modulus as a function of frequency.

18. The complex compliance $J^*(i\omega)$ is defined as the complex operator relating the steady-state harmonic stress and strain through

$$\hat{\gamma}e^{i\omega t} = J^*(i\omega)\hat{\sigma}e^{i\omega t}$$

where $\hat{\gamma}$ and $\hat{\sigma}$ are the amplitudes of components of strain and stress respectively. Using the relationship $J^* = [G^*]^{-1}$ find the real and imaginary parts of the complex compliance for the material in Problem 17.

19. When $G(t) = G_1 t^{-p}$, where G_1 and p are positive constants ($0 < p < 1$) we have a power-law material. Show that $J(t) = t^p \sin(p\pi)/G_1 p\pi$. For a sinusoidal strain find the values of $|G^*|$ and $\tan\delta$.

20. Find the response of a thin semi-infinite rod of Maxwell model material which is subjected to a suddenly applied tensile stress $[\sigma_0 H(t)]$ on the end $x = 0$.

21. Suppose an incompressible linear viscoelastic fluid has been sheared for a long time at a shear rate $\dot{\gamma}_0$. Then motion ceases, at $t = 0$. Find the shear stress in the fluid for $t \gtrless 0$.

3
VISCOMETRIC AND ELONGATIONAL FLOWS

3.1 Introduction

IN THIS chapter we consider some special, but very common, classes of flows applicable to a very wide group of liquids. This approach contrasts with that used in the classical theory of linearly viscous (Newtonian) fluids, in which a special kind of fluid description is used to examine a very wide class of flow fields. We first discuss those motions called viscometric, which are motions substantially equivalent to steady simple shearing, such as Poiseuille and Couette flows, but we do not make any very restrictive prior assumption about the nature of the relation between stress and deformation in the fluids considered. We then consider unsteady shearing flows, again without much restriction on fluid response, and then a class of simple elongational flows, in which the stress system is easily deduced from symmetry. We also consider the problem of wall slip and adherence. The fluids are supposed to be incompressible, and no thermal effects are considered. Non-isothermal flows are considered in Chapter 9. The viscometric motions discussed here can be visualized in terms of sheaf of inextensible material surfaces in relative sliding motion, but the elongational flows are not so readily defined in geometrical terms.

3.2 Kinematics of viscometric flows

In viscometric fluid motions, each fluid element is undergoing a steady simple shearing motion, plus possibly a rigid translation and rotation. The primary example of such a motion is steady simple shearing itself, in which the velocity field has the form

$$\mathbf{v} = \dot{\gamma}(\mathbf{b} \cdot \mathbf{x})\mathbf{a}. \tag{3.1}$$

Here $\mathbf{a}$ and $\mathbf{b}$ are orthogonal unit vectors and $\dot{\gamma}$ is a constant, the shear rate. In such a motion, the fluid in a plane $\mathbf{b} \cdot \mathbf{x} =$ constant forms a material surface that is called a slip surface. Each slip surface moves as a rigid body, and the motion can be visualized as the relative sliding motion of a sheaf of slip surfaces. The vector $\mathbf{a}$ is tangential to the slip surface and represents the direction of relative sliding motion, while $\mathbf{b}$ is the direction normal to the slip surface. It is convenient to define a third unit vector $\mathbf{c}$, orthogonal to $\mathbf{a}$ and $\mathbf{b}$ and thus tangential to the slip surfaces. We call $\mathbf{a}$, $\mathbf{b}$, and $\mathbf{c}$ the shear axes [Fig. 3.1(a)]. The velocity field (3.1) is easily shown to be divergence-free, and hence it is possible in an incompressible fluid.

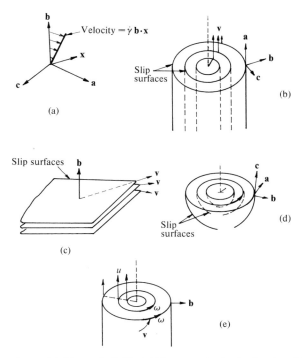

Fig. 3.1   Various viscometric kinematic fields. (a) Shear axes in simple shearing. (b) Steady parallel flow. (c) skew rectilinear flow (d) circular flow. (e) helical flow.

All viscometric flows can be visualized as the relative sliding motion of a sheaf of inextensible material surfaces. In the examples most useful in practice, the slip surfaces move steadily as rigid bodies, and each slip surface occupies the same locus in space at all times. It is known from kinematical studies that the relative motion of two rigid bodies contacting over an area (a 'lower pair') is described by combinations of sliding, turning and screw actions (Fig. 3.2). We can use this information when thinking about viscometric motions with rigid slip surfaces. In these kinds of viscometric motion the slip surfaces must be general cylinders, surfaces of revolution, or helicoids. We now list the main categories of these restricted viscometric flows without as yet considering the dynamical admissibility of any of them; the incompressibility condition is always satisfied.

3.2.1   Steady parallel flows

In steady flows with velocity fields of the form

$$\mathbf{v} = w(x, y)\mathbf{k}, \tag{3.2}$$

where $\mathbf{k}$ is a unit vector in the z-direction, the slip surfaces are the general cylinders $w(x, y) = $ constant [Fig. 3.1(b)]. The shear direction $\mathbf{a}$ is $\mathbf{k}$, and the

Name	Sketch
Screw pair $f=1$	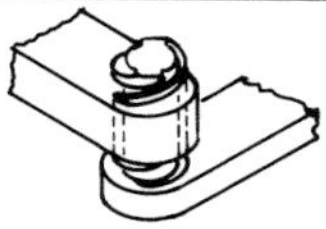
Hinge $f=1$	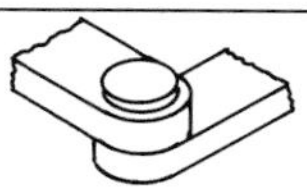
Prism pair $f=1$	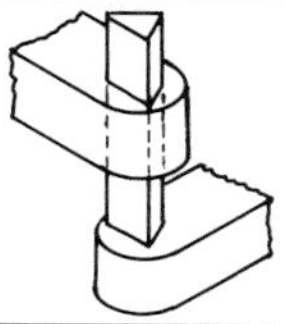
Cylinder pair $f=2$	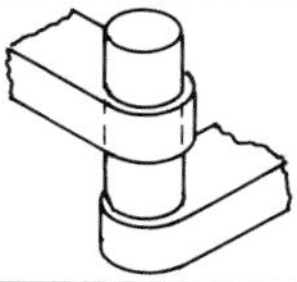
Spherical pair $f=3$	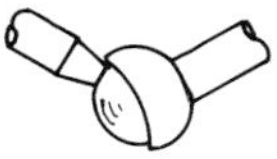

Fig. 3.2 Lower kinematic pairs in mechanisms, showing contact over an area. f is the number of degrees of freedom.

normal direction $\mathbf{b}$ is parallel to ∇w. The shear rate $\dot{\gamma}$ is equal to $|\nabla w|$, and it is constant in time at each particle, as required. We note that

$$\nabla \mathbf{v} = \nabla w \mathbf{k} = \dot{\gamma}\mathbf{ba}, \quad \text{or} \quad (\partial/\partial x_i)v_j = \dot{\gamma}b_i a_j. \tag{3.3}$$

The result can also be expressed as follows. The only non-zero components of $\nabla \mathbf{v}$ are $\partial w/\partial x$ and $\partial w/\partial y$. The shear rate $\dot{\gamma}$ is $\sqrt{(\partial w/\partial x)^2 + (\partial w/\partial y)^2}$ and is equal to $|\nabla w|$. The $\mathbf{a}$ direction coincides with $\mathbf{k}$ and $\mathbf{b}$ is the direction of ∇w, by definition. Thus we can write $\partial v_j/\partial x_i = (\partial w/\partial x_i)k_j$; resolving the gradient with respect to the $\mathbf{a}$, $\mathbf{b}$, and $\mathbf{c}$ vectors as axes we find the coordinate form in eqn (3.3), where k_j, b_i and a_j are the components of $\mathbf{k}$, $\mathbf{b}$ and $\mathbf{a}$ respectively, and $\mathbf{a} = \mathbf{k}$.

3.2.2 Skew rectilinear flows

Another class of flows in which all particles move steadily in straight lines are those with velocity fields of the form

$$\mathbf{v} = u(z)\mathbf{i} + v(z)\mathbf{j}. \tag{3.4}$$

These flows have parallel plane slip surfaces $z = $ constant moving steadily in skew directions. The normal direction $\mathbf{b}$ is parallel to $\mathbf{k}$ [Fig. 3.1(c)]. By taking the gradient of (3.4) we find

$$\mathbf{L}^T = \boldsymbol{\nabla}\mathbf{v} = \boldsymbol{\nabla}z\mathbf{v}'(z) = \mathbf{k}\mathbf{v}'(z). \tag{3.5}$$

To give (3.5) the form $\dot{\gamma}\mathbf{ba}$ as in the preceding example, then $\dot{\gamma}\mathbf{a}$ is given by

$$\dot{\gamma}\mathbf{a} = \mathbf{v}'(z) = u'(z)\mathbf{i} + v'(z)\mathbf{j}. \tag{3.6}$$

Then $\mathbf{a}$ is orthogonal to $\mathbf{b}$ as it must be, and $D\dot{\gamma}/Dt = 0$, where D/Dt means the material or particle-following time-derivative. We notice that the shear direction $\mathbf{a}$ is generally not parallel to the direction of motion $\mathbf{v}$. There is a broader class of skew rectilinear flows in which the slip surfaces are non-parallel planes, tangential to a general ruled surface, with each plane moving parallel to the line where it is tangent to the ruled surface (Yin and Pipkin 1970). Although these flows satisfy the kinematical requirements for viscometric flow, we do not consider them further.

3.2.3   Steady circular flows

In steady flows with coaxial circular streamlines, the velocity field is given in terms of cylindrical coordinates by

$$\mathbf{v} = r\omega(r, z)\mathbf{i}_\theta(\theta), \tag{3.7}$$

where $\mathbf{i}_\theta$ is the field of unit vectors in the azimuthal direction [Fig. 3.1(d)]. The slip surfaces are the surfaces of constant angular velocity $\omega(r, z)$. Evidently the shear direction $\mathbf{a}$ is $\mathbf{i}_\theta$, and the normal direction $\mathbf{b}$ is parallel to $\boldsymbol{\nabla}\omega$. At a given particle, the shear axes rotate as the particle circles the axis. An observer moving with the shear axis system for a given fluid element would see a motion locally equivalent to a steady simple shearing, but the velocity gradient observed from a fixed frame of reference is more complicated:

$$\boldsymbol{\nabla}\mathbf{v} = \boldsymbol{\nabla}r\omega\mathbf{i}_\theta + r\boldsymbol{\nabla}\omega\mathbf{i}_\theta + r\omega\boldsymbol{\nabla}\theta\mathbf{i}'_\theta(\theta)$$

$$= r\boldsymbol{\nabla}\omega\mathbf{i}_\theta + \omega(\mathbf{i}_r\mathbf{i}_\theta - \mathbf{i}_\theta\mathbf{i}_r). \tag{3.8}$$

The right-hand side of (3.8) contains an antisymmetric part that corresponds to a rigid rotation at the angular velocity ω. The remainder, $r\boldsymbol{\nabla}\omega\mathbf{i}_\theta$, is identified as $\dot{\gamma}\mathbf{ba}$, and we find that the shear rate is $r|\boldsymbol{\nabla}\omega|$. Since this is constant on each circle, then $D\dot{\gamma}/Dt = 0$, as required in order for the motion to be a locally-steady shearing.

3.2.4 Helical flows

Coaxial circular cylinders can move parallel to their common axis and simultaneously rotate about it, to produce a velocity field of the form [Fig. 3.1(e)]

$$\mathbf{v} = r\omega(r)\mathbf{i}_\theta(\theta) + w(r)\mathbf{i}_z. \tag{3.9}$$

The normal direction is evidently the radial direction, $\mathbf{b} = \mathbf{i}_r$. As in the case of skew rectilinear flows, the direction of shearing $\mathbf{a}$ is not obvious; applying the gradient operator we obtain

$$\nabla\mathbf{v} = \mathbf{i}_r(r\omega'\mathbf{i}_\theta + w'\mathbf{i}_z) + \omega(\mathbf{i}_r\mathbf{i}_\theta - \mathbf{i}_\theta\mathbf{i}_r). \tag{3.10}$$

The antisymmetric term proportional to ω is evidently the contribution due to the rotation of the shear axes at a particle as it moves around the axis. The remainder has the form $\dot{\gamma}\mathbf{ba}$ if

$$\dot{\gamma}\mathbf{a} = r\omega'(r)\mathbf{i}_\theta + w'(r)\mathbf{i}_z. \tag{3.11}$$

Then $\mathbf{a}$ is perpendicular to $\mathbf{b}$ as required, and the value of $\dot{\gamma}$ at each particle is conserved.

3.2.5 Helicoidal flows

The helical streamlines in the preceding example have a rise per turn $2\pi w/\omega$ which is constant over each cylinder but may vary from one cylinder to another. If all helices have the same rise per turn, the slip surfaces need not be cylindrical but can be general helicoids. In these helicoidal flows, the velocity field has the form

$$\mathbf{v} = (r\mathbf{i}_\theta + c\mathbf{i}_z)\omega(r, z - c\theta). \tag{3.12}$$

The shear direction $\mathbf{a}$ is the same as the direction of motion, and the normal direction $\mathbf{b}$ is parallel to $\nabla\omega$, orthogonal to the helicoids $\omega = $ constant.

Since the velocity gradient transpose has the form

$$\nabla\mathbf{v} = \nabla\omega(r\mathbf{i}_\theta + c\mathbf{i}_z) + \omega(\mathbf{i}_r\mathbf{i}_\theta - \mathbf{i}_\theta\mathbf{i}_r), \tag{3.13}$$

by identifying the first term as $\dot{\gamma}\mathbf{ba}$ we find that the shear rate is given by

$$\dot{\gamma}^2 = (r^2 + c^2)\nabla\omega \cdot \nabla\omega. \tag{3.14}$$

By computing its material derivative, we find that the shear rate is constant along streamlines and is thus constant in time for each particle.

3.2.6 General kinematics of viscometric flows

The preceding examples are of the kind that Lodge (1964) calls steady curvilinear shearing motions. These examples include all viscometric flows that have so far proved to be of any practical interest, but they do not by any means exhaust the kinematical possibilities.

The definition we use is that the history of deformation of a fluid element is viscometric if its motion, viewed from a system of possibly rotating axes, is a steady simple shearing motion. Let $\mathbf{a}$, $\mathbf{b}$, and $\mathbf{c}$ be the (time-dependent) shear axes at a given particle. Consider infinitesimal material elements of length $\mathrm{d}x$ along the directions of the shear axes at time t, $\mathbf{a}(t)\mathrm{d}x$, $\mathbf{b}(t)\mathrm{d}x$, and $\mathbf{c}(t)\mathrm{d}x$. The configuration of these material fibres at time t' defines the deformation of the infinitesimal volume element containing them between time t and time t'. The deformation is viscometric if, for all t', the directions and lengths of the three fibres are given by

$$\mathbf{a}(t)\,\mathrm{d}x \rightarrow \mathbf{a}(t')\,\mathrm{d}x, \quad \mathbf{c}(t)\mathrm{d}x \rightarrow \mathbf{c}(t')\,\mathrm{d}x,$$
$$\mathbf{b}(t)\,\mathrm{d}x \rightarrow [\mathbf{b}(t') + (t' - t)\dot{\gamma}\mathbf{a}(t')]\,\mathrm{d}x, \qquad (3.15)$$

where $\dot{\gamma}$ is a constant for the particle considered.

This form of the definition of a viscometric deformation history does not require that all particles in a given flow be in viscometric motion in order to call the motion of one of them viscometric, but we are generally concerned with cases in which in fact all particles are in viscometric motion. Some of the global geometrical features of such flows are immediately apparent from the definition. Consider the trajectories of the field $\mathbf{a}$, the curves with the vectors $\mathbf{a}$ as tangent vectors. Call these a-lines. The definition states that a material element along an a-line at time t is still along an a-line at time t, and its length is not changed. Thus, a-lines are material lines, convected by the motion, and they move without stretching. Similarly, c-lines are inextensible material lines.

It is also true, but much more difficult to prove (Yin and Pipkin 1970) that a-lines and c-lines must mesh to form material surfaces, the slip surfaces. If this is granted, then it is evident that the slip surfaces are material surfaces that move without stretching. In the examples already given, the slip surfaces not only do not stretch during the motion but also do not change shape. The latter feature is not a general kinematical property of globally viscometric flows. Velocity fields in which the slip surfaces curl up during the course of the motion can be constructed mathematically (Yin and Pipkin 1970). In such cases the slip surfaces are necessarily ruled surfaces, and the straight rule lines are material lines that remain straight when the surface curls.

It is natural to conjecture that viscometric flows must be steady with respect to some global frame of reference, but this, too, is not a general kinematical property. Motions with curling slip surfaces are unsteady in every frame, and motions with rigid slip surfaces are in general unsteady as well, even though the steady shear rate is constant in time at every particle. For details see Yin and Pipkin (1970).

3.3 Stresses in steady viscometric flows

When a viscoelastic liquid is brought from rest into a state of steady shearing motion, the stress is time-dependent because of the transient effects of the past history of deformation. However, if the fluid is sheared at a constant rate, these transients die out in the course of time, and the shearing stress approaches a steady-state value that depends only on the shear rate. It is also assumed that the material properties are not changing with time due to structural breakdown (see Section 1.5).

For the simplest special case, with a velocity field of the form $\mathbf{v} = \dot{\gamma} y \mathbf{i}$, the zx- and zy-components of stress must be zero by symmetry, and the shearing stress σ_{xy} is some odd function of the shear rate,

$$\sigma_{xy} = \dot{\gamma} \eta(\dot{\gamma}), \quad \eta(-\dot{\gamma}) = \eta(\dot{\gamma}). \tag{3.16}$$

In drawing conclusions from symmetry as we have done here, we implicitly assume that the fluid is isotropic, so that any directional properties it may have are induced by the flow itself. The ratio $\sigma_{xy}/\dot{\gamma} = \eta$ is the viscosity function, or simply, the viscosity.

The normal stress components σ_{xx}, σ_{yy}, and σ_{zz} are even functions of the shear rate by a symmetry argument; reversal of shearing cannot affect these components. Consequently, the differences between these components must be zero in any theory depending linearly on shear rate; this does not apply to visco-elastic liquids. These symmetry arguments for deducing the form of the stress field in shear flows are due to Weissenberg (Russell 1946).

The two independent differences,

$$\sigma_{xx} - \sigma_{yy} = N_1(\dot{\gamma}) \quad \text{and} \quad \sigma_{yy} - \sigma_{zz} = N_2(\dot{\gamma}), \tag{3.17}$$

are functions of the shear rate that vanish when $\dot{\gamma} = 0$.

We call these two functions of the shear rate the first and second normal stress differences. Under the classical conception of a fluid as a material that cannot indefinitely remain at rest when under the action of even a very small shearing stress, there can be no shearing stress on any surface element in the case $\dot{\gamma} = 0$, since here the statement that $\dot{\gamma}$ is zero means that the fluid has been at rest for so long that all transients have disappeared. This implies not only that σ_{xy} must vanish, but also that all normal stress components must be equal when $\dot{\gamma} = 0$. It is often convenient to make this explicit by writing

$$N_1 = \dot{\gamma}^2 \Psi_1(\dot{\gamma}) \quad \text{and} \quad N_2 = \dot{\gamma}^2 \Psi_2(\dot{\gamma}). \tag{3.18}$$

The functions Ψ_1 and Ψ_2 are even functions of the shear rate. We call them the first and second normal stress coefficients.

3.3.1 Qualitative behaviour of the viscometric functions for polymer solutions

Most of the results to be discussed do not depend on any special assumptions about the forms of the viscometric functions η, Ψ_1, and Ψ_2; indeed, for the most part we discuss experimental methods of determining the forms of these functions. However, for the sake of concreteness it is useful to understand their general nature in the case of polymer solutions. Figure 3.3 (full lines), shows the three viscometric functions N_1, N_2, and $\tau(\equiv \eta\dot{\gamma})$ for a 6.8 per cent solution of a polyisobutylene (Oppanol B-100) in cetane at $24\,^{\circ}$C; note that the SI unit of viscosity (1 Pa-s) equals ten Poises.

The viscosity function typically decreases, as the shear rate increases, from a limiting value η_0 at zero shear rate to a much lower value η_∞ in the limit of high shear rates. (See Fig. 1.2.) In Fig. 3.3 these limiting values were not reached experimentally. Special forms of the viscosity function were discussed in Chapter 1; an especially simple and useful form is the power-law approximation $\eta = k\dot{\gamma}^{n-1}$, with $n-1$ the slope of a doubly logarithmic plot. This kind of approximation is useful for the region of intermediate shear rates in which the viscosity is dropping from η_0 toward η_∞. For shear-thinning (pseudo-plastic) fluids the power n is less than unity. For any fluid, n must be greater than zero in order to satisfy the stability requirement that the shearing stress increases with the shear rate (see Chapter 10); occasionally (rheopectic) fluids have n greater than one.

The first normal stress difference is positive, at least for polymeric fluids, and the coefficient Ψ_1 has roughly the same form as η. The slope in the power-law region is generally greater than that for η, and no limiting value at high shear rates has ever been reached to the author's knowledge. Theoretical models based on microstructural ideas (Chapter 5) indicate that $\Psi_1 \rightarrow 0$ as $\dot{\gamma} \rightarrow \infty$.

Early measurements with polymeric systems always found N_1 (and Ψ_1) to be positive, but liquid crystal systems can show negative N_1. Kiss and Porter (1978) seem to have been the first to report this phenomenon. In such systems one can have a region (or regions) of positive N_1 interspersed with a region of negative N_1.

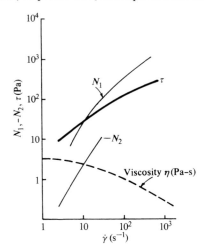

Fig. 3.3   The rheological functions τ, N_1 and $-N_2$ shown as functions of shearing rate for a 6.8 per cent polyisobutylene in cetane solution at $24\,^{\circ}$C. The viscosity function is also shown.

The second normal stress difference is negative for concentrated polymer solutions. It appears that Ψ_2 is generally smaller than Ψ_1, and is often less than 20 per cent of Ψ_1. Partly for this reason Ψ_2 is more difficult to measure than Ψ_1, and less is known about the form of this function. Weissenberg's hypothesis that $N_2 = 0$ is not correct, but for some purposes it is qualitatively adequate. In broadest terms, the effect of normal stress differences is that there is an extra tension N_1 in the direction of shearing, and this extra tension is not small. Further discussion of material properties is given in Section 3.8 below; data for N_2/N_1 are given in Section 3.8.3.

3.3.2 Stress fields

We are only considering situations in which the fluid can be treated as incompressible. The pressure is then a reaction to the constraint of no volume change, and it cannot be specified directly as a function of the shear rate. Instead, the momentum equation and the boundary conditions are used to determine the pressure required to enforce the constraint. Both because of this arbitrariness and because normal stresses in different directions are not equal, the term 'pressure' must be used with great caution.

In steady simple shearing motion, with $\mathbf{v} = \dot{\gamma} y \mathbf{i}$, the shear axes $\mathbf{a}$, $\mathbf{b}$, and $\mathbf{c}$ correspond to the Cartesian unit vectors $\mathbf{i}$, $\mathbf{j}$, and $\mathbf{k}$. It is convenient to write the stress in dyadic notation as

$$\sigma = -\bar{p}\mathbf{I} + \dot{\gamma}\eta(\mathbf{ab} + \mathbf{ba}) + N_1\mathbf{aa} - N_2\mathbf{cc}. \tag{3.19}$$

This is only a formal statement of the symmetry of the stress field discussed above.

Here the reaction pressure $\bar{p}$ is the pressure normal to the slip surfaces (direction $\mathbf{b}$), and it can be written as $-\mathbf{b} \cdot \sigma \cdot \mathbf{b}$. The stress tensor can be written in a form that is more useful in further developments by letting $p = \bar{p} + N_2$ be a new, equally arbitrary pressure:

$$\sigma = -p\mathbf{I} + \dot{\gamma}\eta(\mathbf{ab} + \mathbf{ba}) + (N_1 + N_2)\mathbf{aa} + N_2\mathbf{bb}. \tag{3.20}$$

In this case the pressure p represents the normal pressure in the neutral direction, $-\mathbf{c} \cdot \sigma \cdot \mathbf{c}$ (or σ_{cc}).

In any viscometric flow, each fluid element is performing a steady simple shearing motion. Consequently, the stress in each element is given by an expression of the same form as that derived for steady simple shearing, with the same response functions η, Ψ_1, and Ψ_2, but with shear axes and shear rate appropriate to the particle considered. For the special flows considered in Sections 3.2.1 to 3.2.5, we can immediately write down expressions for the stress by substituting the expressions for the shear axes and shear rates found in those sections into eqn (3.20):

Steady parallel flows:

$$\boldsymbol{\sigma} = -p\mathbf{I} + \eta(\mathbf{k}\boldsymbol{\nabla}w + \boldsymbol{\nabla}w\mathbf{k}) + \dot{\gamma}^2(\Psi_1 + \Psi_2)\mathbf{kk} + \Psi_2\boldsymbol{\nabla}w\boldsymbol{\nabla}w$$
$$\dot{\gamma}^2 = \boldsymbol{\nabla}w \cdot \boldsymbol{\nabla}w. \tag{3.21}$$

Skew rectilinear flows:

$$\boldsymbol{\sigma} = -p\mathbf{I} + \eta(\mathbf{v}'\mathbf{k} + \mathbf{k}\mathbf{v}') + (\Psi_1 + \Psi_2)\mathbf{v}'\mathbf{v}' + \Psi_2\dot{\gamma}^2\mathbf{kk}$$
$$\dot{\gamma}^2 = \mathbf{v}' \cdot \mathbf{v}'. \tag{3.22}$$

Steady circular flows:

$$\boldsymbol{\sigma} = -p\mathbf{I} + \eta r(\mathbf{i}_\theta\boldsymbol{\nabla}\omega + \boldsymbol{\nabla}\omega\mathbf{i}_\theta) + \dot{\gamma}^2(\Psi_1 + \Psi_2)\mathbf{i}_\theta\mathbf{i}_\theta + \Psi_2 r^2\boldsymbol{\nabla}\omega\boldsymbol{\nabla}\omega$$
$$\dot{\gamma}^2 = r^2\boldsymbol{\nabla}\omega \cdot \boldsymbol{\nabla}\omega. \tag{3.23}$$

Helical flows:

$$\boldsymbol{\sigma} = -p\mathbf{I} + \eta(\dot{\gamma}\mathbf{a}\mathbf{i}_r + \mathbf{i}_r\dot{\gamma}\mathbf{a}) + (\Psi_1 + \Psi_2)(\dot{\gamma}\mathbf{a})(\dot{\gamma}\mathbf{a}) + \dot{\gamma}^2\Psi_2\mathbf{i}_r\mathbf{i}_r,$$
$$\dot{\gamma}^2 = (r\omega')^2 + (w')^2, \quad \dot{\gamma}\mathbf{a} = r\omega'\mathbf{i}_\theta + w'\mathbf{i}_z. \tag{3.24}$$

Helicoidal flows:

$$\boldsymbol{\sigma} = -p\mathbf{I} + \eta[(r\mathbf{i}_\theta + c\mathbf{i}_z)\boldsymbol{\nabla}\omega + \boldsymbol{\nabla}\omega(r\mathbf{i}_\theta + c\mathbf{i}_z)]$$
$$+ (\Psi_1 + \Psi_2)\boldsymbol{\nabla}\omega \cdot \boldsymbol{\nabla}\omega(r\mathbf{i}_\theta + c\mathbf{i}_z)(r\mathbf{i}_\theta + c\mathbf{i}_z) + (r^2 + c^2)\Psi_2\boldsymbol{\nabla}\omega\boldsymbol{\nabla}\omega,$$
$$\dot{\gamma}^2 = (r^2 + c^2)\boldsymbol{\nabla}\omega \cdot \boldsymbol{\nabla}\omega. \tag{3.25}$$

It remains to consider whether these flows are dynamically possible or not.

3.4   Controllable viscometric flows

For the experimental determination of the viscometric functions η, Ψ_1, and Ψ_2 for a particular fluid, it is desirable to be able to set up experiments in which the velocity field is completely or at least partly known from the outset. The analysis of experimental data is then not complicated by the simultaneous determination of an unknown velocity field. A few specific velocity fields have such a high degree of symmetry that they satisfy the momentum equation exactly, or with neglect of inertia, no matter what forms the viscometric functions may take. We call such flows completely controllable. All such flows are known.

Homogeneous steady simple shearing, with a velocity field $\mathbf{v} = \dot{\gamma}y\mathbf{i}$, is a completely controllable flow. Since there is no acceleration, the momentum equations reduce to equilibrium equations. If the pressure p is uniform, then all stress components are independent of position, and the equilibrium equations are satisfied identically, no matter what forms the viscometric functions may have. If the momentum equation involves a body force term, then in the case of simple shearing the pressure p is the hydrostatic pressure plus an arbitrary constant. Since a conservative field of body force can be equilibrated by a hydrostatic pressure, we ordinarily omit both the body force and the hydrostatic pressure in discussing solutions of problems; it can be added in easily if it is needed. Unfortunately, simple shearing, because of edge effects, is in practice difficult to realize.

Steady shearing between tilted plates is a completely controllable flow in which the shear rate varies with position. If one plate is at rest and the other, tilted at an angle θ_0 to it, moves with a constant speed U parallel to the line where the plates would intersect, then the velocity field $\mathbf{v} = (U\theta/\theta_0)\mathbf{k}$ is kinematically admissible. The slip surfaces are the planes $\theta = $ constant in cylindrical coordinates. The stress field, found by setting $w = U\theta/\theta_0$ in eqn (3.21) is

$$\boldsymbol{\sigma} = -p\mathbf{I} + \eta\dot{\gamma}(\mathbf{k}\mathbf{i}_\theta + \mathbf{i}_\theta\mathbf{k}) + (N_1 + N_2)\mathbf{k}\mathbf{k} + N_2\mathbf{i}_\theta\mathbf{i}_\theta, \tag{3.26}$$

where the shear rate (see Appendix) is $\dot{\gamma} = U/r\theta_0$. Since the shear rate depends only on r, then whatever forms the viscometric functions may take, their values in this flow can depend only on r. We will ignore body forces in the problem. From (3.26) $\sigma_{r\theta}$ and σ_{rz} are zero, and from the form of the velocity field it is clear that the particle accelerations are also zero. Hence, from the Appendix, noting that all quantities are independent of z, we can deduce that $dp/d\theta = 0$, since N_2 is independent of θ; we also know that N_2 is independent of θ from (3.26). Hence two of the equilibrium equations are satisfied. The remaining equilibrium equation is

$$r\frac{dp}{dr} + N_2 = 0. \tag{3.27}$$

Now $\dot{\gamma} = U/r\theta_0$, and $N_2 = \Psi_2\dot{\gamma}^2$, where Ψ_2 is only a function of $\dot{\gamma}$. By changing to $\dot{\gamma}$ as a variable instead of r in (3.27), it is easy to verify that the momentum equation is satisfied if p has the form

$$p = p(r_0) + I_2(U/r\theta_0) - I_2(U/r_0\theta_0), \tag{3.28}$$

where $I_2(\dot{\gamma}) = \int_0^{\dot{\gamma}} \Psi_2\dot{\gamma}d\dot{\gamma}$. Despite the completeness of the solution, practical realization of this flow is difficult, due to end and edge effects.

There is a flow somewhat like Couette flow that formally satisfies that conditions for a controllable viscometric flow. The angular velocity is $\omega = \dot{\gamma}\log(r/r_0)$, where the constant $\dot{\gamma}$ is the shear rate. Since the shear rate is independent of position, then so are the physical components of stress with respect to the shear axis system, except for p, which is not directly determined by the shear rate. By using eqn (3.23) in the momentum equation we find that it is satisfied if p has the form

$$p = 2\dot{\gamma}\eta\theta - N_1 \log(r/r_0) + \tfrac{1}{2}\rho r^2(\omega^2 - \dot{\gamma}\omega + \tfrac{1}{2}\dot{\gamma}^2) + \text{constant}. \tag{3.29}$$

Because the pressure has a term proportional to θ, such a flow can be sustained only in a sector and not in the full annular gap of a Couette viscometer. The flow is of interest as an example of an inhomogeneous flow with a uniform shear rate.

The preceding examples, together with others obtained from them by super-imposing rigid-body motions, exhaust the list of completely controllable flows (Yin and Pipkin 1970). In the hope of finding more useful examples, we loosen the requirements slightly by considering flows that are completely con-trollable if inertia can be neglected. Since the Reynolds number is often extremely low in experiments on complex fluids, neglect of inertia is often a reasonable approximation.

There are only two new kinds of flows that are controllable if inertia is neglected. One is an intrinsically unsteady motion with slip surfaces that curl up (Yin and Pipkin 1970), which appears to have no experimental value. The other is the class of helicoidal flows with right helicoidal slip surfaces for which the velocity field has the form

$$\mathbf{v} = k(z - c\theta)(r\mathbf{i}_\theta + c\mathbf{i}_z). \tag{3.30}$$

The stress for this case is given by eqn (3.25), in which

$$\omega = k(z - c\theta) \quad \text{and} \quad \dot{\gamma} = k(r + c^2/r). \tag{3.31}$$

Since the shear rate depends only on r, then so do the values of the viscometric functions, and it is found that the momentum equation is satisfied with neglect of inertia if the pressure has the form

$$p(r) = p(r_0) + \int_{r_0}^{r} \left[r^{-1}(2ck\eta - \dot{\gamma}^2 \Psi_2) - k\dot{\gamma}\Psi_1 \right] dr. \tag{3.32}$$

Among these flows, there is one special case that is convenient experimentally, the case $c = 0$ (torsional flow). We consider this particular kind of flow in detail in Section 3.4.1.

3.4.1   Torsional flow in parallel-plate geometry

The torsional flow in a parallel-plate viscometer is the only completely con-trollable flow that is used in practical viscometry. In the flow between a fixed disc and a disc rotating with angular velocity ω_0, the velocity field has the form $\mathbf{v} = (\omega_0 rz/h)\mathbf{i}_\theta$, where h is the separation between the plates. The slip surfaces are the parallel planes $z =$ constant, rotating with angular velocity $\omega = \omega_0 z/h$ about the z-axis. The shear rate is $\dot{\gamma} = r\omega_0/h$. Consequently, at any given shear rate centrifugal force can be made arbitrarily small by decreasing the angular velocity ω_0 and the gap width h in the same proportion. Thus inertial effects, proportional to $\rho r_0 \omega_0^2$, can be made negligible.

If the fluid is in contact with the plates out to the radius r_0, then the moment required to turn the rotating disc or hold the fixed plate is

$$M = 2\pi \int_0^{r_0} \dot\gamma \eta r^2 \, dr. \tag{3.33}$$

In terms of the variables

$$m = M/2\pi r_0^3 \quad \text{and} \quad \dot\gamma_0 = r_0 \omega_0 / h \tag{3.34}$$

this relation is

$$m = \dot\gamma_0^{-3} \int_0^{\dot\gamma_0} \eta \dot\gamma^3 \, d\dot\gamma. \tag{3.35}$$

Thus, the relation among the four measurable quantities $M, r_0, \omega_0,$ and h is reduced to a relation between only two variables, m and $\dot\gamma_0$. Data that cannot be reduced to a single curve of m versus $\dot\gamma_0$ usually indicates that the fluid is not adhering to the plates, contradicting our implicit assumption of a no-slip condition, or that the plates are not truly set up, or that edge effects disturb the flow.

The viscosity function can be determined by numerical differentiation of a plot of m versus $\dot\gamma_0$. For, on differentiating eqn (3.35) we obtain

$$\eta(\dot\gamma_0) = \frac{m}{\dot\gamma_0} \left[3 + \frac{d(\log m)}{d(\log \dot\gamma_0)} \right]. \tag{3.36}$$

The pressure p is given by eqn (3.32) in which we set $c = 0$ and $k = \omega_0/h$. Since the r-direction is the direction of the **c**-vector (neutral direction) in this flow, the radial stress $\sigma_{rr} = -p$. If the fluid is held in the gap by surface tension or by the tension along streamlines, so that the outer boundary of the fluid is approximately the cylinder $r = r_0$, then $-\sigma_{rr}$ there is approximately atmospheric pressure, which we set as zero gauge pressure. (Surface tension can also be included if required; it will add a pressure of σ/r_0 to the atmospheric pressure at the rim where σ is the surface tension coefficient; if the outer boundary is of a more complex shape, then this needs to be considered, too. This term is only of importance experimentally if it *changes* with shear rate.)

Then eqn (3.32) gives

$$p(r) = \int_{\dot\gamma}^{\dot\gamma_0} \dot\gamma (\Psi_1 + \Psi_2) \, d\dot\gamma. \tag{3.37}$$

The pressure increases as r decreases because of the squeezing caused by the extra tension along the circular streamlines. The axial stress is, from eqn (3.25),

$$\sigma_{zz}(r) = -p + N_2. \tag{3.38}$$

Measurement of the distribution of normal thrust can give information about a combination of the functions Ψ_1 and Ψ_2. However, if these pressure measurements are made by attaching pressure gauges to small holes in the surface, the holes introduce a large systematic error (see Chapter 4). Since this was discovered only in 1968, earlier data involve such errors. By using flush-mounted pressure transducers, one could use (3.38) to measure N_2 directly, since, at least ideally, $p = 0$ at the rim of the flow.

The total thrust on the plates can be measured without interference from any hole error. The total thrust is, from eqn (3.38), noting that we assume $p = 0$ at $r = r_0$,

$$F = 2\pi \int_0^{r_0} (p - N_2) r \, dr = -\pi \int_0^{r_0} (2rN_2 + r^2 p') \, dr. \tag{3.39}$$

By letting $\dot\gamma = r\omega_0/h$ be the integration variable and using the expression for $p'(r)$ found from eqn (3.37), we obtain

$$f = F/\pi r_0^2 = \dot\gamma_0^{-2} \int_0^{\dot\gamma_0} \dot\gamma (N_1 - N_2) \, d\dot\gamma. \tag{3.40}$$

The combination $N_1 - N_2$ can be determined from a plot of the force per unit area f versus the rim shear rate $\dot\gamma_0$. By differentiating eqn (3.40) we find that

$$N_1 - N_2 = f \left(2 + \frac{d \log f}{d \log \dot\gamma_0} \right). \tag{3.41}$$

Thus careful total thrust measurements may be used to find $N_1 - N_2$.

3.5   Partially controllable flows

Among controllable flows, in which the velocity field is fully known in advance of any knowledge of the forms of the viscometric functions, only torsional flow has been used as a practical method of viscometry. Many other flows are used for viscosity measurements. In these flows the velocity field is not completely specified at the outset, but its general nature is known.

There is a category of flows in which the shapes of the slip surfaces are known in advance, but their speeds depend on the form of the viscosity function, which is to be determined. These flows have such symmetry that the normal stress distribution does not influence the velocity distribution. Whatever forms the normal stress functions may have, the distribution of normal stress differences can be equilibrated by an appropriate distribution of pressure. Such flows are called partially controllable. All practical methods of viscosity measurement involve flows that are at least partially controllable. All such flows are known (Yin and Pipkin 1970). Aside from the completely controllable cases already mentioned in Section 3.4, the only flows that are partially controllable are the skew rectilinear motions, some Poiseuille flows, the helical flows, and the motion in a cone-and-plate viscometer.

We now examine the various types of partially controllable viscometric flows and indicate how to determine the velocity fields in detail.

3.5.1 Skew rectilinear motions

Skew rectilinear flows can be visualized in terms of the motion of a fluid between two plates in parallel translation, with a pressure gradient in the fluid that is generally not parallel to the direction of relative motion of the plates. These flows can be used as approximations to flows between two cylinders of almost equal radii, in which, for example, one cylinder is rotating with respect to the other and there is also an axial pressure gradient. If the gap width is small in comparison to the radii, so that the curvature can be ignored, the flow may be treated as a skew rectilinear flow.

Since there is no acceleration, the momentum equation reduces to the form $\nabla \cdot \boldsymbol{\sigma} = \mathbf{0}$. The stress is given by eqn (3.22). By using this expression in the momentum equation, we obtain, since there is no variation of $\mathbf{v}'$ in the x and y directions

$$\nabla p = (\mathrm{d}/\mathrm{d}z)(\eta \mathbf{v}' + \mathbf{k}N_2). \tag{3.42}$$

Since the right-hand member is independent of x and y, the partial derivatives of p in these directions must be constants. Thus, the part of the pressure gradient parallel to the plates is a constant vector, $-\mathbf{G}$ say, and we obtain

$$\mathrm{d}(\eta \mathbf{v}')/\mathrm{d}z = -\mathbf{G}, \tag{3.43}$$

and

$$\nabla p = -\mathbf{G} + \mathbf{k} \, \mathrm{d}N_2/\mathrm{d}z. \tag{3.44}$$

The equation for $\mathbf{v}(z)$ involves the viscosity function. However, no matter what the velocity turns out to be, and no matter what form the function N_2 may have, the equation for p can be integrated to give

$$p = -\mathbf{G} \cdot \mathbf{x} + N_2(\dot{\gamma}) + \text{constant}. \tag{3.45}$$

Normal stress differences affect the pressure distribution but do not affect the velocity field, so the motion is of the kind that we call partially controllable.

From eqn (3.43) we obtain

$$\eta \mathbf{v}' = -\mathbf{G}z + \mathbf{C}, \tag{3.46}$$

where $\mathbf{C}$, like $\mathbf{G}$, has no z-component. It is convenient to use the inverse of the viscosity function, called the fluidity, $\phi(\tau)$, at this point. ϕ is defined by

$$\dot{\gamma} = \tau\phi(\tau), \tag{3.47}$$

where τ is the shear stress. Then we have, since the magnitude of the shear stress τ is given by $|\eta \mathbf{v}'|$,

$$\mathbf{v}'(z) = (-\mathbf{G}z + \mathbf{C})\phi(\tau), \qquad \tau = |\mathbf{G}z - \mathbf{C}|. \tag{3.48}$$

If we consider the plate in the plane $z = 0$ to be at rest, then

$$\mathbf{v}(z) = -\mathbf{G} \int_0^z \phi(\tau)z \, dz + \mathbf{C} \int_0^z \phi(\tau) \, dz. \tag{3.49}$$

The integration constant $\mathbf{C}$ is to be determined by using the boundary condition at the other plate, say $\mathbf{v}(h) = \mathbf{U}$. Ordinarily this cannot be done exactly except by numerical methods. However, when ϕ is proportional to τ^2, corresponding to the fairly realistic law $\eta = k\dot\gamma^{-2/3}$, the integration is easy. (See Problem 3.1.)

3.5.2 Poiseuille flows

The parallel flows described in Section 3.2.1 are partially controllable if the slip surfaces are parallel planes or coaxial circular cylinders. The former case corresponds to plane Poiseuille flow and the latter to Poiseuille flow in a tube of circular cross-section or in the annular gap between two coaxial tubes. In both cases the boundaries may be in relative motion parallel to the direction of the pressure gradient.

Before considering these special cases, let us examine steady parallel flows more generally, in order to understand why other cases such as the flow in a square duct are not partially controllable. Since the velocity $\mathbf{v}$ is of the form $w(x, y)\mathbf{k}$ there is no acceleration, and the momentum equation becomes $\partial\sigma_{ij}/\partial x_j = 0$. By using eqn (3.21) we obtain

$$\nabla p = \nabla \cdot (\eta\nabla w)\mathbf{k} + \nabla \cdot (\Psi_2\nabla w)\nabla w + \Psi_2\nabla w \cdot \nabla(\nabla w). \tag{3.50}$$

Since the right-hand member is independent of the axial coordinate z, then $\partial p/\partial z$ is a constant, $-p'$, say, where p' is the magnitude of the pressure drop per unit length in the axial direction. Then

$$p = -p'z + p^+(x, y). \tag{3.51}$$

The momentum equation can now be split into axial and non-axial components to give

$$\nabla \cdot (\eta\nabla w) = -p', \tag{3.52}$$

and

$$\nabla p^+ = \nabla \cdot (\Psi_2\nabla w)\nabla w + \Psi_2\nabla w \cdot \nabla(\nabla w). \tag{3.53}$$

The former is used with the boundary conditions to determine w, and we see that the form of w will depend on the form of the viscosity function but not on the normal stress coefficients.

The latter equation is to be used to determine p^+ when w is already known. However, there is no solution of this equation unless the right-hand member is the gradient of some function; that is, it must be irrotational. The final term is equal to $\Psi_2\dot{\gamma}\nabla\dot{\gamma}$, so this term is the gradient of a function of $\dot{\gamma}$ and hence is irrotational. Since the first term on the right is parallel to ∇w, it is irrotational if and only if $\nabla \cdot (\Psi_2\nabla w)$ is constant over surfaces $w = \text{constant}$; that is, constant on each slip surface. To see this denote $\nabla \cdot (\Psi_2\nabla w)$ by $f(x, y)$. Then if this term is the gradient of some function ψ, we have

$$\frac{\partial\psi}{\partial x} = f\frac{\partial w}{\partial x} \qquad \text{and} \qquad \frac{\partial\psi}{\partial y} = f\frac{\partial w}{\partial y}.$$

For consistency we need that

$$\frac{\partial}{\partial y}\left(f\frac{\partial w}{\partial x}\right) = \frac{\partial}{\partial x}\left(f\frac{\partial w}{\partial y}\right). \tag{3.54}$$

This is equivalent to

$$\frac{\partial f}{\partial y}\bigg/\frac{\partial f}{\partial x} = \frac{\partial w}{\partial y}\bigg/\frac{\partial w}{\partial x}, \tag{3.55}$$

or the ∇f is parallel to ∇w everywhere. When this is true the contours of constant f will coincide with the contours of w, which are the slip surfaces.

Thus, in general the equations may or may not be satisfied exactly depending on what the boundary conditions are and on the relation of the function Ψ_2 to the function η. It is possible to invent various special forms of η and Ψ_2 that allow certain special velocity fields to satisfy the equations exactly. However, in order to be assured in an experiment that the flow is indeed rectilinear as assumed, without any prior knowledge of the forms of the functions η and Ψ_2, the flow must be such that $\nabla \cdot (\Psi_2\nabla w)$ is automatically constant on each slip surface, regardless of what form the function Ψ_2 may take. These are the special cases that we call partially controllable. It can be shown (Yin and Pipkin 1970) that the partial controllability conditions are satisfied only if the slip surfaces are parallel planes or coaxial circular cylinders. It is elementary to verify that eqn (3.53) can be integrated in these cases. In the case of plane Poiseuille flow, with $w = w(y)$ and $\dot{\gamma} = w'(y)$, we obtain

$$p^+ = \dot{\gamma}^2\Psi_2(\dot{\gamma}) + \text{constant} = N_2(\dot{\gamma}) + \text{constant}. \tag{3.56}$$

For flows in tubes or annular gaps, with $w = w(r)$ and $\dot{\gamma} = w'(r)$ we obtain

$$p^+ = N_2(\dot{\gamma}) + \int_0^r N_2 \, dr/r + \text{constant}. \tag{3.57}$$

To illustrate the calculation of the speed $w(y)$ or $w(r)$, we first consider the case of plane Poiseuille flow between fixed walls in the planes $y = \pm h$. We suppose that the fluid adheres to the walls, so that $w(\pm h) = 0$. In this case, eqn (3.52) yields

$$\eta w' = -p'y, \qquad (3.58)$$

where we have set the constant of integration equal to zero in anticipation of a velocity profile symmetrical about the middle plane $y = 0$. Then, on inverting eqn (3.58) in terms of the fluidity ϕ, we obtain

$$w'(y) = -\tau \phi(\tau), \qquad \tau = p'y. \qquad (3.59)$$

Integration yields $w(y)$, which must satisfy $w(h) = 0$.

The quantity of primary importance in experiments is the volumetric discharge rate Q, which is found by integration by parts to be

$$Q \equiv \int_{-h}^{h} w(y)\,\mathrm{d}y = -\int_{-h}^{h} yw'\,\mathrm{d}y = (2/p'^2) \int_0^{\tau w} \tau^2 \phi(\tau)\,\mathrm{d}\tau. \qquad (3.60)$$

Here $\tau_w = p'h$ is the wall shear stress; this is a more fundamental quantity than the axial pressure gradient to use when describing tube or channel flow, although the latter is the directly measurable quantity.

Example

Suppose the shear-stress – shear rate relation is given by

$$\tau + \tau^3/\tau_0^2 = \eta_0 \dot\gamma,$$

where η_0 is the zero-shear viscosity and τ_0 is a constant. For small shear rates this relationship gives

$$\tau = \eta \dot\gamma \sim \eta_0 \dot\gamma [1 - (\lambda \dot\gamma)^2 + O(\dot\gamma^4)],$$

where $\lambda \equiv \eta_0/\tau_0$, and for high shear rates the result approaches a power-law

$$\tau = \eta \dot\gamma = \tau_0 (\lambda \dot\gamma)^{\frac{1}{3}} \left[1 - \frac{1}{3}(\lambda \dot\gamma)^{-2/3} + O(\lambda \dot\gamma)^{-1} \right].$$

In this case, the fluidity function ϕ is clearly given $(1 + (\tau/\tau_0)^2)/\eta_0$, and integrating the velocity equation (3.59) with the boundary condition $w(h) = 0$ gives the result

$$w(y) = \frac{p'h^2}{2\eta_0} \left[1 - \left(\frac{y}{h}\right)^2 + \frac{1}{2}\left(\frac{p'h}{\tau_0}\right)^2 \left\{ 1 - \left(\frac{y}{h}\right)^4 \right\} \right],$$

which returns to the Newtonian profile when τ_0 becomes very large. Calculation of the discharge rate Q gives

$$Q = \frac{2p'h^3}{3\eta_0}\left[1 + \frac{3}{5}\left(\frac{p'h}{\tau_0}\right)^2\right].$$

Equation (3.56) can be used to find the pressure distribution across a section and (3.21) gives the complete stress distribution.

For flow in a circular tube, eqn (3.52) yields

$$\sigma_{rz} = \eta w' = -p'r/2, \tag{3.61}$$

and thus,

$$w' = \frac{dw}{dr}(r) = -\tau\phi(\tau), \qquad \tau = p'r/2. \tag{3.62}$$

Integration of (3.62) with the no-slip boundary condition will give $w(r)$. As an important example consider the power-law fluid where

$$\eta = k\left|\frac{dw}{dr}\right|^{n-1}. \tag{3.63}$$

The absolute value of dw/dr is relevant here; p' is positive, being the magnitude of $\partial p/\partial z$, and η is also positive, hence dw/dr is negative; τ is the (positive) magnitude of the shear stress. The actual shear stress σ_{rz} is negative in pipe flow, from eqn (3.61) when w is positive (in annular flow, τ changes sign between the walls).

From (3.62) we have $\tau = k|dw/dr|^n$, and hence

$$\frac{dw}{dr} = -\left(\frac{\tau}{k}\right)^{1/n} = -\tau\phi(\tau) = -\left(\frac{p'r}{2k}\right)^{1/n}. \tag{3.64}$$

Integrating, and setting $w = 0$ at $r = R_0$, we find

$$w = \left(\frac{n}{n+1}\right)\left(\frac{p'}{2k}\right)^{1/n} R_0^{1+1/n}\left\{1 - \left(\frac{r}{R_0}\right)^{1+1/n}\right\}. \tag{3.65}$$

Dimensionless velocity profiles $w(r)/\bar{w}$ for various values of n are given in Fig. 3.4, where the mean velocity $\bar{w}$ is given by $\bar{w} = Q/\pi R_0^2$. The rate of discharge is given by

$$Q = \frac{\pi n}{3n+1}\left(\frac{p'}{2k}\right)^{1/n} R_0^{(3+1/n)}. \tag{3.66}$$

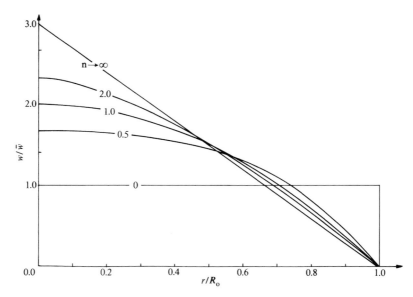

Fig. 3.4 Power-law fluid profiles in a fully-developed circular tube flow as a function of n. The shear-thinning fluids ($n < 1$) have flatter profiles, the shear-thickening ($n > 1$) steeper profiles, than the Newtonian case ($n = 1$).

In general, the discharge from a tube of radius R_0 is found to be

$$Q = 2\pi \int_0^{R_0} wr \, dr. \qquad (3.67)$$

Poiseuille flow in a circular tube was one of the most important experiments not only in the early history of Newtonian fluid dynamics but also in early investigations of non-Newtonian flow. By 1929 the key formula relating measurable quantities to the viscosity function (or fluidity) had been discovered by Weissenberg (Rabinowitsch 1929), and it was extended to allow for slip at the tube walls shortly thereafter by Mooney (1931). For the case of no slip manipulation of the result (3.67) yields the required relation. This result reduces the relation among the three measurable quantities Q, R_0 and p' to a relation between only two quantities, $\tau_w (= p' R_0 / 2)$ and the reduced discharge, $q = Q / \pi R_0^3$. If data from tubes of various sizes cannot be reduced to a single curve of q versus τ_w, then wall slip or some other wall effect is indicated; we will consider wall slip in Section 3.9.

To show the relation between τ_w and q, we note that the shear stress magnitude τ at a radius r is $rp'/2$, and at the wall ($r = R_0$) $\tau_w = p' R_0 / 2$. If (3.67) is integrated by parts, and it is assumed that no slip at the wall occurs, so that $w(R_0) = 0$, then

(3.67) becomes

$$Q = -\pi \int_0^{R_0} r^2 w'(r) \, dr.$$
(3.67a)

Using the fluidity function $\phi(\tau)$ and the transformation $r = 2\tau/p'$ (3.67a) becomes, recognizing that $dw/dr = w' = -\tau\phi(\tau)$,

$$Q = (8\pi/p'^3) \int_0^{\tau_w} \tau^3 \phi(\tau) \, d\tau.$$
(3.67b)

Using the definitions of q and τ_w, division of (3.67b) by πR_0^3 gives the result

$$q = \tau_w^{-3} \int_0^{\tau_w} \tau^3 \phi(\tau) \, d\tau.$$
(3.68)

By differentiation with respect to τ_w we find

$$\frac{dq}{d\tau_w} = \phi(\tau_w) - \frac{3q}{\tau_w}$$
(3.69)

Now $\phi(\tau_w) = 1/\eta(\tau_w) = \dot{\gamma}_w/\tau_w$, hence (3.69) can be rearranged as

$$\dot{\gamma}_w = q(\tau_w)\left[3 + \frac{d(\log q)}{d(\log \tau_w)}\right].$$
(3.70)

Thus, the shear rate at the stress τ_w has been obtained from a plot of q versus τ_w. Although this relation is usually convenient, it does require differentiation of the data, and numerical schemes for direct inversion of eqn (3.68) have been devised (Tanner and Williams 1970).

In some cases results are simply presented in terms of a nominal shear rate equal to $4\bar{w}/R_0$ (or $8V/D$ as it is often written), so that (3.70) is not used. It is easy to show that this nominal shear rate assumes a parabolic (Newtonian) velocity field, and that it is equal in fact to $4Q/\pi R_0^3$ or $4q$; it can be seen that for highly shear-thinning fluids gross errors can occur if the nominal instead of the true shear rate is used. It is easy to derive corresponding formulae for the plane case from eqn (3.60); defining q as $Q/2h$, the inversion formula is identical to eqn (3.70) except that the 3 is replaced by 2.

Practical precautions that need to be observed in the Poiseuille experiment are connected with inlet and outlet effects, temperature control, generation of heat by viscous dissipation increase of viscosity with pressure, and transition to turbulent flow. Many of these factors are discussed later in Chapters 8–10.

Here we will mention the end-effects which affect the determination of the shear stress. The flow in the instrument will consist of an inlet region, an exit region, and the flow in the fully-developed central tube region on which the analysis above is based. Couette suggested that two lengths of tube be used and that the difference in pressure drop over the difference in length be used to find p', thus eliminating end effects. Some commercial viscometers use this technique, normally the pressure is measured in a reservoir upstream of the capillary tube entrance, and the exit pressure is assumed to be zero.

Although the effective exit length is short (a few radii at most), the extra pressure drop at inlet is significant, and can amount to an apparent extra length of tens of radii, so care is needed. Bagley (1957) proposed a modification of Couette's method and the so-called 'Bagley correction' is often used to deal with entry and exit pressure losses. Bagley assumed that the entry pressure drop is independent of the capillary length for a given discharge rate and tube diameter. Tests with several capillary tube lengths (L), following Couette, enable one to deduce the entry losses as an equivalent extra tube length, (NR_0) and the 'true' shear stress can then be computed to be

$$\tau_w = \Delta p R_0 / 2(L + NR_0) \tag{3.62a}$$

where NR_0 is the entry (and exit) correction. Unfortunately, N is usually a function of flow rate, temperature and other variables for many non-Newtonian fluids. Some instruments use two tubes, one with a 'zero' length, which should enable one to directly eliminate end effects, but there must be some doubt that such short dies truly mimic the end effects in longer tubes (see Chapter 8). Despite these problems, the capillary rheometer is widely used because it often enables very high shear rates ($> 10^5\,\mathrm{s}^{-1}$) and stresses to be reached relative to those values attainable with other instruments.

These problems are well known from Newtonian fluid mechanics, but are less severe in that case. When adequate precautions have been taken, it seems to be possible to obtain values of the viscosity function with errors of order of one per cent from Poiseuille flows; accuracy is greatest with nearly Newtonian fluids.

No information on normal stress differences is obtained from this test. Some attempts have been made to measure the actual normal thrust on the walls as a function of distance from the exit. Linear extrapolation appears to show that a non-zero stress $(-\sigma_{rr})_w$ exists at the end of the tube. This is consistent with the qualitative idea that with an extra tension in the direction of shearing, the fluid will tend to contract in the axial direction and thus swell in the radial direction on leaving the tube, as viscoelastic fluids are observed to do. However, this is a dangerous argument, as the flow outside the tube and inside the tube near the exit is not viscometric and cannot be described by using the viscometric constitutive equation. (See Chapter 8.)

Axial flow in the annular region between two cylinders is of interest as an arrangement that allows measurement of the second normal stress difference. Usually a pressure gradient drives the flow, but the use of a moving inner cylinder is also possible. Whether the inner cylinder is moving or not, we find directly from the equations of motion and the symmetry of the flow that the difference between the radial thrusts at the outer radius r_o and the inner radius r_i is related to the normal stress difference N_2 by

$$[-\sigma_{rr}(r_o)] - [-\sigma_{rr}(r_i)] = \int_{r_i}^{r_o} N_2 \, dr/r. \tag{3.71}$$

There are at least two difficulties associated with this method of testing. First, the basic velocity profile is no longer governed by eqn (3.62) but by

$$\eta w' = -p'r/2 + C/r, \tag{3.72}$$

in which the constant of integration C is to be determined by using the boundary conditions at the two walls. This generally needs numerical solution. For example, one can attempt to use the power-law model, and it is immediately clear that there are simpler and more convenient methods of viscosity measurement.

The second difficulty is that pressure-hole errors must be considered (see Chapter 4). From the early sets of data using annular flows it appeared that the second normal stress difference was positive. However, when the data were corrected for pressure-hole errors, the sign of N_2 was reversed, and the N_2 is negative, which is now believed to be the case. But since the corrections are larger than the measured quantities, the process is of dubious accuracy. Tests with flush-mounted (hole-free) pressure transducers are possible, but, due to the wall curvature, these are not easy experiments. See Tanner and Walters (1998) for past attempts to measure N_2, and the history of pressure-hole errors.

3.5.3   Couette flow and helical flows

The helical flows listed in Section 3.2.4 are all partially controllable. We will not give a detailed verification of this, but merely note that with three unknown functions, w, ω, and p at our disposal, it is always possible to satisfy the momentum equations. These flows, with coaxial cylindrical slip surfaces, can be visualized as flows between two coaxial cylinders which may be in relative motion. Cases in which the motion is purely axial have already been discussed in Section 3.5.2. Cases that would require an azimuthal pressure gradient, such as the example in Section 3.4, can be considered. Such instruments are not discussed further here; there are some practical problems (Kraynik *et al.* 1984).

If the gap between the inner and outer cylinders is small in comparison to the radius of either one, by ignoring the curvature one can treat a helical flow approximately as a skew rectilinear flow (Section 3.5.1). For the present we confine our attention to the special case of Couette flow. In the Couette visco-meter, the flow is driven by steady rotation of one or both of the cylinders, with no applied pressure gradient. Since there is no angular acceleration, the moment M per unit axial length on each slip surface must be the same, and by expressing this moment in terms of the shear rate we obtain a first integral of the azimuthal component of the momentum equation

$$M = 2\pi r^2 \dot{\gamma} \eta(\dot{\gamma}). \tag{3.73}$$

Recalling that the shear rate is $r\omega'(r)$ and expressing the relation in terms of the fluidity ϕ, we obtain

$$r\omega'(r) = \tau\phi(\tau), \qquad \tau = M/2\pi r^2. \tag{3.74}$$

Then, integration yields

$$\omega(r) - \omega(r_i) = \frac{1}{2}\int_\tau^{\tau_i} \phi \, d\tau, \qquad \tau_i = M/2\pi r_i^2, \tag{3.75}$$

where r_i is the radius of the inner cylinder. The difference in angular velocities between the cylinders is

$$\Omega = \frac{1}{2}\int_{\tau_o}^{\tau_i} \phi \, d\tau, \qquad \tau_0 = M/2\pi r_o^2, \tag{3.76}$$

where r_o is the radius of the outer cylinder. This relation among the four measurable quantities Ω, M, r_i, and r_o shows that the data from cylinders of various sizes can be reduced to a relation among only three quantities, Ω, τ_i, and τ_o. Data that cannot be reduced in this way indicate that the hypothesis of no slip at the wall may have failed or some other effect may be present.

In order to obtain η (or ϕ) from the measured quantities, it is necessary to invert eqn (3.76). In the case of a gap $h = r_o - r_i$ that is small in comparison to either radius, the shear rate is nearly uniform at the value $\dot{\gamma} = \bar{r}\Omega/h$, where $\bar{r}$ is the average radius, and there is no difficulty in finding $\eta(\dot{\gamma})$. Using this approximation is equivalent to using the midpoint rule of numerical integration. Many approximate schemes of inversion are available (Coleman *et al.* 1966); a direct numerical approach is straightforward.

There are several commercial viscometers that use this configuration. Commonly the inner cylinder rotates and the speed and torque on this cylinder are measured. Problems due to end effects and Taylor instability (see Chapter 10) can be minimized by using a guard-ring design similar to Couette's original proposal. The small quantity of fluid needed, compared to that needed for Poiseuille viscometry, is an attractive feature of this type of instrument.

Measurable normal stress effects occur in the Couette configuration. The stress is given by eqn (3.24), with w set equal to zero. Since the angular velocity is a function of radius alone, then whatever forms the normal stress functions may have, they too are functions of r in the present problem. Consequently, the momentum equation is satisfied if the pressure p in eqn (3.24) is a function of r determined by the radial component of the momentum equation (noting

$\sigma_{zz} = -p$ here):

$$\frac{dp}{dr}(r) = \rho\omega^2 r - N_1/r + dN_2/dr. \tag{3.77}$$

Here ρ is the density of the fluid. We omit the hydrostatic pressure, which can be added to all normal stresses when the computation is otherwise complete. The radial stress, from eqn (3.24), is

$$\sigma_{rr} = -p + N_2. \tag{3.78}$$

Consequently, on integrating eqn (3.77) we find that the difference between the normal thrusts on the outer and inner cylinders is

$$[-\sigma_{rr}(r_o)] - [-\sigma_{rr}(r_i)] = \int_{r_i}^{r_o} [\rho r\omega^2 - (N_1/r)] \, dr. \tag{3.79}$$

The difference in thrust due to centrifugal force is positive, of course. The remaining thrust difference is negative if the first normal stress difference is positive, and thus the radial pressure at the inner cylinder can exceed that at the outer cylinder.

Although this test is sound in principle, and there are some examples in the literature, there is the serious pressure-hole difficulty associated with its practical execution; we note that the pressure-hole errors are found to be of the same order of magnitude as the quantities measured.

The axial component of stress σ_{zz} is equal to $-p$. In a flow with gravity acting in the negative z-direction, by adding the hydrostatic pressure (ρgz) we obtain

$$-\sigma_{zz} = -\rho gz + N_2 + \int_{r_i}^{r} [\rho r\omega^2 - (N_1/r)] \, dr + \text{constant}. \tag{3.80}$$

If the upper surface of the fluid is open to the atmosphere, the shape of the free surface can be found approximately by using this relation to find the value of z at which $-\sigma_{zz}$ is equal to atmospheric pressure:

$$\rho gz = N_2 + \int_{r_i}^{r} [\rho r\omega^2 - (N_1/r)] \, dr + \text{constant}. \tag{3.81}$$

The effects of centrifugal force and the extra tension N_1 are opposite to one another, if N_1 is positive as expected. If N_2 is relatively small and the effect of centrifugal force is small, then the surface stands highest at the inner cylinder. This is the classical Weissenberg effect (Fig. 1.3). The result (3.81) is valid only for small surface disturbances; a large climbing effect disturbs the flow near the surface.

Example: The Weissenberg rod-climbing effect
The shear stress at radius r in a Couette flow is given by [eqn (3.74)] $\tau = M/2\pi r^2$, where M is a constant. If we assume $N_1 = a\tau^m$, where m (>0) is often nearly 2 (Fig. 3.21), then we can integrate (3.81) to find

$$\rho gz = N_2(r) + \int_{r_i}^{r} \rho r\omega^2 \, dr + \frac{1}{2m}[N_1(r) - N_1(r_i)] + \text{constant}.$$

If inertia is negligible, and z is set at zero when r is very large, then

$$\rho gz = N_2(r) + N_1(r)/2m.$$

Hence one requires $N_1 + 2mN_2 > 0$ for rod climbing. If $N_1 > 0$ and $N_2 < 0$, then if $m = 2$, $|N_2/N_1|$ must be less then 0.25 to get climbing.

The inertia term will always be negative and can be found easily if the viscosity is constant; in this case the second-order model [eqn(4.30c)], with $m = 2$, gives, when the outer radius is very large

$$\rho gz = N_2(r) + \tfrac{1}{4}N_1(r) - \tfrac{1}{2}\rho\omega_i^2 r_i^4/r^2,$$

where ω_i is the angular speed of the rod and r_i is its radius.

3.5.4 Cone and plate flow

The cone-and-plate device (Fig. 3.5, with $c = 0$) produces a flow in which the shear rate is very nearly uniform. It is by far the most used normal stress measuring instrument. The flow is partially controllable to the same degree of approximation that the shear rate is uniform. In Fig. 3.5 we show on the left the boundary condition with a 'sea' of fluid and on the right the more common free boundary condition.

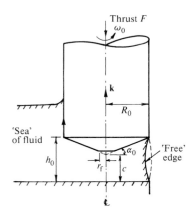

Fig. 3.5 Cone-plate geometry. Normal setting is that extrapolated cone-tip just touches plate ($c = 0$). On the left is shown a drowned edge and on the right the more normal free edge.

The slip surfaces in a cone-and-plate flow are cones rotating about a common axis. Let α be the latitudinal angle, so that $\alpha = 0$ is the equatorial plane and $\alpha = \pi/2$ is the axis of rotation. The fluid is contained in the gap between a plate in the plane $\alpha = 0$ and a cone $\alpha = \alpha_0$. Either the cone or the plate or both may rotate. To be definite we suppose that the plate is fixed and the cone rotates with a constant angular velocity ω_0.

Typically the gap angle α_0 is small, say $4°$ or less (0.07 radians). The shear rate in the fluid depends on the angular variable α, but α itself changes so little that this variation is insignificant. The shear rate at a given radius r is then the linear speed of the cone there, $r\omega_0$, divided by the gap width, $r\alpha_0$, so the shear rate is constant at the value $\dot{\gamma} = \omega_0/\alpha_0$. A more detailed analysis shows that the error in this estimate is $O(\alpha_0^2)$, which is often negligible (Lodge 1964).

If the fluid fills the gap out to the radius R_0, the moment M exerted on slip surfaces out to this radius must be the same for each surface, since there is no angular acceleration, and it is equal to

$$M = 2\pi \int_0^{R_o} \dot{\gamma}\eta(R\cos\alpha)^2 \, \mathrm{d}R = (2\pi/3)R_o^3\dot{\gamma}\eta(\dot{\gamma})[1 + O(\alpha_0^2)]. \tag{3.82}$$

Thus, measurement of the moment required to turn the cone or hold the plate gives practically a direct reading of the shearing stress:

$$\tau = \dot{\gamma}\eta(\dot{\gamma}) = 3M/2\pi R_o^3; \qquad \dot{\gamma} = \omega_0/\alpha_0. \tag{3.83}$$

The expression for the stress given in eqn (3.19) is more convenient than that in eqn (3.20) in the present case. The shear axes $\mathbf{a}$, $\mathbf{b}$, and $\mathbf{c}$ are unit vectors along the θ, α, and R directions, where R is the spherical radius:

$$\boldsymbol{\sigma} = -\bar{p}\mathbf{I} + \dot{\gamma}\eta(\mathbf{i}_\alpha\mathbf{i}_\theta + \mathbf{i}_\theta\mathbf{i}_\alpha) + N_1\mathbf{i}_\theta\mathbf{i}_\theta - N_2\mathbf{i}_R\mathbf{i}_R. \tag{3.84}$$

To the lowest order of approximation in the small parameter α_0 one may replace $\mathbf{i}_\alpha$ by $\mathbf{i}_z$ here, but a corresponding replacement of the spherical radial vector $\mathbf{i}_R$ by the cylindrical radial vector $\mathbf{i}_r$ leads to serious errors in the normal thrust calculation. By using eqn (3.84) in the momentum equation, we obtain

$$\nabla\bar{p} = \mathbf{i}_r\rho r\omega^2 - \mathbf{i}_r(N_1/r) - \mathbf{i}_R(2N_2/R) \tag{3.85}$$

where $\omega = \dot{\gamma}\alpha$. The two latter terms are gradients but the inertial term is not, so the equation has no solution unless the inertial term is neglected. By doing so and also noting that r and R are the same to the order of approximation considered, we obtain

$$\bar{p}(R) = \bar{p}(R_o) - (N_1 + 2N_2)\log(R/R_o). \tag{3.86}$$

In the present problem, $\bar{p}$ is the thrust against the plate and $\bar{p} + N_2$ is the radial pressure $-\sigma_{rr}$. With a free boundary at R_o, we suppose that the radial pressure there is atmospheric pressure, which is our datum line of zero. Then the thrust on the plate at the rim is

$$\bar{p}(R_o) = -N_2. \tag{3.87}$$

By using this in eqn (3.86), we find that the total thrust on the plate out to the radius R_o is

$$F = 2\pi \int_0^{R_o} \bar{p}R \ dR = N_1 \pi R_o^2/2. \tag{3.88}$$

The simplicity of this formula and the uniformity of the shear rate are two of the reasons for the popularity of the cone and plate instrument.

The relation (3.88) is not exact for several reasons. One of these is that the outer radius (the meniscus) of the test fluid is not exactly spherical and thus the flow near the outer boundary is not exactly viscometric. It appears that errors from this source are probably less than 5 per cent (Kaye *et al.* 1968). The error due to neglect of centrifugal force can be corrected in an approximate way. This is best done by calibrating with a Newtonian liquid. A theoretical estimate can be obtained by averaging the centrifugal force in eqn (3.85) across the gap. The result (3.88) is then replaced by

$$F = (\pi R_o^2/2)[N_1 - c\rho(R_o\omega_0)^2], \tag{3.89}$$

with $c = 1/6$. This seems to be somewhat in excess of measured corrections. The value $c = 3/20$ has been found from a more elaborate analysis (Savins and Metzner 1970) and seems to agree with data quite well when inertial effects are small. Larger inertial effects may cause appreciable transverse flow, which makes the analysis much more difficult. Fluid degradation can also cause errors. Finally, the role of surface tension at the edge should be considered; often it can be shown to be a negligible contribution to the thrust.

The second normal stress difference can be found directly by measuring the rim pressure, according to eqn (3.87). Another method that can be used if N_1 is already known is to measure the distribution of normal thrust on the plate; according to eqn (3.86), the slope of a plot of thrust versus log R should be constant at the value $-N_1 - 2N_2$. Errors in the estimation of $\bar{p}(R_o)$ do not affect this slope. Also, if a set of pressure holes is used to measure the thrust distribution (Kaye *et al.* 1968), the correct slope is obtained in spite of pressure-hole errors because the error is the same for each hole; the error depends on the local shear rate, which is uniform.

Miller and Christiansen (1972) have made absolute measurements of the thrust distribution by using flush-mounted pressure gauges, for which there is no hole error. Their evaluations of N_2 from the rim pressure and the thrust slope are consistent, and confirm that N_2 is negative for the fluid tested. This work has been extended by Magda and Baek (1994).

3.5.5 Open channel methods for finding N_2

A direct method of using tube flows to estimate N_2 is to observe the flow under gravity in an open channel and use the free surface as a pressure gauge. Here we consider flows in channels of circular cross-section or deep channels with parallel, straight walls, for which the unperturbed flow is exactly viscometric.

First, consider the viscometric flow in a tube of circular cross-section, radius R. The thrust $-\sigma_{\theta\theta}$ on a diametral plane is the pressure p, and from eqns (3.51) and (3.57) we find that this reaction is

$$p = -Gz + N_2 + \int_0^r N_2 \, dr/r + C. \tag{3.90}$$

When the tube is tilted at an angle β to the horizontal and the flow is driven by gravity rather than an axial pressure gradient G, the term Gz is absent but there is an additional hydrostatic pressure term. In a co-ordinate system x, y, z with z along the tube axis, the x-direction horizontal, and y positive upward, the hydrostatic pressure is $-\rho g y \cos \beta$. Then

$$p = -\rho g y \cos \beta + C + N_2 + \int_0^r N_2 \, dr/r. \tag{3.91}$$

Suppose now that the tube is cut in half and the top half is removed, but normal tractions given by eqn (3.91) are applied over the diametral plane $y = 0$. Since no shearing stresses are required on this surface, the flow will continue uninterrupted in the lower half of the tube. If the second normal stress difference is negative, the pressure p required to hold the surface flat is largest in the middle, and if it is positive, it is largest near the wall. If we now remove the applied traction, we expect the surface to rise where p was largest, until the extra weight of fluid above the surface $y = 0$ supplies the missing force. Thus, if N_2 is negative, the surface will bulge upward in the middle. This agrees with observation (Fig. 3.6) and this seems to give an unambiguous proof that the second normal stress difference is negative for the particular fluids that have been tested by this method.

The surface warping effect can be used for quantitative estimation of N_2. By setting $p = 0$ in eqn (3.91) we find the extra height of fluid $h(r)$ needed to give the required value of p on the plane $y = 0$, and we thus find that the shape of the free surface is given by

$$(\rho g \cos \beta)h = C + N_2 + \int_0^r N_2 \, dr/r. \tag{3.92}$$

Since the shearing traction on this surface is not exactly zero if the flow is exactly the same as the viscometric flow in a tube, the channel flow is not exactly viscometric and the preceding result is correct only to first order in the ratio h/R. The values of the second normal stress difference shown in Fig. 3.7 were obtained

(a)

(b)

Fig. 3.6 A Weissenberg-type effect in an inclined open-channel flow. (a) Curved surface of a (non-Newtonian) 1 per cent solution of polyethylene oxide in water. (b) Flat surface of Newtonian fluid (glycerol). The free surfaces of the fluid reflect the straight edges of the metal bridge.

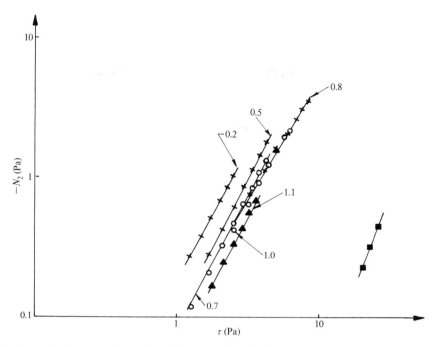

Fig. 3.7 The shear stress dependence of $-N_2$; ○ Polyox WSR = 301 in water; × Separan AP-30 in water; ▲ Oppanol B200 in cetane; ■ NBS Non-linear Fluid No. 1. The percentage of polymer by weight is shown on the curves.

from experiments on open channel flow. Corresponding ratios of $-N_2/N_1$ are given in Table 3.9.

For the flow between vertical parallel plane walls, the shape of the free surface is given by the slightly simpler formula

$$(\rho g \cos \beta)h = C + N_2, \tag{3.93}$$

with similar restrictions. A further set of interesting nearly viscometric flows occuring in twisted tubes of circular cross-section has been investigated by Barnes and Walters (1969). In these flows, transverse circulation is more the rule than the exception, because both streamline tension and centrifugal force tend to produce secondary flow, and for this reason such flows are more difficult to analyse. We refer the reader to the original paper for details.

3.6 Unsteady shearing flows

When one considers the kinematic aspects of the viscometric flows discussed above, there is no reason to restrict the motion to steady shearing rates. For example, it is perfectly easy to conceive an unsteady simple shearing motion in which $\dot{\gamma}$ is a function of time; the flow still continues to be described by the relative sliding of slip surfaces. If we confine our attention to the case of a sample which has been at rest long enough before testing so that stresses due to any previous motions have decayed, then the same stress symmetry occurs in both steady and unsteady shearing. As long as the slip surfaces do not change form, the above considerations give a method of approaching unsteady shearing motion. If there is a switch from one set of shearing axes $(\mathbf{a}, \mathbf{b}, \mathbf{c})$ to another during an experiment, then there is more complexity, even though the entire flow is viscometric.

Generally, in unsteady motions the inertia terms in the equations of motion will not vanish, but sometimes they are small enough to be ignored. The problem of inertia is not special to non-Newtonian flows, and since inertia forces tend to obscure effects due to non-linear material behaviour, it is often more interesting to consider situations where inertial effects are minimal. Thus we shall ignore inertia wherever this is realistic, and in particular in our present study of simple shearing; in Section 6.8.3 we discuss some aspects of inertia effects.

For inertia-less unsteady shearing, the response functions $\tau(t)$, $N_1(t)$ and $N_2(t)$ define the situation completely for a given $\dot{\gamma}(t)$. Of the infinite number of possibilities for $\dot{\gamma}(t)$, the following are the main ones which have been tried experimentally

(a) Oscillatory shear $\dot{\gamma} = \omega\hat{\gamma} \sin \omega t$.

(b) Sudden imposition of shearing,

$$\begin{cases} \dot{\gamma} = 0 & t < 0 \\ \dot{\gamma} = \dot{\gamma}_0 & t \gtrless 0. \end{cases}$$

(c) Cessation of shearing (or stress relaxation)

$$\begin{cases} \dot{\gamma} = \dot{\gamma}_0 & t \lessgtr 0 \\ \dot{\gamma} = \dot{\gamma}_0 & t > 0. \end{cases}$$

(d) Step displacement of *shear*. If γ is the shear strain then a shear strain is suddenly applied (or removed) at certain times; $\dot{\gamma}(t)$ is formally a set of impulses or delta functions.

(e) Combinations of a steady shearing with the above.

Clearly, in (e) steady shear axes can be oriented parallel to the unsteady flow axes, or in some other direction. Most combined flows use the same axes but Tanner and Williams (1971) have superposed an oscillatory shear at right angles to a simple shear, thus producing an unsteady skew motion similar to that discussed in Section 3.2.2.

Each of these flows will be discussed later in Section 3.8; many results for $\tau(t)$ and $N_1(t)$ are available, but few for $N_2(t)$. In addition to the above flows where $\dot{\gamma}(t)$ is imposed, it is also possible to perform creep and constrained elastic recovery tests where the shear stresses are imposed and the response is $\dot{\gamma}(t)$. Thus we can add to the above list

(f) Creep tests:

$$\begin{cases} \tau = 0 & t \lessgtr 0 \\ \tau = \tau_0 & t > 0 \end{cases}$$

(g) Constrained elastic recovery tests:

$$\begin{cases} \tau = \tau_0 & t < 0 \\ \tau = 0 & t \gtrless 0. \end{cases}$$

Sketches of typical responses for $\tau(t)$ and $N_1(t)$ are shown in Fig. 3.8 for some of the tests. Using these data one can study other unsteady viscometric flows undergoing similar temporal histories.

3.6.1 Pipkin's classification diagram for shearing flows

In connection with these flows, it is useful to discusss the flow diagnosis diagram introduced by Pipkin (1972) for shearing motions; it can also be applied to other motions. Consider a material where the mean relaxation time is λ; for example from Fig. 1.12 we can define λ as $\int_0^\infty tG \, dt / \int_0^\infty G \, dt$. Then the flow regimes may

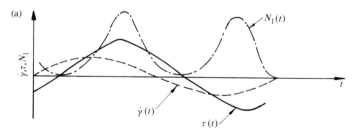

Fig. 3.8(a) Sinusoidal strain-rate ($\dot{\gamma}$). The shear stress response (τ) is not sinusoidal but has the same frequency (ω) as the strain-rate. The normal-stress (N_1) signal is always positive and fluctuates at 2ω.

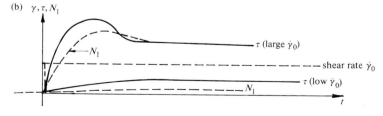

Fig. 3.8(b) Sudden imposition of strain-rate. The shear stress and normal stress difference curves may overshoot as shown at larger shearing rates; at low enough shear rates no overshoot occurs and the curves rise smoothly to their steady-state values.

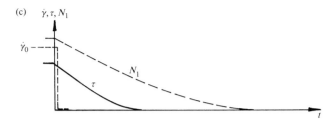

Fig. 3.8(c) Cessation of shearing. Usually N_1 vanishes slower than the shear stress τ. No under-shoot has ever been observed. Both τ and N_1 vanish, typically, more rapidly for higher $\dot{\gamma}_0$ values.

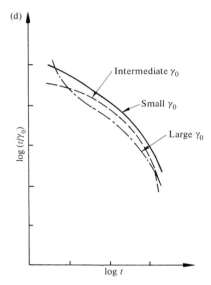

Fig. 3.8(d) Step of shear; here $\dot{\gamma}$ is an impulse or delta function. The curves (τ/γ_0) are proportional to the linear relaxation function $G(t)$ (Fig. 1.15) for small γ_0. At larger strains the stress has a similar, but not identical shape, and the scale factor is not linear in γ_0.

be classified by considering a typical rate of change as being fast or slow relative to λ, and by the maximum shear strain amplitude A; in the case of steady flows we take A to be the amount of shearing in time λ, hence $A = \dot{\gamma}\lambda$ for this case. Suppose that a characteristic rate ω exists for the variation of the kinematics of a particle in time. We then plot A against the dimensionless product $\omega\lambda$ (Fig. 3.9).

This diagram thus plots as abscissa a *Deborah number* $\omega\lambda$, denoted henceforth by (De); the ordinate $\lambda\dot{\gamma}$ is an example of *Weissenberg number*, denoted henceforth as (Wi). (Some writers use We for this dimensionless quantity, but We is already used for the ratio of the inertia to surface tension, the so-called *Weber number*.) The regimes of the three classical bodies discussed in Chapter 2 are then seen to lie along the $\omega\lambda$ axis; for very slow changes of motion one is near the steady viscometric flow regime just discussed, and for very fast applications of

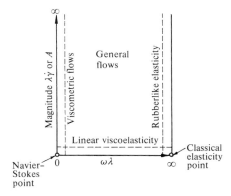

Fig. 3.9 Pipkin flow diagnosis diagram for a fluid with a characteristic time λ. Here ω is a characteristic strain frequency, $\dot{\gamma}$ is a characteristic strain rate, and A is a characteristic strain magnitude. The scales $(0 - \infty)$ are highly non-linear. The product $\omega\lambda$ is a Deborah number (De) and $\lambda\dot{\gamma}$ is a Weissenberg number (Wi).

strain ($\omega\lambda \to \infty$) the material behaves in a rubberlike manner, as typified by 'bouncing putty'. In the centre of the diagram lie the large-amplitude unsteady shearing flows. It is here that further studies of relevant constitutive relations are needed; the edge strips are fairly well-understood regions and the constitutive relations that are relevant there are known.

3.7 Elongational flows

Suppose that a rod of material is being extended homogeneously along its x-axis, so that each part of the rod is stressed uniformly. At each section we suppose that the rate of elongation $\partial u/\partial x(\equiv \dot{\varepsilon})$ is independent of x and is a function of time at most. Mass conservation and axial symmetry then demand that $\partial v/\partial y = \partial w/\partial z = -\dot{\varepsilon}/2$. All shearing-stress components are zero and $\sigma_{yy} = \sigma_{zz}$, by symmetry. For an incompressible liquid the stress response is then completely defined by the dependence of $\sigma_{xx} - \sigma_{yy}$ on the rate of extension $\dot{\varepsilon}$ and the time t elapsed since the stetching began:

$$\sigma_{xx} - \sigma_{yy} = \dot{\varepsilon}\eta_E(\dot{\varepsilon}, t), \tag{3.94}$$

where η_E is the stretching viscosity; generally it is a function of the rate of stretching $\dot{\varepsilon}$ and time. The initial response of a polymer melt is elastic for many materials and might more appropriately be described in terms of the dependence of the stress on the strain ε. For this reason the stress generally increases as time progresses during the initial stages of the motion (Fig. 3.10). If the rate of extension is large, this initial stage may end in fracture of the specimen. For example, it is a simple matter to fracture some silicone liquids ('silly putty') in one's own hands.

When rupture does not occur, the stretching viscosity η_E may eventually approach a limiting value, $\eta_E(\dot{\varepsilon}, \infty)$, which we will usually denote simply by $\eta_E(\dot{\varepsilon})$.

The limiting value is called the steady extensional viscosity, or Trouton viscosity. Trouton (1906) found that the extensional viscosity of mixtures of pitch and tar is independent of $\dot{\varepsilon}$ and about equal to $3\eta_0$, the value for an incompressible Newtonian fluid with shear viscosity η_0.

Concern with the properties of the Trouton viscosity began in the 1930s in connection with the important problem of spinning synthetic fibres from molten liquid. Useful compilations of material on elongational flows are the book by Petrie (1979), and the article of Meissner (1992). For polymer solutions the work of Sridhar and co-workers (Orr and Sridhar 1996) should be consulted.

Spinning experiments can rarely be used for the unambiguous determination of the steady state viscosity η_E, because each fluid element experiences a highly unsteady stretching and there is usually not enough time for transient elastic effects to die away. Ballman's (1965) work on polystyrene was the first in which the strain rate was kept constant during the motion. A tensile test was carried out on a bar of very viscous liquid, with the ends of the specimen moving apart at an exponentially increasing rate so as to keep the velocity gradient $\partial u / \partial x$ constant. Essentially the same technique was used by Stevenson (1972), who obtained the results shown in Figs 3.10–3.13. Usually the weight of the specimen is supported by floating it in a liquid of the same or slightly higher density, as Trouton (1906) did.

With this technique only limited extensions can be achieved, since the length of the specimen increases in proportion to $\exp(\dot{\varepsilon}t)$, and the rate of extension cannot be much larger than about 1–$10\,\text{s}^{-1}$. Meissner (1971) introduced a substantially improved method of testing. Each end of a strand of liquid is drawn between a pair of gears that rotate at constant angular velocity. The two ends are pulled in opposite directions, so that the axial velocities are U and $-U$, say, at

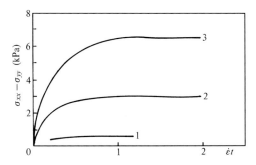

Fig. 3.10 Elongation stress versus strain for Butyl 035 at $100\,^\circ\text{C}$ (low elongation rates). Curve 1: $\dot{\varepsilon} = 2.85 \times 10^{-4}\,\text{s}^{-1}$; Curve 2: $\dot{\varepsilon} = 1.48 \times 10^{-3}\,\text{s}^{-1}$; Curve 3: $\dot{\varepsilon} = 2.82 \times 10^{-3}\,\text{s}^{-1}$. Note that the elongational stress difference $\sigma_{xx} - \sigma_{yy}$ attains a steady state.

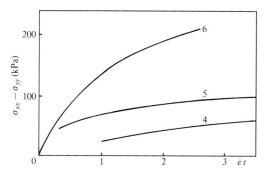

Fig. 3.11 Elongational stress versus strain for Butyl 035 at 100 °C (higher elongation rates). Curve 4: $\dot{\varepsilon} = 2.48 \times 10^{-2}\,\text{s}^{-1}$; Curve 5: $\dot{\varepsilon} = 4.51 \times 10^{-2}\,\text{s}^{-1}$; Curve 6: $\dot{\varepsilon} = 0.135\,\text{s}^{-1}$. Note that no steady stress state is apparent at the higher rates.

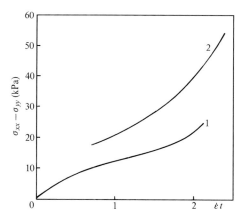

Fig. 3.12 Elongational stress versus strain for Natsyn 410 at 80 °C. Curve 1: $\dot{\varepsilon} = 6.37 \times 10^{-4}\,\text{s}^{-1}$; Curve 2: $\dot{\varepsilon} = 1.57 \times 10^{-3}\,\text{s}^{-1}$. Note s-shape of response curves.

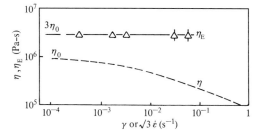

Fig. 3.13 Steady elongational (η_E) and shear (η) viscosity data for Butyl 035 at 100 °C. The abscissa is the square root of the second invariant of the rate-of-strain tensor.

the locations of the gears, $x = L$ and $x = -L$. The axial velocity in the strand is then $u = Ux/L$, giving a constant extension rate $\dot{\varepsilon} = U/L$. With this method, one can draw samples out to about 1000 times their original length and achieve extension rates of $10\,\mathrm{s}^{-1}$. The data in Fig. 3.14 were obtained by Meissner (1971); they portray an unsteady flow.

In order to measure the steady-state viscosity $\eta_E(\dot{\varepsilon})$, the fluid must be subjected to a constant rate of extension for so long that the stress reaches a constant value, or a constant stress for so long that the rate of extension becomes constant. Cases in which it is certain that a steady state was reached involve fluids of very high viscosity and rates of extension not much greater than about $1.0\,\mathrm{s}^{-1}$. In this limited range, the Trouton viscosity has usually been found to be either nearly constant or, at the highest rates of extension, an increasing function of $\dot{\varepsilon}$. Although these rates of extension are very low, the steady-shearing viscosities $\eta(\dot{\gamma})$ of the same materials are usually significantly lower at a shear rate $\dot{\gamma} = 0.1\mathrm{s}^{-1}$ than they are in the limit of zero shear rate. Consequently, the ratio of the extensional viscosity to the shearing viscosity is an increasing function of the strain rate even when the Trouton viscosity is still constant, as in the data of Stevenson (1972) shown in Fig. 3.13.

3.7.1 Unsteady extensions

It is fairly easy to obtain much higher rates of extension, but the experiments in which this is done have always involved flows with a non-constant rate of extension (in space) unlike the spatially homogeneous unsteady flow shown in Fig. 3.14.

Spinning experiments typically involve extension rates of the order of 0.1 to $10^2\,\mathrm{s}^{-1}$. Convergent die-entry flows involve extension rates of the order of 1 to $100\,\mathrm{s}^{-1}$. Jet-thrust experiments involve extension rates of the order of 10 to $10^3\,\mathrm{s}^{-1}$. In two jets colliding head-on, or the reverse, the rate of extension has been

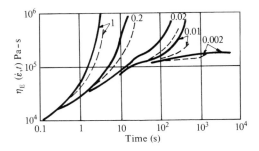

Fig. 3.14 Growth of unsteady elongational viscosity $\eta_E(\dot{\varepsilon}, t)$ due to elongation at a constant rate $(\dot{\varepsilon})$ suddenly applied at time $t = 0$. Dashed lines: data of Meissner (1971) for low-density branched polyethylene ('Melt I' at 150 °C: density at 20 °C 918 kg/m³; melt index = 1.33; molecular weight $(M_w) = 4.82 \times 10^5 = 28.1 M_n$). Full curves are response curves calculated from the Lodge rubberlike liquid [see eqn (5.150)] with a relaxation function chosen to fit the data at $\dot{\varepsilon} = 10^{-3}\,\mathrm{s}^{-1}$. The values of $\dot{\varepsilon}(\mathrm{s}^{-1})$ are shown on the curves.

estimated as $50\,s^{-1}$ for a melt (Mackley and Keller 1973) and $8 \cdot 10^3\,s^{-1}$ for a polymer solution (Frank *et al.* 1971). See the surveys by Gupta and Sridhar (1998) and James and Walters (1993) for further discussion of experimental methods.

Except in the case of spinning, analyses of these flows make use of approximations and assumptions that are subject to doubt. Furthermore, it is usually clear that the 'viscosity' in these flows is not the steady-state Trouton viscosity. However, the general trend is toward higher values of viscosity as the rate of extension increases, and the rise is often so drastic that no error of approximation could explain it. The results are often stated in terms of the ratio of the extensional viscosity to the steady-shearing viscosity at a shear rate equal to the rate of extension. Ratios of order 3×10^4 have been reported (Metzner and Metzner 1970).

It is generally agreed that the observed increases in extensional viscosity are caused by alignment of the long-chain polymer molecules along the direction of stretching. This may explain the data of Cogswell (1969) on polypropylene, from which he deduced that the extensional viscosity decreases as the rate of extension increases, because this polymer has large side groups that may inhibit orientation.

Since polymer molecules are long, flexible chains, the degree of extension and orientation in a solution of such molecules depends on the rate of extension of the solution. Both extension and orientation grow larger as the rate of extension increases, and both effects increase the extensional viscosity of the solution. We consider these matters in Chapter 5.

3.7.2 Biaxial and other related flows

Stretching motions of the form (3.94) with $\dot{\varepsilon}$ negative can be produced by stretching a sheet of material, so that $-\dot{\varepsilon}/2$ represents the stretching rate in the plane of the sheet.

In inflation of a circular sheet with clamped edges, the flow at the centre of the sheet is an equal biaxial extension. Denson and Gallo (1971) have proposed a method of achieving a nearly constant rate of extension at the centre. This method has been used by Maerker and Schowalter (1974) who found that the Trouton viscosity at first decreases as the rate of extension increases. Maerker and Schowalter also found that, at higher rates of extension, the viscosity passes through a minimum and then increases sharply. Dobraszczyk (1997) has used this method to explore dough rheology.

In any steady velocity field of the form $v_i = \dot{\varepsilon}_i x_i$ (no sum on i), with the sum of the extension rates equal to zero for incompressible flow, the only relevant material properties are the dependence of the two independent normal stress differences on $\dot{\varepsilon}_i$ and t. In the flows that we have discussed, $\dot{\varepsilon}_2 = \dot{\varepsilon}_3$ and thus $\sigma_{yy} = \sigma_{zz}$ by symmetry. In strip biaxial tests, a sheet is stretched in the x-direction while its width in the $z-$direction is held constant, so that $\dot{\varepsilon}_3 = 0$ and $\dot{\varepsilon}_2 = -\dot{\varepsilon}_1 = -\dot{\varepsilon}$ say. The normal stress difference $\sigma_{xx} - \sigma_{yy}$, divided by $\dot{\varepsilon}$, defines the viscosity for this test. The difference $\sigma_{xx} - \sigma_{zz}$ can also be measured.

Strip biaxial tests have been reported by Peng and Landel (1974), who also report tests on equal biaxial extension. They find that the viscosity decreases as the rate of extension increases, with very much the same form of dependence in both tests. Denson and Crady (1974) have used inflation of rectangular strips, in which the strip becomes more or less cylindrical, and they report that the strain-rate along the axial direction is less than 1 per cent of that along the circumferential direction. At the lowest rates of extension they find that the extensional viscosity is of the order $4\eta_0$, the value for an incompressible Newtonian liquid, and they find that the viscosity decreases as the rate of extension increases; see also Walters (1984) for another method of testing.

This set of homogeneous stretching flows has been explored experimentally by Meissner (1992). Consider the rate of deformation tensor **d** which, in the absence of shearing, is diagonal. Thus we can write it as

$$\mathbf{d} = \dot{\varepsilon}_0 H(t) \begin{pmatrix} 1 & 0 & 0 \\ 0 & m & 0 \\ 0 & 0 & -(1+m) \end{pmatrix} \tag{3.95}$$

where $\dot{\varepsilon}_0$ is a constant and $H(t)$ is the unit step function. Equation (3.95) preserves volume (tr $\mathbf{d} = 0$) and needs two defining parameters —$\dot{\varepsilon}_0$ and m. The parameter m defines the flow type:

(i) $m = -0.5$ gives simple uniaxial elongation;

(ii) $m = 0$ gives planar elongation;

(iii) $m = 1$ gives biaxial elongation (or negative elongation);

and other values give 'elliptical' flows.

These flows are generally characterized by two stress differences—defined so that the first axis is along the direction of largest extension rate ($\dot{\varepsilon}_1$) and setting $\dot{\varepsilon}_0 \equiv \dot{\varepsilon}_1 > \dot{\varepsilon}_2 \overline{>} \dot{\varepsilon}_3$. There appear to be differences in the responses depending on the value of m.

3.8   Some experimental data

We have already given some results for fluids above and we have discussed experimental methods briefly. More extensive discussions of experimental methods are given by Walters (1975), Petrie (1979), Gupta and Sridhar (1988), and Meissner (1992).

As to other experimental methods, we refer to the books by Janeschitz-Kriegl (1983) and Fuller (1995) for details of optical measurements.

Here we shall give some results for low-density polyethylene melts, less viscous solutions of polymers, and some other materials. It is strictly necessary to compare the behaviour of melts with concentrated polymer solution data as it is often argued that solutions behave similarly to melts, and that results for solutions can be applied qualitatively to melt behaviour. This will be assumed in the rest of this book: it is an idea supported by microstructural evidence (Chapter 5).

3.8.1 Low density polyethylene

A working party (Meissner 1975) has prepared a document giving a useful survey of properties of three similar commercial low density polyethylene samples, designated A, B, and C. Although differing considerably from the processing point of view they were quite similar rheologically, as judged by measurements made in six independent laboratories. Besides the properties given below, information on other quantities of interest, such as melt flow index, thermal stability and molecular parameters is also given in Meissner's report. It was concluded that the samples A, B, and C were similar except for slight differences in molecular weight distribution, and some differences in extrudate swelling (see Chapter 8) and thermal stability.

We will now consider the results given by Meissner; the reader should refer to the original report to obtain an idea of how difficult it is to obtain consistent results from different laboratories.

Zero-shear viscosity. The zero-shear viscosities η_0 are tabulated in Table 3.1. Below and at 150 °C material C has a value of η_0 about 10 per cent lower than A and B, which were essentially equal. At 170 °C and 190 °C the three specimens cannot be compared because of different thermal stability properties. By plotting $\log \eta_0$ as a function of $1/T$ (reciprocal of absolute temperature in degrees Kelvin) one obtains a straight line. Thus, the Arrhenius form holds, and

$$\eta_0 = A \exp\left(E_0/RT\right), \tag{3.96}$$

where A is a constant, E_0 is an activation energy, and R is the gas constant (8.3143 J/mole K). Thus E_0 is 57.0 ± 0.9 kJ/mole, hence E_0/R is about 6850 K. This agrees with the activation energy given for low-density polyethylene (LDPE) derived by other methods.

Linear stress relaxation. Both (a) stress relaxation after a sufficiently small step of shear strain and (b) stress relaxation after cessation of a sufficiently low shearing rate were investigated by the IUPAC group. The first test determines $G(t)$, and the second generates a related function. At 150 °C the three samples differed little

Table 3.1 Zero-shear viscosities and activation energies E_0

Temp. °(C)	A	B	C
112	260	263	240×10^3(Pa-s)
130	105	118	97
150	55	55	50
170	23.8	25.5	24.3
190	14	(Unstable)	15.5
E_0	57.8	56.5	56.1 kJ/mole

(Fig. 3.15, Table 3.2). Comparison of results obtained by the two methods (a) and (b) is given in Table 3.3. It can be concluded that the samples behave in a similar, but not identical manner.

Frequency-dependent linear viscoelastic material functions. For the dynamic mechanical measurements, three different types of apparatus were used. Table 3.4 presents three different sets of data obtained at 150 °C by three operators using two different instruments. The data obtained by different operators (α and β) with the same instrument are, in general, in excellent agreement. The agreement between the data obtained with different instruments is less than perfect but is still reasonable. It is quite obvious from the results reported in the table that no significant differences exist between the three samples as far as linear viscoelastic behaviour is concerned. At low frequencies (α) finds systematically lower values for G' and G'' in the case of sample A compared with samples B and C. Lower values for sample A in the range of lower frequencies are also found by (β) but contrary to the findings of (α), the difference between the values of G' for samples B and C is found by (β) to be of the same order of magnitude as the difference between A and B. All sets of data agree with respect to the effect of increasing frequency which tends to diminish the observed differences between the samples.

The differences between the data obtained with different instruments also seem to be frequency-dependent: with increasing frequency, the difference between

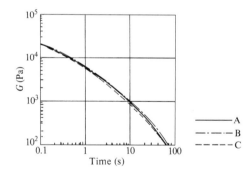

Fig. 3.15 Shear relaxation modulus $G(t)$ from stress relaxation after a step function shear strain. Temperature 150 °C. Three samples A, B, and C of low-density polyethylene are shown.

Table 3.2 Shear relaxation moduli $G(t)$; $T = 150$ °C

Time	Sample		
	A	B	C
$t = 0.1$s	2.0	2.04	1.97×10^4 Pa
1s	5.4	5.7	5.3×10^3 Pa
10s	8.6	9.6	7.8×10^2 Pa

Table 3.3 Comparison of the linear viscoelastic shear relaxation modulus $G(t)$, measured directly in the relaxation experiment (a) and calculated from relaxation after steady shear flow (b)

Time, s	Type of test	A	B	C
0.1	(a)	20 000	20 400	19 700
	(b)	20 800	19 700	18 900
1.0	(a)	5 400	5 700	5 300
	(b)	5 050	4 950	5 250
10	(a)	860	960	780
	(b)	860	880	810
100	(a)	—	—	—
	(b)	52	65	47
1000	(a)	—	—	—
	(b)	0.94	1.5	0.83

(Values in Pa).

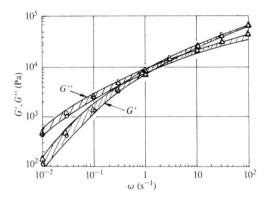

Fig. 3.16 Frequency-dependent storage (G') and loss (G'') Moduli at 150 °C for LDPE samples (A○,B△,C□). The hatched areas correspond to the range of data in Table 3.4.

G' values decreases. The difference between the G'' values appears to change sign for frequencies higher than $10\,\mathrm{s}^{-1}$. Taking account of the limited accuracy of the oscillatory measurements, the general behaviour of G' and G'' as a function of frequency (Fig. 3.16) is in good agreement with the results of the relaxation measurements.

Generally speaking, there are practically (that is, within 10 per cent) no differences in the linear viscoelastic behaviour of the three samples. The differences in G' and G'' may be ascribed to measurement problems.

The temperature dependence of η_0 (expressed by the activation constant E_0) is the same for A, B, and C. If the molten samples can be treated as thermo-rheologically simple materials it follows from the time–temperature

Table 3.4 Storage modulus G' and loss modulus G'' at 150 °C in Pa

$\omega\ [s^{-1}]$	Sample	Investigator		
		(α)	(β)	(γ)
Storage modulus G'				
0.01	A	1.01×10^2	1.05×10^2	1.12×10^2
	B	1.46×10^2	1.40×10^2	1.16×10^2
	C	1.53×10^2	1.09×10^2	8.5×10^1
0.1	A	1.39×10^3	1.20×10^3	1.26×10^3
	B	1.81×10^3	1.40×10^3	1.30×10^3
	C	1.70×10^3	1.27×10^3	1.18×10^3
1	A	8.10×10^3	7.0×10^3	7.0×10^3
	B	8.70×10^3	7.3×10^3	7.3×10^3
	C	8.60×10^3	7.35×10^3	7.0×10^3
10	A	2.70×10^4	2.6×10^4	2.45×10^4
	B	2.85×10^4	2.8×10^4	2.49×10^4
	C	2.80×10^4	2.6×10^4	2.60×10^4
100	A	6.80×10^4	7.0×10^4	
	B	7.20×10^4	7.1×10^4	
	C	7.00×10^4	7.1×10^4	
Loss modulus G''				
0.01	A	4.76×10^2	4.2×10^2	4.7×10^2
	B	5.50×10^2	4.5×10^2	4.5×10^2
	C	5.70×10^2	4.55×10^2	4.25×10^2
0.1	A	2.65×10^3	2.24×10^3	2.25×10^3
	B	2.90×10^3	2.41×10^3	2.38×10^3
	C	2.83×10^3	2.38×10^3	2.30×10^3
1	A	8.90×10^3	8.15×10^3	7.7×10^3
	B	9.00×10^3	8.40×10^3	8.20×10^3
	C	8.80×10^3	8.50×10^3	8.30×10^3
10	A	1.98×10^4	2.10×10^4	1.95×10^4
	B	2.05×10^4	2.20×10^4	2.0×10^4
	C	2.07×10^4	2.20×10^4	1.92×10^4
100	A	3.70×10^4	4.80×10^4	
	B	3.70×10^4	4.65×10^4	
	C	3.70×10^4	4.70×10^4	

superposition principle that the validity of the conclusion concerning the identical linear viscoelastic behaviour for 150 °C can be extended to other temperatures in the molten state. (See Chapter 9.)

The viscosity function. The viscosity function $\eta(\dot{\gamma})$, commonly used to represent the non-Newtonian behaviour of polymer melts, was determined by using rotational viscometers at low rates of shear and at higher shear rates by means of capillary viscometers using two types of corrections: (a) the Weissenberg correction (eqn 3.70) yields the true shear rate at the die wall for non-Newtonian liquids independent of the velocity distribution within the die, (b) the Bagley correction (3.62a) provides the true pressure gradient from which the true shear stress at the die wall is calculated.

The measurements at higher shear rates with the capillary viscometer were performed at 150 and 190 °C. All the viscosity data at these two temperatures are given in Fig. 3.17. The data cover the remarkably wide range of shear rates from 10^{-4} to 10^{+3} s^{-1}. At first sight, the results for the three samples measured by five different instruments coincide very well at 150 °C, but there are clear differences at 190 °C, not for the different samples but for the different participants; that is, for different test methods. This fact again demonstrates the importance of thermal stability if reliable data are to be obtained. For any one participant, the three samples again show practically identical curves, for 190 °C as well as for 150 °C.

For a discussion in more detail, the results at 150 °C were interpolated graphically and the viscosity data tabulated for fixed decades of shear rate (Table 3.5). At low shear rates, the coincidence of the data within ±10 per cent is

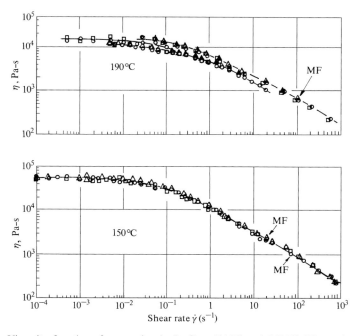

Fig. 3.17 Viscosity functions for samples A, B, C at 190 °C and 150 °C. The various symbols indicate the various instruments used: a Weissenberg rheogoniometer, a Kepes rotational rheometer, and two capillary instruments. MF indicates the onset of melt fracture or instability in the capillaries.

Table 3.5   Shear viscosity $\eta(\dot{\gamma})$ at $T = 150\,^{\circ}\text{C}$

$\dot{\gamma} \cdot s^{-1}$	From	A	B	C	
0.0001	IVa	5.5	5.5	5.0	10^4 Pa-s
0.001	Ia	4.2	5.2	—	10^4 Pa-s
	II	4.7	4.8	4.9	10^4 Pa-s
	IVa	5.5	5.5	5.0	10^4 Pa-s
0.01	Ia	4.2	5.2	4.95	10^4 Pa-s
	II	4.4	4.5	4.7	10^4 Pa-s
	IVa	4.4	4.85	4.35	10^4 Pa-s
0.1	Ia	3.17	3.4	3.0	10^4 Pa-s
	II	3.1	3.2	3.2	10^4 Pa-s
	IVa	2.90	2.90	2.82	10^4 Pa-s
	IVb	2.85	3.0	3.0	10^4 Pa-s
1.0	Ia	1.23	1.23	1.22	10^4 Pa-s
	II	1.1	1.18	1.18	10^4 Pa-s
	IVb	1.18	1.21	1.20	10^4 Pa-s
10	Ib	4.0	3.95	3.88	10^3 Pa-s
	II	2.9	3.2	3.1	10^3 Pa-s
	IVb	3.6	3.65	3.41	10^3 Pa-s
100	Ib	10.3	10.3	9.6	10^2 Pa-s
	IVb	8.8	8.8	8.8	10^2 Pa-s
1000	Ib	2.2	2.2	2.11	10^2 Pa-s
	IVb	2.0	1.93	1.92	10^2 Pa-s

Obtained by graphical interpolation to the fixed shear rates $\dot{\gamma}$ indicated. Ia uses a shear rate-temperature shift method in order to determine the viscosity function at $150\,^{\circ}\text{C}$ from data measured at $130\,^{\circ}\text{C}$. Ia and IVa used a Weissenberg rheogoniometer, II used a Kepes cone/plate device, Ib and IVb used capillaries.

again evident. At $0.1\ \text{s}^{-1}$, a similar coincidence is evident; the coincidence at $1\ \text{s}^{-1}$ is even better. At higher shear rates, the two capillary instruments give good agreement (difference about 10 per cent). Only at $10\ \text{s}^{-1}$ is there a more pronounced difference between the rotational data and the two sets of capillary data. However, this may be due to the pronounced time-dependence of viscous flow at this relatively high shear rate.

Comparing the three samples, the 10 per cent difference at the lowest shear rates denotes the largest difference between the viscosity data measured. At higher shear rates, the magnitude of this difference is reduced below 10 per cent to such an extent that the viscosity functions of the three samples at $150\,^{\circ}\text{C}$ can be assumed to be identical. The identity of the viscous behaviour was one criterion for the selection of the three samples for this IUPAC program. The results confirm that this criterion was fulfilled. Some processing differences were found with these melts and it becomes clear that neither the linear viscoelastic nor the purely viscous behaviour of the melts accounts for these differences.

Steady-state normal stress differences. The definitions of the two normal stress differences N_1 and N_2 have been given in eqn (3.17).

Using a cone-plate device the three IUPAC LDPE samples yielded the steady-state results shown in Table 3.6 at $130\,°C$ (there were thermal stability problems at $190\,°C$).

The comparison of data for samples A, B, and C, can be summarized as follows: the shear stresses are practically the same, the difference between A, B, and C being often much less than ten per cent. The results for N_1 are, on the average, also equal for A and C. Sample B, however, seems to have a slightly higher normal stress difference which differs from that for A by about 0–20 per cent, except at the highest shear rate, $\dot{\gamma} = 0.85\,s^{-1}$, at which the normal stress difference for B is lower than that for A. Experimental difficulties must be kept in mind, however, and the above data represent small, if any, differences in the behaviour of the three samples A, B, and C.

The measurement of N_2 is more difficult. Meissner (1992) describes the use of a two-part cone-plate rheometer to deduce N_1 and N_2 for polyethylene at $105\,°C$. An average of the results showed $N_2 \approx -0.24N_1$.

Time-dependence of τ and N_1 at constant shear rate. These studies were performed using an instrument modified to obtain the transient behaviour of polyethylene melts correctly (Meissner 1972). The measurements were made at $150\,°C$. and the constant shear rates $\dot{\gamma} = 0.1 - 1 - 10\,s^{-1}$. The time-dependent behaviour differs from one shear rate to another, as was shown by detailed studies previously presented for a sample A (Meissner 1972).

The small differences found in the rheogoniometer data for A, B, and C raise the question of the reproducibility of the results. For this purpose, the reproducibility was checked with sample A always using a new specimen for each

Table 3.6 Shear stress (τ) and first normal stress difference (N_1) at $130\,°C$ in kPa. Results for three LDPE samples (A, B, C)

$\dot{\gamma}$	A		B		C	
$[s^{-1}]$	τ	N_1	τ	N_1	τ	N_1
0.0043	0.353	–	0.356	–	–	–
0.0085	0.707	–	0.83	–	0.714	–
0.0135	1.11	–	1.21	–	1.02	–
0.0269	1.99	2.13	2.00	2.32	1.85	2.00
0.0425	2.72	3.78	2.92	4.30	2.73	3.93
0.085	4.52	9.66	4.84	9.89	4.40	8.59
0.135	6.00	12.6	6.45	16.2	5.81	14.5
0.269	9.01	25.2	9.36	29.7	9.11	15.7
0.425	11.6	37.1	11.3	38.0	10.9	34.7
0.85	15.6	55.3	15.2	52.7	–	–

measurement performed under equal experimental conditions. These reproducibility tests resulted in a large scatter-band for the transient functions. For N_1 the band-width increases with decreasing shear rate and amounts to 23 per cent at $\dot{\gamma} = 0.1\,\mathrm{s}^{-1}$, probably because of the low relative sensitivity of the normal stress-measuring system at this low shear rate. The 10 per cent bandwidth for shear stress τ is surprisingly high.

The time-dependence of τ and N_1 is shown for $1\,\mathrm{s}^{-1}$ shear rate, for sample A, in Fig. 3.18. The general shape of the curves and the magnitude of the bandwidths for repeated measurements are comparable for the three samples.

The maxima for τ and N_1 are approximately located at a constant shear strain $\gamma = \dot{\gamma}t$ independent of shear rate $\dot{\gamma}$.

The main result of this work is that at all times the shear stress τ is practically equal for A and B. τ is about 10 per cent lower for C, whereas differences in N_1 can be measured around the maxima of these time-dependent functions, provided the shear rate is low enough.

Relaxation of stress after cessation of flow at constant shear rate. At the cessation of flow with constant shear rate, the relaxation of stress (that is, of τ and N_1) was determined. The scatter for repeated tests is rather high, especially for long times. Therefore, definite conclusions concerning the different behaviour for the three samples A, B, and C in this type of relaxation test cannot be presented.

In spite of this difficulty, the following conclusions can be drawn from the data obtained. (a) With increasing magnitude of the preceding shear rate $\dot{\gamma}$, the relaxing signals for τ and N_1 decay more rapidly. (b) Normal stress differences relax more slowly than shear stresses, as follows from theoretical reasoning (Lodge 1964). (c) Because of the scatter of the data already mentioned, conclusions to be drawn as

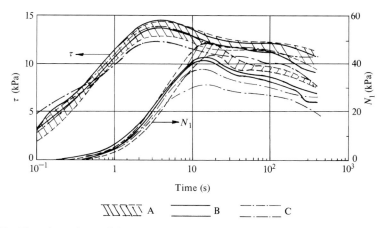

Fig. 3.18 Time-dependence of shear stress (τ) and normal stress difference (N_1) at constant shear rate $\dot{\gamma} = 1.0\,\mathrm{s}^{-1}$ for LDPE at 150 °C. The hatched bands indicate the range for eight specimens of sample A.

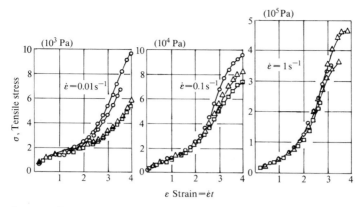

Fig. 3.19 Stress–strain relation $\sigma(\varepsilon)$ for homogeneous elongated test specimens at 150 °C and various constant strain rates $\dot{\varepsilon}(s^{-1})$ (A○, B△, C□).

to well-established differences in the relaxation behaviour of A, B, and C are rather limited.

Elongational behaviour. For homogeneous deformations, large total strains[†] up to $\varepsilon_H = \ln \Lambda = 4$ were achieved [Λ denotes (length at time t)/(initial length)]. At any chosen point on the stress–strain diagram, the specimen was cut into short pieces which shrank from the original length L_A, between the scissors performing the cutting operation, to the length L_R, after complete recovery. From the equation

$$\varepsilon_R = \ln(L_A/L_R) \tag{3.97}$$

the recoverable portion ε_R of the total tensile strain ε, obtained at the end of the extensional operation, can be calculated (see also Meissner 1971). At low tensile strain rates, samples A and C were difficult to extend to the maximum total strain $\varepsilon = 3$, whereas sample B completed this total tensile strain without rupture. Tests with $\dot{\varepsilon} = 0.001\,\mathrm{s}^{-1}$ gave rise to thermal stability problems. In Fig. 3.19 stress–strain relations for the three samples are given for the strain rates $\dot{\varepsilon} = 0.01-0.1-1\,\mathrm{s}^{-1}$. The temperature of the measurement was 150 °C; it should be added that, in general, the reproducibility of the results is good (see Meissner 1971), except for sample A at 0.01 and sample B at 1 s^{-1}. In Fig. 3.19 the upper and lower limiting curves show the range in which stress/strain responses were located in these two cases.

From Fig. 3.19 it follows that, for $\dot{\varepsilon} = 0.01$ and $0.1\,\mathrm{s}^{-1}$, the stress/strain relations show pronounced differences between the three samples. However, these differences start to develop at total strains of $\varepsilon > 2$, and they are large for $\varepsilon > 3$.

[†] The logarithmic strain measure ε_H here is the Hencky (1929) strain.

In this region of highest total strain, sample A shows the highest tensile stresses σ, whereas B and C show a relative early flattening of the strain-hardening range of the stress–strain diagram (at $\dot{\varepsilon} = 0.1\,\mathrm{s}^{-1}$), or a later onset of this strainhardening region (at $\dot{\varepsilon} = 0.1\,\mathrm{s}^{-1}$). Comparing B and C, it should be noted that at $\dot{\varepsilon} = 0.01\,\mathrm{s}^{-1}$, the curve for B lies between the curves for A and C, whereas, at $\dot{\varepsilon} = 0.01\,\mathrm{s}^{-1}$, B and C have identical stress–strain curves. At $\dot{\varepsilon} = 1\,\mathrm{s}^{-1}$, however, no difference can be detected between the samples. The curves for A and C fall within the error band for sample B. It should be added that, at this tensile rate, not all tests could be performed up to $\varepsilon = 4$ because of rupture of the specimens: for example, for sample C, three runs gave practically the same stress–strain curve, with a sudden rupture at $\varepsilon \approx 2.6$. Thus, a higher total strain than this limit did not seem to be possible with that sample.

It is notable that the differences in elongational behaviour of the samples occur at $\varepsilon > 2$ and that they become minor with increasing strain rate $\dot{\varepsilon}$. The recoverable tensile strain ε_R for sample A is listed in Table 3.7. Laun and Münstedt (1978) and Wagner *et al.* (1979) have extended material A to larger strains. In the latter paper no steady stress state was reached.

Stress–strain relations $\sigma(\varepsilon)$ measure at constant tensile strain-rate $\dot{\varepsilon}$ at 150 °C show differences between A, B, and C, but only at large total strains ($\varepsilon > 2$) and at relatively low strain rates: at $\dot{\varepsilon} = 0.01\,\mathrm{s}^{-1}$, sample A has a value of stress σ which is much higher (at these large total strains) than the values for B and C, which are nearly equal. At $\dot{\varepsilon} = 0.1\,\mathrm{s}^{-1}$, the difference between A and B or C is smaller, and at $\dot{\varepsilon} = 1\,\mathrm{s}^{-1}$, the difference is practically zero. In spite of these

Table 3.7 Elastic recovery ε_R after cessation of tensile tests at total (logarithmic) tensile strain ε

$\dot{\varepsilon}_0\,[\mathrm{s}^{-1}]$	ε	ε_R
0.001	1	0.17
	2	0.22
	3	0.29
0.01	1	0.38
	2	0.70
	3	1.0
	4	1.24
0.1	1	0.64
	2	1.2
	3	1.66
	4	1.88
1	1	0.83
	2	1.62
	3	2.12

The tensile tests were performed at constant tensile strain rate $\dot{\varepsilon}_0$ and a temperature of 150 °C. Data for sample A.

differences in the stress–strain relations, the recoverable tensile strain values ε_R do not differ markedly between the three samples, even at low strain-rates.

This set of data is most valuable; the data were obtained under practically every test condition available at present (except measurement of the temperature dependence of the non-linear viscoelastic properties which were mainly determined at 150 °C only). The results of the melt rheology measurements are accurate and reproducible (for one material and the same measurement) within an order of magnitude of 10 per cent. This is typical considering the very different types of measurements performed.

For this research programme, the three samples A, B, and C were selected in such a way that no differences appear in the usual characterization procedure for LDPE products (however, light scattering tests indicate that sample A has a small portion of molecules with a very high molecular weight). Similarly, melt rheology yields indistinguishable behaviour for the three samples if:

(a) the deformation is in the linear viscoelastic range, or

(b) the viscosity function (flow curve) is measured.

However, there are differences in the processing behaviour of the three samples which are not reflected by the above-mentioned characterization, and there are also differences to be found in melt rheology:

(c) In shear flow, these differences are connected with melt elasticity in the non-linear viscoelastic range. It is remarkable that the (relative) differences are the higher the lower is the shear rate; for example, at $0.1 \, \text{s}^{-1}$ shear rate, the difference in N_1 between samples A and C is about 40 per cent.

(d) In elongational flow, differences between A, B, and C occur at large total strains only, and (as in shear flow) are more pronounced at low strain rates.

The conclusion which follows immediately from (a) to (d) is that the behaviour in linear viscoelastic flow does not unambiguously reflect the behaviour in non-linear viscoelastic flow.

3.8.2   A polymer solution

As a second example, we consider a solution which consisted of 2.5 per cent by weight polyisobutylene ($M_w \sim 1.2 \times 10^6$) dissolved in a solvent of 51.3 per cent polybutene ($M_w \sim 950$) and 48.7 per cent decalin. Results were given by Ooi and Sridhar (1994) for this so-called S1 test liquid, and Byars et al. (1997) did extensive data fitting. Other groups were given the fluid, but the samples were not identical (over 20 per cent differences in zero shear viscosity were recorded at 20 °C, probably due to sample evaporation) and so we will refer to the excellent data of Ooi and Sridhar who conducted shear, oscillatory, and constant strain-rate elongational tests (Fig. 3.20). It is difficult to measure the steady-state elongational behaviour here [Fig. 3.20(c)].Tanner and Walters (1998) show

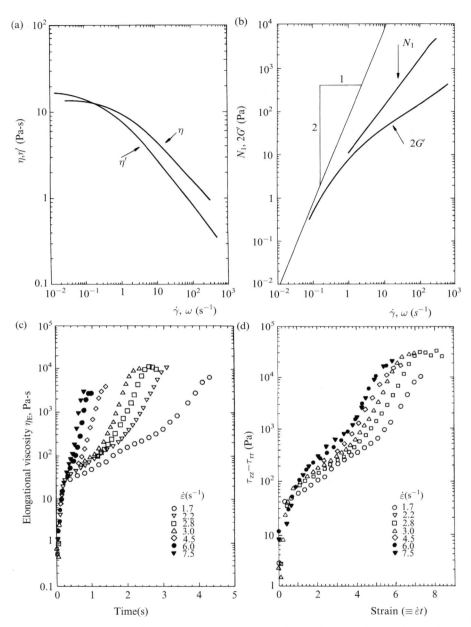

Fig. 3.20 Data for polyisobutylene solution S1 at 21 °C. (a) Steady (η) and dynamic (η') shear viscosities. (b) First normal stress difference (N_1) and storage modulus (G'). (c) Uniaxial extensional stress growth versus time at various stretch rates. (d) As for (c) except abscissa is strain, $\dot{\varepsilon}t$ (Ooi and Sridhar 1994).

(p.149) the immense range of results for η_E when various approximate techniques and theories are used.

These data were fitted to several models by Byars *et al.* (1997) and we shall refer to the data again later. The linear viscoelastic properties as fitted by Byars *et al.* (1997) are given in Table 3.8. One can see that the response is not very different in character from the polymer melt described in the previous section. No results for shear start-up were given but a similar S1 sample which did not shear-thin (M1, see Hudson and Ferguson 1990) showed much more stress overshoot in start-up of shear than Meissner's IUPAC polyethylene discussed above. Remarkably, with the S1 and M1 solutions, the graph of N_1 versus τ is quadratic, $N_1 \infty \tau^2$, and is practically independent of temperature (Fig. 3.21). This surprising behaviour was emphasized earlier by Vinogradov and Malkin (1980). (See also Chapter 9 of this book.) The second normal stress difference is discussed below.

3.8.3 *Other fluid properties*

The results discussed above are seen to be typical of molten polymers and concentrated solutions (there are fewer measurements, except shear viscosity, G' and G'', for dilute solutions). The book by Bird *et al.* (1977) contains results for several polyacrylamide solutions. Ferry (1970) shows many results for polyisobutylene solutions, especially for the small-strain dynamic properties and shear viscosities. Middleman (1968) also shows some data for polyisobutylene solutions. Tanner (1973) correlated data for polyisobutylenes. Elongational behaviour is discussed by Stevenson (1972), Petrie (1979), and Meissner (1992), and double-step relaxation experiments were reported by Zapas (1974), Osaki *et al.* (1981) and Wagner and Ehrecke (1998). Laun (1986) developed a rule to compute N_1 from G' and G''; he suggests

$$N_1(\omega) = 2G'(\omega)[1 + (G'/G'')^2]^{0.7}.$$

The reader is referred to the sources for details.

Other systems behave somewhat differently and these will be mentioned from time to time. We have in mind dough (see Fig. 1.9) where no steady shear viscosity

Table 3.8 Linear viscoelastic parameters for S1 polyisobutylene solution at 21 °C

Mode No.	λ_i (s)	η_i (Pa−s)	g_i ($\equiv \eta_i/\lambda_i$) (Pa)
1	6.48	4.22	0.651
2	0.705	5.06	7.18
3	0.116	3.24	27.9
4	0.0078	1.01	129.5
Solvent	–	0.15	–

[See eqn (2.105) for notation].

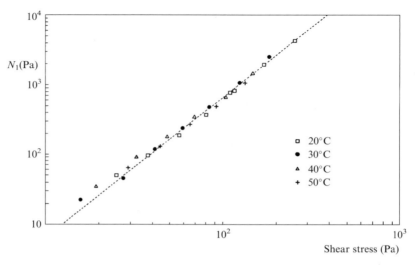

Fig. 3.21   First normal stress difference (N_1) versus shear stress (τ) for polyisobutylene fluid S1 at different temperatures. Dashed line is a quadratic fit to the data (Binding *et al.* 1994).

exists, suspensions (Chapter 5), soft tissues from animals (Fung 1993), and liquid crystals (Kiss and Porter 1978). Generally, shear-thinning is in evidence; in some cases (dough, for example), the linear viscoelastic regime occupies only an exceedingly narrow range of strain (of order 0.1 per cent or less) in contrast to some polymeric fluids, where a 10 per cent strain may still appear linear.

Second normal stress differences

We have not discussed the second normal stress difference, N_2, in any detail. This quantity is more difficult to measure than N_1, and results are scarce. Keentok (1997) has reviewed much of the data available for various liquid systems.

   Table 3.9 gives a selection of the data. In most cases an increase of shear rate decreased the value of $|N_2/N_1|$, but it is often sufficient, in absence of good data, to assume a constant value. Occasionally (Keentok 1997) a positive value of N_2/N_1 can be found; generally, however, N_2/N_1 is negative, as shown. Often it is small in magnitude.

   For melts, the optical result of Wales (1976) showed a range of N_2/N_1 around -0.1 for polystyrene. The fluid of Boger (1978) shows a very small N_2 value of the order of 1 per cent or less of N_1. (It also shows a remarkably constant shear viscosity over a wide range of shear rates, and has been used in experiments to separate shear-thinning from elastic effects.)

   In summary, we shall often assume that N_2/N_1 is negative and fairly small for polymer melts and solutions; this is not necessarily the case for other systems.

Table 3.9 N_2/N_1 for various fluids

Fluid	Concentration by weight	$-N_2/N_1$	Source
Polyox WSR 301/water	1%	0.075 ± 0.01	Keentok *et al.* (1980)
	0.7%	0.065 ± 0.01	
Separan AP-30/water	0.8%	0.1 ± 0.01	Keentok *et al.* (1980)
	0.5%	0.085 ± 0.015	
	0.2%	0.065 ± 0.01	
Oppanol-B 200 polyiso-butylene in dekalin	1.1%	0.11 ± 0.01	Keentok *et al.* (1980)
NBS Nonlinear fluid No.1	7.1%	0.3 ± 0.04	Keentok *et al.* (1980)
Silicone 1000 cSt ($\dot{\gamma} = 100\,\mathrm{s}^{-1}$)	–	0 ± 0.2	Keentok (1997)
Silicone 12,500 cSt ($\dot{\gamma} = 60\,\mathrm{s}^{-1}$)	–	0.5 ± 0.1	Keentok (1997)
Silicone 30,000 cSt ($\dot{\gamma} = 60\,s^{-1}$)	–	0.5 ± 0.1	Keentok (1997)
Polyisoprenes	various	0.17–0.30	Lee *et al.* (1992)
Liquid MI ($\dot{\gamma} \sim 20\,\mathrm{s}^{-1}$)	0.244%	0.07	Binding *et al.* (1990)
Polyisobutylene($\dot{\gamma} = 2\,\mathrm{s}^{-1}$)	0.03%	~ 0.01	Magda *et al.* (1991)
Polystyrene solutions	various	0.28	Magda and Baek (1994)
Shell Barbatia grease ($\dot{\gamma} \sim 10\,\mathrm{s}^{-1}$)	–	0.5	Binding *et al.* (1976)
Suspensions of spheres	–	1–0.5	Ohl and Gleissle (1992) Laun (1994)
Polyethylene	–	0.24	Meissner (1992)
Polystyrene	–	0.1	Wales (1976)

3.9   Wall slip

It is nearly always assumed in Newtonian fluid mechanics that fluid particles stick to a solid surface, thereby driving (or retarding) the flow. However, early workers in the subject did question the point, and Bingham (1922) in his book said (pp. 29–35):

These results seem to make it quite certain that, whether the liquid wets the solid or not, there is no measurable difference between the velocity of the solid and the liquid in contact with it. . .

By 1929, rheologists were actively considering slip, especially for rubbery materials. Mooney (1931), who knew that rubbers did not adhere fully to solid walls, extended Weissenberg's capillary analysis to include slip, as follows.

Suppose we return to eqn (3.67) and integrate by parts, finding, if slip does not vanish at the wall

$$Q = \pi w(R) R^2 - \pi \int_0^R r^2 \frac{\mathrm{d}w}{\mathrm{d}r}\,\mathrm{d}r, \tag{3.67c}$$

or, with $q = Q/\pi R^3$, we find

$$q = w(R)/R + \tau_{\mathrm{w}}^{-3} \int_0^{\tau_{\mathrm{w}}} \tau^3 \phi(\tau)\,\mathrm{d}\tau, \tag{3.67d}$$

where $w(R)$ is the slip velocity. Clearly, slip increases the discharge, other things being equal.

Early ideas made $w(R)$ a linear function of τ_{w} (Navier's wall condition), but while this may be a useful model for those cases where a distinct film of solvent appears next to the wall, due perhaps to a lubricant, it appears that in many cases there is an onset stress at which slip suddenly begins, and the relation between τ_{w} and $w(R)$ is not linear.

Following Mooney (1931) we assume that the slip speed $w(R) = \beta\tau_{\mathrm{w}}$, where β, the slip coefficient, may also be a function of τ_{w}. Then (3.67d) becomes

$$q = \beta\tau_{\mathrm{w}}/R + \tau_{\mathrm{w}}^{-3} \int_0^{\tau_{\mathrm{w}}} \tau^3 \phi\,\mathrm{d}\tau. \tag{3.67e}$$

Mooney (1931) suggested plotting q against τ_{w} for various capillary radii—if they coincide, then no slip is occuring, but if they do not, one can plot q against $1/R$ for constant τ_{w}, find $[\partial q/\partial(1/R)]_{\tau=\mathrm{const}}$ and hence find $\beta\tau_{\mathrm{w}}$ as a function of τ_{w}. This is clearly a difficult procedure, but the analysis has been used (Hatzikiriakos and Dealy 1992).

That slip does occur suddenly is shown in the capillary results of Vinogradov *et al.* (1972) shown in Fig. 3.22. These results suggest that slip begins at a wall shear stress of about 0.2 MPa, with a consequent jump in discharge rate. While it is commonplace to see slip in plastic flow of metals and in the flow of powders, the slip of fluids is less easy to believe in spite of the interesting early work by Mooney (1931).

However, Migler *et al.* (1993) measured directly the local speed of a sheared polymer melt within 10^{-4} mm from the solid surface. They reported that in the absence of strong polymer adsorption, polydimethylsiloxanes slip at all shear stresses, no matter how small. For strong chain adsorption there seems to be a critical shear stress at which a transition from a weak to a strong slip takes place. These seem to be the first direct, as opposed to inferred, measurements of slip.

One can distinguish several possibilities concerning slip:

(i) In solutions, it is clear that the presence of a solid wall may alter the concentration of solute near the wall, thereby inducing a low-viscosity layer and apparent slip. In this case, the solvent does not slip at the wall. This possibility is discussed in detail in the recent survey of Barnes (1995); see also Buscall *et al.* (1993) for an example involving colloids.

(ii) In a molten polymer, the wall will alter the configuration space available to molecules near the wall, even if slip does not occur, thereby altering the material behaviour next to the wall.

(iii) Slip may occur, separately from or in addition to (ii) above.

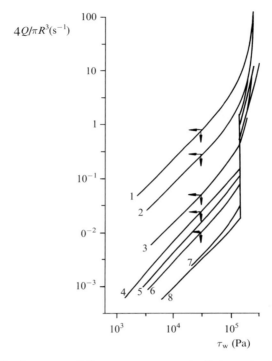

$4Q/\pi R^3 (\text{s}^{-1})$

Fig. 3.22 Plots of volume rate of flow versus pressure (wall shear stress) for polyisoprenes extruded from a capillary of radius 0.48 mm and length 24.8 mm. Each curve corresponds to a different polymer. Note the jump in Q at around $\tau_\text{w} = 0.1$ MPa. (After Vinogradov *et al.* 1972.)

Effect (i) is relatively well understood and need not be discussed further.

Effect (ii) must certainly occur, but so far only simulations can detect it and the altered layer seems generally not to penetrate far into the bulk material.

In connection with effect (iii), Pearson and Petrie(1968) made a basic contribution to the subject, although the work was not widely noticed, and the subject languished until Ramamurthy (1986) showed, unequivocally, that real slip can occur in extrusion.

Pearson and Petrie (1968) consider the length scales involved near the wall, nominating L as the scale of the apparatus, (for example, a diameter), ℓ as the scale of the roughness on the surface of the apparatus, and ℓ_p as the scale of the microstructural entities making up the fluid. It is the relative scales of these lengths, they said, that determines, to a large extent, the boundary conditions. In the various cases:

1. $L \gg \ell \gg \ell_p$ is the typical condition for small-molecule fluids (water, light oils). In this case the small scale of the asperities means that any flow over them is at very low Reynolds number, and viscous forces dominate near the boundary. Here, the argument of Richardson (1971) shows that actual

adhesion to the wall is not necessary in order to see, on the gross scale (L), an apparent no-slip condition, even if slip occurs on the ℓ_p scale.

2. $L \gg \ell_p \gg \ell$. This condition is typical of coarse powders in smooth containers, where slip at the wall is often seen.

3. $L \gg \ell \approx \ell_p$. This is often the case for large molecule fluids near smooth walls, and the question of slip may well depend on chemical and/or mechanical adhesion, in contrast to the cases above.

In summary, with the benefit of both theoretical and experimental interest, we can now conclude that at least three factors are of importance:

(i) The Pearson–Petrie scaling argument means that no effective slip can occur when molecular size is smaller than the wall roughness scale. This really explains all the classical evidence on liquids (and non-rarefied gases) and shows that a no-slip condition is appropriate in these cases; chemical adhesion is not needed to avoid slip; see also recent work by Sarkar and Prosperetti (1996).

(ii) For large molecules (relative to the wall roughness scale), the temperature and adherence (chemical) properties may be of great significance in setting the critical shear stress at which slipping occurs.

(iii) Normal pressure may assist in reducing slip.

We have discussed the wall slip situation as if slip uniformly appears everywhere. In fact, since the wall shear stresses are non-uniform at inlet and exit, it is likely that slip will begin at these points before the constant stress region begins to slip (Phan-Thien 1988; Tanner 1994). Thus the idea of partial slip arises. Further, since there is now evidence (Denn 1992; Hatzikiriakos and Dealy 1991) that the normal force on the wall is a factor in slip, the matter is clearly not very simple. An example which takes account of these factors is given below.

Some idea of the magnitudes of slip speed is shown in Fig. 3.23. The onset of slip is often in the region of $\tau_w = 0.1 - 0.2$ MPa, and beyond that a power-law fit has often been used (Hatzikiriakos and Dealy 1991). Slip speeds of 10–20 mm/s were quoted by these workers for $\tau_w = 0.3$ MPa using high-density polyethylene (HDPE). Figure 3.23 shows data obtained with a sliding plate viscometer (Hatzikiriakos and Dealy 1991) showing an onset of slip at about 0.1 MPa in HDPE. These workers also showed that slip was affected by surface treatment with chemicals (mainly fluoropolymers), confirming Ramamurthy's (1986) result. Further, Hatzikiriakos (1993) states that the slip speed is a function of temperature, pressure and various molecular parameters. In addition, one also has the wall condition (additives, roughness) and the shear stress. According to this paper, slip velocity increases with temperature, molecular weight, and shear stress. Slip decreases with increased polydispersity and pressure. Many of these factors can be understood in the light of the Pearson–Petrie arguments given above, relating the relative sizes of roughnesses and molecules. Slip speed is often

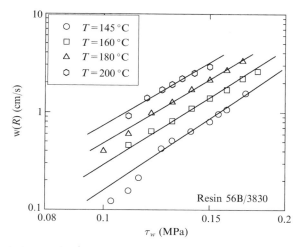

Fig. 3.23 Slip velocity as a function of wall shear stress on a log–log plot to show power-law for slip velocity (Courtesy Dr S. Hatzikiriakos, University of British Columbia).

modelled using a truncated power-law expression, possibly also with a pressure factor. Person and Denn (1997) used the form

$$
\begin{aligned}
w_s &= a\,\tau_w^m \exp\left(-\beta_p\right) \quad (\tau_w > \tau_c) \\
w_s &= 0 \qquad\qquad\qquad\;\; (\tau_w \le \tau_c).
\end{aligned}
\tag{3.98}
$$

Here τ_c is the onset (or critical) stress, a, m, and β are constants, and p is the pressure. Estimates of β vary quite widely, from $10^{-1} - 10^{-2}(\text{MPa})^{-1}$ for the same Linear Low Density Polyethylene (LLDPE).

Figure 3.23 shows some data for high density polyethylene (HDPE) obtained by Hatzikiriakos and Dealy (1991); they show a critical shear stress of about 0.1 MPa and the temperature effect is evident. In this case, using a sliding plate viscometer, pressure effects were essentially zero. Hatzikiriakos (1993) fitted data for LLDPE with $a = 7\,(\text{MPa})^{-4}\,\text{m/s}$ and $m = 4$. One sees that an abrupt drop to zero slip is required at τ_w so Hatzikiriakos (1993) used a modified Eyring rate theory (see Glasstone *et al.* 1941) to avoid this step change. He proposed a slip speed of the form

$$
w_s = f(T)\sinh\left[\frac{b}{T}\left(\frac{\tau_w}{\tau_c} - 1\right)\right].
\tag{3.99}
$$

Thus, the onset stress is the point of rapid slip increase here; no slip occurs if $\tau_w \lessgtr \tau_c$.

The exact values of the function f and the constant b depend on polymer properties, notably molecular weight. In this formula the critical stress is related

Table 3.10 Critical wall shear stresses for mono-disperse polyethylenes [from Hatzikiriakos (1993)]

Polymer	Type	M_w	τ_c(MPa)
56B	HDPE	178,000	0.090
60% 56B	HDPE	129,000	0.120
20% 56B	HDPE	80,000	0.150
2910	HDPE	56,000	0.180
GRSN/7047	LLDPE	114,000	0.100

to the work of adhesion (W_a), which is a function of surface properties, and b is proportional to W_a. The search for mechanisms of slip continues; see also Wang et al. (1996).

Some values of the critical shear stress τ_c are given in Table 3.10; they are all of order 0.1 MPa.

In addition to the above steady-state slip there appears to be a time-dependence of slip, so that a given stress applied for a very long time will cause slip at a lower critical stress than a sharp pulse will. Also, it is well known that extra roughness can inhibit slip. For polyvinyl chloride (PVC) this effect has been studied by Knappe and Krumböck (1986).

Work continues in this area; however, it is clear that one cannot ignore slip when high shear stresses (~ 0.1 MPa or greater) exist in fluids or soft solids; for other systems (rubber, granular materials) slip is also important.

Finally, one needs to consider the effect of 'failure' within the fluid or at the fluid–solid interfaces and some discussion is provided in Section 9.2.1.

Example: Uhland's analysis of slip
In this analysis (Uhland 1976) it is assumed that the shear stress on the wall is related to the normal stress as it is in the Amontons–Coulomb friction law:

$$\tau_w = p\mu,$$

where μ is a friction coefficient and p is the local pressure, equal to the normal stress at the wall. If the local pressure gradient in the (circular) tube is $+\,\mathrm{d}p/\mathrm{d}z$, then ($\tau_w$ is a positive magnitude, $+\mathrm{d}p/\mathrm{d}z$ is negative)

$$\tau_w = -\frac{R}{2}\left|\frac{\mathrm{d}p}{\mathrm{d}z}\right| = \mu p, \qquad (3.100)$$

and one can integrate to find the pressure. If the entry pressure is p_0 ($z = 0$), and the exit pressure is assumed to be p_L ($z = L$), then

$$p = p_L \exp\frac{2\mu}{R}(L - z). \qquad (3.101)$$

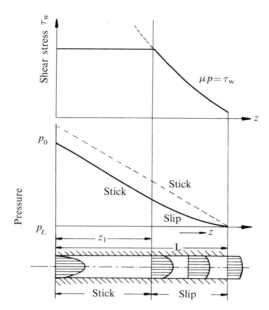

Fig. 3.24   Uhland's slip analysis.

and

$$\tau_w = \mu p_L \exp \frac{2\mu}{R}(L - z).$$ (3.102)

In this example the wall shear stress is not constant along the tube, and it is easy to see that if no slip occurs for a large enough pressure, then this model often gives partial slip. Figure 3.24 gives a sketch of some results. Michaeli (1992) discusses computations of velocity profiles with this analysis. However, this analysis is clearly not of universal applicability, since at $z = L$, it is necessary to have non-zero pressure in order to maintain a shear stress at exit. Since this is not always the case it is clear that this simple model is not adequate to describe slip. A serious problem thus arises in numerical analysis, since the question of slip (or no-slip) relates to boundary conditions on the flow problem. We shall return to these questions in Chapter 8.

3.10   Summary

There is now a broad agreement on the type of response common to polymer melts and concentrated solutions, even though it is sometimes difficult to find accurate results for a given substance in the literature. One sees clearly the problems of obtaining accurate data, and the accuracy which can be expected, from the data for polyethylene and the S1 solution. Certainly more work is needed to clarify elongational behaviour at very high rates of strain and at negative strain rates (biaxial extension: see Section 3.7.2). One also needs to look carefully at step and superposed shear experiments and to investigate dilute solution behaviour, but the main kind of response is now reasonably well-documented, at least for isothermal conditions, and it follows the pattern for polyethylene. We shall refer again to the effect of temperature in Chapter 9. Other materials have been studied less intensely, but often they follow, broadly, the patterns given above for polymers.

Hence the response of fluids in the simpler flows can be discussed from both the theoretical and experimental viewpoints; it is not always necessary to use complex constitutive equations. This should not be forgotten; there are many practical cases where the flow is essentially of a simple kind (viscometric, usually) and useful results can be found with a small amount of analysis. We shall consider more difficult cases in the following chapters.

References

Bagley, E. G. (1957). *J. Appl. Phys.*, **28**, 624.
Ballman, R. L. (1965). *Rheol. Acta*, **4**, 137.
Barnes, H. A. (1995). *J. Non-Newtonian Fluid Mech.*, **56**, 221.
Barnes, H. A. and Walters, K. (1969). *Proc. R. Soc.*, **A 314**, 85.
Binding, D. M., Hutton, J. F., and Walters, K. (1976). *Rheol. Acta*, **15**, 540.
Binding, D. M., Jones, D. M., and Walters, K. (1990). *J. Non-Newtonian Fluid Mech.*, **35**, 121.
Binding, D. M., Maia, J., and Walters, K. (1994). *J. Non-Newtonian Fluid Mech.*, **52**, 137.
Bingham, E. C. (1922). *Fluidity and plasticity*, McGraw-Hill, New York.
Bird, R. B., Armstrong, R. C., and Hassager, O. (1977). *Dynamics of polymeric liquids*, Vol. 1, *Fluid dynamics*, Wiley, New York.
Boger, D. V. (1978). *J. Non-Newtonian Fluid Mech.*, **3**, 87
Buscall, R., McGowan, J. I., and Morton-Jones, A. J. (1993). *J. Rheology*, **37**, 621.
Byars, J. A., Binnington, R. J., and Boger, D. V. (1997). *J. Non-Newtonian Fluid Mech.*, **72**, 219.
Cogswell, F. N. (1969). *Rheol. Acta*, **8**, 187.
Coleman, B. D., Markovitz, H., and Noll, W. (1966). *Viscometric flows of non-Newtonian fluids*. Springer-Verlag, Berlin.
Denn, M. M. (1992). In *Theoretical and applied rheology* (eds. P. Moldenaers and R. Keunings), p. 45. Elsevier, Amsterdam.
Denson, C. D. and Crady, D. L. (1974). *J. Appl. Polym. Sci.*, **18**, 1611.
Denson, C. D. and Gallo, R. J. (1971). *Polymer Eng. Sci.*, **11**, 174.
Dobraszczyk, B. J. (1997). *Cereal Foods World*, **42**, 516.
Ferry, J. D. (1970). *Viscoelastic properties of polymers*. Wiley, New York.
Frank, F. C., Keller, A., and Mackley, M. R. (1971). *Polymer*, **12**, 467.
Fuller, G. G. (1995). *Optical rheometry of complex fluids*. Oxford University Press.
Fung, Y. C. (1993). *Biomechanics, 2nd edn*. Springer-Verlag, New York.
Glasstone, S., Leidler, K., and Eyring, H. (1941). *The theory of rate processes*. McGraw-Hill, New York.
Gupta, R. K. and Sridhar, T. (1988). In *Rheological measurement* (eds. A. A. Collyer and D. W. Clegg), p. 211. Elsevier Applied Science, London.

Hatzikiriakos, S. (1993). *Int. Poly. Processing*, **8**, 135.
Hatzikiriakos. S. and Dealy, J. M. (1991). *J. Rheol.*, **35**, 497.
Hatzikiriakos. S. and Dealy, J. M. (1992). *J. Rheol.*, **36**, 703.
Hencky, H. (1929). *Ann. Physik* (5), **2**, 617.
Hudson, N. E. and Ferguson, J. (1990). *J. Non-Newtonian Fluid Mech.*, **35**, 159.
James, D. F. and Walters, K. (1993). In *Techniques in rheological measurement* (ed. A. A. Collyer), p. 33. Chapman & Hall, London.
Janeschitz-Kriegl, H. (1983). *Polymer melt rheology and flow birefringence*, Springer-Verlag, Berlin.
Kaye, A., Lodge, A. S., and Vale, D. G. (1968). *Rheol. Acta*, **7**, 368.
Keentok, M. (1997). Ph.D. Thesis, University of Sydney.
Keentok, M. V., Georgescu, A. G., Sherwoood, A. A., and Tanner, R. I. (1980). *J. Non-Newtonian Fluid Mech.* **6**, 303.
Kiss, G. and Porter, R. S. (1978). *J. Polym. Sci. Polym. Symposium*, **65**, 193.
Knappe, W. and Krumböck, E. (1986). *Rheol. Acta*, **25**, 296.
Kraynik, A. M., Aubert, J. H., and Chapman, R. N. (1984). *Proc. IX Intl. Congr. on Rheology*, Mexico **4**, 77.
Laun, H. M. (1986). *J. Rheol.*, **30**, 459.
Laun, H. M. (1994). *J. Non-Newtonian Fluid Mech.*, **54**, 87.
Laun, M. and Munstedt, H. (1978). *Rheol. Acta*, **17**, 415.
Lee, C. S., Magda, J. J., Devries, K. L., and Mays, J. W. (1992). *Macromolecules*, **25**, 4744.
Lodge, A. S. (1964) *Elastic liquids*, Academic Press, New York.
Mackley, M. R. and Keller, A. (1973). *Polymer*, **14**, 16.
Maerker, J. M. and Schowalter, W. R. (1974). *Rheol. Acta*, **13**, 627.
Magda, J. J. and Baek, S. G. (1994). *Polymer*, **35**, 1187.
Magda, J. J., Lou, J., Baek, S. G., and Devries, K. L. (1991). *Polymer*, **32**, 2000.
Meissner, J. (1971). *Rheol. Acta*, **10**, 230.
Meissner, J. (1972). *Trans. Soc. Rheol.*, **16**, 405.
Meissner, J. (1975). *Pure and Appl. Chem.*, **42**, 551.
Meissner, J. (1992). In *Applied polymer analysis and characterization* (ed. J. Mitchell). p. 201, Hanser, Munich.
Metzner, A. B. and Metzner, A. P. (1970). *Rheol. Acta*, **9**, 174.
Michaeli, W. (1992). *Extrusion dies for plastics and rubber, 2nd edn*, Hanser, Munich.
Middleman, S. (1968) *The flow of high polymers*, ch. 2. Interscience Publishers, New York.
Migler, K. B., Hervet, H., and Leger, L. (1993). *Phys. Rev. Letters*, **70**, 287.
Miller, M. J. and Christiansen, E. B. (1972). *Am. Inst. chem. Engrs. J.* **18**, 600.
Mooney, M. (1931). *J. Rheology*, **2**, 210.
Ohl, J. and Gleissle, W. (1992). *Rheol. Acta*, **31**, 294.
Ooi, Y. W. and Sridhar, T. (1994). *J. Non-Newtonian fluid Mech.*, **52**, 153.
Orr, N. and Sridhar, T. (1996). *J. Non-Newtonian Fluid Mech.*, **67**, 77.
Osaki, K., Kimura, S., and Kurata, M. (1981). *J. Rheol.*, **25**, 549.
Pearson, J. R. A. and Petrie, C. J. S. (1968) In *Polymer systems: deformation and flow* (eds. R. E. Wetton and R. W. Whorlow), p. 163. Macmillan, London.
Peng, St. T. J. and Landel, R. F. (1974) *Rheol., Acta*, **13**, 548.
Person, T. and Denn, M. M. (1997). *J. Rheol.*, **41**, 249.
Petrie, C. J. S. (1979). *Elongational flows*. Pitman, London.
Phan-Thien, N. (1988). *J. Non-Newtonian Fluid Mech.*, **26**, 377.
Pipkin, A. C. (1972). *Lectures on viscoelasticity theory*, Springer-Verlag, New York.
Pipkin, A. C. and Tanner, R. I. (1972). *Mechanics Today*, **1**, 262.
Rabinowitsch, B. (1929). *Z. Physik. Chem.*, A **145**, 1.

Ramamurthy, A. V. (1986). *J. Rheol.*, **30**, 337.

Richardson, S. (1971). *J. Fluid Mech.*, **59**, 707.

Russell, R. J. (1946). Ph.D. Thesis, University of London.

Sarkar, K. and Prosperetti, A. (1996). *J. Fluid Mech.*, **316**, 223.

Savins, J. G. and Metzner, A. B. (1970). *Rheol. Acta*, **9**, 365.

Stevenson, J. F. (1972). *Am. Inst. Chem. Engrs. J.*, **18**, 540.

Tanner, R. I. (1973). *Trans. Soc. Rheol.*, **17**, 365.

Tanner, R. I. (1994). *Ind. Eng. Chem. Res.*, **33**, 2434.

Tanner, R. I. and Walters, K. (1998). *Rheology: an historical perspective*, Elsevier, Amsterdam.

Tanner, R. I. and Williams, G. (1970). *Trans. Soc. Rheol.*, **14**, 19.

Tanner, R. I. and Williams, G. (1971). *Rheol. Acta*, **10**, 528.

Trouton, F. T. (1906). *Proc. R. Soc.*, A **77**, 426.

Uhland, E. (1976). *Rheol. Acta*, **15**, 30.

Vinogradov, G. V., Malkin, A. Ya., Yanovskii, Yu., G., Borisenkova, E. K., Yarlykov, B. V., and Berezhnaya, G. V. (1972). *J. Polymer Sci.*, A-2, **10**, 1061.

Wagner, M. H. and Ehrecke, P. (1998). *J. Non-Newtonian Fluid Mech.*, **76**, 183.

Wagner, M. H., Raible, T., and Meissner J. (1979). *Rheol. Acta*, **18**, 427.

Wales, J. L. S. (1976). *The application of flow birefringence to rheological studies of polymer melts*. Delft University Press.

Walters, K. (1975). *Rheometry*. Chapman & Hall, London.

Walters, K. (1984). *Proc. IX Intl. Congr. Rheology*, Mexico, **1**, 31.

Wang, S.-Q., Drda, P. H. and Inn, Y.-W. (1996). *Rheol.*, **40**, 875.

Yin, W. L. and Pipkin, A. C. (1970). *Archs. ration. Mech. Anal.*, **37**, 111.

Zapas, L. J. (1974). In *Deformation and fracture of high polymers* (ed. H. Kausch), p. 381. Plenum Press, New York.

Problems

1. Consider the flow between two large parallel planes, $z = \pm h$. Suppose the upper plane moves at a speed U in the x-direction; and the lower moves at a speed U in the opposite direction. A pressure gradient $G(\equiv -\partial p/\partial y)$ exists in the y direction. Suppose the space between the planes is filled with a 'power-law' fluid so that in simple shearing

$$\eta = k \mid \dot{\gamma} \mid^{-2/3}$$

Formulate an expression for the fluidity ϕ. Using this and knowledge of the shear stresses τ_{xz}, τ_{yz}, compute the velocity field in terms of the shear stress $\tau_{xz}(\equiv \tau)$ and G. Also find the rate of flow per unit width (Q) in the y-direction.

2. Reconsider the problem when the viscosity function is the same but one uses the full viscometric equation (3.22) with $\Psi_1 = a_1 \mid \dot{\gamma} \mid^{-4/3}$, $\Psi_2 = a_2 \mid \dot{\gamma} \mid^{-4/3}$. Find the stress and pressure fields. Assume a_1 and a_2 are constants.

3. Show that the flow $\mathbf{v} = c(\theta/\theta_0)\mathbf{k}$ (in r, θ, z coordinates) is a completely controllable flow between two inclined planes. Find the stress field and the stresses on the planes.

4. Consider a Couette flow. Show that in a Reiner–Rivlin fluid ($\Psi_1 = 0$ in eqn (3.24)) a Weissenberg rod-climbing effect may occur if $\Psi_2 > 0$. Include inertia in the problem.

5. Consider the Pochettino viscometer (Fig. 3.25) which consists of a solid cylinder A of radius R_i sliding axially inside a concentric fixed cylinder B of radius R_0. Find the relation between the speed w of A and the force F on it for the following viscosity laws.

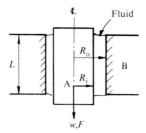

Fig. 3.25 Problem 5.

(a) a power-law fluid [eqn (1.11)]

(b) a Bingham body [eqn (1.9)].

6. In a cone-plate flow assume the velocity field changes in a linear fashion across the gap. Compute an approximate correction for fluid inertia by averaging the inertia forces across the gap. Show that this procedure gives the value $c = \frac{1}{6}$ in eqn. (3.89).

7. Find the response of a linear Maxwell fluid (eqn 2.81 and 2.105) to the sudden start and stop of shearing, and also the response to a double step of shear (Fig. 4.10). Neglect inertia.

8. Compute the average instantaneous inertia forces in the elongational flow of a cylinder of radius R and length L undergoing a steady extensional flow of rate $\dot{\varepsilon}$ by finding the mean pressure due to inertia on the ends $x = \pm L/2$

9. A purely viscous non-Newtonian fluid is inelastic, but can under some circumstances show 'elastic' behaviour. Suppose a sample of a 'power-law' fluid is contained between two plates distance h apart, moving relative to one another at a speed $\{h\dot{\gamma}_0 + \varepsilon\omega \sin \omega t\}$, so that we have superposed simple and sinusoidal shearing. Ignoring fluid inertia, compute the shear stress on the plates. Write the response in a power series in the small parameter ε: $\tau = \tau_0 + \varepsilon\tau_1 + \varepsilon^2\tau_2 + \cdots$. Using the first order terms in stress and strain only find the values of G' and G'' for the fluid. Use the second-order terms in ε^2 to find the new 'steady-state' viscosity.

10. Work out the velocity, pressure and stress fields in an annulus through which a second-order fluid (η, Ψ_1 and Ψ_2 constant) is flowing. Assume the inner tube is half the radius of the outer tube.

11. Suppose that the boundary condition at the wall of a circular tube is the Navier condition.

$$w(R) = \beta\tau_w,$$

where β is a constant, $w(R)$ is the slip speed at the wall and τ_w is the wall shear stress. If the fluid is Newtonian, with a true, but unknown, viscosity η_0, show that the material appears to have a viscosity of $\eta_0/(1 + \beta\eta_0/R)$. How could you find η_0 and β?

4

CONTINUUM-DERIVED THEORIES AND EXPERIMENTAL DATA

IN THE previous two chapters we examined the behaviour of materials in various simple flows—at small strains, viscometric flows and elongational flows, both steady and transient. In some cases (for example, steady viscometric flows) we were able to derive some results in special classes of flows (for example, viscometric flows) without the need for an explicit general constitutive equation connecting stress and deformation. Note that eqn (3.20) is not a constitutive equation; it is a restatement of the known stress field. In many cases, however, one is forced to assume a constitutive equation and the question of how one arrives at such a descriptor for general flows needs to be considered.

Two types of approach have been widely used: one based on continuum mechanics, and one based on microstructural ideas. We will consider the former in this chapter and the latter in the next.

When a constitutive relation is proposed, it is appropriate to find out what it predicts in various simple flow classes, and in particular to note behaviour which it cannot describe. So far we have discussed the Boltzmann linear viscoelastic constitutive relation (eqn 2.81) which is completely successful for small strains, and now we consider other, non-linear equations.

4.1 Reiner–Rivlin and purely viscous fluids

A natural generalization of the Navier–Stokes fluid (eqn 2.69) is to permit the stresses to depend in a non-linear way on the rate of deformation; systematic observations of viscosity changes with shear rate go back at least to Schwedoff (1890). Considerable work continued over the next forty years until Hohenemser and Prager (1932) showed how to write the viscosity as a function of the invariants of the rate of deformation tensor. Finally, Reiner (1945) and Rivlin (1948) considered the general concept of stress in an incompressible body as a function of the rate of deformation tensor. They showed that if the extra stress tensor $\sigma_{ij} + p\delta_{ij}(\equiv \tau_{ij})$ is an isotropic function of the rate-of-deformation tensor (d_{ij}) only, then it follows that

$$\sigma_{ij} = -p\delta_{ij} + 2\eta d_{ij} + 4\Psi_2 d_{ik} d_{kj}, \tag{4.1}$$

where η and Ψ_2 are functions of the second and third invariants of d_{ij}, see eqns (2.10–2.11) for the definition of these invariants. (The first invariant of d_{ij} is zero for incompressible materials.) The form (4.1) can be deduced readily if it is supposed that

$$\sigma_{ij} = -p\delta_{ij} + f_{ij}(d_{ij}), \tag{4.2}$$

where f_{ij} can be expanded in a series of powers of d_{ij}. Successive applications of (2.13c) and (2.13d) remove all higher powers of d_{ij}, leaving the result (4.1).

It is easily shown that eqn (4.1) is objective (see Section 2.5). An alternative approach which does not assume that f_{ij} can be developed in a series form is given by Leigh (1968), who also gives a more general treatment showing that neither the velocity nor the vorticity can be included as independent variables in a constitutive equation that is objective.

In equation (4.1) the quantities η and Ψ_2 are functions of the invariants of d_{ij}. The first invariant $I_d = d_{ii}$ is zero for the incompressible fluids being considered. The third invariant (det d_{ij}) is zero in plane flows. If we consider a simple (plane) shearing flow where $\mathbf{v} = \dot{\gamma}y\mathbf{i}$, say, then the only non-zero components of d_{ij} are $d_{xy} = d_{yx} = \frac{1}{2}\dot{\gamma}$. The second invariant II_d is just $-\dot{\gamma}^2/4$. Hence, it is convenient to consider, for all flows, that the equivalent (positive) shear rate is defined as

$$\dot{\gamma} = 2\sqrt{-II_d} = \sqrt{2d_{ij}d_{ij}}. \tag{4.3}$$

For example, in an elongational flow, where $d_{xx} = \dot{\varepsilon}$, $d_{yy} = d_{zz} = -\dot{\varepsilon}/2$, and the other components are zero, the second invariant II_d is $-3\dot{\varepsilon}^2/4$, and so the equivalent shear rate is $\dot{\gamma} = \sqrt{3}\dot{\varepsilon}$. (In this flow the third invariant is non-zero and is equal to $\dot{\varepsilon}^3/4$.) Thus one can replace the dependence on II_d by a dependence on $\dot{\gamma}$, and vice versa.

Returning to the Reiner–Rivlin fluid model (4.1), computation of the response (4.1) in simple shearing ($\mathbf{v} = \dot{\gamma}y\mathbf{i}$) shows that

(a) the shear stress component can be modelled exactly because η is an arbitrary function of $\dot{\gamma}$.

(b) N_2, ($\equiv \sigma_{yy} - \sigma_{zz}$) can be modelled exactly as Ψ_2 is an arbitrary function of $\dot{\gamma}$;

(c) $N_1 = 0$. This is contrary to experimental evidence (Fig. 3.3).

Because of the failure to model the dominant normal stress difference, it is not possible to use this equation in any flow where this factor is important. It is also not a realistic description of the transient response of polymer fluids since when motion ceases τ_{ij} immediately vanishes which is contrary to observation. In fact, the only flow which is described accurately by this equation is a steady elongational flow. It is nevertheless often used when the flow is dominated by a variable viscosity. The function Ψ_2 is then usually set equal to zero and the viscosity taken as a function of the second invariant of d_{ij} only. One is then back to a simple non-Newtonian fluid, often called a generalized Newtonian fluid, in which the viscosity varies with the shear rate. The only flows of polymer fluids which are strictly suitable for modelling in this way are those which are very close to steady viscometric; for fluids where elastic effects are minimal it is more generally useful. It would be futile to attempt to catalogue all solutions to flow problems which use this limited and atypical model of fluid behaviour. Most often the viscosity function is described by a 'power-law' behaviour; for a selection of the literature see Truesdell and Noll (1965), p. 479, and the book by Bird *et al.* (1977), Chapter 5. Many solutions can be generated for the 'power-law' case where

$$\sigma_{ij} = -p\delta_{ij} + 2k|\mathrm{II}_d|^{(n-1)/2}d_{ij} \tag{4.4}$$

and k and n are constants. In this case there is no characteristic time in the constitutive model and it is possible to look for similarity solutions of the equations. The creeping flow into a cone where streamlines are straight is an example of a similarity solution. However, it must be reiterated that the inelastic fluid model often omits important phenomena, and hence it is not reasonable to devote great efforts to the analytical solution of this system of equations. Numerical solution is readily available if required using several commercially available codes, and eqn (4.4) is frequently used in practice. Some examples of solutions are given in Chapter 8.

The techniques used in Chapter 1 for deriving the many purely viscous models can be extended to three-dimensional flows; an interesting attempt to analyse shear and elongational behaviour using a Reiner–Rivlin model has been given by Debbaut and Crochet (1988). Except for occasional problems in securing numerical convergence in extreme shear-thinning cases, the solution of purely viscous fluid flows is now essentially routine.

4.2 Materials with a yield stress

Many materials appear to show a yield stress. In them no detectable flow arises until the local stresses exceed a critical value. Some authors have denied that a true yield stress exists, but here we take the view that such a model is often of great utility. We have already mentioned several such models in Section 1.5.1 in a one-dimensional context.

Although the Bingham body and its power-law extension, the Herschel–Bulkley body, (Fig. 4.1) have been used since the 1920s to model some non-Newtonian behaviour, it is only recently that effective computational methods have become available to solve the resulting flow problems. Figure 4.1(a) shows the fit to a propellant dough; Fig. 4.1(b) shows the relation between the rigid-plastic, Bingham, and Herschel–Bulkley models on a linear plot of shear rate. The steep η_0 curve is part of the bi-viscosity model, discussed below.

The advent of constant-stress rheometers has made it possible to scrutinize the behaviour of materials at very low shear rates, and this has led some writers (Barnes and Walters 1985; Barnes 1992), to doubt, in large measure, that a true yield stress exists in many soft materials. The point made by Barnes and Walters is that while a yield stress (τ_y) may be extrapolated to $\dot{\gamma} = 0$ from a finite shear rate [Figs 4.2(a–c)] measurements closer to $\dot{\gamma} = 0$ may show just a variable

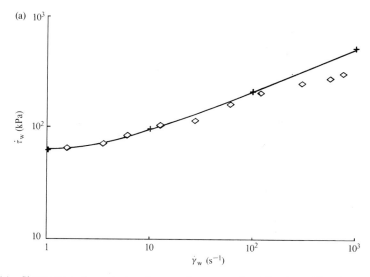

Fig. 4.1(a) Shear stress–shear strain-rate curve for a propellant dough. $\diamond$ Experimental points; + Herschel–Bulkley fit.

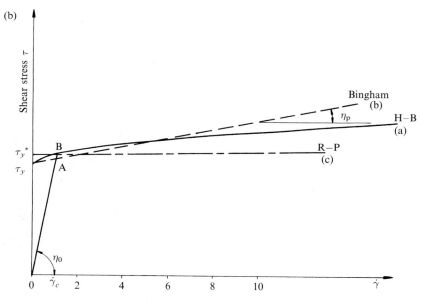

Fig. 4.1(b) $\dot{\gamma}$, τ are on a linear scale (a) Herschel–Bulkley (H–B) model $\tau = 5 + 0.5\dot{\gamma}^{0.5}$, full line. (b) Bingham model $\tau = 5 + 0.2\dot{\gamma}$, dashed line. (c) Rigid plastic (R–P) model, yield stress τ_y^* — — — — . The bi-viscosity models (with $\eta_0 = 5$) are also sketched; at $\dot{\gamma}_c$ there is a switch from stiff, Newtonian behaviour to Bingham or H–B curves.

viscosity behaviour, and no yield stress will be seen [Fig. 4.2(d)]. Barnes (1992) did later modify this view to admit that a true yield stress (with possible elastic behaviour up to the yield point) may exist in certain concentrated suspensions. Thus a variety of models is possible, as shown in Fig. 4.1, and in those models with a true yield stress (τ_y) one often assumes an elastic, viscoelastic, or rigid behaviour for $\tau < \tau_y$.

Provided one is not concerned with exceedingly low shear rates, the bi-viscosity models and the Bingham and Herschel–Bulkley models will give very similar results in simple flows, for example, Poiseuille flow (Fig. 4.3). A similar dilemma occurs in metal flow: we often think of elastic and plastic behaviour, but often a small amount of creep occurs, which is usually ignored except at higher temperatures. Creep also occurs in soft materials (Fig. 4.4) often indicating visco-elastic behaviour before yield. In the bi-viscosity models, some (viscous) creep

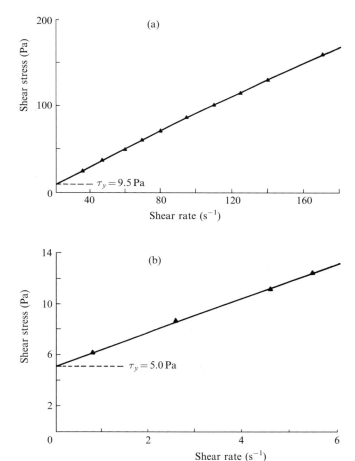

Fig. 4.2(a–b)

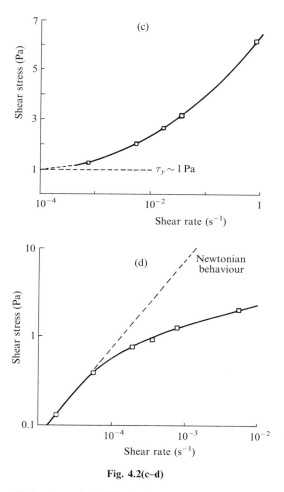

Fig. 4.2(c–d)

Fig. 4.2 Barnes and Walters's results (1985) showing the problems of defining τ_y for a PVA latex sample. As the shear rate is progressively reduced, the yield stress reduces, and finally vanishes for this sample.

occurs at all stress levels. Provided the zero-shear viscosity (η_0 in Fig. 4.1(b)) is high enough, so that the creep rates at stresses near τ_y are negligible, one might well decide to consider certain regions in the flow as effectively rigid. This means that rates of deformation (or rates of shear) less than τ_y/η_0 are assumed to be 'rigid' regions. Some care is evidently needed in using this concept for the bi-viscosity model where maximum stresses are not far from τ_y, and we shall discuss this later. Usually, η_0 is set to about 10^4 times other characteristic viscosities in the system.

In addition to the models given above, a cruder model still is the rigid-plastic model of material behaviour, Fig. 4.1(b). In this model the shear stress does not

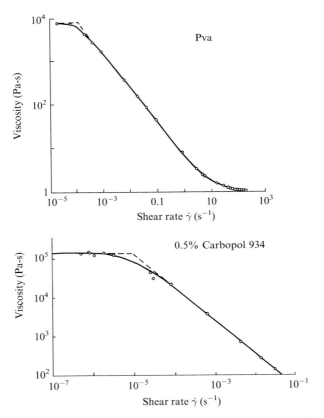

Fig. 4.3 Barnes and Walters's (1985) results showing the strict non-existence of yield stress for PVA latex and Carbopal samples. The dotted lines show biviscosity Herschel–Bulkley fits; η_0 and power-law slopes, of -0.78 and -0.87 respectively, are evident.

vary with shear rate, and the stress τ_y^* is chosen to give a best fit over rates of shear of interest. The model is most useful when small rates of shear are of interest, so that τ_y^* and τ_y are not far apart. There is a vast literature on rigid-plastic flow (for example, Hill 1950; Chakrabarty 1987) and sometimes one can use these ideas in a rheological context. In addition, the elastic-plastic model, an improvement on the rigid-plastic case, has an elastic component for small strains, with stresses not exceeding τ_y. Another, purely viscous, approach is that of Papanastasiou (1987), where a very smooth viscous stress-strain rate function is assumed. We refer the reader to the literature for details and further discussion; in this case unyielded regions are often not well defined.

So far we have only considered one-dimensional rheology. The generalization to three dimensions (we will only consider the isotropic incompressible case, implying a Poisson's ratio of 0.5 in any elastic region) is immediate, provided one

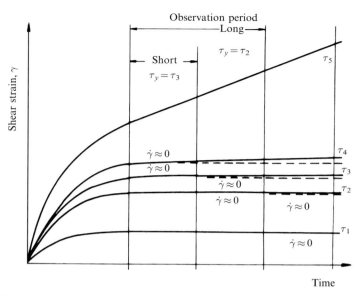

Fig. 4.4   From Cheng (1986) showing the apparent time-dependence of τ_y and sample creep. The viscoelastic response before yielding is evident.

assumes that yield occurs via a von Mises type of yield criterion. We suppose the yield stress in simple shear is τ_y, as above. The extra (or deviatoric) stress $\boldsymbol{\tau}$ in the cases considered is defined by

$$\boldsymbol{\tau} = p\mathbf{I} + \boldsymbol{\sigma}, \tag{4.5}$$

where $\boldsymbol{\sigma}$ is the total stress tensor, p is the pressure and $\mathbf{I}$ is the unit tensor. In the case of incompressible inelastic flow, the pressure p is given by

$$p = -\frac{1}{3}(\sigma_{xx} + \sigma_{yy} + \sigma_{zz}). \tag{4.6}$$

The criterion of yield is then some function of J_2 and J_3, the second and third invariants of $\boldsymbol{\tau}$ respectively. There is a considerable literature on metal yield, but rheologists have effectively always used the von Mises criterion

$$J_2 = \tau_y^2 = \frac{1}{3}Y^2 \tag{4.7}$$

as the criterion of yield, in the absence of any other information. J_2 can be written in various forms; in Cartesian components we get

$$J_2 = \frac{1}{2}\tau_{ij}\tau_{ij} = \frac{1}{3}Y^2. \tag{4.8}$$

In eqn (4.7), Y is the yield stress in simple tension, τ_y is the yield stress in shear, as before. No sufficiently accurate tests seem to be available to comment on the applicability of eqn (4.8) to most soft materials.

Besides the yield criterion, a flow rule is needed. For plasticity theory, neglecting elastic strains, the rule is

$$d_{ij} = \dot{\lambda}\tau_{ij} \qquad (4.9)$$

where $\dot{\lambda}$ is a rate parameter, to be found as part of the solution; it is not a material property. For the Bingham body, we have (Chakrabarty 1987)

$$
\begin{aligned}
d_{ij} &= 0 \qquad\qquad (J_2 \ll \tau_y^2) \\
\tau_{ij} &= 2\left\{ \frac{\tau_y}{\dot{\gamma}} + \eta_p \right\} d_{ij} \quad (\tau_y^2 > J_2)
\end{aligned}
\qquad (4.10)
$$

where $\dot{\gamma} = \sqrt{2d_{ij}d_{ij}}$, the equivalent shear rate we have defined in eqn (4.3). Hence near the yield point where the term $2\eta_p d_{ij}$ is small, the Bingham and rigid-plastic materials behave similarly [Fig. 4.1(b)]. The extension to Herschel–Bulkley behaviour follows. It is to be understood that in all cases a value of τ_y (or τ_y^*) is to be chosen so that an appropriate range of shear is covered. The various bi-viscosity fluids also have similar constitutive laws; in this case one can write the equivalent of eqn (4.10) as

$$\tau_{ij} = 2\eta(J_2)d_{ij} \quad (J_2 > \tau_y^2), \qquad (4.11)$$

where the effective viscosity is a function of J_2; for small strain rates $\eta = \eta_0$. In discussion of the bi-viscosity model, one refers to unyielded regions as being these regions where $J_2 < \tau_y^2$; the rates of deformation in these regions will be small, of order τ_y/η_0, as absolute rigidity is not expected here. In the Bingham body and similar cases, truly rigid unyielded regions can exist. We now consider these unyielded regions.

4.2.1   Rigid regions in flows with yield stress

Lipscomb and Denn (1984) said that 'Bingham fluids cannot contain yield surfaces (that is, rigid areas) in complex confined geometries like those used in molding and lubrication. Rather, yielding and flow must occur everywhere, although shear-free plug regions may sometimes be approximated'. While they had the rectification of certain common errors in mind in the paper, notably the erroneous depiction of unyielded zones found from classical lubrication type theory in squeeze-film flow (Chapter 6), it is believed that the above statement is too sweeping (O'Donovan and Tanner 1984). The authors recognized the necessity for continuity in the Bingham velocity field, which gives some constraints on kinematics. The plastic Bingham viscosity (η_p) serves to diffuse the sharp slip surface of rigid-plastic theory without altering greatly the general nature of the flow, so one may expect some general guidance from plasticity theory in the Bingham case. Lipscomb and Denn pointed out, in different language, that rigid regions may occur in flows where the stress field is known (statically determined problems); for example, in Poiseuille flow. Their analysis of flow commencement in a Bingham squeeze-film flow is, however, incorrect. There is generally no determinate stress state in a rigid body, and one cannot assume a uniform stress state is imposed before yield takes place. To avoid this problem, either an elastic (or linear viscoelastic or viscous) initial response can be assumed. In these cases a singularity occurs at the outer rim of the platens and yield begins there, moving inwards. (A similar argument for a punch penetrating a block is given by Chakrabarty (1987) in the elastic-plastic case, where yielding begins at the punch corners.) Lipscomb and Denn also discuss the relation between a true Bingham body and the bi-viscosity model. Up to the yield point, as mentioned above, an incompressible elastic solid undergoing strain and a linear viscous fluid of the same shape undergoing strain-rate, may have the same stress patterns with small deformations. Hence the stress field before yield can be established; in a strictly rigid body it is generally not known.

We now consider some other work. Sherwood et al. (1991) studied Bingham-type materials for which the yield stress varies through the sample. They used a bi-viscosity model in their analysis.

Atapattu et al. (1990) compared their experimental results with the computations of Beris et al. (1985) and others, with reasonable agreement on the extent of the yielded cavity zone.

Wilson (1993a) considered the squeeze-film flow, generally agreed with Lipscomb and Denn's conclusions, and presented further analysis. He emphasized that non-shear stress components could not be neglected in the squeeze-film problem, as is traditional in lubrication type analyses. He developed the analysis for the bi-viscosity model in terms of δ(gap/radius ratio $= h/a$) and $\varepsilon = \eta_p/\eta_0$, with the following conclusions:

When $h\eta_0/a\eta_p$ is small, a usual lubrication-type theory holds. However, the strict Bingham limit ($\eta_0 \to \infty$) shows that lubrication theory fails, and no simple theory exists.

Wilson (1993b) found unyielded regions in the flow of a Bingham body in a closing wedge flow. He claims that '...all the flows for which yield surfaces are known to exist beyond doubt are in a sense kinematically trivial; that is, they have straight or circular streamlines or a superposition, etc;' but one needs only to consider the flow over a deep slot to find a counter example to this statement. It was concluded that yield surfaces can exist in the closing wedge problem in the Bingham limit but only when the yield stress τ_y is suitably large.

Szabo and Hassager (1992) looked at the eccentric tube flow discussed by Walton and Bittleston (1991) and analysed the plug flow regions using both a bi-viscosity and a true Bingham model. The agreement in the extent of the unyielded regions using both models was excellent.

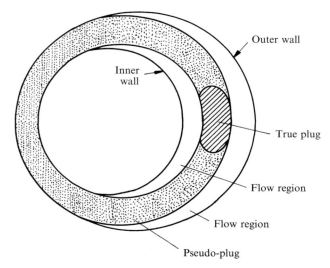

Fig. 4.5   Axial flow between eccentric cylinders showing true plug or unyielded regions and regions
of small shear rate (pseudo-plugs).

Beverly and Tanner (1992) also analysed the Walton–Bittleston flow, (Fig. 4.5)
with results comparable to those mentioned above; it is also clear that unyielded
cores will tend to vanish when the inner cylinder is rotated. Atkinson and El-Ali
(1992) considered several corner flows in a Bingham body and found unyielded
regions.

   Beris *et al.* (1985) used a true Bingham model in studying the flow around a
sphere. They showed small unyielded regions around the poles of the sphere, in
the stagnation areas.

   Axisymmetric plasticity problems are not hyperbolic (Hill 1950) and the
velocity and stress fields are continuous in these cases. If we focus on the axis of
symmetry, then the shear stress is zero, $\sigma_{\theta\theta} = \sigma_{rr}$ so $\sigma_{rr} - \sigma_{zz} = \pm\sqrt{3}\tau_y$ there.
Similar considerations can be applied to the true Bingham body; the problem is
not hyperbolic, and the velocity and stress fields are continuous. The lack of
definition of stress in a strictly rigid region continues, and we will now assume an
elastic response before yielding. If we look at the field near an axis of symmetry,
and close to a solid boundary, we find a stagnation point. Because of the no-slip
condition at the wall, the displacement $u = 0$ and $\partial u/\partial r = 0$ on the wall surface,
and hence, using the incompressibility condition, $\partial w/\partial z$ is also zero at the wall
surface. Hence the stagnation point is a point of zero deviatoric stress, and the
material is clearly unyielded there. As one moves away from this point, say along
the axis of symmetry, the stresses must change in a continuous manner, governed
in any unyielded regions by the elastic response, until the yield criterion is
exceeded. Hence one expects unyielded regions near stagnation points, and these
will be larger or smaller depending on the general stress field in the body. The
computations of Beris *et al.* (1985) with a true Bingham body support these

conclusions. In summary, for both Bingham and bi-viscosity models, one sees that there are many small regions, often near stagnation points, where no yield takes place, in addition to the known cases of cores in simpler flows like the Poiseuille flow.

4.2.2   The rigid-plastic limit

In order to help understand these flows we cite here a few rigid-plastic examples and compare them with other analyses and experiments. An important example (Hill 1950) is extrusion in plane strain (Fig. 4.6). In plastic-rigid solutions, where one assumes maximum friction at the walls, there are 'dead' or rigid zones in the corners N of the die. Similar 'dead zones' are found under the centre of a punch problem (Chakrabarty 1987).

This 'nose' is similar to that found by Beris *et al.* (1985) near the poles of a sphere in their true Bingham solution. In the extrusion flow there is slip at the wall, then a complex yielded flow near the exit, then a plug flow downstream. Beverly and Tanner (1989) found a similar behaviour at the exit from a tube in a viscoelastic case. Magnin and Piau (1992) have reported a static zone remaining in the corner of the contraction in axisymmetric experiments.

For the rigid-plastic case corresponding to a plane squeeze-film problem we can use the Prandtl solution as given by Hill (1950) and Chakrabarty (1987), Fig. 4.7. There are unyielded zones near the stagnation points at the centreline which continually lose material to the adjacent yielding zones. Near the outer edge there is also an unyielded zone, but in the axisymmetric case this edge behaviour will not take place. O'Donovan and Tanner (1984) showed a result, for

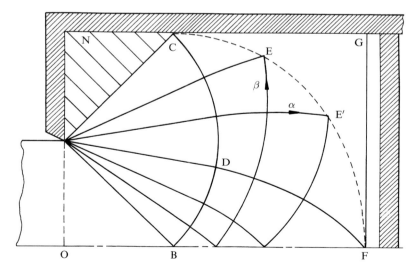

Fig. 4.6   Plane extrusion of a rigid-plastic material. The shaded area is a rigid (dead metal) zone.

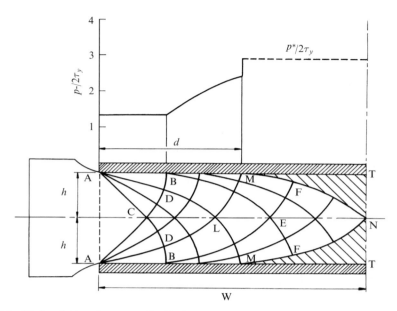

Fig. 4.7 Slipline field and pressure distribution of plane compression of a rigid-plastic material between perfectly rough platens. In this case the shaded zones are rigid; the outer zone ACA is also rigid in this plane case.

an axisymmetric geometry, of the squeeze film; this result was computed using the bi-viscosity model. The unyielded zones there (that is, $J_2 < \tau_y^2$) do not meet at the gap centre, in contrast to the rigid-plastic plane case (Fig. 4.7). This may be attributed to the broadening and diffusive effect of the post-yield viscous element in the model and to the axisymmetric geometry. For further discussion of squeeze films see Chapter 6.

Thus we are, in conclusion, able to get some guidance as to Bingham-type behaviour from the rigid-plastic case; one expects unyielded zones in regions of low stress, especially in reentrant corners like N (Fig. 4.6) and near stagnation points. However, the ability to develop slip zones in the plane strain rigid-plastic analysis means that care must be taken in these analogies when considering axisymmetric cases and Bingham materials, in which only continuous velocity fields are allowed.

So far we have regarded the material as a complicated generalized Newtonian fluid. It is possible to pursue this line of argument much further so that the 'yield' material can have many of the constitutive laws previously discussed; one simply adds the stress $2\tau_y d_{ij}/\dot{\gamma}$ to the polymeric stresses τ_{ij} derived previously. In this way one obtains a viscoelastic–plastic material description. An alternative approach to this subject has been given by White and Tanaka (1981); see also Bird *et al.* (1982) for a useful survey.

In the case of viscoelastic–plastic flow, much work remains to be done, both on the constitutive model and with computation. Early work by Beverly and Tanner (1989) shows that a dominant yield stress gives results not too different from those discussed above for inelastic fluids. Finally, we affirm the utility of the bi-viscosity model: if the yield areas are small then the model returns a stress field of a viscous nature (often the same as a linear elastic stress field); once appreciable size yield areas appear, then the flow field will be the same as the true Bingham model to within differences of order η_p/η_0, which can be made very small $[O(10^{-3}, 10^{-4})]$. Clearly, some careful judgement may be needed in interpreting results, but this is normal in numerical work. One may also, in view of the work of Barnes and Walters (1985), question whether the true Bingham model is in any case more appropriate for many real materials than a bi-viscosity model.

4.3 Oldroyd's developments

A form of the linear viscoelastic relation can be developed from a mechanical model which may be imagined as a Maxwell model [Fig. 1.16(c)] with a second viscous element in parallel. We find

$$\lambda_1 \frac{\partial \boldsymbol{\tau}}{\partial t} + \boldsymbol{\tau} = \eta_0 \left(1 + \lambda_2 \frac{\partial}{\partial t} \right) \mathbf{A}_1, \tag{4.12}$$

where $\mathbf{A}_1$ is just $2\mathbf{d}$ (see Chapter 2, eqn 2.35a).

Oldroyd (1950) realized that this (linear) model did not obey the material objectivity principle he laid down in this landmark paper, and suggested that for large strains in flows the rates of change of the stress $\boldsymbol{\tau}$ and the deformation rate $\mathbf{d}$ needed to be objective and hence non-linear. (See Section 2.5 for a short discussion of the principle of material objectivity.) As an example of the use of the principle he formulated the following generalization of (4.12) where if

$$\sigma_{ij} + p\delta_{ij} = \tau_{ij},$$

then τ_{ij} obeys the following equations

$$\begin{aligned}
\tau_{ij} + \lambda_1 &\left\{ \frac{\mathrm{D}}{\mathrm{D}t} \tau_{ij} + \frac{\partial v_m}{\partial x_i} \tau_{mj} + \frac{\partial v_m}{\partial x_j} \tau_{mi} \right\} + \frac{\mu_0}{2} \tau_{kk} A_{ij}^{(1)} \\
&- \frac{\mu_1}{2} \left(\tau_{im} A_{mj}^{(1)} + \tau_{jm} A_{mi}^{(1)} \right) + \tfrac{1}{2} a_1 \left(\tau_{km} A_{km}^{(1)} \right) \delta_{ij} \\
&= \eta_0 \left[A_{ij}^{(1)} + \lambda_2 A_{ij}^{(2)} - \mu_2 A_{ik}^{(1)} A_{kj}^{(1)} + \tfrac{1}{2} a_2 (A_{km}^{(1)} A_{km}^{(1)}) \delta_{ij} \right].
\end{aligned} \tag{4.13}$$

Here

$$\frac{\mathrm{D}}{\mathrm{D}t} \equiv \frac{\partial}{\partial t} + v_k \frac{\partial}{\partial x_k},$$

is the material derivative. In direct notation eqn (4.13) becomes

$$\boldsymbol{\tau} + \lambda_1 \left\{ \frac{D\boldsymbol{\tau}}{Dt} + \mathbf{L}^T \boldsymbol{\tau} + \boldsymbol{\tau}\mathbf{L} \right\} + \frac{\mu_0}{2} \operatorname{tr}(\boldsymbol{\tau}\mathbf{A}_1)$$

$$- \frac{\mu_1}{2}(\boldsymbol{\tau}\mathbf{A}_1 + \mathbf{A}_1\boldsymbol{\tau}) + \frac{1}{2}a_1 \operatorname{tr}(\boldsymbol{\tau}\mathbf{A}_1)\mathbf{I}$$

$$= \eta_0 \left[\mathbf{A}_1 + \lambda_2 \mathbf{A}_2 - \mu_2 \mathbf{A}_1^2 + \tfrac{1}{2}a_2 \operatorname{tr}(\mathbf{A}_1^2)\mathbf{I} \right]. \tag{4.13a}$$

Here $\mathbf{A}_2$ is the second Rivlin–Ericksen tensor (2.39a). One can test this equation for objectivity by first noting that all terms except those multiplied by λ_1 and λ_2 are clearly objective. The tensor $\mathbf{L}$ is not objective, because we have, from (2.63)

$$\mathbf{x}' = \mathbf{c}(t) + R(t)\mathbf{x}; \quad x_i' = c_i + R_{ik}x_k. \tag{2.63}$$

Differentiating (2.63) to find the velocity one finds

$$\dot{\mathbf{x}}_i' = \mathbf{v}_i' = \dot{\mathbf{c}}_i' + R_{ik}\dot{x}_k + \dot{R}_{ik}x_k. \tag{4.14}$$

Now $L_{ij}' \equiv \partial v_i'/\partial x_j' = (\partial v_i'/\partial x_m)(\partial x_m/\partial x_j')$, so using the results $\partial x_i/\partial x_j' = R_{ji}$, one finds, using the chain rule;

$$L_{ij}' = R_{ik}L_{km}R_{jm} + \dot{R}_{ik}R_{jk} \quad \text{or} \quad \mathbf{L}' = \mathbf{R}\mathbf{L}\mathbf{R}^T + \dot{\mathbf{R}}\mathbf{R}^T, \tag{4.15}$$

which confirms that the velocity gradient tensor $\mathbf{L}$ is not objective. In the D/Dt operator, $v_k \partial/\partial x_k$ is a scalar, and hence this term is objective. The term $\partial \boldsymbol{\tau}'/\partial t$ is not objective. One has

$$\frac{\partial \boldsymbol{\tau}'}{\partial t} = \frac{\partial}{\partial t}(\mathbf{R}\boldsymbol{\tau}\mathbf{R}^T) = \dot{\mathbf{R}}\boldsymbol{\tau}\mathbf{R}^T + \mathbf{R}\frac{\partial \boldsymbol{\tau}}{\partial t}\mathbf{R}^T + \mathbf{R}\boldsymbol{\tau}\dot{\mathbf{R}}^T. \tag{4.16}$$

Taking the operator $(\partial \boldsymbol{\tau}'/\partial t) + \mathbf{L}'^T\boldsymbol{\tau}' + \boldsymbol{\tau}'\mathbf{L}'$ and putting in terms of the un-primed frame, one finds it becomes

$$\dot{\mathbf{R}}\boldsymbol{\tau}\mathbf{R}^T + \mathbf{R}\dot{\boldsymbol{\tau}}\mathbf{R}^T + \mathbf{R}\boldsymbol{\tau}\dot{\mathbf{R}}^T + \mathbf{R}\mathbf{L}^T\boldsymbol{\tau}\mathbf{R}^T + \mathbf{R}\dot{\mathbf{R}}^T\mathbf{R}\boldsymbol{\tau}\mathbf{R}^T + \mathbf{R}\boldsymbol{\tau}\mathbf{L}\mathbf{R}^T + \mathbf{R}\boldsymbol{\tau}\mathbf{R}^T\dot{\mathbf{R}}\mathbf{R}^T. \tag{4.17}$$

Now the orthogonal tensor $\mathbf{R}$ obeys $\mathbf{R}\mathbf{R}^T = \mathbf{I}$, and by differentiation, we have

$$\dot{\mathbf{R}}\mathbf{R}^T + \mathbf{R}\dot{\mathbf{R}}^T = \mathbf{0}. \tag{4.18}$$

Using this result we find that the first and third terms in (4.17) can be rewritten as $-\mathbf{R}\dot{\mathbf{R}}^T\mathbf{R}\boldsymbol{\tau}\mathbf{R}^T$ and $-\mathbf{R}\boldsymbol{\tau}\mathbf{R}^T\dot{\mathbf{R}}\mathbf{R}^T$, and they cancel out the fifth and seventh terms.

Hence one finds

$$\frac{D\tau'}{Dt} + \mathbf{L}^{T'}\tau' + \tau'\mathbf{L}' = \mathbf{R}\left[\frac{D\tau}{Dt} + \mathbf{L}^T\tau + \tau\mathbf{L}\right]\mathbf{R}^T \tag{4.19}$$

so that the bracketted quantity is objective. In a similar way one can show that $\mathbf{A}_2$ is objective, and in fact the $\mathbf{A}_n$ for all n are objective (Huilgol and Phan-Thien 1997).

There are eight constants in this version of the equation which must be found. λ_1 is the relaxation time, and λ_2 is the retardation time; these and η_0 can be found approximately from small-strain experiments. It is not difficult to compute the response in simple steady shearing. Oldroyd (1958) showed that the viscometric functions η, Ψ_1, and Ψ_2 are given by

$$\eta = \eta_0\left[\frac{1 + k_2\dot{\gamma}^2}{1 + k_1\dot{\gamma}^2}\right] \tag{4.20}$$

where k_1 and k_2 are constants; explicitly

$$k_1 = \lambda_1(\mu_1 - a_1) + \mu_0(\mu_1 - \lambda_1 - 3a_1/2) + \mu_1(\lambda_1 + a_1 - \mu_1) \tag{4.21}$$

$$k_2 = \lambda_1(\mu_2 - a_2) + \mu_0(\mu_2 - \lambda_2 - 3a_2/2) + \mu_1(\lambda_2 + a_2 - \mu_2) \tag{4.22}$$

$$\Psi_1 = 2\eta_0\lambda_1\left[\frac{\eta}{\eta_0} - \frac{\lambda_2}{\lambda_1}\right] \tag{4.23}$$

$$\Psi_2 = -(2\lambda_1 - \mu_1)\eta + (2\lambda_2 - \mu_2)\eta_0. \tag{4.24}$$

In simple shearing of real fluids, the shear stress curve usually seems to rise monotonically with shear rate. If the model is to reflect this behaviour, we must restrict the values of k_1 and k_2, and hence η_∞/η_0, so that

$$k_2 > \tfrac{1}{9}k_1, \quad \text{or} \quad \eta_\infty > \tfrac{1}{9}\eta_0.$$

This is too restrictive to describe real materials over a wide range of shear rates (cf. Fig. 1.2 where the viscosity changes a thousand fold over five decades of shear rate). The elongational response is also often not typical of real fluids (see Problem 4.3 at the end of this chapter).

Thus despite the complexity of this equation, it is only illustrative; it is not an accurate description of real material behaviour. The most popular form of this equation (termed here the upper convected Maxwell (or UCM) model) can be obtained by setting $\mu_0 = \lambda_2 = \mu_2 = a_1 = a_2 = 0$, and also $\mu_1 = 2\lambda_1$; we shall return to it below. Another popular variant of this model is the so-called Oldroyd-B model, see Section 4.3.2. Other subsets of 4.13 have been used, but are mostly in worse accord with experimental behaviour than the UCM and Oldroyd-B models. See Petrie (1979) and Tanner and Walters (1998) for further details.

No special analytical methods of solution have been developed for this set of equations. However, numerous perturbation solutions have been obtained. These are sometimes useful in showing the direction in which viscoelasticity affects flow.

4.3.1   The (upper) convected Maxwell model

In the (upper) convected Maxwell (UCM) case the equation (4.13) simplifies to the form

$$\lambda_1 \frac{\Delta \boldsymbol{\tau}}{\Delta t} + \boldsymbol{\tau} = 2\eta_0 \mathbf{d}, \tag{4.25}$$

and the (upper) convected derivative $\Delta/\Delta t$ is defined as

$$\frac{\Delta \tau_{ij}}{\Delta t} \equiv \frac{\partial \tau_{ij}}{\partial t} + v_k \frac{\partial \tau_{ij}}{\partial x_k} - L_{jk}\tau_{ki} - L_{ik}\tau_{kj}, \tag{4.26}$$

where L_{ij} is the velocity gradient tensor. Forms of $\Delta/\Delta t$ in Cartesian and cylindrical coordinates are given in the Appendix.

This equation has proved to be very popular for testing computational schemes (Chapter 8) and in many other theoretical studies but it is not very realistic for the description of the flow of many polymers. For example, one sees that $k_1 = k_2 = 0$ in (4.20) so that the shear viscosity is constant at the value η_0 and does not depend on shear rate. Also $\Psi_1 = 2\eta_0\lambda_1$ and $\Psi_2 = 0$, so the second normal stress difference is zero. The steady elongational behaviour can be computed by setting $\mathbf{v} = \dot{\varepsilon}(x\mathbf{i} - \frac{y}{2}\mathbf{j} - \frac{z}{2}\mathbf{k})$ (see Chapter 3), in which case (4.25) reduces to two equations for the non-zero stress components $\tau_{xx}, \tau_{yy}(= \tau_{zz})$

$$\tau_{xx} = 2\eta_0\dot{\varepsilon}/(1 - 2\lambda\dot{\varepsilon}); \quad \tau_{yy} = -\eta_0\dot{\varepsilon}/(1 + \dot{\varepsilon}\lambda). \tag{4.26}$$

The significant rheological quantity here in the incompressible case is the elongational viscosity $\eta_E = (\tau_{xx} - \tau_{yy})/\dot{\varepsilon}$, which is found to be

$$\eta_E = 3\eta_0/(1 + \dot{\varepsilon}\lambda)(1 - 2\lambda\dot{\varepsilon}). \tag{4.27}$$

This gives unrealistic predictions for $\dot{\varepsilon}$ values of $1/2\lambda$ and $-1/\lambda$ and values falling outside these limits.

4.3.2   The Oldroyd-B model

In this case one uses the result

$$\boldsymbol{\tau} = \boldsymbol{\tau}_1 + \boldsymbol{\tau}_2, \tag{4.28}$$

where $\boldsymbol{\tau}_1$ obeys eqn (4.25) (the UCM equation) and $\boldsymbol{\tau}_2$ is a viscous component, equal to $2\eta_s\mathbf{d}$. This is a special case of (4.13) where $\mu_1 = 2\lambda_1, \mu_2 = 2\lambda_2$, $\mu_0 = a_1 = a_2 = 0$. λ_2 is the relaxation time, the total viscosity in this case is $\eta_0 + \eta_s$, and $\lambda_2 = \lambda_1\eta_s/(\eta_s + \eta_0)$.

Hence when $\eta_s = 0$, we recover the UCM model above, and when $\eta_0 = 0$ the behaviour is Newtonian.

The disadvantages of the UCM model persist, but the Oldroyd-B model has been used to represent some (real) Boger (1977) fluids and is often used as a simple model in computations, occasionally with multiple relaxation times.

4.4 Simple and non-simple fluids

This book is mainly concerned with what Truesdell and Noll (1965) have termed 'simple' fluids. The history of this concept in relation to the pioneering work of Oldroyd (1950) is discussed by Tanner and Walters (1998). This idea of a simple fluid can be stated in various terms. One tangible consequence of having a 'simple' fluid is that in a viscometric flow the local velocity gradient governs the viscosity. Higher spatial derivatives of velocity (such as $\partial\dot{\gamma}/\partial r$ in a Poiseuille flow) are not relevant in such a flow. In terms of the microstructure, one can define a simple fluid as one in which the microstructure is small compared with the distance over which the stress state changes, and it also assumes a restriction to local interaction between particles of the fluid. The restriction to local action is important and permits us to replace the complete description of the history of all the particle motions by the history of the single descriptor $\mathbf{C}$. Reference to (2.31–32) shows that this is the quantity governing the change of distance between particles which are near the point $\mathbf{r}$ at time t' and near $\mathbf{x}$ at time t. From these considerations we find the definition of a simple (incompressible) fluid, set out below.

A 'simple' fluid sets the extra stress $\boldsymbol{\sigma} + p\mathbf{I}$ as a *functional* of the history of the strain tensor $\mathbf{C}$ defined in eqn (2.32); that is, $\boldsymbol{\sigma} + p\mathbf{I}$ depends on the entire past history of $\mathbf{C}$, or, formally

$$\boldsymbol{\sigma} + p\mathbf{I} = \underset{t'=-\infty}{\overset{t'=t}{\mathscr{F}}}[\mathbf{C}(t')]. \qquad (4.29)$$

As it stands, (4.29) can only be used to deduce certain symmetries of $\boldsymbol{\sigma}$ given the symmetry of $\mathbf{C}$; in fact, not much can be deduced that cannot be found in simpler ways. It may be shown that (4.29) is an isotropic fluid (Leigh 1968). Expansions of (4.29) have proved somewhat useful, and are discussed next.

4.5 Rivlin–Ericksen expansions

For a sufficiently smooth motion, one can replace $\mathbf{C}(t')$ in (4.29) by the Rivlin–Ericksen expansion (2.40). In the viscometric flows it so happens that the series terminates exactly; all terms $\mathbf{A}^{(3)}$ and higher vanish. In general flows this will not usually occur. We recognize that if the material has a time of memory or time constant λ, then each term of (2.40) can be rearranged in the dimensionless form $(\lambda^n/n!)\mathbf{A}^{(n)}\{(t - t')/\lambda\}^n$ and in flows where all higher terms than the first n can be neglected (or vanish), the history functional of $\mathbf{C}$ over t' (4.29) may be replaced by a function of the first n coefficients in the expansion so that

$$\boldsymbol{\sigma} + p\mathbf{I} = \frac{\eta_0}{\lambda} \mathbf{f}(\lambda \mathbf{A}_1, \lambda^2 \mathbf{A}_2, \ldots, \lambda^n \mathbf{A}_n). \tag{4.30}$$

The function $\mathbf{f}$ is made dimensionless by the factor η_0/λ. If it is expanded in powers of λ, a series of order-fluid descriptions results

$$\boldsymbol{\sigma} + p\mathbf{I} = \mathbf{0} \text{ (zero-order or Euler flow).} \tag{4.30a}$$

Here we recognize that $\mathbf{f}(0) = \mathbf{0}$, to agree with the hydrostatic assumption.

$$\boldsymbol{\sigma} + p\mathbf{I} = \eta_0 \mathbf{A}_1 \text{ (first-order or Newtonian flow).} \tag{4.30b}$$

Here the first-order expansion is made to begin with $\lambda \mathbf{A}_1$, so that η_0 has its usual meaning. The next order fluid, the second-order fluid, will contain terms in $\lambda \mathbf{A}_1$, $\lambda^2 \mathbf{A}_1^2$ and $\lambda^2 \mathbf{A}_2$. We shall write it as

$$\boldsymbol{\sigma} + p\mathbf{I} = \eta_0 \mathbf{A}_1 + (\Psi_1(0) + \Psi_2(0))\mathbf{A}_1^2 - \tfrac{1}{2}\Psi_1(0)\mathbf{A}_2, \tag{4.30c}$$

where the Ψ_1 and Ψ_2 have the meaning assigned to them in eqn (3.18); they are the first and second normal stress coefficients, and are determinable from viscometric measurements. The third-order fluid is of the form

$$\begin{aligned}
\boldsymbol{\sigma} + p\mathbf{I} = {} & \eta_0(1 + \lambda^2\phi_1 \operatorname{tr} \mathbf{A}_1^2)\mathbf{A}_1 + \{\Psi_1(0) + \Psi_2(0)\}\mathbf{A}_1^2 \\
& - \tfrac{1}{2}\Psi_1(0)\mathbf{A}_2 + \phi_2\lambda^3(\mathbf{A}_2\mathbf{A}_1 + \mathbf{A}_1\mathbf{A}_2) + \phi_3\lambda^3\mathbf{A}_3.
\end{aligned} \tag{4.30d}$$

Here the viscosity begins to change with shear rate, in contrast to the lower-order fluids. The rapid proliferation of coefficients is also noticeable.

These results were first obtained by Rivlin and Ericksen (see Pipkin 1972) who started from the assumption that the stress at a given instant is determined by the gradients of the velocity, acceleration and higher time derivatives at the same instant, and used invariance (or objectivity, see Section 2.5) arguments to show that the dependence on these gradients must be through the combinations $\mathbf{A}_n$. Under the assumption that the fluid is isotropic, they showed that the function $\mathbf{f}$ must be an isotropic function of its arguments. Instead of expanding as we did above, for dependence on $\mathbf{A}_1$ and $\mathbf{A}_2$ alone they proved that the function $\mathbf{f}$ could be expressed in the form

$$\mathbf{f} = \sum_0^2 \sum_0^2 \phi_{mn}(\mathbf{A}_1^m \mathbf{A}_2^n + \mathbf{A}_2^n \mathbf{A}_1^m), \tag{4.31}$$

in which the scalar coefficients ϕ_{mn} are functions of the traces of various products of $\mathbf{A}_1$ and $\mathbf{A}_2$. The form (4.31) is a consequence of the assumption of isotropy, and it was obtained under that assumption alone, with no other prior assumptions about the nature of the function $\mathbf{f}$, nor about the shear rate magnitude. This procedure is useful because in the viscometric flows it is easy to show that all the Rivlin–Ericksen tensors except $\mathbf{A}_1$ and $\mathbf{A}_2$ are zero and the expression (4.31) for the stress is then exactly equivalent to eqn (4.29) in so far as the dependence of the stress on the deformation history is concerned. Exact solutions of these problems (neglecting centrifugal force in the case of torsional flow), accounting for the normal stress differences explicitly were obtained. These appear to be the first exact, general solutions of viscometric flow problems. The form (4.31) is not the most convenient one, however.

At about the time of this work it was becoming usual to think of the stress in a material element as a functional of its strain history. Green and Rivlin (1957) developed this notion for finite deformations and non-linear response, and obtained canonical forms incorporating the consequences of isotropy. It was later shown that for simple fluids the stress can be expressed as an isotropic functional of the strain history $\mathbf{C}(t')$ [see eqn (2.32)] and that within the context of assumptions then currently accepted, a fluid would necessarily be isotropic (Truesdell and Noll 1965).

Ericksen, in a paper delivered in 1958 seems to have been the first to discuss viscometric flows explicitly in terms of the strain history. He also observed that in the known examples of flows for which $\mathbf{A}_n = 0$ for $n > 3$, the tensors $\mathbf{A}_1$ and $\mathbf{A}_2$ could always be put into a certain form. Criminale, Ericksen, and Filbey (1958) observed that if $\mathbf{A}_1$ and $\mathbf{A}_2$ have this form, then all of the products in eqn (4.31) can be expressed in terms of $\mathbf{I}$, $\mathbf{A}_1$, $\mathbf{A}_2$, and $\mathbf{A}_1^2$ alone. The expression for the stress then reduces to the convenient form

$$\boldsymbol{\sigma} = -p\mathbf{I} + \eta\mathbf{A}_1 - \tfrac{1}{2}\Psi_1\mathbf{A}_2 + (\Psi_1 + \Psi_2)\mathbf{A}_1^2, \tag{4.32}$$

where η, Ψ_1 and Ψ_2 are functions of

$$2\dot{\gamma}^2 = \operatorname{tr}\mathbf{A}_1^2 = \operatorname{tr}\mathbf{A}_2. \tag{4.33}$$

We observe that in viscometric flows, following Section 3.3,

$$\dot{\gamma}(\mathbf{ab} + \mathbf{ba}) = \mathbf{A}_1, \quad \dot{\gamma}^2\mathbf{bb} = \tfrac{1}{2}\mathbf{A}_2, \quad \text{and} \quad \dot{\gamma}^2\mathbf{aa} = \mathbf{A}_1^2 - \tfrac{1}{2}\mathbf{A}_2, \tag{4.34}$$

and all higher Rivlin–Ericksen tensors vanish as stated above. Consequently, the expression for the stresses that we have given in eqn (3.20) is equivalent to that given by Criminale, Ericksen, and Filbey (1958).

The work of Criminale, Ericksen, and Filbey has had effects more important than the particular result obtained. Their constitutive equation was presented as a relation applying to a special class of motions, and not as a model defining some fictitious class of materials. In order to understand the physical applicability of the mathematical models that are used to describe viscoelastic fluids, it is essential to be aware that none of these models may be applicable to all states of motion of any particular fluid. Many different forms of constitutive equations can be valid descriptions of the behaviour of one fluid, depending on the circumstances, and one form of constitutive equation may describe the behaviour of many different fluids in one particular kind of motion.

Higher Rivlin–Ericksen fluid expansions arise in non-viscometric motions. However, the number of fluid parameters rises rapidly and it has not even been possible to find all six parameters experimentally [eqn (4.30d)] for a third-order fluid. Thus although there has been a number of perturbation calculations with higher-order fluids they are of limited applicability. Even if it were possible to find the parameters needed, the range of applicability of these order fluids is too narrow to be very interesting as we shall see below, and it does not seem useful to pursue this path. Here we will concentrate on second-order representations [eqn (4.32)] for which only three functions need be found experimentally.

4.5.1 Behaviour of Rivlin–Ericksen fluids in unsteady shearing motions

To see that the second-order fluid [eqn (4.30c)] is not suitable for discussing unsteady motions consider a fluid disturbed so that $\mathbf{v} = [u(y, t), 0, 0]$. We see that continuity is satisfied. From the constitutive equation (4.30c) we find

$$\sigma_{xy} = \eta \frac{\partial u}{\partial y} - \tfrac{1}{2}\Psi_1 \frac{\partial^2 u}{\partial y \partial t}. \tag{4.35}$$

The equation of motion for shearing motions (with $\partial/\partial x = \partial/\partial z = 0$) is

$$\rho \frac{\partial u}{\partial t} = \eta \frac{\partial^2 u}{\partial y^2} - \tfrac{1}{2}\Psi_1 \frac{\partial^3 u}{\partial y^2 \partial t}. \tag{4.36}$$

Suppose we have a channel (width h) in which the walls are stationary. Let $u = \sum_{m=1}^{\infty} \alpha_m(t)\phi_m(y)$. Substituting each component in (4.36), we find

$$\rho \frac{d\alpha_m}{dt} \phi_m = \eta \alpha_m \phi_m''(y) - \tfrac{1}{2}\Psi_1 \frac{d\alpha_m}{dt} \phi_m''(y) \tag{4.37}$$

which is not directly solvable.

By letting

$$\phi_m = a_m \sin \frac{m\pi y}{h}$$

we can represent any initial velocity profile by a Fourier expansion in y. Setting this in (4.37) we find, for the mth harmonic

$$\left[\rho - \tfrac{1}{2}\Psi_1 \frac{m^2\pi^2}{h^2}\right] \frac{d\alpha_m}{dt} = -\frac{m^2\pi^2}{h^2}\eta\alpha_m. \tag{4.38}$$

Solving,

$$\alpha_m(t) = \alpha_m(0) \exp\{\eta t / (\tfrac{1}{2}\Psi_1 - \rho h^2 / m^2 \pi^2)\}. \tag{4.39}$$

If the exponent is positive, we have a very unrealistic result as the material is clearly unstable in the sense that any initial disturbance of the form $\Sigma_m a_m \sin(\pi m y)/h$ will amplify itself. In (4.39) η, Ψ_1 and $\rho h^2/m^2\pi^2$ are all (usually) positive. Therefore the exponent is positive if $\tfrac{1}{2} > (\rho h^2/m^2\pi^2\Psi_1)$. For fixed ρ, h, Ψ_1 we can always find m so that we have an instability, so that some profiles amplify; if $\rho h^2/\Psi_1 \to 0$, then all profiles amplify; if $\Psi_1 = 0$, (Navier–Stokes case) then all disturbances die away. This shows the unsuitability of the second-order-fluid for unsteady flow description. If we consider only unsteady shearing disturbances, $u = y\dot\gamma(t)$, then the expression for σ_{xy} becomes $\sigma_{xy} = \eta\dot\gamma - \tfrac{1}{2}\Psi_1(\mathrm{d}\dot\gamma/\mathrm{d}t)$. Neglecting inertia, and keeping σ_{xy} constant, we can solve for $\dot\gamma$ to find

$$\dot\gamma = \frac{\sigma_{xy}}{\eta} + A \exp(2\eta t/\Psi_1), \tag{4.40}$$

where A is a constant of integration. Clearly, eqn (4.40) is unstable. This analysis can be extended to other models of the order type, and by elementary stability criteria we find they are all unstable. Thus all unsteady flows are too 'fast' for the second-order constitutive model, and also for higher-order fluids of this type (Joseph 1981).

4.5.2 Steady elongational flow response for viscometric constitutive models

A steady elongational flow, where $\mathbf{v} = \dot\varepsilon\{\mathbf{i}x - \tfrac{1}{2}\mathbf{j}y - \tfrac{1}{2}\mathbf{k}z\}$ is a flow which is very different from the viscometric flows (Section 3.2). If we apply the flow field to the viscometric fluid model (4.32) we find, neglecting fluid inertia,

$$\sigma_{xx} - \sigma_{yy} = 3\dot\varepsilon\eta[1 + \dot\varepsilon\{\Psi_2 + \tfrac{1}{2}\Psi_1\}/\eta]. \tag{4.41}$$

For small values of $\dot\varepsilon\{\Psi_2 + \tfrac{1}{2}\Psi_1\}/\eta$, the result (4.41) is in agreement with observation. The group $(\Psi_2 + \tfrac{1}{2}\Psi_1)/\eta$ is positive, as far as present measurements show, and hence it is likely that at some negative $\dot\varepsilon$ (4.41) will predict a stress difference and strain rate of opposite sign, which is completely unacceptable. When η, Ψ_1 and Ψ_2 are constants this is inevitable, and care must be taken in attempting to apply the viscometric model to highly non-viscometric flows. The plane stretching flow, where $\mathbf{v} = \dot\varepsilon\{\mathbf{i}x - \mathbf{j}y\}$, gives the result $\sigma_{xx} - \sigma_{yy} = 4\dot\varepsilon\eta$ for the second-order model and the problem of stress reversal does not arise.

4.6 Green–Rivlin expansions

We have mentioned the failure of the second-order fluid model in unsteady flows (Section 4.5.1). Green and Rivlin (1957) suggested an alternative expansion scheme to the slow-flow expansions discussed in Section 4.5. When specialized to incompressible fluids (Pipkin 1964) the expansion becomes

$$\boldsymbol{\sigma} = -p\mathbf{I} + \int_{-\infty}^{t} \mu_1(t - t_1)\mathbf{G}(t_1) \, dt_1$$
$$+ \int_{-\infty}^{t} \int_{-\infty}^{t} \mu_{22}(t - t_1, t - t_2)\mathbf{G}(t_1)\mathbf{G}(t_2) \, dt_1 \, dt_2 + \cdots \tag{4.42}$$

where the strain tensor $\mathbf{G}$ is defined as $\mathbf{C} - \mathbf{I}$. The structure of (4.42) may be understood by referring to the discussion in Chapter 1 [Section 1.5.2(e)] where the single integral form represents the effect of strain increments added at various times t' in the past; the effect of an increment at one time has no influence on the effect of another increment at a later time. The possibility of such an effect is recognized in the double integral form which estimates the contribution to the stress from two contributions to the strain added at different times. Higher order integral terms may be interpreted in a similar way. From the experimental point of view, the memory function $\mu_1(t)$ is not difficult to find, since it is the derivative of the linear relaxation function $G(t)$. However, the kernel μ_{22} and higher kernels are not so easy to find and experimental results are rare. To begin with we shall concentrate on the single integral form. In simple shearing, where the velocity field is of the form $\mathbf{v}(\mathbf{x}, t) = \dot{\gamma}y\mathbf{i}$, where $\mathbf{x} = (x, y, z)$ and $\dot{\gamma}$ is the constant shear rate we can compute the strain tensor $\mathbf{C}$ and hence $\mathbf{G}$. We have, in the present case, from Section 2.3.6, eqn (2.41);

$$r_2 = y, \quad r_3 = z. \tag{4.43}$$

and

$$r_1 = x + \dot{\gamma}y(t' - t). \tag{4.44}$$

Using the definition of the deformation gradient $\mathbf{F}$ (2.28), we find the value of $\mathbf{C}$ as

$$\mathbf{C} \equiv \mathbf{F}^T\mathbf{F} = \begin{bmatrix} 1 & \dot{\gamma}(t' - t) & 0 \\ \dot{\gamma}(t' - t) & 1 + \dot{\gamma}^2(t' - t)^2 & 0 \\ 0 & 0 & 1 \end{bmatrix}. \tag{2.43}$$

We can now compute the viscometric functions using the first integral of (4.42). The results are, with $s = t - t'$:

$$\left.\begin{aligned} \sigma_{xy} &= -\dot{\gamma} \int_0^{\infty} s\mu_1(s) \, ds = \eta_0\dot{\gamma} \\ N_1 &= \sigma_{xx} - \sigma_{yy} = -\dot{\gamma}^2 \int_0^{\infty} s^2\mu_1(s) \, ds \\ N_2 &= \sigma_{yy} - \sigma_{zz} = -N_1 \end{aligned}\right\} \tag{4.45}$$

σ_{xy} and N_1 are positive, as they should be. However, it is usual to find $|N_2/N_1| \ll 1$ (Section 3.8.3), so the above result is not a good representation of real fluid response. Addition of the double integral corrects this problem. Clearly the unsteady shear response is at least reasonable but for steady elongational flows another problem arises. In this case we find $\mathbf{C}$ to be a diagonal tensor (see Section 2.3)

$$\mathbf{C} = \mathrm{diag}[\exp -2\dot{\varepsilon}(t - t'), \ \exp \dot{\varepsilon}(t - t'), \ \exp \dot{\varepsilon}(t - t')]. \tag{4.46}$$

Computing the stress difference of interest, we find

$$\sigma_{xx} - \sigma_{yy} = \int_0^\infty \mu_1(s)[\exp(-2\dot{\varepsilon}s) - \exp(\dot{\varepsilon}s)] \, ds. \tag{4.47}$$

In the typical case when we take $\mu_1(s)$ to be the derivative of a Maxwell form like (2.105), we find

$$\mu_1(s) = -\sum_{n=1}^{m} \frac{g_n}{\lambda_n} \exp(-s/\lambda_n). \tag{2.106}$$

It is clear that the integral (4.47) only converges for

$$-\tfrac{1}{2} < \lambda_1 \dot{\varepsilon} < 1, \tag{4.48}$$

where λ_1 is the longest relaxation time in (2.106). In practice, one can exceed these limits, and hence this model is not a good description of fast elongational flows. The multiple integral terms do nothing to alleviate this problem.

4.7   Elastic behaviour for rapid deformations

When a viscoelastic fluid is strained very rapidly, the initial response may be elastic in character, see Section 3.6.1. In this case the functional of the history of $\mathbf{C}$ [eqn (4.29)] becomes a function $\mathbf{g}$ of the strain suddenly applied, or

$$\sigma + p'\mathbf{I} = \mathbf{g}(\mathbf{C}), \tag{4.49}$$

where p' is the pressure.

Note that in this formulation the strain is computed from the stressed or deformed configuration as reference, rather than from the unstressed state as is usual in elasticity. For the present application, the formulation (4.49) is convenient. We can expand $\mathbf{g}$ using the result for strains corresponding to eqn (2.15) to rewrite (4.49) in the form

$$\sigma + p'\mathbf{I} = g_0\mathbf{I} + g_1\mathbf{C} + g_2\mathbf{C}^2, \tag{4.50}$$

where g_0, g_1 and g_2 are functions of the strain invariants I_c and II_c; III_c is just unity for the incompressible materials considered here. From the corresponding

Cayley–Hamilton result (2.13b), after premultiplying by $\mathbf{C}^{-1}$, we obtain

$$\mathbf{C}^2 = \mathrm{I}_c\mathbf{C} - \mathrm{II}_c\mathbf{I} + \mathbf{C}^{-1}, \tag{4.51}$$

so that (4.50) can be replaced by

$$\boldsymbol{\sigma} + p\mathbf{I} = (g_1 + g_2\mathrm{I}_c)\mathbf{C} + g_2\mathbf{C}^{-1}, \tag{4.52}$$

where the isotropic stress terms have been incorporated in the pressure. Finally, we note that in incompressible materials the second invariant of $\mathbf{C}(\mathrm{II}_c)$ is equal in magnitude to the first invariant of $\mathbf{C}^{-1}(\mathrm{I}_{c-1})$, and (4.52) can be put in the convenient form

$$\boldsymbol{\sigma} = -p\mathbf{I} + h_1(\mathrm{I}_c, \mathrm{I}_{c^{-1}})\mathbf{C} + h_{-1}(\mathrm{I}_c, \mathrm{I}_{c^{-1}})\mathbf{C}^{-1}. \tag{4.53}$$

The functions h_1 and h_{-1} are connected to a potential in elasticity; we have $h_1 = -\partial W/\partial \mathrm{I}_c$ and $h_{-1} = \partial W/\partial \mathrm{I}_{c^{-1}}$ where $W = W(\mathrm{I}_c, \mathrm{I}_{c^{-1}})$ is the strain-energy function. It is, of course, not expected that this constitutive relation will be generally useful in flow problems, but it does correspond to the very fast deformations on the right-hand boundary of the Pipkin diagram, Fig. 3.9. Care must be taken with the use of these ideas.

Example: shear deformation of a Mooney material
One of the simplest models of ideal rubberlike behaviour is the Mooney material (Ogden 1984) where h_1 and h_{-1} in eqn (4.53) are constants. We consider a deformation where the amount of shear is γ. Then comparison with eqn (2.43) shows that the $\mathbf{C}$ tensor has components $C_{11} = C_{33} = 1$, $C_{22} = 1 + \gamma^2$, $C_{12} = C_{21} = -\gamma$, and the rest are zero. The tensor $\mathbf{C}^{-1}$ has $C_{11}^{-1} = 1 + \gamma^2$, $C_{12}^{-1} = \gamma$, with the rest zero or one. The homogeneous stress state satisfies equilibrium; mass is conserved.

One finds $\sigma_{12} = \tau = G\gamma$, $\sigma_{11} - \sigma_{22} = N_1 = G\gamma^2$, where the shear modulus $G = h_{-1} - h_1$, and $N_2 = h_1\gamma^2$. Because $h_1 = -\partial W/\partial \mathrm{I}_c$, N_2 is negative. Thus typical normal stresses, N_1 positive, N_2 negative, and a constant shear modulus are found for the Mooney model.

If a polymer solution containing a small-molecule solvent is considered, then the solvent will generally respond in a viscous manner to any signal separately from the elastic component due to the dissolved polymer.

Equation (4.53) does not contain this viscous component, which will dominate the initial response. There are many results for simple deformations in non-linear elasticity recorded in the literature; see Truesdell and Noll (1965) and Ogden (1984).

For rubbers there is evidence (Roland 1989) that the picture of ideal rubber behaviour indicated in (4.53) is not accurate for high-speed deformations, and quite slow deformation is needed to get to the ideal form. With polymer melts also, as we will see below, (4.53) is too 'elastic' to describe materials in many cases.

4.8   The Kaye–Bernstein–Kearsley–Zapas model

Kaye (1962) and Bernstein, Kearsley, and Zapas (1963) independently proposed a constitutive model which appears to have been inspired by the theory of rubberlike elasticity. When specialized for the incompressible case this constitutive model is

$$\sigma = -p\mathbf{I} + \int_{-\infty}^{t} \left\{ \frac{\partial U}{\partial \mathbf{I}_{c^{-1}}} \mathbf{C}^{-1} - \frac{\partial U}{\partial \mathbf{I}_c} \mathbf{C} \right\} dt'. \tag{4.54}$$

Here $\mathbf{C}^{-1}$ is the inverse of the strain tensor $\mathbf{C}$ and $\mathbf{I}_{c^{-1}}$ and $\mathbf{I}_c$ are the traces (first invariants) of these tensors. The potential U is a function of various quantities.

$$U = U(\mathbf{I}_c, \mathbf{I}_{c^{-1}}, t - t'). \tag{4.55}$$

The bracketted form in (4.54) is the response for an incompressible elastic solid to *strain* (Section 4.7), provided the time-dependence in U is suppressed. Thus (4.54) bears the same relation to finite deformation elasticity as linear viscoelasticity theory bears to classical elasticity.

This form of constitutive model gives fairly good results for all of the flows discussed in Chapter 3. Since sudden shear strain applications are a severe test of the equation and provide a convenient way of finding out something about the U-function, we shall consider these in detail. We will initially be concerned with the shear stress response. It can be shown that if the magnitude of the step of shear strain is γ, and the strain is applied at time zero, then the shear components of $\mathbf{C}$ and $\mathbf{C}^{-1}$ are $-\gamma$ and γ respectively, and $\mathbf{I}_{c^{-1}} = \mathbf{I}_c = 3 + \gamma^2$, when $t' < 0$; for $t > t' > 0$ the shear strains are zero. Hence

$$\tau = \int_{-\infty}^{0} \gamma \left[\frac{\partial U}{\partial \mathbf{I}_{c^{-1}}} + \frac{\partial U}{\partial \mathbf{I}_c} \right] dt', \tag{4.56}$$

where we have denoted σ_{xy} by τ. Considering the square bracket, it is only a function of γ^2 and $t - t'$, and if we denote $t - t'$ by s, and call the integrand in (4.56) γK^+ for short, then

$$\tau(\gamma, t) = \int_{t}^{\infty} \gamma K^+ (\gamma^2, s) \, ds, \tag{4.57}$$

where K^+ is a function of s and γ^2. Differentiation with respect to t gives the result

$$-\gamma K^+ (\gamma^2, t) = \dot{\tau}(\gamma, t). \tag{4.58}$$

Thus K^+ can be determined from a series of stress relaxation experiments at different strains; at small enough strains K^+ is $-dG/dt$, where G is the linear relaxation function. Einaga *et al.* (1971) have done tests of this kind (Fig. 4.8) and find that for polystyrene K^+ can be factorized into a time-dependent and a

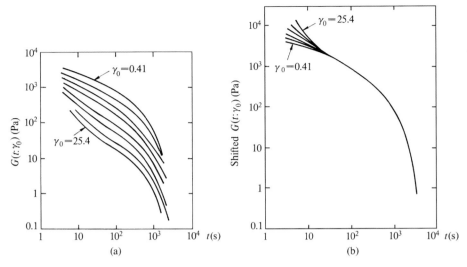

Fig. 4.8 The stress relaxation modulus $[G(t:\gamma_0)]$ for 20 per cent polystyrene (narrow distribution $M_w = 1.8 \times 10^6$) in Aroclor. In (a) the value of γ_0, the step strain magnitude, increases as one moves down from the top curve ($\gamma_0 = 0.41$ and 1.87) to the bottom curve ($\gamma_0 = 25.4$). The intermediate values are $\gamma_0 = 3.34$, 5.22, 6.68, 10.0, 13.4, and 18.7 respectively. In (b) the data are superposed by vertical shifting to show the similarity of $G(t:\gamma_0)$ at large times irrespective of the magnitude of γ_0. Note that $\gamma_0 = 25.4$ is now the highest curve at the left.

strain-dependent part. Hence we find, approximately

$$\tau(t, \gamma) = \gamma f(|\gamma|)G(t). \tag{4.59}$$

In this case G is the linear relaxation function and f is the non-linear factor. The value of $f(\gamma)$ is given for a polyisobutylene solution in Fig. 4.9.

Now consider a two-step strain as in Fig. 4.10. γ_2 may be smaller or larger than γ_1 but in either case the position $\mathbf{r}$ of the particles in the shear direction, relative to the current particle position $\mathbf{x}$, is given by $r_2 = y$, $r_3 = z$, as before, and

$$
\left.
\begin{array}{ll}
\left.
\begin{array}{ll}
r_1 = x - \gamma_1 y, & t' < 0 \\
r_1 = x, & t' > 0
\end{array}
\right\} t_1 > t > 0 \\[2mm]
\left.
\begin{array}{ll}
r_1 = x - (\gamma_2 - \gamma_1)y, & t_1 > t' > 0 \\
r_1 = x - \gamma_2 y, & t' < 0 \\
r_1 = x, & t' > t_1
\end{array}
\right\} t > t_1
\end{array}
\right\}. \tag{4.60}
$$

Hence, the strains can be computed as γ_1, 0, $\gamma_2 - \gamma_1$, γ_2, 0 in the five zones listed in (4.60). For $0 < t < t_1$, the response is the same as for a single step; if we denote the shear stress response at time t to a step γ_s imposed at time t_s by $\tau(\gamma_s, t - t_s)$,

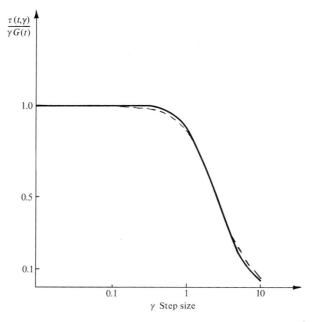

Fig. 4.9   Non-linear part of KBKZ kernel from single step strain data. ---- $[1 + \gamma^2/6.4]$ eqn (4.82). —— Experimental results for polyisobutylene solution. Note linear portion for γ less than about 0.5.

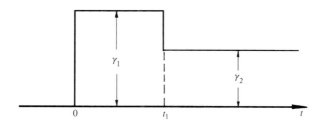

Fig. 4.10   Double-step strain.

then the response in the present case, for $0 < t < t_1$, will be denoted by $\tau(\gamma_1, t)$. For times greater than t_1, we have

$$\tau(t) = \int_t^\infty \gamma_2 K^+(\gamma_2^2, s)\, \mathrm{d}s + \int_{t-t_1}^t (\gamma_2 - \gamma_1) K^+[(\gamma_2 - \gamma_1)^2, s]\, \mathrm{d}s. \qquad (4.61)$$

From (4.57) we see that (4.61) can be put in the form

$$\tau(t) = \tau(\gamma_2, t) + \tau(\gamma_2 - \gamma_1, t - t_1) - \tau(\gamma_2 - \gamma_1, t). \qquad (4.62)$$

Note that this reduces to the linear viscoelastic result $\gamma_1 G(t) + (\gamma_2 - \gamma_1)G(t - t_1)$ for small strains. A series of experiments on single steps can be used to find the $\tau(\gamma, t)$ function, and then one can test (4.62). Zapas (1974) and Einaga *et al.* (1971) have carried out such test sequences for two cases; in the first case a small step ($\gamma_1 = 0.532$) was followed by a larger step ($\gamma_2 = 1.77$), after a delay of 2.8 s. The experimental agreement in this case was good. However, in the second case, where $\gamma_1 = 1.77$ and $\gamma_2 = 0.85$ ($t_1 = 1.62$ s), the agreement is apparently poor, the sign being incorrect for some of the time. Nevertheless, this theory is certainly a much more accurate predictor of material behaviour than equations discussed previously. The problem of specifying the kernel U in flows more general than shear flows is also addressed below.

The need for considering a 'potential' U follows from the observation that for any set of sudden strains which returns the sample to its original shape, we must avoid the production of work. This is guaranteed by the introduction of U, since the model is then rubberlike in rapid deformations (Section 4.7).

There have also been similar proposals to the KBKZ theory from Tanner and Simmons (1967) who discussed a 'network-rupture' theory. It is, however, best viewed from the microstructural viewpoint (Chapter 5).

Another approach to the KBKZ model is to go back to the functional of **C** (4.29) which, as far as is known, is a sufficiently general description of the isotropic materials being considered, and note that the Green–Rivlin expansion (4.42) single integral results are not very satisfactory due to the form of strain chosen. Instead of considering a functional of **C** and then expanding in integral form, one can consider any other strain measure, which is a function of **C** [say **H(C)**] and then generate an integral expansion. If **H(C)** is an isotropic form, then by using similar results to (4.49)–(4.53) we can find the result, ignoring double and higher integral terms

$$\sigma + p\mathbf{I} = \int_{-\infty}^{t} \mu(t - t')\mathbf{H}(t')\,\mathrm{d}t', \qquad (4.63)$$

where

$$\mathbf{H} = h_1(\mathbf{I}_c, \mathbf{I}_{c^{-1}})\mathbf{C} + h_{-1}(\mathbf{I}_c, \mathbf{I}_{c^{-1}})\mathbf{C}^{-1} \qquad (4.64)$$

is the new strain measure, chosen to be optimal in respect of a certain set of tests. This form of equation was first proposed by White and Tokita (1967), without the time-strain separation implied in (4.63). When h_1 and h_{-1} are chosen so that there is an elastic potential we find that eqn (4.63) is a special case of the KBKZ model, eqn (4.54), with separated time and strain parts in the kernels; this behaviour is suggested by the results of Einaga *et al.* (1971), (Fig. 4.8).

4.8.1   Elongational flows and KBKZ-type models

Wagner *et al.* (1998) have fitted the results of uniaxial ($m = -0.5$) biaxial ($m = 1$) and planar extensional ($m = 0$) steady flows [see eqn (3.95) for the definition of m]

using an equation of the form

$$\sigma(t) = -p\mathbf{I} + \int_{-\infty}^{t} \mu(t - t')\mathbf{H}(t')\,dt' \tag{4.65}$$

where $\mu(t) = -dG(t')/dt$ and $G(t')$ is the linear relaxation function [eqn (1.24)]. The form of the strain tensor $\mathbf{H}$ is taken as being of the form (4.64). When $h_1 = 0$ and $h_{-1} = 1$ we get Lodge's (1964) rubber-like model. In the class of extensional flows being considered, two normal stress differences can be measured

$$\sigma_1(t) = \sigma_{11} - \sigma_{33} = \int_{-\infty}^{t} \mu(t - t')h_1^*(C_{11}^{-1} - C_{33}^{-1})\,dt'$$

$$\sigma_2(t) = \sigma_{22} - \sigma_{33} = \int_{-\infty}^{t} \mu(t - t')h_2^*(C_{22}^{-1} - C_{33}^{-1})\,dt'. \tag{4.66}$$

Wagner *et al.* (1998) discuss the choice of the h_1^* and h_2^* functions to fit available data for high-density-polyethylene (HDPE). (The paper does not deal with simple shearing.) If we focus on the simpler case of start-up of uniaxial elongation, so $\sigma_2 = h_2 = 0$, then

$$\sigma_1(t) = \int_{-\infty}^{t} \mu(t - t')h_1^*(\exp 2\varepsilon - \exp(-\varepsilon))\,dt', \tag{4.66a}$$

where $\varepsilon = \dot{\varepsilon}_0(t - t')$. The 'damping' function h_1^* is a function of ε only, since

$$I_{c^{-1}} = \exp 2\varepsilon + 2\exp(-\varepsilon)$$

and

$$I_c = \exp(-2\varepsilon) + 2\exp\varepsilon.$$

The data for LDPE (Wagner 1979) were fitted by setting h_1^* (a function of ε) so that

$$h_1^*(\varepsilon) = [0.0025\exp 2\varepsilon + 0.9975\exp 0.3\varepsilon]^{-1}, \tag{4.67}$$

and eight time-constants of the form [eqn (2.105)] were used to represent $\mu(t)$. While the fitting for these (transient) elongational flows was successful at strain rates of $0.1\,\text{s}^{-1}$, there were problems in the recoil of these models, as described below.

4.8.2 Recoil after elongation

In a recoil experiment, the stress is released at a certain time, usually by cutting the tensile specimen with a knife. If the material is subjected to a strain history from a time $t = 0$ up to a time t_0, at which time the tensile stress σ is set equal to zero, then the recoverable strain can be computed.

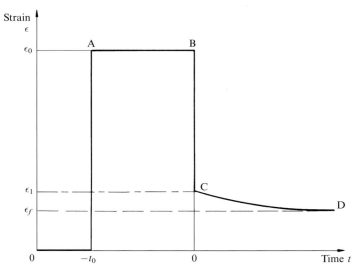

Fig. 4.11 Elongational recoil history.

As an illustration, we will use the history investigated by Zdilar and Tanner (1992), Fig. 4.11. Here a tensile specimen is extended rapidly to a strain ε_0 at $t = -t_0$, held at a constant strain until $t = 0$, and then cut, releasing the stress. Typically, one sees an immediate elastic contraction, followed by a slower return to a final strain ε_f after a long time.

If we were dealing with a linear model, then the form of the curve depends only on $G(t)$. For a Maxwell model with a single relaxation time [Fig. 1.16(c)] the stress history can be found readily. Let the model for the tensile stress σ be described by

$$\lambda \frac{d\sigma}{dt} + \sigma = 2\eta_0 \dot{\varepsilon}, \tag{4.68}$$

where η_0 is the constant viscosity, equal to $G_0\lambda$, and λ is the relaxation time. Application of a sudden strain ε_0 at $t = -t_0$ will cause a stress $G_0\varepsilon_0$ in the system, which then decays so that the stress is equal to $G_0\varepsilon_0 \exp(-t_0/\lambda)$ when $t = 0$. A sudden release of stress means that the specimen must suddenly retract by an amount $\varepsilon_0 \exp(-t_0/\lambda)$, reaching point C in Fig. 4.11, and thereafter stays at the same length, so that

$$\varepsilon_f = \varepsilon_0(1 - \exp(-t_0/\lambda)). \tag{4.69}$$

For two relaxation times λ_1 and λ_2 in parallel we have the stress at point A as $(G_1 + G_2)\varepsilon_0$ where G_1 and G_2 are the rigidities associated with λ_1 and λ_2 respectively. At point B the total stress becomes

$$\sigma(B) = \varepsilon_0(G_1 e^{-t_0/\lambda_1} + G_2 e^{-t_0/\lambda_2}). \tag{4.70}$$

Sudden removal of stress again brings one to point C, with a strain recoil of $\sigma(B)/(G_1 + G_2)$. However, this leaves one spring-dashpot combination in tension and the other in compression, so that a continued creep occurs from C to D while both parts of the model reduce their stresses to zero. To see this, if one forms the equation for the total stress σ in terms of the strain-rate $\dot{\varepsilon}$, one finds, since for $t > 0$ $\sigma = \dot{\sigma} = \ddot{\sigma} = 0$, as an equation for ε:

$$\left[\eta_1 \left(\lambda_2 \frac{\mathrm{d}}{\mathrm{d}t} + 1 \right) + \eta_2 \left(\lambda_1 \frac{\mathrm{d}}{\mathrm{d}t} + 1 \right) \right] \dot{\varepsilon} = 0. \tag{4.71}$$

(The r.h.s. of (4.71) is in general equal to $(\lambda_1(\mathrm{d}/\mathrm{d}t) + 1)(\lambda_2(\mathrm{d}/\mathrm{d}t) + 1)\sigma$, but here σ and its derivatives are zero.) The general solution of (4.71) is $(t > 0)$

$$\varepsilon(t) = A + B\exp(-t/\lambda_3), \tag{4.72}$$

where $\lambda_3 = (\lambda_1\eta_2 + \lambda_2\eta_1)/(\eta_1 + \eta_2)$. By considering the state of the system just after $t = 0$, and evaluating $\dot{\varepsilon}(0^+)$ and $\varepsilon(0^+)$ one finds

$$A = \varepsilon_0 - \varepsilon_1 - B$$

and

$$B = \frac{\varepsilon_0 G_1 G_2 (\lambda_1 - \lambda_2)}{(G_1 + G_2)(\eta_1 + \eta_2)} \{ e^{-t_0/\lambda_1} - e^{-t_0/\lambda_2} \}. \tag{4.73}$$

where $\eta_1 = G_1\lambda_1, \eta_2 = G_2\lambda_2$. The value of ε_1 is given by [eqn. (4.70)] as $\sigma(B)/(G_1 + G_2)$. B is positive, so the final decay curve is as sketched in Fig. 4.11. Zdilar and Tanner (1992) worked with PVC samples, and approximating their material by two relaxation times (10 s, 1s) and letting $G_1 = G_2 = 0.2$ MPa, t_0 (hold time) $= 5$ s, one finds, for a unit input strain ε_0, B $= 0.245$, A $= 0.448$. Hence the recoil is ultimately about 45 per cent of the initial strain. Other non-linear models pose complex computing problems for solution and are discussed by Zdilar and Tanner (1992). In the non-linear regime, it was found that the (K)BKZ model, otherwise the best model, was too elastic, and predicted too much recoil for both LDPE and PVC. The same conclusion had been reached much earlier, following a uniform strain rate extension, by Wagner (1979), see Figure 4.12, and in fact all of the equations discussed in Sections 4.3, and 4.6–4.8 are simply too elastic to predict recoil accurately.

4.9   Wagner's irreversible model

To combat this defect, Wagner (1979) introduced his irreversible model. If one writes (4.66a) with the 'damping function' h_1^* as a function of the invariants of $\mathbf{C}$ and $\mathbf{C}^{-1}$, then the values of these invariants vary during the motion. If the strain invariants are increasing, then h_1^* decreases in both the elastic KBKZ model and in the irreversible equation. However, if the strain measures ultimately decrease,

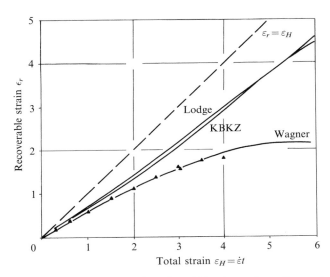

Fig. 4.12   Recoverable strain ε_r as a function of total Hencky strain ε_H for constant strain rate tests with elongation rate $\dot{\varepsilon} = 0.1$: ▲ Experimental (Meissner 1971). Material: LDPE at $150\,^\circ$C. The calculated response of the Lodge, KBKZ and Wagner models is shown. The latter agrees closely with experiment.

then h_1^* increases in the elastic model, while the irreversible model simply maintains h_1^* at its minimum value in the time interval of interest. Applied to recoil, one sees (Fig. 4.12) a dramatic difference in the recovery with the two models. This model thus seems to be suitable for polymer flows, and research continues to be done to broaden its applicability (Wagner *et al.* 1998).

4.9.1   Double-step shear strains

The KBKZ theory is exact in respect of the single-step response and Wagner's (1978) approach improves the double-step response.

We shall consider the response of a (White–Tokita) fluid of the form

$$\boldsymbol{\sigma} + p\mathbf{I} = \int_{-\infty}^{t} \mu(t - t')H(\mathbf{I}_{c^{-1}}, \mathbf{I}_c)\mathbf{C}^{-1}(t')\,\mathrm{d}t', \tag{4.74}$$

which is a special case of (4.54) if one abandons the potential U and simply sets $\partial U/\partial I_{c^{-1}} = \mu H, \partial U/\partial I_c = 0$ in that equation. If we now consider a shear flow where the amount of shear γ is known, then $\mathbf{I}_{c^{-1}}$ and $\mathbf{I}_c$ are simply functions of $\gamma^2(t, t')$. Then (4.74) becomes, for the shear stress $\sigma_{xy}(\equiv \tau)$

$$\tau = \int_{-\infty}^{t} \mu(t - t')\gamma H(\gamma^2)\,\mathrm{d}t'. \tag{4.75}$$

H is regarded as a functional of $\gamma(t, t')$; when $H = 1$ we have the Lodge model, and when H is a function of γ^2 we have a model that has been used extensively with reasonable success many times. To overcome the deficiencies in the double-step response Wagner's irreversible model is used. Here, we recall,

(a) When γ^2 is increasing in time we let $H(\gamma^2) = h(\gamma^2)$, an ordinary (decreasing) function of γ^2. This reflects the increasing loss of network junctions with strain.

(b) When γ is constant, then we still have $H = h$.

(c) When γ^2 decreasing, we do not suppose H is still the function $h(\gamma^2)$.

Rather, it is supposed that junctions lost on a previous increasing period for γ remain lost, and hence $H = \min\{h(\gamma^2)\}$ over the relevant period.

We can now look at the double-step strain for this model. Equation (4.60) gives the displacements and the strains are listed below it. The shear stress results for the KBKZ model ($H = h$ always) are, for the double-step strain shown in Fig. 4.10,

$$\tau = \gamma_2 h(S^2)G(t) + (\gamma_2 - \gamma_1)h((\gamma_2 - \gamma_1)^2)[G(t - t_1) - G(t)]. \tag{4.76}$$

Here $G(t)$ is the linear relaxation function, in the KBKZ model we set $S^2 = \gamma_2^2$, and in the Wagner-modified model $S^2 = \gamma_1^2$. To see what difference is made by this modification use has been made of the results of Zapas (1974) on a poly-isobutylene solution. The relaxation function G of the Zapas material was approximated by the function

$$G(t) = 708e^{-10t} + 600e^{-t} + 63.0e^{-t/10} + 0.3e^{-t/100}, \tag{4.77}$$

which is adequate to demonstrate the point at issue. Similarly, h was approximated as

$$h = 1 - 0.085\gamma^2 \quad (|\gamma| < 2). \tag{4.78}$$

Using these results the complete response curve was computed. Figure 4.13(a) shows that the global response of the two models differs little. The response starting at about $t = 3.0$ s has been published by Zapas (1974). An enlarged view of this experimentally determined region is compared to the present model in Fig. 4.13(b). We see that the experimental curve is much closer to the Wagner than to the KBKZ result, and that the Wagner device pushes the response curve lower for t greater than 1.62 s.

It does not, however, help to explain the discrepancies with two large increasing steps noted by Osaki et al. (1981). However, even these very large steps are represented quite well by the KBKZ equation, and the results are of acceptable accuracy for many purposes. Wagner (1979) has also shown as we saw in the previous section, that the KBKZ type of equation, as modified using the irreversibility idea, can predict accurately the results of various

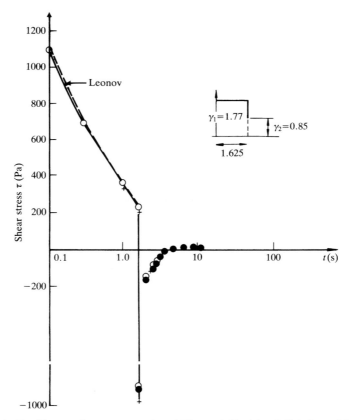

Fig. 4.13(a) Double-step shear response ($\gamma_1 = 1.77$, $\gamma_2 = 0.85$, delay 1.62 s) for polyisobutylene: ● Wagner modification of KBKZ; ○ KBKZ relation; + Leonov relation.

constant–elongation-rate, constant tensile stress, and constant tensile force experiments. Much detail for other tests has also been published by Wagner (1977). Therefore, it is concluded that a constitutive model of the form

$$\sigma(t) = -p\mathbf{I} \int_{-\infty}^{t} \left\{ \begin{array}{l} \mu_1(t - t', \mathbf{I}_{c^{-1}}, \mathbf{I}_c)\mathbf{C}^{-1}(t') \\[2mm] + \mu_2(t - t', \mathbf{I}_{c^{-1}}, \mathbf{I}_c)\mathbf{C}(t') \end{array} \right\} dt' \tag{4.79}$$

will ordinarily be a fairly accurate description of polymer melt and concentrated solution behaviour; often μ_2 (which governs the small second normal stress difference in simple shearing) may be omitted. μ_1 and μ_2 should be derivable from a potential (Larson and Monroe 1987). Further simplifications obtained by splitting the kernel into time- and strain-dependent parts are possible. It should

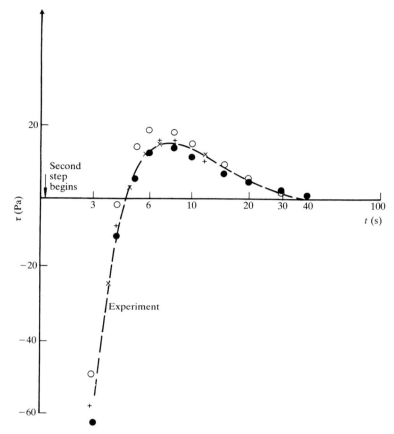

Fig. 4.13(b)   Double-step shear response for polyisobutylene. Same data and symbols as in Fig. 4.13(a). Experimental data (×) from Zapas (1974).

not be forgotten, however, that this model is not extremely accurate for multiple-step strains (Osaki *et al.* 1981).

Equation (4.77) has now been tested over most of the full range of $I_{c^{-1}}$, I_c space. Currie (1982) shows that in order to cover the entire invariant space $(I_c, I_{c^{-1}})$ one needs to perform elongational tests for both positive and negative elongational rates; shear and (positive) elongation alone are insufficient. There are few data for negative rates (biaxial testing); see Larson and Monroe (1987).

A useful tabulation by Laun (1978) and Wagner (1979) of the relaxation function $\mu(t)$ [eqn (4.74)] for a low-density polyethylene similar to the IUPAC sample 'A' described in Section 3.8 is shown in Table 2.1. Figure 4.14 shows the function $G(t)$ graphically. The kernel function $H(\gamma^2)$ in (4.75) for shear flows has been determined to be, for the particular LDPE under discussion,

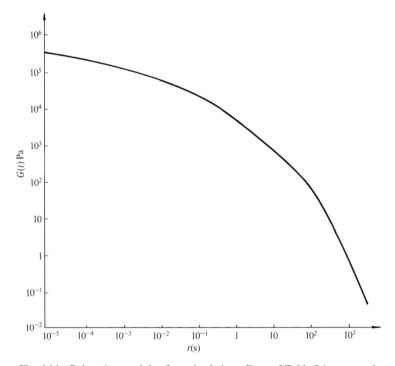

Fig. 4.14   Relaxation modulus for polyethylene. Data of Table 2.1 were used.

(Wagner and Meissner 1980)

$$H(\gamma^2) = 0.57 \exp(-0.31|\gamma|) + 0.43 \exp(-0.106|\gamma|). \tag{4.80}$$

It is difficult to write the general H-function of eqn (4.74). Wagner and Meissner (1980) have used an invariant I equal to $\alpha I_{c^{-1}} + (1 - \alpha)II_{c^{-1}}$ in their discussion of shear and elongational flows. When $\alpha = 0.032$, the H (or damping-) functions for both shear and elongation coincide; both invariants are equal in shear. The kernel is shown in Fig. 4.15.

An earlier paper (Wagner 1979) used the damping function of eqn (4.67).

Some results are shown in Fig. 4.16. Comparison is made both with experimental data and the Leonov model, which is discussed in Section 5.6.9. The shear data are fitted well, but it is clear that to fit the data for both weak and strong flows is not easy. An alternative kernel form has been explored by Papanastasiou *et al.* (1983). They write

$$H = \alpha'/(\alpha' - 3 + \beta' I_{c^{-1}} + (1 - \beta')I_c), \tag{4.81}$$

where α' and β' are constants. It appears possible to fit the Wagner (1979) data for LDPE quite well with this kernel if $\alpha' = 14.38$, $\beta' = 0.018$. For shearing flows

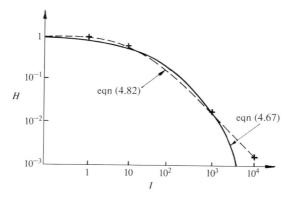

Fig. 4.15 Kernel functions of strain invariant for LPDE Melt I (similar to Sample A of Section 3.8.1). The invariant I used is $\alpha I_{c^{-1}} + (1 - \alpha)I_c$. This is equal to γ^2 in shearing flows. Solid line is eqn 4.67 with $\alpha = 0.032$. Dashed line is eqn (4.82) with $\alpha' = 14.83$ plotted for shearing flows.

one has

$$H(\gamma) = \alpha'/(\alpha' + \gamma^2).   \tag{4.82}$$

While further work on approximate kernels continues, we may point out that eqn (4.81) is more consistent with linear viscoelastic theory than is (4.80) in that it gives a departure from linearity proportional to γ^2, instead of $|\gamma|$. In Fig. 4.15 we compare the two kernels where (4.80) is for shearing only. Equation (4.82) also provides a good fit to experimental data for a polyisobutylene solution. See Larson and Monroe (1987) and Currie (1982) for further discussion on kernel forms.

The form (4.81) has been used frequently in computations, (see Chapter 8) and can provide an excellent fit to LDPE data in shearing and uniaxial extension (Luo and Tanner 1988).

4.10 Other continuum models

Following the nearly-viscometric Rivlin–Ericksen theory (Section 4.5) attempts have been made to construct nearly-elongational approximate theories. Huilgol (1979) has presented results for this case. Since there are no slip planes the theory is not so elegant as in the viscometric case and no really convenient constitutive model emerges. The number of functionals to be found experimentally is five, which is difficult. Several other ideas have appeared, mostly building on work presented above. We shall discuss those with a microstructural motivation in the next chapter; some older continuum models have been reviewed by Walters (1975), Bird *et al.* (1977), Tanner (1968), and Larson (1988). None of these is as uniformly successful as the KBKZ or the Wagner equations or as useful in special cases as the nearly-viscometric equation or as simple as some of the Oldroyd

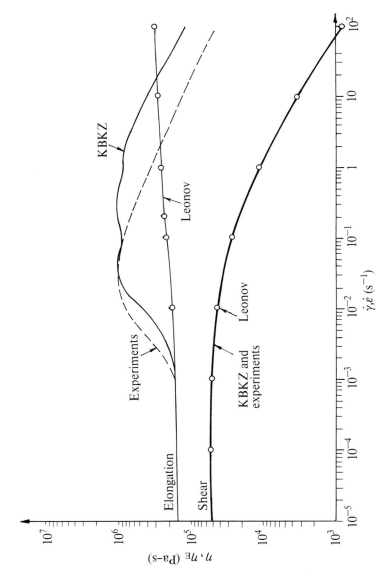

Fig. 4.16 Comparison of Leonov and **KBKZ** models with experimental data for LDPE in steady elongational and shearing flows. Relaxation data of Table 2.1 and kernel of eqn (4.81) were used for **KBKZ** calculations; the Leonov calculation also used data of Table 2.1 with 8 time-constants.

models and hence they will not be discussed. The idea of a co-rotational model, referred to a locally rotating set of axes, has occurred to a number of writers and has been given a full exposition by Bird *et al.* (1977). One case is obtained by setting $\mu_1 = \lambda_1, \mu_2 = \lambda_2$ in (4.13). There are difficulties with the response of these models especially in suddenly applied shearing (and shear) and we refer the reader to the book of Bird *et al.* (1977) for details. Some other, older models can be found in the book by Truesdell and Noll (1965); again, they do not seem to be as good an approximation as the KBKZ model and they will not be discussed.

4.11 Some applications of approximate constitutive equations to experimental measurements

In flows that are close to shearing motions it is possible to use some of the approximate constitutive relations given above in successful flow modelling.

We now give some examples selected from those results which are useful in interpreting experiments. Readers interested in further discussions of constitutive equations may go directly to Chapter 5.

For steady Eulerian flows that are not exactly viscometric, the viscometric constitutive equation, eqn (4.32), in the form given by Criminale, Ericksen, and Filbey (1958) can be used as an approximation. As a further simplifying approximation when using the viscometric equation, it is sometimes useful to consider one or both of the normal stress coefficients Ψ_1 and Ψ_2 to be a constant multiple of the viscosity function η. This is a qualitatively reasonable approximation if the three functions have the same general shape. Reasonable quantitative accuracy is obtained by choosing the constant of proportionality to make the approximation exact at the highest shear rate relevant to the problem at hand. Of course, the result of any approximate computation must be discarded if the predicted flow is not like a viscometric flow, since the approximation scheme is then not even self-consistent.

A still cruder approximation scheme that seems to give qualitatively correct results in flows that are nearly viscometric is to replace all of the viscometric functions by appropriate constants. The best choice of the constants depends on the problem being considered. With constant coefficients, the viscometric equation is formally the same as the second-order approximation for slow viscoelastic flow, eqn (4.30c). Many mathematical problems have been solved within the second-order theory, and although it is difficult to attain shear rates so low that the second-order approximation has relevance, it appears that these solutions may nevertheless be qualitatively correct if the flow is mainly a steady laminar shearing motion. We now consider some solution methods based on the second-order approximation.

4.11.1 Some useful results for second-order flows

In some problems in the second-order approximation, or equivalently the viscometric theory with constant coefficients, the velocity field remains that of a

Newtonian flow, and the non-linearities in the constitutive equation affect only the distribution of normal stress. Furthermore, in such cases it is possible to obtain simple expressions for the pressure analogous to Bernoulli's theorem for inviscid flows.

The first non-trivial theorem of this kind is that the velocity field of any plane, Newtonian creeping flow still satisfies the equations of the second-order approximation. The pressure formula for such flows was derived by Giesekus (1963) and Tanner and Pipkin (1969), independently, and extensions to other kinds of flows have been discussed by Pipkin (1972) and Kearsley (1970). See also Huilgol and Phan-Thien (1997).

The main result is a kinematic identity, independent of any constitutive equation. We consider the velocity field $\mathbf{v}$ of an incompressible fluid motion. Let $\mathbf{A} = \mathbf{A}_1$, and $\mathbf{B} = \mathbf{A}_2$ be the first two Rivlin–Ericksen tensors for this flow. The main hypothesis is that $\mathbf{\nabla} \cdot \mathbf{A}$ is irrotational

$$\frac{\partial A_{ij}}{\partial x_j} = \frac{\partial Q}{\partial x_i}. \tag{4.83}$$

(For example in creeping Newtonian flow the equations of motion have this form, with $Q = p/\eta_0$). Then $\mathbf{\nabla} \cdot (\mathbf{B} - \mathbf{A}^2)$ is also irrotational, and its potential can be exhibited explicitly:

$$\nabla \cdot (\mathbf{B} - \mathbf{A}^2) = \mathbf{\nabla}(\mathrm{D}Q/\mathrm{D}t + \dot{\gamma}^2/2). \tag{4.84}$$

Here the shear rate $\dot{\gamma}$ is defined by $2\dot{\gamma}^2 = \operatorname{tr} \mathbf{A}^2$. The proof of the result (4.84) amounts to writing out both members of eqn (4.84) at length, and then using the hypothesis (4.83) and the definitions of $\mathbf{A}_1$ and $\mathbf{A}_2$ [eqn (2.39)].

The preceding result can be used in the following way. Let us suppose that in a given problem, the equations of Newtonian flow can be put into the form (4.83). If, instead of the Navier–Stokes equations, we use the constant-coefficient (second-order) approximation to the viscometric equation, then in general the velocity field will be different from the Newtonian velocity $\mathbf{v}$. However, we can test the assumption that it remains the same. Under this assumption, the momentum equation becomes (for creeping flow)

$$\mathbf{\nabla}p = \mathbf{\nabla}[\eta Q - \eta T(\mathrm{D}Q/\mathrm{D}t + \dot{\gamma}^2/2)] + \eta T^*\mathbf{\nabla} \cdot (\mathbf{A}^2). \tag{4.85}$$

We have defined $\eta T = \frac{1}{2}\Psi_1$ and $\eta T^* = \frac{1}{2}\Psi_1 + \Psi_2$, for future convenience; recall Ψ_1 and Ψ_2 are the normal stress coefficients [eqn (3.18)]. The condition of integrability is that $\mathbf{\nabla} \cdot (\mathbf{A}^2)$ is irrotational. If this is not true, then the Newtonian velocity field is not satisfactory. If it is true, and the potential of $\mathbf{\nabla} \cdot (\mathbf{A}^2)$ is Q^*, then the Newtonian velocity field $\mathbf{v}$ is still the correct velocity field, but the pressure is changed from its Newtonian value to the value

$$p = \eta Q - \eta T(\mathrm{D}Q/\mathrm{D}t + \dot{\gamma}^2/2) + \eta T^* Q^*. \tag{4.86}$$

The stress, found by using this result in eqn (4.32), is then

$$\boldsymbol{\sigma} = -\mathbf{I}(\eta Q - \eta T \, DQ/Dt) + \eta \mathbf{A} - \eta T(\mathbf{B} - \mathbf{A}^2 - \mathbf{I} \operatorname{tr} A^2/4)$$
$$+ \eta T^*(\mathbf{A}^2 - Q^*\mathbf{I}). \tag{4.87}$$

For plane flow, $\boldsymbol{\nabla} \cdot (\mathbf{A}^2) = \boldsymbol{\nabla}(\dot{\gamma}^2)$ and thus,

$$Q^* = \dot{\gamma}^2 = \tfrac{1}{2} \operatorname{tr} \mathbf{A}^2. \tag{4.88}$$

For parallel flows, if G is the magnitude of the axial pressure gradient and v is the speed, then it is easy to show that

$$\boldsymbol{\nabla} \cdot (\mathbf{A}^2) = -(G/\eta)\boldsymbol{\nabla} v + \dot{\gamma}\boldsymbol{\nabla}\dot{\gamma}, \tag{4.89}$$

and thus

$$Q^* = -(G/\eta)v + \dot{\gamma}^2/2. \tag{4.90}$$

For potential flow, with $\mathbf{v} = \boldsymbol{\nabla}\phi$, we find that $Q = 0$ and that $\boldsymbol{\nabla} \cdot (\mathbf{A}^2) = \boldsymbol{\nabla}(\dot{\gamma}^2)$, whence

$$Q^* = \dot{\gamma}^2. \tag{4.91}$$

In the potential flow case, inertia in the flow can be included by using the total pressure $p + \tfrac{1}{2}\rho v^2$, instead of p, in the calculations.

Results of the same kind can often be obtained when the viscosity function is not treated as a constant but the velocity field satisfies an equation of the form

$$\boldsymbol{\nabla} \cdot (\eta \mathbf{A}) = \boldsymbol{\nabla} Q. \tag{4.92}$$

In such cases one or both of the normal stress coefficients are treated as constant multiples of η, and a pressure formula is obtained provided that the viscosity function is constant along streamlines. We shall make such approximations where necessary in individual cases rather than developing general results that use such approximations even where they are not needed. We give some examples in the following sections.

4.11.2 Pressure-hole errors

A well-established experimental procedure for determining the fluid thrust on a solid wall consists of drilling a small hole in the wall and attaching a pressure-measuring device to the base of the hole (Fig. 4.17) The hole disturbs the primary flow, thus yielding in principle an error in measuring the undisturbed pressure. For Newtonian flow, dimensional analysis shows that the difference between the gauge pressure p_g and the undisturbed thrust p_u must have the form

$$p_e = p_g - p_u = \tau_w f(Re, d/D), \tag{4.93}$$

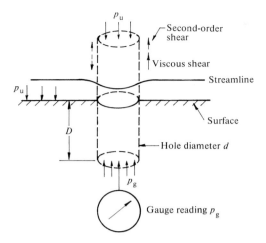

Fig. 4.17 Pressure-hole measurement definition.

where τ_w is the wall shear stress, d is the hole diameter, D is the hole depth, and (Re) is the Reynolds number based on the wall shear stress and the hole diameter:

$$(Re) = \rho\dot{\gamma}d^2/\eta = \rho\tau_w d^2/\eta^2. \tag{4.94}$$

In the limit of zero Reynolds number the flow is reversible and thus there is a symmetrical streamline pattern if the hole is symmetrical. The Navier–Stokes equations then imply that the pressure is constant along the hole centreline, and it follows that the gauge pressure is the same as the pressure outside the hole. This means that the pressure error in a Newtonian flow is an inertial effect, and that the error can be made negligible by reducing the hole diameter until the Reynolds number based on it is sufficiently small.

These facts are well known and it is usual to take care that the diameters of pressure holes are small enough to avoid inertial errors. In connection with viscoelastic fluids, where pressure measurements are needed in order to determine the normal stress functions, various authors assumed that when changes of hole size did not significantly affect their pressure measurements, then the hole sizes must be small enough to give a true pressure reading. In so far as inertial errors are concerned, it is believed that this is a correct inference.

However, the assumption that gauge pressure is equal to undisturbed wall thrust was doubted, and it was suggested by Broadbent *et al.* (1968) that for viscoelastic fluids there is a systematic error not connected with inertia. By assuming that the error is a function of the undisturbed wall shear stress, it has been possible to unify a number of puzzling experimental results, including some anomalous normal stress data. The size of the error is not small in comparison with the shearing stress, and p_e as defined in eqn (4.93) is negative, opposite in sign to the inertial error.

The self-consistency of the data obtained from various viscometers by early workers would lead one to doubt that pressure-hole errors are important. However, it now appears that this consistency is due to the consistency of the pressure-hole error itself. Simple direct measurements (Tanner and Pipkin 1969) have confirmed the existence of such errors and have shown that for sufficiently small holes, the error is independent of hole size. Thus, diminishing the hole size until inertial effects are unimportant does not affect the error due to viscoelasticity.

The cause of the error can be understood in terms of the extra tension in the direction of shearing which occurs in shearing motions of viscoelastic fluids. The streamlines dip slightly toward the hole as they pass it (Fig. 4.17) The direction of shearing is approximately along the direction of flow in such a motion, so that the effect of the first normal stress difference is approximately to produce an extra tension in the streamlines. If we consider a control volume in the shape of a cylinder with one end deep in the hole and the other far out in the undisturbed flow, the pressures on these two ends are respectively p_g and p_u. The shearing stresses on the sides of the cylinder cancel out, and thus in Newtonian flow the equilibrium of the control volume would require that $p_g = p_u$. However, for a viscoelastic fluid the streamline tensions exert an outward force on the control volume and partly balance the thrust on the outside end, p_u. The value of p_g required for equilibrium is then not as large as p_u. This argument suggests that the gauge error should be negative, and that its value should be determined by the first normal stress difference. Since the flows for various hole sizes are geometrically similar (in the creeping flow approximation), we can also understand why the error is independent of size.

An approximate theoretical analysis giving an estimate of the hole error can be based on the results in Section 4.11.1. We consider flow across a deep, narrow slot transverse to the flow rather than a circular cylindrical hole, in order to be able to use the results for plane flows. We suppose that the Reynolds number based on slot width is small enough that the creeping flow approximation is valid. Then the Newtonian velocity field satisfies eqn (4.83) with $Q = p^0/\eta$ where p^0 is the pressure corresponding to Newtonian flow. For viscoelastic response, if we approximate the viscosity and normal stress coefficients by constants, the stress is given by eqn (4.87) with $Q = p^0/\eta$, $Q^* = \dot{\gamma}^2$, $\eta T = \frac{1}{2}\Psi_1$ and $\eta T^* = \frac{1}{2}\Psi_1 + \Psi_2$

$$\boldsymbol{\sigma} = -\mathbf{I}(p^0 - T\mathrm{D}p^0/\mathrm{D}t) + \eta\mathbf{A} - \eta T(\mathbf{B} - \mathbf{A}^2 - \mathbf{I}\dot{\gamma}^2/2)$$
$$+ \eta T^*(\mathbf{A}^2 - \mathbf{I}\dot{\gamma}^2). \tag{4.95}$$

To find the pressure error we consider simple shearing flow along a wall in the plane $y = 0$, for which the undisturbed flow is $\mathbf{v} = \dot{\gamma}y\mathbf{i}$. The disturbance due to the slot mouth is a local effect, so that the Newtonian flow is approximately $\mathbf{v} = \dot{\gamma}y\mathbf{i}$ still, if y is large in comparison to the slot width. The undisturbed thrust p_u is the value of $-\sigma_{yy}$ far from the slot mouth, which is, from (4.95),

$$p_u = p^0 + \tfrac{1}{2}\eta T\dot{\gamma}^2. \tag{4.96}$$

Deep inside the slot, the velocity tends to zero and the stress tends to an isotropic pressure $p_g = p^0$. Since the Newtonian pressure p^0 is constant along the slot centreline, we obtain, since $N_1 = \Psi_1 \dot{\gamma}^2 = 2\eta T \dot{\gamma}^2$,

$$p_e = p_g - p_u = -\tfrac{1}{2}\eta T \dot{\gamma}^2 = -N_1/4. \tag{4.97}$$

Thus, the gauge reads low by one-quarter of the first normal stress difference in the undisturbed flow, and the error is independent of slot width when the slot is narrow enough to neglect inertial error.

When the flow is not globally a simple shearing motion, the formula for the pressure error is the same but the shear rate is that for the undisturbed motion at the site of the slot mouth. This has been verified in detail for Poiseuille flow and flow in an open channel (Tanner and Pipkin 1969).

The theoretical result (4.97) was obtained by treating the viscosity and normal stress coefficients as constants. To compare with experiments we no longer treat N_1 as a constant. In other words, for a given flow we take the constant value of N_1 to be the value of N_1 at the undisturbed shear rate that is ultimately relevant. Under this interpretation, the result (4.97) is in good agreement with direct measurements of the error (Fig. 4.18).

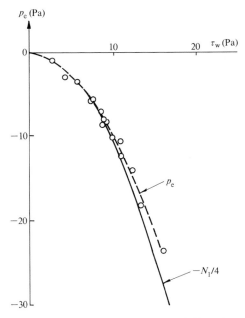

Fig. 4.18 Pressure-error results. The solution is a water-soluble polyacrylamide at 25 °C. The circles are experimental data, and fit the rule $p_e \sim -0.2\,N_1$ closely. The full line is the simple theoretical result $p_e = -0.25\,N_1$.

To obtain these direct measurements, fluid was allowed to flow steadily down an inclined channel under its own weight. In such a flow the stress at the bed of the channel is directly determined by the weight of fluid above it, and thus the undisturbed thrust p_u is known from the outset. Gauges were attached to a slot and to holes in the bottom of the channel to obtain values of p_g. In Fig. (4.18) the measured pressure error is compared with values of $N_1/4$ estimated from the total thrust in cone and plate experiments. See eqn (3.88).

The experiments by Tanner and Pipkin (1969) found no significant difference between the errors for holes and for a slot across the direction of flow. However, Kearsley (1970) pointed out a result that made it clear that there must be a difference that can be significant. If the difference in normal thrusts on the two walls of a Couette viscometer, given by eqn (3.79), is corrected for the pressure errors at both walls by using the formula (4.97), then the difference in gauge readings is zero (apart from the effect of centrifugal force). This is not observed with pressure holes. Kearsley accordingly carried out Couette experiments with slots parallel to the axis rather than holes, and found that at low shear rates the difference in gauge pressures between two slots is indeed much smaller than the difference between two holes. Kearsley also showed that for flow parallel to a slot rather than across it, the pressure error is one-half the second normal stress difference (see Section 4.11.3). This would suggest that for slots in the circumferential direction in a Couette viscometer, the measured difference in gauge pressures should be almost equal to the full value given by eqn (3.79), not corrected for hole error, because the correction based on the second normal stress difference is relatively small. Experiments since made have verified this hypothesis.

Higashitani and Pritchard (1972) have estimated the pressure error for holes, and have found that it is intermediate between those for the two kinds of slots. This means that in the Couette geometry with pressure holes, the observed difference in gauge pressures is neither zero as for slots parallel to the axis nor is it the nearly-full uncorrected value found for slots in the circumferential direction, but rather it is an intermediate value not as large as the uncorrected value (3.79).

In our notation, the pressure error formulas given by Higashitani and Pritchard (1972) are, for flow across a slot:

$$p_e = -\frac{1}{2} \int_0^\tau N_1(d\tau/\tau), \tag{4.98}$$

for flow along a slot,

$$p_e = \int_0^\tau N_2(d\tau/\tau), \tag{4.99}$$

and for flow past a hole,

$$p_e = -\frac{1}{3} \int_0^\tau (N_1 - N_2)(d\tau/\tau), \tag{4.100}$$

where the viscometric functions are treated as functions of the shearing stress τ in the integration. In the constant–coefficient approximation, the first two of these results agree respectively with the results found by Tanner and Pipkin (1969) and by Kearsley (1970) and the result for holes is

$$p_e = -\tfrac{1}{6}(N_1 - N_2). \tag{4.101}$$

The method used by Higashitani and Pritchard to derive these results is a little less convincing in the case of holes than in the simpler cases involving slots, but there is little doubt that the error for holes should have (approximately) the general form

$$p_e = c_1 N_1 + c_2 N_2, \tag{4.102}$$

and the values $-c_1 = c_2 = 1/6$ are reasonably concordant with what one might expect from the results on slots. If we set N_2 equal to $-0.15\, N_1$ (see Table 3.9) in eqn (4.101), we obtain $p_e = -0.19\, N_1$, which is close to the value $-0.2\, N_1$ found experimentally for both slots and holes. The slight difference suggests that $-c_1$ might be a little larger than $1/6$; no inference about c_2 can be made since the error is relatively insensitive to the value of c_2. Proposals have been made to use the hole-pressure difference as a means of measuring N_1 without disturbing the flow greatly. Accurate calibration of such a device at high shear rates needs care, particularly as the flow is then often observed to be unsymmetric with respect to the hole axis. Further discussion is given by Lodge and de Vargas (1983).

For the plane flow (slot) case, Sugeng et al. (1988) did a numerical computation of the flow field and the stresses, finding the hole error was $-0.24\, N_1$ for the UCM model. Using a modified Phan-Thien–Tanner model the experimental results of Pike and Baird (1984) for polystyrene were well described by the computations. The flow over the slot was not symmetrical as assumed in the approximate second-order analysis given above, probably due to the convective stress terms in the UCM and PTT models.

4.11.3   Flow in non-circular tubes and channels

In Section 3.5.2 we found that steady parallel flows are not partially controllable unless the slip surfaces are parallel planes or coaxial circular cylinders, and thus they are not in general dynamically admissible at all in real fluids. Thus, the flow in a tube of non-circular cross-section will generally not be a parallel flow. More particularly, rectilinear flow is only possible if the second normal stress coefficient (Ψ_2) is zero (the Weissenberg hypothesis) or if it is a constant multiple of the viscosity function, but generally not otherwise. Green and Rivlin (1956) examined the flow in a tube of elliptical cross-section, using a special constitutive equation, and found a weak secondary circulation in each quadrant. Langlois and Rivlin (1963) found that the flow in a tube is rectilinear not only in the Newtonian (first-order) approximation but also in the second-order approximation [eqn (4.30c)] and in the third-order approximation. Transverse flow first appears in the fourth-order slow motion approximation.

Although there are a few experimental papers describing secondary flows in non-circular tubes it is not an easy matter to detect these flows. Symmetry considerations show that the transverse flow is broken into cells bounded by the symmetry axes of the cross-section (Fig. 4.19) and conceivably the actual cells could be smaller than those required by symmetry.

Thus there is only a comparatively small part of the cross-section in which the transverse flow could be appreciable. Observations show that although streamlines are indeed spirals, one turn of the spiral requires an axial distance of many ($\sim 10^2$) diameters along the length of the tube. Thus unless one has very long tubes and correctly positioned measuring equipment, one is quite likely to miss the phenomenon.

If we assume that the transverse circulation is negligible, then the flow is viscometric and the viscometric constitutive equation is applicable. The discussion of Poiseuille flow in Section 3.5.2 is then relevant, and in particular the axial velocity satisfies eqn (3.52), with $w = 0$ at the tube walls if there is no slip. From eqns (3.51) and (3.53) we find that the pressure p is given by

$$p = -Gz + I_2(\dot{\gamma}) + \bar{P}(x, y),\qquad(4.103)$$

where I_2 is the integral of the second normal stress coefficient defined in eqn (3.28). The $\bar{P}$ must satisfy

$$\boldsymbol{\nabla}\bar{P} = \boldsymbol{\nabla}\cdot(\Psi_2\boldsymbol{\nabla}w)\boldsymbol{\nabla}w.\qquad(4.104)$$

This equation generally has no solution because the right-hand member is not irrotational.

However, if we write $\Psi_2 = -T_2\eta +$ remainder and pick the constant coefficient T_2 so as to optimize the first term as an approximation to Ψ_2, then on neglecting the remainder we have $\Psi_2 = -T_2\eta$; rectilinear flow is possible if such a relation is valid. Indeed, with this approximation, when we take eqn (3.53) into account we

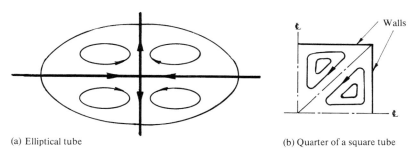

(a) Elliptical tube   (b) Quarter of a square tube

Fig. 4.19 Secondary flows in elliptical and square tubes induced by axial pressure drop.

find that eqn (4.104) becomes

$$\nabla \bar{P} = GT_2 \nabla w, \quad \text{whence} \quad \bar{P} = GT_2 w + C. \tag{4.105}$$

In the final result we replace T_2 by $-\Psi_2/\eta$, evaluated at each point with the shear rate appropriate to that point. By doing so, we obtain

$$p = -Gz + I_2 - G\Psi_2 w/\eta + C. \tag{4.106}$$

By using this result in the expression (4.30c) for the stress, we find that the normal thrust on the tube wall, say p_u, is

$$p_u = -Gz + I_2 - N_2 + C. \tag{4.107}$$

Kearsley (1970) has considered the pressure error for a slot along the direction of flow. The tube, augmented by such a slot, is merely a tube of a different cross-section. We can interpret the preceding result as the undisturbed pressure at the site of the slot mouth. Deep inside the slot, where the velocity and shear rate are negligible, the pressure is $p_g = -Gz + C$. Thus the gauge error for a flow parallel to a pressure slot is

$$p_e = p_g - p_u = N_2 - I_2. \tag{4.108}$$

In the constant $-\Psi_2$ approximation, for which $I_2 = N_2/2$, the pressure error is half the second normal stress difference. This is the result obtained by Kearsley that was mentioned in Section 4.11.2.

In tube and channel flows the first normal stress difference produces an extra tension along the streamlines that tends to pull them straight if they start to curve, but otherwise has no effect. Normal stress effects in such flows are caused by the second normal stress difference, so it is natural to attempt to use such flows to measure N_2. As we have just seen, the error for pressure slot along the direction of flow is directly related to N_2 and more generally the distribution of normal thrust around the tube wall depends on N_2. We have already discussed a method based on measurement of the difference in normal thrusts on the two parts of the boundary in the flow in the annular gap between two coaxial cylinders. Such measurements are subject to correction for pressure-hole errors, of course, if pressure holes are used; note that mounting flush-fitting gauges on curved walls is not easy.

4.11.4   General torsional flows

Steady flows in circles are kinematically admissible viscometric flows, with surfaces of revolution as slip surfaces (Section 3.2.3). These flows generally cannot satisfy the dynamical equations exactly. Both centrifugal forces and normal stress effects cause transverse flow, and the motion is then no longer exactly viscometric. However, when the flow is produced by holding a rotor near a flat plate, as in the parallel plate or cone and plate instruments, this transverse flow is negligibly small when the gap between plate and rotor is sufficiently small.

There are several reasons for considering flows of this kind other than the simple cone and plate and torsional flows. First, analysis of edge effects in the basic flows requires consideration of flows that are not completely controllable. Second, the cone and plate instrument often has a rotor that is not exactly a cone, but a cone with its tip cut off to prevent fouling (Fig. 3.5). Third, in order to estimate the errors that can occur due to incorrect setting of the cone, it is useful to consider the operation of the cone and plate device when the tip of the cone is separated from the plate by a small distance c (Fig. 3.5).

Finally, the cone and plate can be used in attempting to measure the second normal stress difference, by operating the instrument with known, non-zero tip separations.

In all of these cases we suppose that the motion is a steady circular flow (Section 3.2.3). The stress is given by eqn (3.23). We use this in the equilibrium equation (that is, neglecting centrifugal force) and specify that there is no azimuthal pressure gradient.

We suppose that the gap between the plate and the cone or other rotating body has a width $h(r)$ at the radius r, and that h is everywhere much smaller than the outer radius r_0. If ω_0 is the angular velocity of the rotor, then the shear rate is approximately

$$\dot{\gamma} = r\omega_0/h(r), \tag{4.109}$$

and thus it is a function of r alone. In these cases we find (Pipkin and Tanner 1972)

$$p(r) = \int_r^{r_0} (N_1 + 2N_2)(\mathrm{d}r/r) + I_2(\dot{\gamma}) - I_2(\dot{\gamma}_0) + p_0, \tag{4.110}$$

where $\dot{\gamma}_0$ is the shear rate at the rim and p_0 is the radial pressure there; I_2 is defined in (3.28).

If the sample is held in the viscometer by surface tension so that the outer radius r_0 is the boundary of the fluid (the free boundary condition), then p_0 is atmospheric pressure, at least approximately. However, there is some small ambiguity because the shape of the free boundary is not known exactly, as we mentioned previously. If this is ignored, we can compute the stress from eqn (3.23) and (4.110).

Although we believe that this slight error in the free boundary condition is more tolerable than those which occur with other boundary conditions, it has been proposed that the viscometer be operated fully immersed in a sea of liquid so that there is a large body of fluid outside the rim r_0. In this case the estimation of the radial pressure p_0 at the rim requires further analysis. It has been shown that if h is the gap width at the rim (with $h \ll r_0$), the zone in which the shear rate changes from its full operating value inside the gap to negligible values outside the gap, has a width of the order of h. From eqn (4.110) we see that the change of

p across this transition zone, from p_0 inside to p_∞, say, outside, is given by

$$p_0 - p_\infty = I_2(\dot{\gamma}_0) + N_1 O(h/r_0). \tag{4.111}$$

If H is the depth of fluid just outside the gap and we take p_∞ to be the hydrostatic pressure $\rho g H$, and if h/r_0 is sufficiently small, then

$$p_0 = \rho g H + I_2(\dot{\gamma}). \tag{4.112}$$

Thus the radial pressure at the rim is *not* the hydrostatic pressure as often assumed, but depends on the shear rate at the rim. In the constant $-\Psi_2$ approximation, I_2 is equal to $N_2/2$ and thus the pressure just inside the rim is smaller than hydrostatic by half the magnitude of the second normal stress difference (assuming that N_2 is negative). In terms of the thrust against the plate, the result is

$$- \sigma_{zz}(r_0) = p_0 - N_2 = \rho g H + I_2(\dot{\gamma}_0) - N_2(\dot{\gamma}_0). \tag{4.113}$$

With this meaning of pressure, there is an excess of pressure (over hydrostatic) under the rim, equal to $- N_2/2$ if Ψ_2 is treated as constant (Pipkin and Tanner 1972). Thus, neither the radial nor the axial pressure is equal to the hydrostatic pressure.

It has been shown that the pressure near the rim, within distances of the order of the gap width h, is described reasonably well by eqn (4.113). Note that if N_2 is negative, then the rim pressure $-\sigma_{zz}(r_0)$ is greater than hydrostatic, and vice versa. Experiments with flush-mounted pressure gauges show that the excess rim pressure is positive, as expected (Pipkin and Tanner 1972).

To consider the problem of gap-setting errors, we suppose that the gap width $h(r)$ has the form

$$h(r) = h^*(r) + c, \tag{4.114}$$

so that changing c represents changing the spacing.

The moment required to hold the plate is

$$M = 2\pi \int_0^{r_0} \dot{\gamma} \eta r^2 \, \mathrm{d}r. \tag{4.115}$$

To compute the change of M due to a small setting error $c \neq 0$, we first note from eqns (4.109) and (4.114) that

$$\partial(\dot{\gamma}^{-1})/\partial c = 1/\omega_0 r. \tag{4.116}$$

Then

$$\frac{\mathrm{d}M}{\mathrm{d}c} = \frac{2\pi}{\omega_0} \int_0^{r_0} \frac{\mathrm{d}(\dot{\gamma}\eta)}{\mathrm{d}(\dot{\gamma}^{-1})} r \, \mathrm{d}r. \tag{4.117}$$

If the rotor is a cone then the shear rate is uniform when $c = 0$, and the integrands are constants, apart from the explicit factors involving r. The relative error dM/M due to a setting error dc is then found to be

$$\frac{dM}{M} = -\frac{3}{2}\frac{d\log\dot{\gamma}\eta}{d\log\dot{\gamma}}\bigg|_{c=0}\frac{dc}{h_0}, \tag{4.118}$$

where h_0 is the gap at the rim when $c = 0$. For a cone with its tip cut off, so that there is a flat of radius r_f, the derivative dM/dc differs from that shown by a term of relative magnitude $O(r_f^2/r_o^2)$, which is usually negligible.

From eqn (3.23), the thrust against the plate is $p - N_2$, and the total thrust is accordingly

$$F = 2\pi \int_0^{r_o} (p - N_2)r\,dr = \pi r_o^2 p_0 - \pi \int_0^{r_o} (2N_2 r + r^2 p')\,dr. \tag{4.119}$$

For a cone, since $\dot{\gamma}$ is independent of r when $c = 0$, we find that the slope at $c = 0$ is

$$dF/dc = (\pi r_o/\omega_0)[dN_1/d(\dot{\gamma}^{-1}) - \dot{\gamma}N_2]. \tag{4.120}$$

It has been proposed that this slope be used, along with known values of N_1, to determine N_2. However, we notice that since each differentiation with respect to c introduces a new factor $1/r$ into the integrand, the integrand of d^2F/dc^2 is proportional to $1/r$ and thus this derivative is infinite. This means that numerical differentiation to determine dF/dc at $c = 0$ is a practical impossibility, and thus that using eqn (4. 120) to determine N_2 can give highly misleading results. In the case of a cone with its tip cut off, the second derivative is not infinite, but it involves a factor $\log(r_o/r_f)$ that may be substantial.

4.11.5   Eccentric-disc flow

A common method of testing viscoelastic materials is to measure the stress needed to produce a small, sinusoidally oscillating deformation (Chapter 2). The stress can be decomposed into a part in phase with the strain and a part 90° out of phase, in phase with the strain-rate. By dividing these parts by the strain amplitude, one obtains values of the elastic-storage modulus $[G'(\omega)]$ and the loss modulus $G''(\omega)$. The dynamic viscosity is related to the loss modulus by $\eta'(\omega) = G''(\omega)/\omega$. (See Section 2.6.3.)

Gent (1960) described an ingenious method of testing solid polymers for G^* and Maxwell and Chartoff (1965) used the same principle in a device adapted for the testing of polymer melts (Fig. 4.20). In this instrument a *steady* rotational flow is used to measure the frequency-dependent properties of the material, and the measurement of phase angle is replaced by the task of measuring steady forces.

The device consists of parallel discs that rotate at the same angular velocity Ω about axes perpendicular to the discs but not coincident. When the distance h

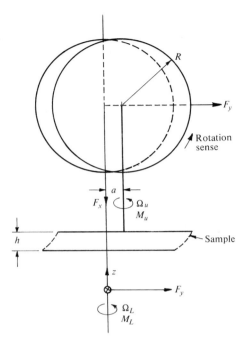

Fig. 4.20 Eccentric disc rheometer showing plan view (*top*) and side view (*bottom*) and a 'free edge' configuration for the sample. In normal operation $\Omega_u \cong \Omega_L = \Omega$.

between the plates is sufficiently small, the fluid can be held in the gap by surface tension. In some cases the lower disc is replaced by a rotating cup.

A fluid element starting at the right-hand side in Fig. (4.20) is sheared from a top-outward position to a top-inward position as the discs rotate through 180°. The elastic force required for the shearing is in the positive y-direction on both sides. The viscous force, related to the rate of shear, is parallel to the x-direction.

Gent (1960) recognized that the stress components in the test specimen would be uniform and constant in time, and he showed that the force needed to hold the top disc in place would be related to the viscoelastic moduli by

$$F_x = A(a/h)G''(\Omega), \tag{4.121}$$

and

$$F_y = A(a/h)G'(\Omega). \tag{4.122}$$

Here A is the contact area, a is the distance between the axes of rotation, h is the gap width, and Ω is the angular velocity. This result requires that the shear amplitude a/h be small enough that non-linear geometrical effects can be neglected.

This analysis also requires the strain amplitude a/h to be in the linear range. Gross and Maxwell (1972) have shown that the measured stress remains proportional to a/h up to values of this strain of the order of 0.50 for some polymers (compare Fig. 4.9). They emphasized that the limit of linear behaviour depends on the strain amplitude, not the strain-rate. This is of interest because some rheological models require a small value of the *strain-rate* amplitude $a\Omega/h$ for linearity (for example, single integral models with a strain-rate dependent kernel).

To obtain the results (4.121–122) we postulate the velocity field ($\mathbf{v} = (u,v,w)$) is (Fig. 4.20)

$$u = -\Omega\left(y - a\frac{z}{h}\right), \quad v = \Omega x \quad w = 0. \tag{4.123}$$

If edge effects and velocity lag (i.e. a difference between Ω_u and Ω_L) are neglected but inertia is taken into account, the velocity field has the form

$$u = -\Omega[y - y_0(z)], \quad v = \Omega[x - x_0(z)], \quad w = 0, \tag{4.124}$$

with a line of centres that is not the straight line $x_0 = 0$, $y_0 = az/h$ assumed above. Abbott and Walters (1970) determined the line of centres exactly for the case of a Newtonian liquid (thus discovering a previously unknown exact solution of the Navier–Stokes equation) and for a linearly viscoelastic liquid with arbitrary moduli. They found that the line of centres departs from straightness by an amount proportional to the offset a and proportional in lowest order to the Reynolds number $\rho\Omega h^2/\eta$. Under ordinary operating conditions this Reynolds number is so small that inertial effects are negligible.

There is an interesting paradox concerning the flow between discs of finite radii. Except near the edges, the rate of energy dissipation is $\sigma_{xz}a\Omega/h$, with σ_{xz} uniform. The total dissipation is then $V\sigma_{xz}a\Omega/h$, with V the effective volume of the flow region. If we define the effective contact area between the fluid and either disc as $A = V/h$, then the total rate of dissipation is seen to be equal to $aF_x\Omega$ with F_x given by (4.121). However, if the effective contact area is a circle centred on the axis of rotation and the stress is uniform in this region, then no work is done by the disc on the fluid. This way of stating the paradox contains its solution; the region in which the stress is tolerably uniform is not centred on the axis of rotation.

In normal operation of the rheometer, one of the discs is driven at a prescribed angular velocity Ω and the other is allowed to rotate freely. Except when $a = 0$, this leads to a small difference $\Delta\Omega$ between the angular velocities of the discs. If the top disc, say, is turned with a moment M and the lower disc rotates freely about its axis, then the power supplied is $M\Omega$. This must be equal to the rate of dissipation $aF_x\Omega$, so that the moment is $M = aF_x$. Then either the effective contact area is not a circle centred on the axis of rotation, or the velocity field does not have the ideal form (4.123).

The eccentric disc device is sometimes a useful method of measuring the moduli G' and G'' especially at low frequencies because one thereby avoids the problem of accurate phase angle measurements. However, instrument structural flexibility can give rise to problems, and the method seems unlikely to supplant time-sinusoidal testing. Further discussion and other variants of this method of measurement are given by Walters (1975); see also Huilgol (1969).

4.12 Summary

From the material in this chapter we see that the central area in Pipkin's diagram (Fig. 3.9) is difficult to describe by the continuum approach to finding constitutive relations. The most notable success is the irreversible KBKZ model, which does not follow the classical pattern of equation development and appears to have been inspired by rubber-elastic theory. We shall now leave the classical approaches and look to microstructural ideas in the next chapter for a further guide to possible types of constitutive models for non-linear viscoelastic materials. At the end of Chapter 5 we shall survey the various constitutive proposals and classify their responses to various test flows (Table 5.4).

References

Abbott, T. N. G. and Walters, K. (1970). *J. Fluid Mech.*, **40**, 205.

Atapattu, D. D., Chhabra, R. P. and Uhlherr, P. H. T. (1990). *J. Non-Newtonian Fluid Mech.*, **38**, 31.

Atkinson, C. and El-Ali, K. (1992). *J. Non-Newtonian Fluid Mech.*, **41**, 339.

Barnes, H. A. (1992). *Proc. XI Int. Congress on Rheology*, (ed. P. Moldenaers and R. Keunings), p. 576. Brussels, Elsevier, Amsterdam.

Barnes, H. A. and Walters, K. (1985). *Rheol. Acta*, **24**, 323.

Beris, A. N., Tsamopoulos, J. A., Armstrong, R. C., and Brown, R. A. (1985). *J. Fluid Mech.*, **158**, 219.

Bernstein, B., Kearsley, E. A., and Zapas, L. J. (1963). *Trans. Soc. Rheol.*, **7**, 391.

Beverly, C. R. and Tanner, R. I. (1989). *J. Rheol.*, **33**, 989.

Beverly, C. R. and Tanner, R. I. (1992). *J. Non-Newtonian Fluid Mech.*, **42**, 85.

Bird, R. B., Armstrong, R. C., and Hassager, O. (1977). *Dynamics of polymeric liquids*, Vol. 1., *Fluid dynamics*. Wiley, New York.

Bird, R. B., Dai, G. C., and Yarusso, B. J. (1982). *Rev. Chem. Eng.*, **1**, 1.

Boger, D. V. (1977). *J. Non-Newtonian Fluid Mech.*, **3**, 87.

Broadbent, J. M., Kaye, A., Lodge, A. S., and Vale, D. G. (1968). *Nature Lond.*, **217**, 55.

Chakrabarty, J. (1987). *Theory of plasticity*. McGraw-Hill, New York.

Cheng, D. C. H. (1986). *Rheol. Acta*, **25**, 542.

Criminale, W. O., Ericksen, J. L., and Filbey, G. L. (1958). *Arch. ration. Mech. Anal.*, **1**, 410.

Currie, P. K. (1982). *J. Non-Newtonian Fluid Mech.*, **11**, 53.

Debbaut, B. and Crochet, M. J. (1988). *J. Non-Newtonian Fluid Mech.*, **30**, 169.

Einaga, Y., Osaki, K., Kurata, M., Kimura, S., and Tamura, M. (1971). *Polymer J. (Japan)* **2**, 550.

Gent, A. N. (1960). *Br. J. appl. Phys.*, **11**, 165.

Giesekus, H. (1963). *Rheol. Acta*, **3**, 59.

Green, A. E. and Rivlin, R. S. (1956). *Q. appl. Math.*, **14**, 229.
Green, A. E. and Rivlin, R. S. (1957). *Archs. ration. Mech. Anal.*, **1**, 1.
Gross, L. H. and Maxwell, B. (1972). *Trans. Soc. Rheol.*, **16**, 577.
Higashitani, K. and Pritchard, W. G. (1972). *Trans. Soc. Rheol.*, **16**, 687.
Hill, R. (1950). *The mathematical theory of plasticity*. Oxford University Press.
Hohenemser, K. and Prager, W. (1932). *ZaMM.*, **12**, 216.
Huilgol, R. R. (1969). *Trans. Soc. Rheol.*, **13**, 513.
Huilgol, R. R. (1979). *J. Non-Newtonian Fluid Mech.*, **5**, 219.
Huilgol, R. R. and Phan-Thien, N. (1997). *Fluid mechanics of viscoelasticity*, Elsevier, Amsterdam.
Joseph, D. D. (1981). *Archs. ration. Mech. Anal.*, **75**, 251.
Kaye, A. (1962). College of Aeronautics, Cranfield, *Note No.* 134.
Kearsley, E. A. (1970). *Trans. Soc. Rheol.*, **14**, 419.
Langlois, W. E. and Rivlin, R. S. (1963). *Rend. Mat.*, **22**, 169.
Larson, R. G. (1988). *Constitutive equations for polymer melts and solutions.* Butterworths, Boston.
Larson, R. G. and Monroe, K. (1987). *Rheol. Acta*, **26**, 208.
Laun, H. M. (1978). *Rheol. Acta.*, **17**, 1.
Leigh, D. C. (1968). *Nonlinear continuum mechanics*. McGraw-Hill, New York.
Lipscomb, G. G. and Denn, M. M. (1984). *J. Non-Newtonian Fluid Mech.*, **14**, 337.
Lodge, A. S. (1964) *Elastic liquids*. Academic Press, London.
Lodge, A. S. and de Vargas, L. (1983). *Rheol. Acta*, **22**, 151.
Luo, X. L. and Tanner, R. I. (1988). *Int. J. Num. Meth. Eng.*, **25**, 9.
Magnin, A. and Piau, J.-M. (1992). in *Theoretical and applied rheology* (ed. P. Moldenaers and R. Keunings), p. 195. Elsevier, Amsterdam.
Maxwell, B. and Chartoff, R. P. (1965). *Trans. Soc. Rheol.*, **9**, 41.
O'Donovan, E. J. and Tanner, R. I. (1984). *J. Non-Newtonian Fluid Mech.*, **15**, 75.
Ogden, R. W. (1984). *Nonlinear elastic deformation*. Ellis Horwood, Chichester.
Oldroyd, J. G. (1950). *Proc. Roy. Soc.*, **A200**, 523.
Oldroyd, J. G. (1958). *Proc. Roy. Soc.*, **A245**, 278.
Osaki, K., Kimura, S., and Kurata, M. (1981). *J. Rheol.*, **25**, 549.
Papanastasiou, A. C., Scriven, L. E. and Macosko, C. W. (1983). *J. Rheol.*, **27**, 387.
Papanastasiou, T. C. (1987). *J. Rheol.*, **31**, 385.
Petrie, C. J. S. (1979). *Elongational flows*, Pitman, London.
Pipkin, A. C. and Tanner, R. I. (1972). *Mechanics today*, **1**, 262.
Pipkin, A. C. (1964). *Rev. mod. Phys.*, **36**, 1034.
Pipkin, A. C. (1972). *Lectures on viscoelasticity theory*. Springer-Verlag, New York.
Reiner, M. (1945). *Am. J. Math.*, **67**, 350.
Rivlin, R. S. (1948). *Proc. R. Soc.*, **A193**, 260.
Roland, C. M. (1989). *Rubber Chem. and Tech.*, **62**, 863.
Schwedoff, T. (1890). *J. Physique*, [2], **8**, 34.
Sherwood, J. D., Meeten, G. H., Farrow, C. A., and Alderman, N. J. (1991). *J. Non-Newtonian Fluid Mech.*, **39**, 311.
Sugeng, F., Phan-Thien, N. and Tanner, R. I. (1988). *J. Rheol.*, **32**, 215.
Szabo, P. and Hassager, O. (1992). *J. Non-Newtonian Fluid Mech.*, **45**, 149.
Tanner, R. I. (1968). *Trans. Soc. Rheol.*, **12**, 155.
Tanner, R. I. and Pipkin, A. C. (1969). *Trans. Soc. Rheol.*, **13**, 471.
Tanner, R. I. and Simmons, J. M. (1967). *Chem. Eng. Sci.*, **67**, 1803.
Tanner, R. I. and Walters, K. (1998). *Rheology: An historical perspective*. Elsevier, Amsterdam.
Truesdell, C. and Noll, W. (1965). *The nonlinear field theories of mechanics*. Springer-Verlag, Berlin.

Wagner, M. H. (1977). *Rheol. Acta*, **16**, 43.
Wagner, M. H. (1978). *J. Non-Newtonian Fluid Mech.*, **4**, 39.
Wagner, M. H. (1979). *Rheol. Acta*, **18**, 681.
Wagner, M. H. and Meissner, J. (1980). *Makromol. Chemie*, **181**, 1533.
Wagner, M. H., Ehrecke, P., Hachmann, P., and Meissner, J. (1998). *J. Rheol.*, **42**, 621.
Walters, K. (1975). *Rheometry*. Chapman & Hall, London.
Walton, I. C. and Bittleston, S. H. (1991). *J. Fluid Mech.*, **222**, 39.
White, J. L. and Tanaka, H. (1981). *J. Non-Newtonian Fluid Mech.*, **8**, 1.
White, J. L. and Tokita, N. (1967). *J. Phys. Soc. Japan*, **22**, 719.
Wilson, S. D. R. (1993*a*). *J. Non-Newtonian Fluid Mech.*, **47**, 211.
Wilson, S. D. R. (1993*b*). *J. Non-Newtonian Fluid Mech.*, **50**, 45.
Zapas, L. J. (1974). *In Deformation and fracture of high polymers* (ed. H. Kausch), p. 381. Plenum Press, New York.
Zdilar, A. M. and Tanner, R. I. (1992). *Rheol. Acta*, **31**, 44.

Problems

1. Compute the response of the Reiner–Rivlin fluid in a steady extensional flow. Show that the extensional viscosity [see eqn (3.94)] is unsymmetric so that $\eta_E(\dot{\varepsilon}) \neq \eta_E(-\dot{\varepsilon})$ in general.

2. Show that for flow of a power-law fluid [(eqn (4.4)] that a similarity solution exists for the (Hamel) flow consisting of flow into a converging plane channel. Derive an equation satisfied by the radial velocity, v_r where v_r is of the form $r^{-1}f(\theta)$ in plane polar coordinates. Omit inertia if needed.

3. Investigate the response of the Oldroyd equation (4.13) in steady elongational flow with $\mu_0 = a_1 = a_2 = \lambda_2 = \mu_2 = 0$ as a function of the two parameters $\lambda_1\dot{\varepsilon}$, and μ_1/λ_1. Find any critical rates of strain (that is, where $\eta_E \to \infty$) as functions of these two parameters.

4. Consider a beam of square cross-section (side a) modelled as an elastic-plastic incompressible material, with yield stress τ_y and Young's modulus E. A constant moment is applied to bend the beam. Let the axis of the beam lie in the z-direction, x and y lie in the cross-section, parallel to the sides. The axis of the applied moment M is the x-axis. Sketch the yielded and unyielded areas as M is increased, using simple beam theory. What is the maximum moment that can be withstood by the beam?

5. Work out the steady shear and elongation results for the co-rotational model. In this model $\mu_1 = \lambda_1$, $\mu_2 = \lambda_2$, and $\mu_0 = a_1 = a_2 = 0$ in eqn (4.13). What do you conclude about its ability to describe typical fluids?

6. Compute the shear viscosity of a third-order fluid as a function of shear rate. Show that there is an upper shear rate beyond which the approximation fails to be realistic.

7. For the second-order fluid model, compute the stresses in a flow into a line sink of strength $2\pi q$ ($q < 0$). (Inertia is to be included.) Suppose the stress σ_{rr} is zero at a finite radius R_i. Plot the results for σ_{rr} and $\sigma_{\theta\theta}$ as a function of radius in creeping flow when $q = 1$, $R_i = 5$.

8. Prove result (4.84).

9. Using the definitions (2.89), (2.106), and (4.45), show that as $\omega \to 0$, $\dot{\gamma} \to 0$, one has the relation $\Psi_1(\dot{\gamma}) = 2G'(\omega)/\omega^2$.

10. For the steady flow of the UCM model [eqn (4.25)] Renardy has argued that near high stress points the constitutive equation can be approximated by $\lambda \Delta \tau / \Delta t = 0$. Show that this has an exact solution of the form $\tau_{ij} = h(\psi) v_i v_j$ where v_i is a velocity component and h is an arbitrary function of the stream function ψ.

11. Show that the result in Problem 7 can be obtained by using (4.87) and (4.91) directly.

12. Consider a slider with a sharp corner (Fig. 4.21) sliding with speed W in the z-direction parallel to the plane $y = 0$; the entire space is filled with a second-order fluid. Compute the normal stress on the surfaces deep within the channel. (i.e. far to the left of A). What difference would a radius (or chamfer) at corner A make?

13. Consider a truncated cone-plate apparatus (Fig. 4.22). The angle α is grossly exaggerated. Compute the torque M and thrust F on the cone as functions of h/H. Assume a second-order fluid model.

14. Repeat Question 13 but with an error in spacing of cone/plate so that h is zero but the cone tip is now separated by a distance c from the plate [eqns (4.118) and (4.120)].

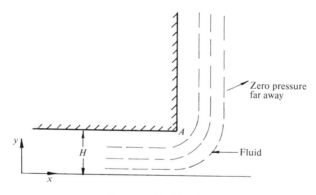

Fig. 4.21 Problem 12.

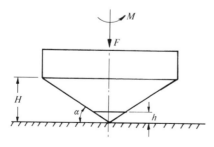

Fig. 4.22 Problem 13.

15. Compute the steady shear response of the double integral form (4.42) assuming $-\mu_1(t) = (\eta_1/\lambda^2)\exp(-t/\lambda)$ and $\mu_{22}(t_1, t_2) = (\eta_2^2/\lambda^4)\exp-(t_1 + t_2)/\lambda$. What is the viscosity function?

16. Calculate $\mathbf{C}^{-1}$ for (2.43). Show that $\mathrm{II}_c = \mathrm{I}_{c^{-1}}$ in this case (this is a general result for incompressible materials).

17. Compute the $\mathbf{C}$ tensor for the eccentric disc flow (4.123).

5
MICROSTRUCTURAL THEORIES

It is clear from Chapter 4 that in many cases it is difficult to find an accurate, useful material description purely from continuum considerations. In nearly-viscometric flows some success has been achieved and KBKZ-type theory has been generally useful. Nevertheless, it is instructive to try and derive constitutive models from micromechanical models, and this is the concern of the present chapter. An ultimate goal is to derive rheological behaviour from the known material parameters and structure, and progress in this direction is being made.

By taking account of the microstructure of substances we expect to gain more insight into the type of constitutive equation needed to describe their mechanical behaviour. At the extremes, we can deal either with dilute or concentrated solutions; in the former, each particle interacts only with the solvent and not with other suspended particles, while in the latter there are particle–particle interactions, which are dominant; in this case molecular 'entanglements' may form. Particles may also be of significant size, leading to suspension theory. To begin we will concentrate on dilute solution theory.

5.1 The polymer molecule

We may dissolve all sorts of substances to form solutions: here we will often think of polymer molecules. Many books (for example, Billmeyer 1964) give details of the chemical structure of polymer molecules, where they occur and how they are made. Note that the linking of many monomer units leads to very elongated structures; for example, poly(ethylene oxide) is formed by linking many units of the form:

$$[-CH_2 - CH_2 - O][-CH_2 - CH_2 - O][- \cdots].$$

This is a linear polymer: others with side branches and groups are quite possible; often 10^5 units can be assembled into a macromolecule. Since the molecular weight (MW) of an element of poly(ethylene oxide) is 44, we can have an overall MW of 4×10^6 or greater. It is possible to make molecular models (for example, Fig. 5.1); note that the molecules at the chain ends have slightly different neighbours from the other molecules and so must behave somewhat differently; this aspect is not considered here. Note also that not all molecules are the same length, although sometimes we assume they are.

Although the chemical form of a linear polymer molecule might suggest that the backbone is a straight line, this is not so, and the spatial configuration of a molecule, at a given instant, is complex and random. Bond angles between pairs of atoms somewhat restrict relative local configurations but due to the very large numbers of atoms in a molecule the possible number of global configurations is immense. Consider only three carbon atoms 1, 2, 3 (Fig. 5.2). We can see that atom 3 has a lot of freedom with respect to atom 1, since each molecule can move on the conical paths suggested in Fig. 5.2. A model of a polymer molecule made in wire looks like a random path in space and represents an instantaneous picture of the molecule. Note that due to Brownian motion the molecule changes continuously from one configuration to another: there is an enormous number of possible configurations. The elasticity of polymer molecules, and hence of solutions and undiluted rubbers, is intimately connected with these random motions and random configurations.

Now the exact solution of the problem of the motion of an actual molecule is clearly a very difficult problem, and we will make simpler models of molecules. For example, we will, in the present elementary treatment, neglect the problems of hindered rotation and excluded volume effects (inability of two molecules to be in the same place). The books by Treloar (1958), Flory (1969), and Yamakawa (1971) have accounts of some of these problems.

(a)

Fig. 5.1(a)   Fisher–Hirschfelder–Taylor models of eight monomer units of a poly(ethylene oxide) chain in two of many possible configurations.

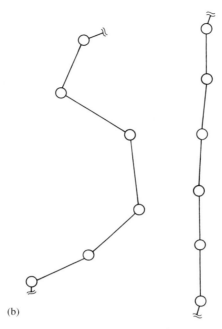

Fig. 5.1(b) Freely jointed chain models duplicating the same configurations as the models above. Normally, however, in the idealization of a macromolecule by a freely jointed chain, each bead-rod unit corresponds to about 10 to 20 monomer units (from Bird *et al.* 1977*a*).

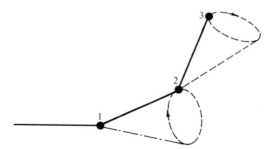

Fig. 5.2 Showing relative motion of adjacent bonds between molecules.

5.1.1 Mean end-to-end length of a polymer molecule

We now regard the chain molecule configuration as a random walk in space so that each 'link' or vector between successive molecules is in a randomly chosen direction relative to its neighbour. Thus we expect the end-to-end distance to be proportional to $a\sqrt{N}$, where N is the number of links and a is the length of a link. This end-to-end distance is commonly taken as a measure of the size of the molecule: we will take an ensemble average over many identical molecules to define this quantity. Hence we define the average end-to-end length r as

$$r^2 = \frac{1}{q}\sum_{j=1}^{q} r_j^2, \tag{5.1}$$

where r_j is the end-to-end length of the jth molecule of the representative set of q molecules. Now suppose the molecule is modelled as shown in Fig. 5.3 being made up of N vectors of length a oriented randomly: then

$$\mathbf{r}_j = \sum_{i=1}^{N} \mathbf{a}_i^{(j)}, \tag{5.2}$$

and hence

$$r^2 = \frac{1}{q}\sum_{j=1}^{q}\left\{\sum_{i=1}^{N} \mathbf{a}_i^{(j)}\right\}^2. \tag{5.3}$$

This is a general result and we have to be more specific about the relative orientation between molecules to get useful results. In general, we have, from eqn (5.3):

$$r^2 = \frac{1}{q}\sum_{i=1}^{q}\left(a_1^{(i)_2} + a_2^{(i)_2} + \cdots a_N^{(i)_2} + \mathbf{a}_1^{(i)}\cdot\mathbf{a}_2^{(i)} + \cdots \mathbf{a}_{N-1}^{(i)}\cdot\mathbf{a}_N^{(i)}\right). \tag{5.4}$$

Now $a_m^{(i)_2} = a^2$ for all links, (they all have the same length) and hence

$$r^2 = Na^2 + \frac{1}{q}\sum_{j=1}^{q}\left(\mathbf{a}_1^{(j)}\cdot\mathbf{a}_2^{(j)} + \cdots \mathbf{a}_{N-1}^{(j)}\cdot\mathbf{a}_N^{(j)}\right). \tag{5.5}$$

The simplest case to consider is when we have an ideal freely-jointed chain; in that case the projection of one bond on another ($\mathbf{a}_p^{(j)}\cdot\mathbf{a}_q^{(j)}$) is, on average, zero, and we find, as anticipated above,

$$r = a\sqrt{N}. \tag{5.6}$$

For other chains (non-freely rotating) we get, depending on constraints (Flory 1969):

$$r = \text{const} \times a\sqrt{N}. \tag{5.7}$$

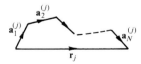

Fig. 5.3   Random-vector molecular model.

Note that the actual length of the molecule is $Na(=R)$ which is enormously $(O(\sqrt{N})$ greater, a thousand times greater, possibly) than the average end-to-end length. The solvent can cause some change in r due to interaction with the molecules: for this problem see Flory (1969).

5.1.2 Distribution function for end-to-end length for ideal chains

For any given ideal chain of N links, the end-to-end vector $\mathbf{r}$ is a statistical quantity and may be characterized by a probability density $P(N, \mathbf{r})$. We may think of one end of the chain being at the origin if we wish. Now consider the addition of one extra link with vector $\mathbf{a}$; clearly

$$P(N+1, \mathbf{r}) = P(N, \mathbf{r}+\mathbf{a}), \tag{5.8}$$

for any given configuration.

If $|\mathbf{a}| \ll |\mathbf{r}|$ we can expand the right-hand side of (5.8) as a Taylor series, noting that a has components a_x, a_y, a_z.

$$P(N, \mathbf{r}+\mathbf{a}) = P(N, \mathbf{r}) + \boldsymbol{\nabla} P \cdot \mathbf{a} + \frac{1}{2!}\left[\frac{\partial^2 P}{\partial x^2}\Big|_{\mathbf{a}=0} a_x^2 + \frac{\partial^2 P}{\partial y^2}\Big|_{\mathbf{a}=0} a_y^2\right.$$
$$\left. + \frac{\partial^2 P}{\partial z^2}\Big|_{\mathbf{a}=0} a_z^2 + 2\frac{\partial^2 P}{\partial x \partial y}\Big|_{\mathbf{a}=0} a_x a_y + \cdots\right] + O(a^3). \tag{5.9}$$

If we average (5.9) over all possible $\mathbf{a}$, noting that $\langle \mathbf{a} \rangle = \mathbf{0}$; then

$$\langle P(N, \mathbf{r}+\mathbf{a}) \rangle = \langle P(N, \mathbf{r}) \rangle + \frac{a^2}{6}\boldsymbol{\nabla}^2 P\Big|_{\mathbf{a}=0} + O(a^3). \tag{5.10}$$

Here we have used the fact that $\langle a_x^2 \rangle = \langle a_y^2 \rangle = \langle a_z^2 \rangle = \frac{1}{3}a^2$; $\langle a_x a_y \rangle = 0$ and so on. Now we denote $\langle P(N, \mathbf{r}) \rangle = P(N, r)$, where we have averaged out the direction of $\mathbf{r}$. Since $\langle P(N+1, \mathbf{r}) \rangle = P(N+1, r)$, then if $N \gg 1$, we can also write

$$P(N+1, r) = P(N, r) + \frac{\partial P}{\partial N} + O(1/N^2). \tag{5.11}$$

Thus, equating $P(N+1, r)$ and $\langle P(N, \mathbf{r}+\mathbf{a}) \rangle$ we have, approximately

$$\frac{\partial P}{\partial N} = \frac{a^2}{6}\boldsymbol{\nabla}^2 P, \tag{5.12}$$

which is a diffusion equation for $P(r, N)$ when N is large. The relevant solution of this equation depends only on r since the undisturbed molecule is spherically symmetric. This solution is the point-source solution of heat-conduction:

$$P = \text{const.} \times N^{-3/2}\exp(-3r^2/2Na^2). \tag{5.13}$$

(Notice that this distribution does not go to zero for $r > Na$, whereas any real molecular distribution function must do this; if $N \gg 1$, P is small for $r > Na$.) The normalizing constant is found from insisting that the end-to-end distance is certainly between 0 and ∞:

$$\int_0^\infty P(r)4\pi r^2 \, dr = 1. \tag{5.14}$$

Hence, we find the complete solution:

$$P(N, r) = \left(\frac{2}{3}\pi a^2 N\right)^{-3/2} \exp\left(-\frac{3r^2}{2Na^2}\right). \tag{5.15}$$

By integration we find

$$\langle r^2 \rangle = \int_0^\infty r^2 P(r)4\pi r^2 \, dr = Na^2. \tag{5.16}$$

Note that the most probable separation of the ends is zero and the molecule is then in a compact spherical configuration; note also that we have neglected the effects of the solvent; its effect is minimal for the so-called θ solvent (Flory 1969).

5.1.3   Tension in a polymer chain

Suppose that we hold the chain ends clamped a distance r apart. Because of the type of ideal chain considered, the internal energy is constant (zero).[†] The entropy is not constant, however, since the number of configurations available to the chain is less when the ends are forcibly held apart. In this development we assume a finite number of configurations exists. From the way we defined P, and the fact that all chain states are equally probable, it follows that the number of configurations available at any given r is proportional to P. Also, from the Boltzmann rule, the entropy (s) is proportional to the logarithm of the number of available configurations. Hence

$$s = k \ln P + \text{const.}, \tag{5.17}$$

where the constant is arbitrary. Thus, from (5.15)

$$s = C - \frac{3kr^2}{2Na^2}, \tag{5.18}$$

where C is an arbitrary constant and k is Boltzmann's constant.[‡] Clearly s is a maximum when $r = 0$, which is the most probable configuration of a free chain.

[†] This approximate result follows from experiments on solid rubber composed of many molecules (Treloar 1958).
[‡] $k = 1.3807 \times 10^{-23} J/K$.

The first law of thermodynamics can now be applied. If the work *output* from the chain is dW, the heat *input* is dQ, and the change of internal energy is dE, then

$$dQ = dW + dE. \tag{5.19}$$

In the present case dE is assumed to be zero (no change in internal energy) and if the force on the free chain ends is $\mathbf{F}$, then in a small displacement $d\mathbf{r}$ of this end, the work output dW is $-\mathbf{F}\cdot d\mathbf{r}$. For mechanical equilibrium, $\mathbf{F}$ and $\mathbf{r}$ are always parallel, and hence $d\mathbf{r}$ and $\mathbf{F}$ are also parallel. Thus the work output dW is just $-F\,dr$, at constant temperature. Hence, since $dQ = T\,ds$ in a reversible process, $dQ = T\,ds = -F\,dr$; and thus $F = |\mathbf{F}| = -T(ds/dr)$ or $\mathbf{F} = (3kT/Na^2)\mathbf{r}$, from eqn (5.18). This gives the average force in the chain when the ends are held a distance r apart. The molecule is equivalent to a Hookean spring of zero length and spring constant $3kT/Na^2$.

This result is only true for small r/Na since our expression for s is only valid under these circumstances, as careful examination will show; if $r/Na \sim 1$, the force must increase indefinitely since the molecule is almost completely straightened out. The actual spring law is an inverse Langevin function L^{-1} (Fig. 5.4), where (Flory 1969)

$$\frac{aF}{kT} = L^{-1}(r/Na), \tag{5.20}$$

and

$$L\left(\frac{aF}{kT}\right) = \coth\left(\frac{aF}{kT}\right) - \frac{kT}{aF} = \frac{r}{Na}. \tag{5.21}$$

A useful approximation to the inverse Langevin function has been derived by Warner (1972). He suggested that one could use, in place of the rule (5.20), the following approximation.

$$F = \frac{3kTr}{Na^2} \Bigg/ \left[1 - \left(\frac{r}{Na}\right)^2\right]. \tag{5.22}$$

This curve is also plotted in Fig. 5.4. In future, we shall often use eqn (5.22) in place of (5.20). The retractile force of individual molecules is responsible for rubber elasticity and for the viscoelastic behaviour of polymer solutions; each molecule can store mechanical energy and subsequently release it.

5.2 Polymer molecules in dilute solution

Two other factors are relevant when a molecule is in solution:

(i) *Hydrodynamic forces.* If the molecule is displaced from its equilibrium state, it will elastically return to the equilibrium configuration; the average motion

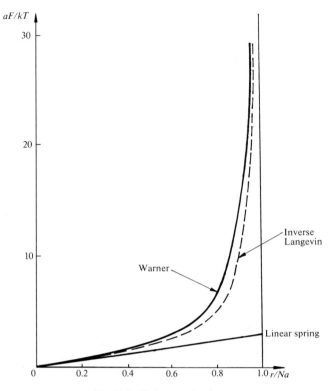

Fig. 5.4 Various spring laws.

through the fluid meets with viscous resistance from the solvent. Since molecules are small, the Reynolds numbers are small, and we can use Stokes's law of resistance. It is usual to idealize the molecule by lumping the viscous resistance at discrete points with frictionless springs in between; each submolecule will be long enough so that the 'spring' between acts as an ideal freely-jointed chain. We now suppose that the viscous resistance force on bead *i* can be written as a force $\mathbf{F}_i^{(v)}$ (see Fig. 5.5):

$$\mathbf{F}_i^{(v)} = \zeta(\mathbf{v} - \dot{\mathbf{r}}_i). \tag{5.23}$$

Here **v** is the local solvent velocity, $\dot{\mathbf{r}}_i$ is the bead velocity referred to the same co-ordinate system, and ζ is a constant friction coefficient.

Each bead will induce a velocity field at every other bead, so that really $\mathbf{F}_i^{(v)}$ contains a component from other, especially nearby, beads; this hydrodynamic interaction leads to the Kirkwood–Zimm (1956) theory. To begin with we will ignore bead–bead interaction. At this stage this seems permissible, since some of the most dramatic effects in solutions occur when the beads are widely separated.

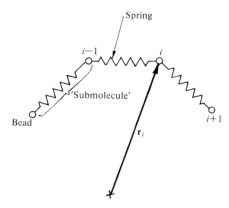

Fig. 5.5   Bead-spring 'molecule' notation.

(ii) *Brownian motion*. This refers to the random action of the solvent molecules on the polymer 'beads'. The theory goes back to Einstein and was very important in establishing the kinetic theory of matter. For a free particle, Einstein's theory imagines that the particle is battered by many random, uncorrelated forces and in consequence undergoes a random walk, composed of many small steps. This is very similar to the spatial random-walk problem of the polymer molecule discussed previously; however, the addition of another 'step' is now equivalent to an extra time step, and the differential equation is formed by replacing N by t; hence $P(\mathbf{r}, N)$ is replaced by $\psi(\mathbf{r}, t)$ and the differential equation becomes,

$$\frac{\partial \psi}{\partial t} = D\mathbf{\nabla}^2 \psi, \tag{5.24}$$

where D is a diffusion coefficient. The solution is, following eqn (5.15)

$$\psi(\mathbf{r}, t) = \frac{1}{(4\pi Dt)^{1.5}} \exp(-r^2/4Dt), \tag{5.25}$$

giving the density of particles at $(\mathbf{r}, t)$, subject to there being a total of one particle over all space.

If a gradient of the ψ-(density of particle) function is present, then this is equivalent to a force $\mathbf{F}^{(b)}$ on a particle. More precisely, on the bead we have a Brownian motion force

$$\mathbf{F}^{(b)} = -kT\frac{\partial}{\partial \mathbf{r}}\ln\{\psi(\mathbf{r}, t)\}. \tag{5.26}$$

Similarly, for N beads, we will have a joint probability density function $\psi(\mathbf{r}_1, \mathbf{r}_2, \ldots, t)$ and the force on the ith bead is, following (5.26)

$$\mathbf{F}_i^{(b)} = -kT\frac{\partial}{\partial \mathbf{r}_i}\ln \psi. \tag{5.27}$$

5.2.1   *Equation of conservation of beads*

Finally, consider many molecules moving in *configuration space* $(\mathbf{r}, t)$; note that we do not work in phase space $(\mathbf{r}, \dot{\mathbf{r}}, t)$ because ψ, will not be assumed to depend on the velocities of the beads. (The use of phase space has been explored by Bird *et al.* (1987*b*)). The number of points in a finite volume of configuration space Ω is, for N beads

$$\int_\Omega \psi \, \mathrm{d}\mathbf{x}_1 \, \mathrm{d}\mathbf{x}_2 \ldots \mathrm{d}\mathbf{x}_N \left(\equiv \int_\Omega \psi \, \mathrm{d}\Omega \text{ say} \right). \tag{5.28}$$

Points $\psi(\mathbf{x}_i, \ldots \mathbf{x}_N)$ in configuration space are not created or destroyed, as each represents a molecule, thus the rate of loss of points in a volume $\Omega(- \partial/\partial t \int_\Omega \psi \, \mathrm{d}\Omega)$ must be equal to the number of points crossing the surface s of $\Omega = \int_s \psi \mathbf{u} \cdot \mathbf{n} \mathrm{d}s$ where $\mathbf{u}$ is a hyper-velocity $(\dot{\mathbf{r}}_1, \dot{\mathbf{r}}_2, \ldots \dot{\mathbf{r}}_N)$ and $\mathbf{n}$ is an outward-pointing unit vector in the space. Thus from a balance of the number of beads, we have

$$\frac{\partial}{\partial t} \int_\Omega \psi \, \mathrm{d}\Omega + \int_s \psi \mathbf{u} \cdot \mathbf{n} \, \mathrm{d}s = 0. \tag{5.29}$$

Now, by Green's transformation $\int_s \psi \mathbf{u} \cdot \mathbf{n} \mathrm{d}s = \int_\Omega \boldsymbol{\nabla} \cdot (\psi \mathbf{u}) \mathrm{d}\Omega$, where $\boldsymbol{\nabla} \cdot (\mathbf{z}) = (\partial/\partial \mathbf{r}_i)\{\mathbf{z}_i\}$; hence, after reduction the integral form (5.29) becomes

$$\frac{\partial \psi}{\partial t} + \frac{\partial}{\partial \mathbf{r}_i} \{\psi \dot{\mathbf{r}}_i\} = 0, \tag{5.30}$$

which is the equation of continuity in configuration space.

5.2.2   *Dilute dumbbell solutions*

Here we will only treat the two-bead model (dumbbell) in detail although the N-bead analysis follows the same pattern; it has been found that the dumbbell qualitatively reproduces many features of the N-bead response. The model molecule is as shown in Fig. 5.6. We will assume the surrounding fluid is at constant temperature, that it is incompressible, and there are n_0 molecules/unit volume.

In a dilute solution molecules do not by definition, interact with one another. The criterion for non-interaction has been given by James and Sridhar (1995) to be that the product of mass concentration (c) and intrinsic viscosity, $[\eta]$, should be less than about 0.77. (The intrinsic viscosity is the quantity $(\eta - \eta_s)/\eta_s c$ in the limit $c \to 0$, where η is the solution viscosity and η_s is the solvent viscosity.) We suppose that this is observed here.

Define

$$\mathbf{r}_0 = \tfrac{1}{2}(\mathbf{r}_1 + \mathbf{r}_2). \tag{5.31}$$

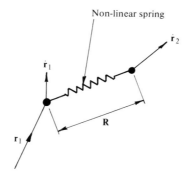

Fig. 5.6 Dumbbell model. Solvent velocity at any point **r** is **Lr**.

Hence, $\mathbf{r}_0$ is the centroidal position, and if **R** is the end-to-end vector (Fig. 5.6),

$$\mathbf{R} = \mathbf{r}_2 - \mathbf{r}_1. \tag{5.32}$$

We will find an equation for $\psi(\mathbf{r}_1, \mathbf{r}_2)$ under the following assumptions:

(a) Dumbbell does not alter local solvent velocity **v(r)**.

(b) Only homogeneous incompressible fluid motions are considered; that is,

$$\mathbf{v} = \mathbf{V}_0 + \mathbf{Lr}, \tag{5.33}$$

where

$\mathbf{L} = \mathbf{L}(t), \mathbf{V}_0$ is the constant speed at the centroid of the molecule and $\operatorname{tr}\mathbf{L} = 0$.(incompressible condition). $\tag{5.34}$

The equations of motion for the beads are then formed by a force balance. Neglecting dumbbell mass-accelerations these become

$$\zeta\{\mathbf{v}(\mathbf{r}_1) - \dot{\mathbf{r}}_1\} + \mathbf{F}^{(c)} - kT\frac{\partial}{\partial \mathbf{r}_1}\ln\psi = 0 \tag{5.35}$$

$$\zeta\{\mathbf{v}(\mathbf{r}_2) - \dot{\mathbf{r}}_2\} - \mathbf{F}^{(c)} - kT\frac{\partial}{\partial \mathbf{r}_2}\ln\psi = 0 \tag{5.36}$$

where $\mathbf{F}^{(c)}$ is the *tension* in the connector (not necessarily a Hookean spring). Subtraction and division by the friction factor ζ gives

$$\dot{\mathbf{R}} = \mathbf{LR} - \frac{2}{\zeta}\mathbf{F}^{(c)} - \frac{2}{\zeta}kT\frac{\partial}{\partial \mathbf{R}}\ln\psi. \tag{5.37}$$

Addition of eqns (5.35) and (5.36) gives

$$\dot{\mathbf{r}}_0 = \mathbf{V}_0 + \mathbf{L}\mathbf{r}_0 - \frac{kT}{2\zeta}\frac{\partial \ln \psi}{\partial \mathbf{r}_0}. \tag{5.38}$$

With the homogeneous velocity gradients assumed, we do not expect the distribution function ψ, to depend on $\mathbf{r}_0$, and hence the last term in eqn (5.38) vanishes. Then the dumbbell centroid moves at the speed of the surrounding fluid. Now the continuity equation is

$$\frac{\partial \psi}{\partial t}(\mathbf{r}_1, \mathbf{r}_2, t) + \frac{\partial}{\partial \mathbf{r}_1} \cdot (\dot{\mathbf{r}}_1 \psi) + \frac{\partial}{\partial \mathbf{r}_2} \cdot (\dot{\mathbf{r}}_2 \psi) = 0. \tag{5.39}$$

Putting eqn (5.39) in terms of $\psi(\mathbf{R}, t)$ and $\mathbf{R}$ we find

$$\frac{\partial \psi}{\partial t} + \frac{\partial}{\partial \mathbf{R}} \cdot (\dot{\mathbf{R}}\psi) = 0. \tag{5.40}$$

Substituting for $\dot{\mathbf{R}}$, using (5.37), we obtain an equation of diffusion for $\psi(\mathbf{R}, t)$:

$$\frac{\partial \psi}{\partial t} + \frac{\partial}{\partial \mathbf{R}} \cdot \left\{ \psi \mathbf{L}\mathbf{R} - \frac{2}{\zeta}\mathbf{F}^{(c)}\psi - \frac{2}{\zeta}kT\frac{\partial \psi}{\partial \mathbf{R}} \right\} = 0. \tag{5.41}$$

Here $\mathbf{L}$ is supposed given; also $\mathbf{F}^{(c)} = (d\chi/d R)\mathbf{R}/|\mathbf{R}|$ where $\chi(R)$ is a spring potential which is supposed known. Thus we can solve for ψ. In case the fluid is at rest ($\mathbf{L} = 0$) then the equilibrium distribution function ψ does not depend on t, is spherically symmetric and may be found by integrating eqn (5.41). Normalizing the result we find

$$\psi = \exp(-\chi/kT) \Big/ \int_0^\infty \left\{ d R 4\pi R^2 \exp -\frac{\chi}{kT} \right\}. \tag{5.42}$$

In the special case of a linear spring, then ψ is a Gaussian function, as already seen in eqn (5.25).

The question of boundary conditions for the diffusion eqn (5.41) arises. The boundary in this case is the limit of the configuration space. For the linear-spring molecule, all points in space can be reached and it is only necessary to ensure that ψ vanishes when $|\mathbf{R}|$ becomes very large. In other cases, one needs to ensure that there is no flux of particles out of the space. If $\mathbf{n}$ is the outward vector from the configuration space, then we must have on the boundary, for the dumbbell case,

$$\psi\dot{\mathbf{R}} \cdot \mathbf{n} = 0. \tag{5.43}$$

A similar result holds for more complex molecules. Initial conditions on ψ will also be needed for unsteady flow problems; these will be supplied as needed.

5.2.3 Calculation of the stress tensor

To calculate the stresses when the function ψ is known we write the total stress as

$$\boldsymbol{\sigma} = -p\mathbf{I} + 2\eta_s\mathbf{d} + \boldsymbol{\tau}^{(m)}, \tag{5.44}$$

where $2\eta_s\mathbf{d}$ is the Newtonian solvent stress contribution and $\boldsymbol{\tau}^{(m)}$ is the molecular contribution to the stresses. There are two factors contributing to $\boldsymbol{\tau}^{(m)}$:

(i) There is a contribution to $\boldsymbol{\tau}^{(m)}$ due to the flux of dumbbells across a fixed element (Bird *et al.* 1987*b*) of amount $-n_0 k\,T\mathbf{I}$ which may be absorbed into the pressure term $-p\mathbf{I}$,

(ii) The main contribution to the stress across a plane is by dumbbells straddling the plane, as sketched in Fig. 5.7.

Each molecule that straddles a plane exerts a tension from outside to inside of the box, parallel to the spring, of $\mathbf{F}^{(c)}$. Resolution of all such forces normal and parallel to the plane will give the stress components. If we consider a unit cube, containing n_0 molecules, then $n_0\,\psi(\mathbf{R})\,dX\,dY\,dZ$ is the number of molecules with an end-to-end vector $\mathbf{R}(=\mathbf{i}X+\mathbf{j}Y+\mathbf{k}Z)$. Then the probability that one of these molecules will cross the face normal to X is just X. Thus the stress $\tau_{xx}^{(m)}$ is $n_0\psi\,dX\,dY\,dZ\,XF_x^{(c)}$ for this group of molecules. Integration over all X, Y, Z gives, where $dV = dX\,dY\,dZ$ and the momentum flux term $n_0\,kT$ has been added:

$$\tau_{xx}^{(m)} = n_0 \int\!\!\int\!\!\int_{-\infty}^{\infty} \psi F_x^{(c)} X\,dV - n_0\,kT. \tag{5.45}$$

Similarly, we find $\tau_{xy}^{(m)} = n_0 \iiint \psi F_y^{(c)} X\,dV$, and generally

$$\boldsymbol{\tau}^{(m)} = n_0 \int\!\!\int\!\!\int_{-\infty}^{\infty} \psi\,\mathbf{F}^{(m)}\mathbf{R}\,dV - n_0\,kT\,\mathbf{I},$$

which can be written as

$$\boldsymbol{\tau}^{(m)} \equiv n_0\langle\mathbf{F}^{(c)}\mathbf{R}\rangle - n_0\,kT\,\mathbf{I}, \tag{5.46}$$

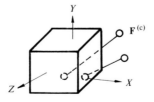

Fig. 5.7 Basis for stress calculation showing connectors crossing planes.

where $\langle\,\rangle$ denotes the average over configuration space. Since $\mathbf{F}^{(c)}$ is parallel to $\mathbf{R}$, clearly the stress component $\boldsymbol{\tau}^{(m)}$ is *symmetric*—an important result in view of the discussion given in Section 2.2.

In most cases we cannot avoid solving the diffusion equation (5.40) and using that value of ψ, to find $\boldsymbol{\tau}^{(m)}$ via integration of (5.46). This is very laborious, and even if done numerically it is often expensive and inaccurate. In some special flows solution is easier, as we now show.

5.2.4   Steady potential flows

Let $\mathbf{v} = \boldsymbol{\nabla}\phi$; then $\phi = \tfrac{1}{2}\mathbf{R}^T\mathbf{L}\mathbf{R}$ if $\mathbf{L}$ is a constant symmetric velocity gradient matrix. In this case it is easy to solve for ψ:

$$\psi = C\exp\left(-\frac{\chi}{kT} + \frac{\zeta\phi}{2kT}\right), \tag{5.47}$$

where C is a normalizing constant and $(\mathrm{d}\chi/\mathrm{d}\mathbf{R}) = \mathbf{F}^{(c)}$ so that χ is the spring potential. Now consider a linear spring law where

$$\mathbf{F}^{(c)} = \frac{3kT}{Na^2}\mathbf{R}; \qquad \chi = \frac{3kT}{2Na^2}R^2, \tag{5.48}$$

and an elongational flow where

$$\mathbf{v} = \left(\dot{\varepsilon}x, \ -\frac{\dot{\varepsilon}y}{2}, \ -\frac{\dot{\varepsilon}z}{2}\right) \qquad \text{and} \qquad \phi = \dot{\varepsilon}\left(\frac{x^2}{2} - \frac{1}{4}(y^2 + z^2)\right), \tag{5.49}$$

then

$$\psi = C\exp\left\{-\frac{3R^2}{2Na^2} + \frac{\zeta\dot{\varepsilon}}{4kT}\left(x^2 - \frac{y^2}{2} - \frac{z^2}{2}\right)\right\} \tag{5.50}$$

and $\boldsymbol{\tau}^{(m)}$ can be found using (5.46). Now if $\zeta\dot{\varepsilon}/4kT > 3/2Na^2$ then $\psi \to \infty$ as $\mathbf{R} \to \infty$, and we will not be able to find C because the integral of ψ over all space does not exist. Physically, the springs have been stretched indefinitely by viscous forces, and no steady solution exists. Generally, we need a molecule of limited length to avoid this problem so that $\chi \to \infty$ at finite $|\mathbf{R}|$. This is an easy but tedious problem in general. For non-potential flows few general results can be obtained, except for linear springs.

5.3   Constitutive equations for dilute dumbbell solutions with Hookean springs

In this case we can find a constitutive equation valid for any flow by an averaging process.

Using index notation and absorbing the $n_0 k T \mathbf{I}$ momentum flux term into the stresses, we can define τ_{ij} using eqn (5.46) to be

$$\tau_{ij} \equiv n_0 k T \delta_{ij} + \tau_{ij}^{(m)} = \frac{3n_0 k T}{Na^2} \langle R_i R_j \rangle. \tag{5.51}$$

If we take the diffusion equation (5.41), with $\mathbf{F}^{(c)} = 3kT\mathbf{R}/Na^2$, multiply it by $R_i R_j$ and then average over configuration space, the unsteady term in (5.41) becomes $\partial/\partial t \langle R_i R_j \rangle$, which is proportional to $\partial \tau_{ij}/\partial t$. Similarly, by the use of the divergence theorem on the other terms, we can show, when ψ vanishes on a large sphere, that the following expression results

$$\frac{\partial}{\partial t} \langle R_i R_j \rangle - L_{jm} \langle R_m R_i \rangle - L_{im} \langle R_m R_j \rangle$$
$$+ \frac{12kT}{\zeta Na^2} \langle R_i R_j \rangle - \frac{4kT}{\zeta} \delta_{ij} = 0. \tag{5.52}$$

Multiplication by $3kTn_0/Na^2$ gives a constitutive equation for τ_{ij}:

$$\frac{\partial \tau_{ij}}{\partial t} - L_{jk} \tau_{ki} - L_{ik} \tau_{kj} + \frac{12kT}{Na^2 \zeta} \tau_{ij} = \frac{12(kT)^2 n_0}{Na^2 \zeta} \delta_{ij}. \tag{5.53}$$

If we let $Na^2 \zeta /12kT$ be λ, a time constant, then if we define $(\Delta(\ )_{ij}/\Delta t)$ by the following [compare eqn (4.26)]:

$$\frac{\Delta(\ )_{ij}}{\Delta t} = \frac{\partial(\ )_{ij}}{\partial t} + v_k \frac{\partial(\ )_{ij}}{\partial x_k} - L_{jk}(\ )_{ki} - L_{ik}(\ )_{kj}, \tag{5.54}$$

and note that spatial gradients of $\tau_{ij}^{(m)}$ are assumed zero here, we can rewrite (5.53) as

$$\lambda \frac{\Delta \tau_{ij}}{\Delta t} + \tau_{ij} = n_0 k T \delta_{ij}. \tag{5.55}$$

Now in the dilute solution the total stress $\boldsymbol{\sigma}$, from (5.44), is made up of three parts, one of which is the pressure term. If we subtract off the momentum flux pressure $n_0 k T \mathbf{I}$ from $\boldsymbol{\tau}$, so that we have (eqn 5.51)

$$\boldsymbol{\tau}^{(m)} = \boldsymbol{\tau} - n_0 k T \mathbf{I} \tag{5.56}$$

then the equation for $\boldsymbol{\tau}^{(m)}$ becomes,

$$\lambda \frac{\Delta \boldsymbol{\tau}^{(m)}}{\Delta t} + \boldsymbol{\tau}^{(m)} = 2\eta^{(m)} \mathbf{d} \tag{5.57}$$

where $\mathbf{d}$ is the rate-of-deformation tensor and the viscosity $\eta^{(m)}$ is just $\lambda \eta_0 k T$. This (upper) *convected Maxwell model* (5.57) is identical to (4.25). The term "convected" derives from the form (5.54); it is often called the UCM model.

5.3.1   An equivalent integral form of the constitutive equation

Consider the equation

$$\tau_{ij}(x_i, t) = \int_{-\infty}^{t} \frac{\eta^{(m)}}{\lambda^2} \exp\{-(t - t')/\lambda\} \frac{\partial x_i}{\partial r_m} \frac{\partial x_j}{\partial r_m} \, \mathrm{d}t'. \tag{5.58}$$

Then r_k are the components of the position vector at time t' of the particle which will be at x_i at time t. The expression $\partial r_i/\partial x_j$ is the displacement gradient F_{ij} [eqn (2.28)] and $\partial x_k/\partial r_i$ is the component of the inverse of $\mathbf{F}$, $(F^{-1})_{ik}$. Thus the term $(\partial x_i/\partial r_m)(\partial x_j/\partial r_m)$ is the inverse of the strain tensor $\mathbf{C}$, [eqn (2.34)]. If we now differentiate (5.58) with respect to time, while following a particle, we find (noting x_i is a function of time t)

$$\frac{\mathrm{D}\tau_{ij}}{\mathrm{D}t} = \frac{\eta^{(m)}}{\lambda^2} \delta_{ij} - \frac{1}{\lambda} \tau_{ij} + \int_{-\infty}^{t} \frac{\eta^{(m)}}{\lambda^2} \exp\{-(t - t')/\lambda\}$$

$$\times \left[\frac{\partial \dot{x}_i}{\partial r_m} \frac{\partial x_j}{\partial r_m} + \frac{\partial x_i}{\partial r_m} \frac{\partial \dot{x}_j}{\partial r_m} \right] \mathrm{d}t'. \tag{5.59}$$

Now we can rewrite $\partial \dot{x}_i/\partial r_m$ as $(\partial v_i/\partial x_k)(\partial x_k/\partial r_m)$, and hence the integral in (5.59) can be converted into a form containing stresses

$$\frac{\mathrm{D}\tau_{ij}}{\mathrm{D}t} = \frac{\eta^{(m)}}{\lambda^2} \delta_{ij} - \frac{1}{\lambda} \tau_{ij} + L_{ik}\tau_{kj} + L_{jk}\tau_{ik}. \tag{5.60}$$

Rearranging (5.60) we recover (5.55). Thus the forms (5.55) and (5.58) are equivalent, and either may be used in calculations; the form (5.57) differs only by an isotropic pressure term; it can also be used conveniently.

5.3.2   Multiple relaxation times

With the form (5.58), we can immediately generalize to the case where there is a number (N) of relaxing species of dumbbells, each with its own value of $\eta^{(m)}$ and λ; the stress contributions are additive, and we find

$$\boldsymbol{\tau} = \sum_{n=1}^{N} \boldsymbol{\tau}_n, \tag{5.61}$$

where

$$\boldsymbol{\tau}_n = \int_{-\infty}^{t} \eta_n \lambda_n^{-2} \exp\{-(t - t')/\lambda_n\} \mathbf{C}^{-1} \, \mathrm{d}t'. \tag{5.62}$$

In the differential form (5.57) each $\boldsymbol{\tau}_n$ satisfies a similar constitutive equation. The integral form will be referred to as the Lodge model, since the polymer stress contribution is identical to that of Lodge's (1964) rubber-like liquid, discussed in Section 5.6.2 below.

A second source of multiple relaxation times arises in molecular models more complex than the dumbbell, where many 'modes' exist (Bird *et al.* 1987*b*), even if all molecules are alike.

Multiple relaxation times in dilute solutions may be difficult to measure. In the absence of a measured spectrum, for monodisperse molecules where all molecules are the same, one may use as an approximation the Zimm spectrum, which includes hydrodynamic interaction between beads in an $N+1$-bead, N-spring model. In this case (Huilgol and Phan-Thien 1997; Bird *et al.* 1987*b*) a good fit to G' and G'' is obtained, and one finds the viscosity difference to be

$$\eta - \eta_s = n_0 \, kT \sum_{j=1}^{N} \lambda_j \tag{5.62 a}$$

where η_s is the solvent viscosity.

Now n_0, the number of molecules per unit volume, can be expressed in terms of the polymer weight concentration c. We have $n_0 = N_A c/M$, where N_A is Avogadro's number (6.02×10^{23}) and M is the molecular weight of the (linear chain) polymer. The viscosity difference (or the intrinsic viscosity $[\eta] \equiv (\eta - \eta_s)/c\eta_s$, vanishingly small c) can be measured, and hence the sum of the relaxation times can be found from (5.62a) once M is known. If the longest relaxation time ($j = 1$) in (5.62a) is λ_1 then the remaining relaxation times scale approximately as $\lambda_{1j}^{-(2+\sigma)}$, where σ varies with the 'goodness' of the solvent for the molecules (Bird *et al.* 1987*b*).

One finds σ varies from 0 to -0.5, and estimates of σ can be found from the literature. Once σ is known, and the number of relaxation times (N) is fixed, then $\sum_{j=1}^{N} j^{-(2+\sigma)}$ can be found, enabling λ_1 to be calculated, and hence the remaining $N-1$ relaxation times.

5.3.3   *Response of the convected Maxwell model*

We will compute the response from the integral form (5.58) thus avoiding some careful limit processes for step strains. Consider a simple unsteady shearing motion where the velocity vector is

$$\mathbf{v} = y\dot{\gamma}(t)\mathbf{i}, \tag{5.63}$$

where $\dot{\gamma}(t)$ is arbitrary. Then we can easily compute $\mathbf{C}^{-1}$, following Section 2.3.6

$$\mathbf{C}^{-1}(t') = \begin{bmatrix} 1+s^2 & s & 0 \\ s & 1 & 0 \\ 0 & 0 & 1 \end{bmatrix} \tag{5.64}$$

where the amount of shear

$$s = \int_{t'}^{t} \dot{\gamma}(t')\,dt' = \gamma(t) - \gamma(t') \tag{5.65}$$

is independent of position $\mathbf{x}$. Computing the shear response τ_{xy} from (5.58) we get,

$$\tau_{xy}(t) = \frac{\eta}{\lambda^2} \exp -t/\lambda \int_{-\infty}^{t} [\gamma(t) - \gamma(t')] e^{t'/\lambda} \, dt', \tag{5.66}$$

or, integrating by parts, we find the linear viscoelastic form [cf. eqns (1.24) and (2.83)]

$$\tau_{xy}(t) = \frac{\eta^{(m)}}{\lambda} \int_{-\infty}^{t} \exp(-(t - t')/\lambda)\dot{\gamma}(t') \, dt' \tag{5.67}$$

The results for stress relaxation and sinusoidal shearing are the same as (2.105) and (2.94) respectively; the viscosity is constant in steady shearing for all shear rates. For an impulse of strain-rate (step-strain of magnitude γ_0 at $t = 0$) we have

$$\tau_{xy}(t) = \frac{\eta^{(m)}}{\lambda} \gamma_0 \, e^{-t/\lambda} \quad (t > 0). \tag{5.68}$$

Lodge (1964) has also shown that N_1 relaxes at the same rate as the shear stress for one relaxation time. He also shows that in a sinusoidal shearing N_1 has a frequency twice that of the shear stress. For a sudden-start of shearing, we find

$$\tau_{xy} = \eta^{(m)}\dot{\gamma}_0(1 - \exp -t/\lambda), \tag{5.69}$$

and the normal stress $N_1 = \tau_{xx}^{(m)} - \tau_{yy}^{(m)}$ is ($t > 0$),

$$N_1(t) = 2\eta^{(m)}\lambda\dot{\gamma}^2 \left[1 - \left(1 + \frac{t}{\lambda}\right) \exp(-t/\lambda)\right], \tag{5.70}$$

while $N_2 = 0$ at all times. Thus there is no overshoot of either shear stress or normal stress difference.

Finally we consider steady elongational flow. Let the rate of elongation be $\dot{\varepsilon}$ for all t, and the velocity field be $\mathbf{v} = (\dot{\varepsilon}x, -\dot{\varepsilon}y/2, -\dot{\varepsilon}z/2)$. Then we obtain the Trouton viscosity η_E, already given in Chapter 4 [eqn (4.27)]

$$\frac{\tau_{xx} - \tau_{yy}}{\dot{\varepsilon}} \equiv \eta_E = \frac{3\eta^{(m)}}{(1 - 2\lambda\dot{\varepsilon})(1 + \lambda\dot{\varepsilon})}. \tag{5.71}$$

For $\dot{\varepsilon} = 0$ we find $\eta_E = 3\eta^{(m)}$ as expected. It is curious that η_E has a minimum when $\dot{\varepsilon}\lambda = -0.25$ of $1.6\eta^{(m)}$ and rises sharply so that no solution exists [cf. eqn (5.50)] for $\lambda\dot{\varepsilon} \gtrsim \frac{1}{2}$, $\lambda\dot{\varepsilon} \lesssim -1$. The hydrodynamic forces expand the dumbbells to infinite separation in these cases, and the model is useless in this region.

For the two-dimensional elongational field (sometimes called pure shearing or strip biaxial extension) where the velocity field is $(\dot{\varepsilon}x, -\dot{\varepsilon}y, 0)$, the corresponding value of η_E is given by

$$\eta_E = 4\eta^{(m)}/(1 - \lambda^2 \dot{\varepsilon}^2). \tag{5.72}$$

In this case the minimum η_E is just $4\eta^{(m)}$ at $\dot{\varepsilon} = 0$.

5.4 Weak and strong flow classification via dumbbell mechanics

It is clear from the preceding section that the Lodge model shows great differences in response to shearing and elongational flows. The obvious question that arises is to ask what happens in other cases, and is it possible to use dumbbell mechanics to set up a flow classification scheme which enables one to decide whether a given flow is shear-like or not? Such a classification scheme is complementary to that of Pipkin's (Section 3.6.1), which deals with 'fast' or 'slow' flows within a single category (for example shearing flows).

Suppose that a dilute solution of dumbbells is subjected to a steady, homogeneous flow field, so that the velocity gradient tensor $\mathbf{L}$ is constant. After a long time, the dumbbells will either reach some equilibrium average end-to-end distance or they will be extended beyond bound. We shall call the former *weak* flows and the latter *strong*.

We make use of previous calculations (Section 5.3) and we see from eqn (5.51) that if $\mathbf{R}$ is the dumbbell end-to-end vector,

$$\langle R_i R_j \rangle = \text{constant} \times \tau_{ij}, \tag{5.73}$$

and hence $\langle R_i^2 \rangle = \text{constant} \times \tau_{ii}$. The question of whether the flow is weak (or not) depends on the convergence (or non-convergence) of the integral in eqn (5.58). Equivalently, we can go back to eqn (5.37), which gives, for the linear-spring case with relaxation time λ,

$$\dot{\mathbf{R}} = \mathbf{L}\mathbf{R} - \mathbf{R}/2\lambda + \text{Brownian motion}. \tag{5.74}$$

Suppose we consider the injection of some (possibly hypothetical) tracer particles into our flow field. If they are simply small spheres or other compact particles, they will trace out the fluid particle path lines but will not tell us anything about the expected response of a viscoelastic material element in such a flow. If, however, we consider injecting dumbbell models, consisting of two very small spheres linked by a linear-law spring whose rest length is zero, then the response of these models in a given flow will resemble the response of a dilute solution in the same flow.

It will be demonstrated below that a dilute solution of dumbbell models and a plausible network model of a concentrated solution behave in an identical manner (that is, have an identical constitutive equation for the polymer contribution to the stress tensor) and that this constitutive law is useful for at least qualitative discussion of real viscoelastic flows of concentrated polymer fluids.

Hence the flow classification will also be useful for concentrated solutions and melts. In the present classification scheme for flows we are concerned with the solutions of eqn (5.74), and the question of interest is the *boundedness* of these solutions: this depends on the eigenvalues of (5.74) and not upon the random excitations. Hence we ignore Brownian motion here; it is not necessary for modelling the transition between weak and strong flows. There is no need of Brownian motion to prevent total collapse of the hypothetical test particles; at time zero they are inserted into the fluid with a definite spacing between the beads. Thus, because of the nature of the response it seems reasonable to use our hypothetical dumbbell 'behaviour' to characterize the kinematics of a flow field. The vague term 'behaviour' must mean something related to the length of the vector joining the beads for a single dumbbell; it will actually turn out to be more convenient to consider a swarm of dumbbells and to compute ensemble averages of the quantities of interest. By studying the dumbbell tracer response, we will demonstrate a method of flow classification and characterization. We assume the velocity field is homogeneous in space around a given dumbbell, so that at point $\mathbf{x}$ the velocity is

$$\mathbf{v} = \mathbf{L}(t)\mathbf{x}, \tag{5.75}$$

where the components of the velocity gradient matrix $\mathbf{L}$ may in general be functions of time (t) but not of space ($\mathbf{x}$); no loss of generality occurs due to the suppression of the motion of the origin in this calculation. To justify these simplifications, we assume that the scale of length appropriate to our test particles is much smaller than any length scale appropriate to the flow field. It is also assumed that the centroids of the test particles are convected at the local fluid velocity without slip. It then follows that $\mathbf{L}$ may depend on time even in a Eulerian steady flow and that the variations of the velocity gradients at the particle due separately to unsteadiness and transportation in such a fixed frame are irrelevant; it is only the material or particle-following rate of change that is significant. This point is emphasized here because it seems to be largely independent of the details of the rheology of the fluid. The above formulation is, of course, identical, except for the omission of Brownian motion, to the above formulation for a dilute solution of linear-spring dumbbell molecules when hydrodynamic interaction between the spheres and internal viscosity are neglected (Section 5.3). In most cases, the main functions of the Brownian motion are to maintain the molecules in a non-collapsed condition in a quiescent fluid and to maintain a random orientation in the same condition. In many so-called strong flows its contribution to the force balance on the molecule is minimal except in very small velocity gradients, and except in these conditions, our test dumbbells show molecular behaviour, as we will discuss below. Defining dimensionless quantities in (5.74) we have

$$t^* = t/\lambda, \quad \mathbf{L}^* = \lambda\mathbf{L}, \tag{5.76}$$

so that (5.74) becomes

$$\frac{d\mathbf{R}}{dt^*} = \mathbf{L}^*(t^*)\mathbf{R} - \tfrac{1}{2}\mathbf{R}. \tag{5.77}$$

For any given dumbbell, we suppose that an initial value of $\mathbf{R}$ is given at $t = 0$:

$$\mathbf{R} = \mathbf{R}_0 \text{ at } t^* = 0. \tag{5.78}$$

Thus, since $\mathbf{R}$ depends only on $\mathbf{L}^*$ when $\mathbf{R}_0$ is fixed, if we can solve (5.77) and compute the end-to-end vector $\mathbf{R}$ as a function of time, we can characterize our flow field.

The solution of (5.77) is

$$\mathbf{R} = e^{-t^*/2}\mathbf{F}_0(t^*)\mathbf{R}_0, \tag{5.79}$$

where $\mathbf{F}_0(t)$ is the displacement gradient $(\partial\mathbf{R}/\partial\mathbf{R}_0)$ referred to the initial configuration at $t^* = 0$. So, for any given dumbbell, the value of $R^2(t)$ may be computed formally as

$$R^2 = e^{-t^*}\mathbf{R}_0^{\mathrm{T}}\mathbf{C}\mathbf{R}_0, \tag{5.80}$$

where $\mathbf{C}$ is the relevant strain tensor. We now suppose that our seeding of dumbbells was initially random in orientation and that all dumbbells have the same initial length h_0.

If we average over the orientation space, we find the average of R^2 as a measure of dumbbell distortion by the flow field. Carrying out this process, we find

$$h^2 = \langle R^2 \rangle = \frac{h_0^2}{3} e^{-t^*} \operatorname{tr}\mathbf{C}, \tag{5.81}$$

where $h_0 = |R_0|$. The suggestion is that

$$\mathrm{Sg} \equiv \left(\frac{h}{h_0}\right)^2 = 1/3\,e^{-t^*}\{\operatorname{tr}\mathbf{C}_0(t^*)\} \tag{5.82}$$

is a suitable criterion for judging the severity of structural distortion in a time t of a structure being convected by a flow field. We shall call Sg the dimensionless stretching parameter.

While the above results are exact, they are not very useful owing to the need to compute $\operatorname{tr}\mathbf{C}$; this is not a trivial operation in a general flow field. We now consider the simpler case of constant velocity gradients.

5.4.1 Constant velocity gradients: strong and weak flows

Before going to the finite time case, we can consider the steady-state case when $t \to \infty$, and the test particles spend a very long time in a region of constant velocity gradient. From eqn (5.80), the value of Sg in (5.82) will depend only on whether the largest real part of any of the eigenvalues of $\mathbf{L}^*$ exceeds, equals, or is less than 0.5. In the first case, Sg is infinite, and we have a strong flow; in the two latter cases it is not. Thus, all the flows may be divided into strong and weak flow classes. A convenient way to consider this is to form the matrix $\mathscr{L} = \mathbf{L}^* - \frac{1}{2}\mathbf{I}$ [see eqn (5.77)], where $\mathbf{I}$ is the unit matrix. Then, if any eigenvalue of $\mathscr{L}$ has a positive real part, we have a strong flow. Forming the determinantal equation explicitly, noting that tr $\mathbf{L} = 0$ and tr $\mathscr{L} = -3/2$, we find that the eigenvalues χ of $\mathscr{L}$ satisfy

$$\chi^3 + \frac{3}{2}\chi^2 + \frac{\chi}{2}\left[\frac{9}{2} - \text{tr }\mathscr{L}^2\right] - \det \mathscr{L} = 0 \tag{5.83}$$

Thus, they depend only on the two parameters tr $\mathscr{L}^2$ and det $\mathscr{L}$. We can use the Hurwitz–Routh (Bellman 1960) criterion to obtain the result that weak flows exist if

$$\det \mathscr{L} < 0 \tag{5.84}$$

and

$$\tfrac{27}{16} - \tfrac{3}{4} \text{ tr } \mathscr{L}^2 + \det \mathscr{L} > 0. \tag{5.85}$$

Figure 5.8 shows a chart where these regions of strong and weak flows are delineated in terms of the two parameters det $\mathscr{L}$ and tr $\mathscr{L}^2$. We may note that the above criterion gives precisely the same results for strong and weak flows as other criteria in which Brownian motion is considered. The result for plane flows with a velocity gradient matrix of the form

$$\mathbf{L}^* = \begin{pmatrix} a & b & 0 \\ c & -a & 0 \\ 0 & 0 & 0 \end{pmatrix} \tag{5.86}$$

is

$$\det \mathscr{L} = \tfrac{1}{2}(a^2 + bc - \tfrac{1}{4}) < 0, \tag{5.87}$$

and

$$\tfrac{27}{16} - \tfrac{3}{4} \text{ tr } \mathscr{L}^2 + \det \mathscr{L} = 1 - a^2 - bc > 0. \tag{5.88}$$

In the case of a simple elongational flow, where $\mathbf{L}^* = \text{diag }(\dot{\varepsilon}, -\dot{\varepsilon}/2, -\dot{\varepsilon}/2)$, we have shown that the flow is weak if

$$-1 < \dot{\varepsilon} < \tfrac{1}{2}. \tag{5.89}$$

If we compute (5.84) and (5.85), we find, for a weak flow

$$(\dot{\varepsilon} + 1)^2(2\dot{\varepsilon} - 1) < 0, \tag{5.90}$$

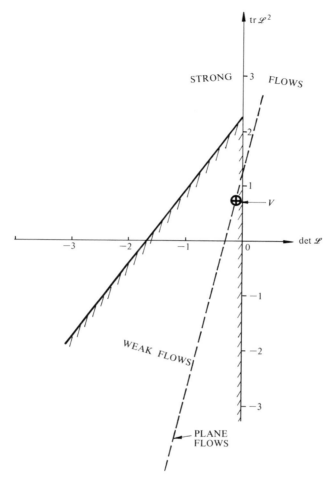

Fig. 5.8 Regions of strong and weak flow in terms of the invariants of the modified velocity gradient matrix, where $\mathscr{L} = \lambda\mathbf{L} - \frac{1}{2}\mathbf{I}$. Dashed line is locus of all plane flows and also the eccentric disc rheometer flow (det $\mathscr{L} < 0$). The point V represents fluids at rest and all viscometric flows.

and

$$\frac{1}{4}\dot{\varepsilon}^3 - \frac{3}{4}\dot{\varepsilon}^2 + 1 = \left(\frac{\dot{\varepsilon}}{2} - 1\right)^2 (\dot{\varepsilon} + 1) > 0. \tag{5.91}$$

Thus, both criteria are significant here. In summary, we see that in these simple flows, in which the Deborah number (*De*) is zero, we find from the present approach that there are two dimensionless numbers which are related to the Weissenberg number (*Wi*); both det $\mathscr{L}$ and tr $\mathscr{L}^2$ play the part of significant dimensionless numbers in classifying steady homogeneous flows.

Neither det $\mathscr{L}$ nor tr $\mathscr{L}^2$ can be regarded as a Weissenberg number. This may be seen by looking at simple shearing or at any other viscometric flow, where we find

$$\det \mathscr{L} = -1/8 \quad \text{and} \quad \text{tr} \, \mathscr{L}^2 = 3/4, \tag{5.92}$$

which are independent of shear rate. To define a Weissenberg number, we may go back to the idea of Pipkin (Section 3.6.1) for shear flows, where a significant parameter is the amount of shear in one relaxation time or we may take the value of the stretching parameter after one relaxation time, which may be found from (5.82), or, as is more usual, we simply define $(Wi) = \lambda \sqrt{2 \, \text{tr} \, \mathbf{d}^2}$, following (4.3).

5.4.2   The Deborah number

The concept of the Deborah number has been mentioned. We now approach the definition of such a number through the microstructural ideas discussed above. In some early uses of the Deborah number a flow time was compared to the fluid characteristic time in problems involving starting-up from a state of rest. Similarly, in sinusoidal motions it is easy to define a characteristic frequency when the Deborah number is formed. In general, we need to identify a characteristic rate of change of flow conditions to define the Deborah number. It is natural that the stress state be taken as the relevant flow quantity. According to Lodge (1974), who considers convected coordinates and computes the stress state relative to the particle trajectory, there are scarcely any rheologically steady states. Even spatially-steady homogeneous flows are not in general deemed rheologically steady from his point of view because the axes of principal stress rotate relative to the particle path framework. Thus, elongational flow is not considered steady in this sense. From the microscopic view, this definition appears to be too restrictive, because the stresses are not changing in magnitude along the particle trajectory. Thus, here we propose only to consider measures of rheological unsteadiness depending on the invariants of the stress states; in a homogeneous flow field these quantities are constant.

Let us suppose the flow field has been divided up into regimes in time and space which one is prepared to consider as regions of homogeneous steady flow. Then we may consider whether or not each regime is strong or weak and define a Weissenberg number for each regime as outlined above. It is then natural to define the Deborah number for a regime simply as

$$(De) = \lambda/(\text{residence time in regime}). \tag{5.93}$$

We will set out for clarity the definitions that have been used in this section; λ is the fluid relaxation time.

1. Test for weak flow. Flow is weak if [see Fig. (5.8)]

$$\det(\lambda \mathbf{L} - \tfrac{1}{2}\mathbf{I}) < 0, \tag{5.84}$$

and

$$\tfrac{27}{16} - \tfrac{3}{4}\operatorname{tr}\left(\lambda\mathbf{L} - \tfrac{1}{2}\mathbf{I}\right)^2 + \det\left(\lambda\mathbf{L} - \tfrac{1}{2}\mathbf{I}\right) > 0. \tag{5.85}$$

2. Definition of the Weissenberg number

$$(Wi) = \lambda\sqrt{2\operatorname{tr}\mathbf{d}^2} \tag{5.94}$$

3. Definition of (De) in a steady homogeneous flow regime

$$(De) = \lambda/(\text{time of dwell in regime}) \tag{5.93}$$

These are the definitions of 'strong' and 'weak' flows, and the distinction between the Weissenberg and Deborah numbers, which will be used later.

5.5 Dumbbells with limited extension

The behaviour of the linear dumbbell in elongational flows is unrealistic and it is desirable to improve this by limiting the extension of a dumbbell to the total molecular length, Na. In general, no closed form results like (5.57) can be found for these cases, and approximation or numerical solution is necessary.

Solutions of the diffusion eqn (5.41) for the distribution function ψ in strong flows often show [compare eqn (5.50)] that ψ is a highly localized function, and this fact can be made use of to produce an approximate constitutive equation. The high localization of the beads means that the diffusive Brownian forces are weak compared to the other forces on the beads. Consider eqn (5.46) and suppose we normalize the stress by the factor $n_0 kT$ so that

$$\mathbf{S} \equiv \boldsymbol{\tau}/n_0 kT \equiv (\boldsymbol{\tau}^{(\mathrm{m})} + n_0 kT\mathbf{I})/n_0 kT = 3N\langle K\mathbf{rr}\rangle, \tag{5.95}$$

where $\mathbf{r}$ is the dimensionless end-to-end vector

$$\mathbf{r} = \mathbf{R}/R_{\mathrm{m}}, \tag{5.96}$$

R_{m} is the maximum molecule length, $R_{\mathrm{m}} = Na$, and K is the dimensionless spring constant giving the deviation from linearity for the spring. If the spring force is $\mathbf{F}^{(\mathrm{c})}$ at an extension $\mathbf{R}$, then we can define the dimensionless spring constant K by

$$\mathbf{F}^{(\mathrm{c})}/\{3NkT/R_{\mathrm{m}}\} = \mathbf{r}K(r). \tag{5.97}$$

For a Hookean spring, $K = 1$, for the Warner spring (5.22) $K = (1 - r^2)^{-1}$.

We now suppose that the distribution function ψ is highly localized and that we can approximate it as a delta-function, so that

$$\psi = \delta(\mathbf{r} - \mathbf{b}). \tag{5.98}$$

where **b** is an unknown vector. By using this approximation in (5.95) and taking the trace of this equation, we find ($b = |\mathbf{b}|$)

$$\operatorname{tr} \mathbf{S} = 3NK(b)b^2. \tag{5.99}$$

We now suppose that K is a once-differentiable single-valued function of r so that we can solve (5.99) for b in terms of tr **S**. Then, from (5.95) we have

$$3N\langle \mathbf{rr} \rangle = 3N\,\mathbf{bb} = \mathbf{S}/K(\operatorname{tr} \mathbf{S}). \tag{5.100}$$

If we now return to the averaged diffusion eqn (5.52), noting now that we cannot remove the $3kT/Na^2$ spring constant from the fourth term, we can rewrite this equation as

$$\frac{\Delta \langle \mathbf{RR} \rangle}{\Delta t} + \frac{4}{n_0 \zeta} \boldsymbol{\tau} = \frac{4kT}{\zeta} \mathbf{I}, \tag{5.101}$$

where $\Delta/\Delta(t)$ is defined in (5.54). Using (5.51) we can rewrite (5.101) as the Giesekus (1966) relation, valid for all dumbbells:

$$\boldsymbol{\tau}^{(m)} = -\frac{n_0 \zeta}{4} \frac{\Delta \langle \mathbf{RR} \rangle}{\Delta t}. \tag{5.102}$$

Normalizing and replacing $\langle \mathbf{RR} \rangle$ in (5.102) by using (5.100), we find a constitutive relation for **S**:

$$\lambda \Delta[\mathbf{S}/K]/\Delta t + \mathbf{S} = \mathbf{I}. \tag{5.103}$$

In terms of $\boldsymbol{\tau}$, we can write (5.103) as

$$\lambda \frac{\Delta(\tau_{ij}/K)}{\Delta t} + \tau_{ij} = n_0\, kT\, \delta_{ij}, \tag{5.104}$$

where $K = K(\tau_{ii}/n_0 kT)$. When $K = 1$, we recover eqn (5.55). For the Warner spring [eqn (5.22)],

$$K = 1 + \frac{1}{3N} \operatorname{tr} \mathbf{S}, \tag{5.105}$$

and (5.103) becomes, as a constitutive equation for $\boldsymbol{\tau}$ (Tanner 1975)

$$\lambda \frac{\Delta}{\Delta t} [\tau_{ij}/(1 + \tau_{ii}/3Nn_0\, kT)] + \tau_{ij} = n_0\, kT\, \delta_{ij}. \tag{5.106}$$

A variant of this equation is described by Bird *et al.* (1987*b*). In general, this equation needs to be solved numerically. We can replace $n_0 kT$ by the macroscopic parameter $(N + 1)\eta_0/\lambda N$, where η_0 is the viscosity at zero shear rate, and obtain a constitutive model in terms of measurable quantities and the parameter N.

The replacement of the true distribution function by a delta-function is an approximation that has been tested against more accurate theoretical predictions. It is expected to be a much better approximation for strong flows than for weak flows and in order to demonstrate how it performs in the weak flows we need now to generate some test cases from (5.106).

When the fluid is at rest, the stress is just an isotropic tension of amount $n_0 kT$. Hence at very slow deformation rates tr $\boldsymbol{\tau} \sim 3n_0 kT$, and thus (5.106) becomes the same as the convected Maxwell case (5.55) with λ replaced by $\lambda N/(N+1)$. Since the zero-shear viscosity in (5.55) is $\lambda n_0 kT$, it follows that the zero-shear viscosity for eqn (5.106) is $N/(N+1)$ times this value. We use this modified zero-shear viscosity as a normalizing factor in what follows. Clearly, there are also modifications to G' and G'', obtained by replacing λ by $\lambda N/(N+1)$. Since $N \gg 1$ often, these are not large changes. Similarly, the value of Ψ_1 at zero-shear rate is reduced by $N^2/(N+1)^2$ from the pure Maxwell case. N_2 is zero for all steady shear rates; this result is valid for all springs (Problem 5.4).

At high rates it may be shown that (Problem 5.7)

$$\eta/\eta_0 \sim \dot{\gamma}^{-2/3}, \frac{\Psi_1}{\Psi_{1_0}} \sim \dot{\gamma}^{-4/3}, \tag{5.107}$$

both of which are fairly typical of real fluid behaviour. Figure 5.9 shows the computed values for various shear rates using the approximate constitutive model (5.106) compared with the exact calculations of Warner (1972) and

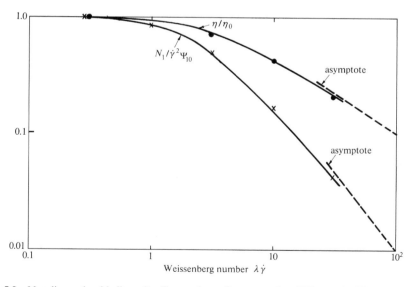

Fig. 5.9 Non-linear dumbbell results. Comparison of exact results of Warner (1972)—and results of constitutive equation (5.106) for $N = 3\frac{1}{3}$ ($\times$ and $\bullet$). The asymptotic forms correspond to [eqn (5.107)].

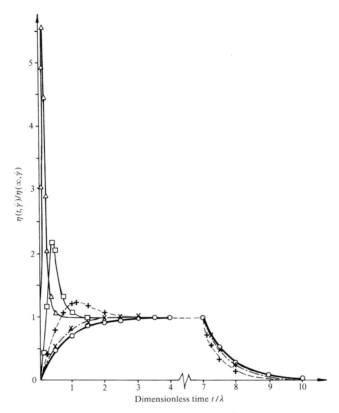

Fig. 5.10 Transient shear flow from [eqn (5.106)]. $N = 3\frac{1}{3}$ in this calculation. Weissenberg numbers $(\lambda\dot{\gamma})$ are

○	$\lambda\dot{\gamma} = 0.1$	□	$\lambda\dot{\gamma} = 10.0$
×	$\lambda\dot{\gamma} = 1.0$	△	$\lambda\dot{\gamma} = 31.62$
+	$\lambda\dot{\gamma} = 3.162$		

Fan (1984). The Warner–Fan calculations solve first for the distribution function and then use eqn (5.46). The comparison in the weak flow being considered is very good and provides support for the delta-function approximation. Figure 5.10 shows the shearing viscosity curves as a function of time after starting and also the decay of stress, and Fig. 5.11 shows the elongational flow response after start-up and cessation of elongation. The sudden application of a step strain is inevitably accompanied instantaneously by corresponding bead movement; a certain maximum strain which can be applied is obtained, and up to this point one obtains a 'stiffer' response as strains get larger. It is not possible to apply a step strain greater than this critical value. However, because of the Newtonian solvent, to which a step strain cannot be applied, this is not a serious problem with the model. In summary, the approximate constitutive model is a great

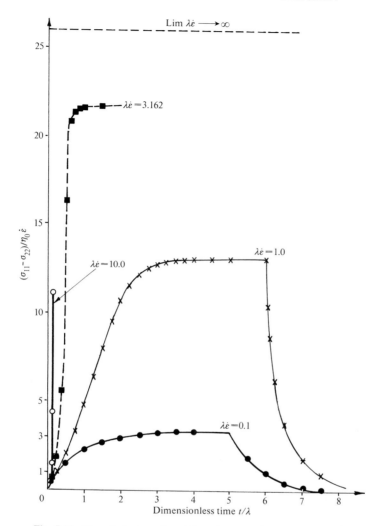

Fig. 5.11   Transient elongational flow from eqn (5.106) for $N = 3\frac{1}{3}$.

improvement on the Maxwell model, especially as regards steady elongational and shearing viscosities, and may be used when the Maxwell model is grossly inadequate.

5.5.1   *Chilcott–Rallison and Giesekus models*

Chilcott and Rallison (1988) produced a related model, setting

$$\lambda \frac{\Delta \mathbf{A}}{\Delta t} = -f(R)(\mathbf{A} - \mathbf{I}), \tag{5.108}$$

where the structure tensor $\mathbf{A} = \langle \mathbf{RR} \rangle$ here; $\Delta/\Delta t$ is the (upper) convected derivative (5.54), λ is the relaxation time, and

$$f(R) = 1/(1 - R^2/L^2), \qquad (5.109)$$

where $R^2 = \mathrm{tr}\,\mathbf{A}$, and L is the fully-extended length of a dumbbell. To compute the (dimensionless) stress they set

$$\sigma = -p\mathbf{I} + 2\eta_s\,\mathbf{d} + \frac{\phi}{\lambda}f(R)\mathbf{A}, \qquad (5.110)$$

where the stress and pressure have been made dimensionless using $\eta_s U/a$, where U/a is a characteristic shear rate, and η_s is the solvent viscosity; ϕ is the volume concentration of dumbbells; $\mathbf{d}$ is made dimensionless using U/a. The difference between (5.108) and the equation (5.104) is in the right-hand side. The right-hand side of (5.104) is equivalent to $-f\mathbf{A} + \mathbf{I}$ in the notation of (5.108). The steady-shear viscosity of the Chilcott–Rallison model is constant, equal to $\eta_s(1 + \phi)$, in contrast to the result shown in Fig. 5.9. The first normal stress difference is quadratic at $\lambda\dot{\gamma} \to 0$ and approaches linearity in $\lambda\dot{\gamma}$ for large $\lambda\dot{\gamma}$.

The elongational viscosity η_E rises to a finite value proportional to L^2 for large elongation rates. Originally this model was proposed to fit Boger fluid behaviour (Boger 1977) with a constant viscosity; it clearly differs greatly from (5.104) in behaviour, despite the structural similarity.

The second normal stress difference is zero in the models (5.104) and (5.108), and Giesekus (1982) introduced the idea of a non-isotropic drag on the beads. Instead of the scalar drag coefficient (ζ) in eqns (5.35) and (5.36) he introduced a tensor drag coefficient. The resulting constitutive equation may be written, in a notation like (5.57)

$$\lambda \frac{\Delta \boldsymbol{\tau}^{(m)}}{\Delta t} + \boldsymbol{\tau}^{(m)} + \frac{\alpha}{G}(\boldsymbol{\tau}^m)^2 = 2\eta\,\mathbf{d}, \qquad (5.111)$$

where α is a constant, G is a rigidity, and $\eta = \lambda G$. The range of α is from 0 to 1; when $\alpha = 0$ one recovers the UCM model (5.57). With (5.111) the second normal stress difference is non-zero (negative) in shearing and the elongational viscosity η_E asymptotes to a constant value for large $\lambda\dot{\varepsilon}$. Thus this model avoids many of the problems of the UCM model (5.57). We shall compare it with actual data for polymers later.

5.5.2 Exact models and approximations

The models discussed above, although reasonably accurate descriptions of steady-state flows, often give excessive overshoot (Fig. 5.10) in shear transient flows, and are generally too 'elastic' in recovery. Many of these effects are due to the closure schemes (like 5.98) used to form the constitutive equations. Although early workers had computed steady state probability distributions (see above) it was left to Lielens *et al.* (1998) to systematically compute the exact transient response for elastic dumbbells and compare this with the results of various closure approximations. Lielens *et al.* considered the dilute solution of dumbbells with a Warner spring (5.22) in one-dimensional transient flows; these have been shown by Keunings (1997) to be quantitatively and qualitatively relevant to real three-dimensional elongations, such as the experimental work of Doyle *et al.* (1998). In the one-dimensional (dimensionless) approximation, the equations (5.22), (5.41), and (5.46) become scalars, of the form

$$F(R) = R/(1 - R^2/L^2) \tag{5.22a}$$

$$\tau = \langle FR \rangle - 1, \tag{5.46a}$$

and the equation for the probability $\psi(R, t)$ is

$$\frac{D\psi}{Dt} = \frac{1}{2}\frac{\partial^2 \psi}{\partial R^2} - \frac{\partial}{\partial R}(\dot{\varepsilon}(t)R\psi) + \frac{1}{2}\frac{\partial}{\partial R}(F\psi) \tag{5.41a}$$

where $\dot{\varepsilon}(t)$ is the unsteady, spatially uniform extension rate. Numerical solution of (5.41a) is feasible for various transient flows, and was reported by Keunings (1997). To look at various closure schemes Lielens *et al.* (1998) define the quantities

$$A = \langle R^2 \rangle, \quad B = \langle R^4 \rangle, \quad A^c = 2\left\langle \frac{R^2}{1 - R^2/L^2} \right\rangle \quad \text{and}$$

$$B^c = 4\langle R^4/(1 - R^2/L^2) \rangle.$$

Admissible values of B were found to lie between the values A^2 and L^2A^2. If a δ-function [Fig. 5.12(a)] distribution is assumed, as in (5.98), with delta functions at $\pm b$, then one finds $A = b^2$, $A^c = 2b^2/(1 - b^2/L^2)$ and so

$$A^c = \frac{2A}{1 - A/L^2} \tag{5.112}$$

which is the FENE-P (finitely extensible non-linear elastic-Peterlin) δ-function approximation used above (eqn (5.98)) (Assumptions of more than one location for the delta functions were made by Leilens *et al.* (1998) without significant improvement in performance. A reasonable approximation where two peaks at b and another between 0 and b were assumed, has also been found.) The evolution equation for τ using delta functions at $\pm b/2$ [Fig 5.12(a)] corresponds to

$$\frac{DA}{Dt} = 1 + 2\dot{\varepsilon}(t)A - \left(\frac{A}{1 - A/L^2}\right) \tag{5.113}$$

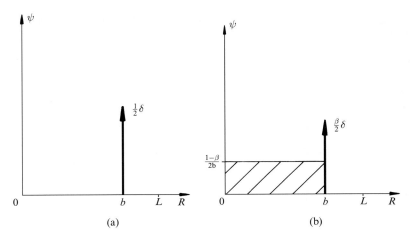

Fig. 5.12 Distribution functions ψ. (a) Two delta functions at $\pm b$ (Tanner 1975). (b) Flat distribution plus delta (L-shape) functions (Lielens *et al.* 1998).

$$\text{and} \qquad \tau = \frac{A}{(1 - A/L^2)} - 1 \equiv \frac{1}{2} A^c - 1, \tag{5.114}$$

which is essentially the same result as (5.100) above. A marked improvement in response followed from the inclusion of a uniform distribution with the delta functions [Fig. 5.12(b)].

This leads to the equations (β is the fraction of dumbbells in the delta function)

$$A \approx \beta b^2 + (1 - \beta)b^2/3$$
$$B = \beta b^4 + (1 - \beta)b^4/5$$

$$A^c \approx \frac{2\beta b^2}{1 - b^2/L^2} + 2(1 - \beta)L^2 \left[\frac{L}{b}\ln\left(\frac{L + b}{L - b}\right) - 1\right]$$

$$B^c \approx \frac{4\beta b^4}{1 - b^2/L^2} + 4(1 - \beta)L^4 \left[\frac{L}{b}\ln\left(\frac{L + b}{L - b}\right) - 1 + \frac{b}{3L}\right] \tag{5.115}$$

Here $B \leq 1.8A^2$, and the size of the delta-function parameter β is found to be

$$\beta = \left[9A^2 - 5B + \sqrt{(9A^2 - 5B)9A^2}\right] \Big/ 10B \tag{5.116}$$

and also

$$b = \sqrt{\frac{5B}{3A + \sqrt{9A^2 - 5B}}}.$$

With these results the so-called FENE-L evolution equations (so-called from the L-shaped distribution function) emerge:

$$\frac{DA}{Dt} = 1 + 2\dot{\varepsilon}(t)A - \frac{1}{2}A^c$$
$$\frac{DB}{Dt} = 6A + 4\dot{\varepsilon}(t)B - \frac{1}{2}B^c \tag{5.117}$$

with τ given by (5.114) as before.

The results of comparing the numerical (stochastic) simulation, the simple result (5.113–5.114) and the system 5.117 are shown in Figs 5.13 and 5.14. The constraint $1.8A^2 \geq B$ is enforced in the solution of 5.117. The results are given for a complex flow history:

$$\dot{\varepsilon}(t) = 100t(1 - t)\exp -4t \quad \text{for} \quad 0 \leq t \leq 1$$
$$= 0 \text{ otherwise.} \tag{5.118}$$

The results in Fig. 5.14 show that good agreement with the simulation is given by the FENE-L model, and it also shows the hysteresis characteristic of the

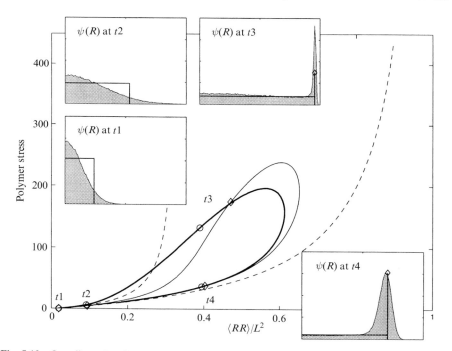

Fig. 5.13 One-dimensional complex flow: polymer stress versus mean square extension ($L^2 = 50$). —; dashed curve, FENE-P; thin curve, FENE-L; thick curve, FENE. Comparison between FENE and FENE-L distribution functions at four selected items, defined in insets. The symbols $\diamond$ and $\circ$ mark these times on the FENE-L and FENE hystereses, respectively.

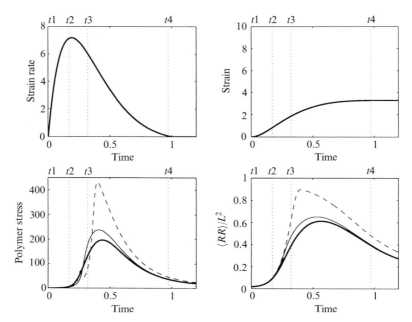

Fig. 5.14 One-dimensional complex flow: evolution of strain rate and strain. Comparison between FENE-P, FENE-L and FENE results for the polymer stress and mean square extension ($L^2 = 50$). Upper dashed curve, FENE-P, thin curve, FENE-L; thick curve, FENE (numerical).

numerical results and the rapid decay of elongational stress. This may be seen from studying the response of the various models to a step pulse of strain rate (Fig. 5.15). The FENE-P closure does not show hysteresis; the numerical and FENE-L results do. Thus the closure chosen has a great impact on the results, and molecular simulations can be used to devise improvements.

More complex models can be contemplated, but their worth has not been established, particularly with respect to numerical solutions. We will return to numerical simulation methods in Chapter 8.

5.5.3   The approach of Acierno and co-workers

Acierno *et al.* (1976) have proposed a Maxwell-type model of the following form for the polymeric stress tensor $\boldsymbol{\tau}$;

$$\boldsymbol{\tau} = \sum_i \boldsymbol{\tau}_i \tag{5.119}$$

$$\frac{1}{G_i}\boldsymbol{\tau}_i + \lambda_i \frac{\Delta}{\Delta t}\left(\frac{1}{G_i}\boldsymbol{\tau}_i\right) = 2\lambda_i \mathbf{d}, \tag{5.120}$$

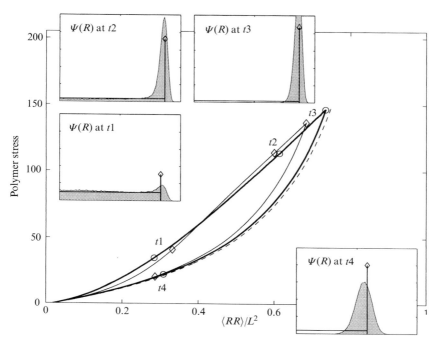

Fig. 5.15 Start-up elongation followed by relaxation: polymer stress versus mean square extension ($L^2 = 50$ and $\lambda\dot{\gamma} = 2$). Dashed curve (right), FENE-P; thin curve, FENE-L; thick curve, FENE. Comparison between FENE and FENE-L distribution functions at four selected times, also defined. The symbols $\diamond$ and $\circ$ mark these times on the FENE-L and FENE hystereses, respectively.

where the modulus G_i and the relaxation time λ_i are related to structural evolution parameters x_i by the rules

$$G_i = G_{0i}x_i; \quad \lambda_i = \lambda_{0i}x_i^{1.4}, \tag{5.121; 5.122}$$

and an evolution equation

$$\frac{dx_i}{dt} = \frac{1}{\lambda_i}(1 - x_i) - ax_i\frac{1}{\lambda_i}\sqrt{\left(\frac{E_i}{G_i}\right)}, \tag{5.123}$$

where $E_i = \frac{1}{2}\text{tr}\,\boldsymbol{\tau}_i$, and a is a parameter.

The values of G_{0i} and λ_{0i} are not disposable, being the equilibrium values for the fluid at rest. Hence a is the only adjustable parameter. The values of a that have been used are in the range 0.4–1.2. Though the theory is quite similar in structure to those just described, it is not intended only for dilute solutions and most applications have been to concentrated solutions and melts. The results from this model are quite realistic in many respects but the *ad hoc* nature of

(5.121) and (5.122) is to be noted. The results for suddenly applied strain are not realistic. In this case the structure has no time to 'evolve' away from equilibrium after the strain step before relaxation begins; it must therefore initially relax in the same way as the simple Maxwell model, which is contrary to the experimental results of Einaga *et al.* (1971) (for molten polymers) shown in Fig. 4.8. There is also a large overshoot of elongational stress when starting up, which is at variance with some experimental results; see, however, Wagner *et al.* (1979). The general pattern of (5.119)–(5.123) is interesting and similar to that of Hinch and Leal (see Section 5.8) and the Lielens *et al.* (1998) results given above: one has a structural dynamics equation plus an equation for computing the stresses. This seems to be a common feature of many complex constitutive models.

5.5.4   Further developments

Many examples of non-linear spring calculations are given by Bird *et al.* (1987*b*), Bird and DeAguiar (1983), and DeAguiar (1983).

The Maxwell model has been generalized to many beads and (linear) springs (chain molecule), and the result is precisely of the form (5.61) and (5.62). Many-bead non-linear spring models have been discussed briefly by Bird (1982) and by Bird *et al.* (1987*b*). Some idea of the difficulties in this case are given by Wiest and Tanner (1989). Further developments of approximate constitutive models for dumbbells with variable friction, internal viscosity and hydrodynamic interaction (Phan-Thien and Tanner 1978) have also been studied. In most of these cases the results are similar in kind to those derived above, and the reader is referred to the original papers for details. See also Fan and Bird (1984).

Verhoef *et al.* (1999) considered a dumbbell model with a viscous component to the connector force. They assumed a connector force of the form $F_i^{(c)} = f(R)R_i + g(R_j R_k d_{jk})R_i$ where **R** is the connector vector and d_{jk} is the rate of deformation tensor. They were able to use pre-averaging of **R** to obtain good agreement with their experiments. Once again, a reduction of the model's elasticity was sought, but the physics of the model is not completely clear.

5.5.5   Rigid dumbbells and rods

In many cases macromolecules are not flexible (for example, some proteins, DNA in a helix configuration, tobacco mosaic virus) and a rigid dumbbell model, albeit crude, is more relevant than a flexible bead-spring model. Figure 5.16(a) shows a multi-bead and rigid rod arrangement; when there are two beads and one rod, of length L, one has the rigid dumbbell. We will only give detailed results for this case; for many other bead-rod-spring models see Bird *et al.* (1987*b*).

In the case of the dumbbell model [Fig. 5.16(b)], with two beads spaced a distance R apart, the beads are not supposed to interact hydrodynamically in the present calculation; this may not always be a realistic assumption.

From Fig. 5.16(b), $\mathbf{r}_1 - \mathbf{r}_2 = R\mathbf{u}$ where **u** is a unit vector directed from bead 2 to bead 1. The solution is supposed to be dilute; in a similar way to the elastic

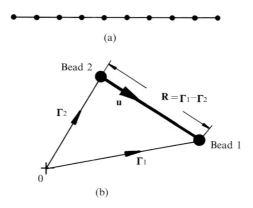

Fig. 5.16 (a) Bead-rod assembly. (b) Rigid dumbbell molecule.

dumbbell [eqn. (5.38)] the centroid moves with the fluid. The restraint of a constant length (R) can be incorporated by recognizing that two unrestrained beads have a relative velocity of $\mathbf{v} = \mathbf{LR} \equiv R\mathbf{Lu}$. The velocity component along the $\mathbf{u}$-vector is $\mathbf{v} \cdot \mathbf{u} = RL_{ij}u_ju_i$, or, using the notation $R\mathbf{L}:\mathbf{uu}$, we find the relative speed along the line of $\mathbf{R}$ to be $R(\mathbf{L}:\mathbf{uu})\mathbf{u}$. If this is subtracted from $R\mathbf{Lu}$, the beads, when constrained by a rigid link, have a relative velocity $\dot{\mathbf{R}}$ given by

$$\mathbf{v} = \dot{\mathbf{R}} = R\dot{\mathbf{u}} = R(\mathbf{Lu} - (\mathbf{L}:\mathbf{uu})\mathbf{u}). \tag{5.124}$$

Brownian motion is also present, and produces rotary diffusion in this case.

Let the Brownian force be $\mathbf{F}^{(b)}$, equal to $-kT\,\partial \ln\psi/\partial\mathbf{u}$, where ψ is the probability density for the state $\mathbf{u}$ at a given time. Then the constraint of rigidity needs us to consider only the effect of that part of $\mathbf{F}^{(b)}$ which is not aligned with $\mathbf{u}$. To obtain this, the part of $\mathbf{F}^{(b)}$ parallel to $\mathbf{u}$ is $(\mathbf{F}^{(b)} \cdot \mathbf{u})\mathbf{u}$, or $F_k^{(b)}u_ku_i$ and subtracting this from $\mathbf{F}^{(b)}$ one finds a reduced Brownian force of $(\mathbf{I} - \mathbf{uu}) \cdot \mathbf{F}^{(b)}$. Using this result by adding the resulting motion to (5.124) yields

$$\dot{\mathbf{u}} = \mathbf{L} \cdot \mathbf{u} - \mathbf{L}:\mathbf{uuu} + \frac{1}{\zeta R}(\mathbf{I} - \mathbf{u}) \cdot \mathbf{F}^{(b)}. \tag{5.125}$$

The connector force $\mathbf{F}^{(c)}$ can be found from the force balance (5.37)

$$\mathbf{F}^{(c)} = \frac{1}{2}\mathbf{F}^{(b)} + \zeta(\mathbf{LR} - \dot{\mathbf{R}})/2. \tag{5.126}$$

From (5.125) and (5.126) we have

$$\mathbf{F}^{(c)} = \frac{\zeta R}{2}[L:\mathbf{uuu}] + \frac{1}{2}(\mathbf{uu} \cdot \mathbf{F}^{(b)}). \tag{5.127}$$

The Kramers–Bird expression for the polymer stress $\boldsymbol{\tau}^{(m)}$ is given by (5.46) and using (5.127) in (5.46) we find

$$\boldsymbol{\tau}^{(m)} = n_0 \left\langle \frac{\zeta R^2}{2} (\mathbf{L} : \mathbf{u}\mathbf{u}\,\mathbf{u}\mathbf{u}) + \frac{R}{2} \mathbf{u}\mathbf{u}\mathbf{u} \cdot \mathbf{F}^b \right\rangle - n_0 kT \mathbf{I} \qquad (5.128)$$

where $\langle \cdot\cdot \rangle$ denotes the average $\iint\int dV \psi (\dots)$.

From a result of Huilgol and Phan-Thien (1997) we have

$$\langle \mathbf{u}\mathbf{u}\mathbf{u} \cdot \mathbf{F}^{(b)} \rangle = \frac{6kT}{R} \langle \mathbf{u}\mathbf{u} \rangle \qquad (5.129)$$

and hence (5.128) becomes

$$\boldsymbol{\tau}^{(m)} = \frac{1}{2} n_0 \zeta R^2 \mathbf{L} : \langle \mathbf{u}\mathbf{u}\mathbf{u}\mathbf{u} \rangle + 3 n_0 kT \langle \mathbf{u}\mathbf{u} \rangle - n_0 kT \mathbf{I}. \qquad (5.130)$$

By using the Giesekus form (5.102) we derive an evolution equation for $\mathbf{u}$:

$$\lambda [\Delta \langle \mathbf{u}\mathbf{u} \rangle / \Delta t + 2\mathbf{L} : \langle \mathbf{u}\mathbf{u}\mathbf{u}\mathbf{u} \rangle] + \langle \mathbf{u}\mathbf{u} \rangle = \frac{1}{3} \mathbf{I}, \qquad (5.131)$$

where $\lambda = \zeta R^2 / 12\, kT$, the time constant, has the same form as the elastic dumbbell result (5.52) if Na^2, the mean square and end–end distance is replaced by R^2.

Generally, (5.131) will need a closure approximation for the fourth-order product in terms of $\langle \mathbf{u}\mathbf{u} \rangle$. Various closure schemes are discussed by Feng *et al.* (1998), and in Section 5.8.1 below.

Alternatively, one may solve a diffusion equation for ψ. This has the form given by Bird *et al.* (1987a) and is not detailed here.

To find ψ one can either solve this equation (see Bird *et al.* 1987a for many examples) or, one can avoid it by using a simulation based on a stochastic formulation (Chapter 8). Typical results for the rheology of rigid dumbbells include (Bird *et al.* 1987a).

(i) Shear thinning with $\tau^{(m)}(\dot\gamma) \sim (\dot\gamma)^{1/3}$ for very large shear rates:

(ii) $N_2 = 0$, as in all simple dumbbell models:

(iii) $N_1 \propto \dot\gamma^{2/3}$ at very large shear rates:

(iv) A finite extensional viscosity at all elongational rates., equal to 6 η_0 as $\lambda\dot\varepsilon$ becomes very large.

Extensions to various 'molecules' including those of the type shown in Fig. 5.16(a) with or without hydrodynamic interactions and internal viscosity, have been made. See Bird *et al.* (1987b). No explicit, exact constitutive equation seems to be available for this model.

Dilute solution theories for a variety of other 'solutes' are described by Bird *et al.* (1987b). Rallison and Hinch (1988) discussed a 'molecule' with rigid bonds on links of fixed length jointed together; typically they used 100 links and studied the unfolding numerically.

5.5.6 *Experimental polymer molecule dynamics in dilute extensional flow*

Perkins *et al.* (1997) studied optically the details of the elongation of single molecules of a DNA in a sugar solution, molecule by molecule. Hence, there is no doubt that their results are for dilute solutions. They were able to study the uncoiling of identical molecules (all the DNA molecules were the same) in strong elongational flows. The configurations adopted by the molecules were many and varied; few were stretched to full length despite a relatively long exposure to elongation compared with the relaxation time. Figure 5.17 shows a partial taxonomy of the stretched molecules. For lower extensional rates ($\dot{\varepsilon}\lambda_{max} \leq 0.5$) the molecules remained coiled. The behaviour of individual molecules is very diverse, some deforming only slightly, while others reached a steady-state extension comparable to the total molecular length. For $\dot{\varepsilon} = 0.86\,\mathrm{s}^{-1}$, with a relaxation time of $3.9\,\mathrm{s}$, and a long dwell time, one would expect most molecules to be highly extended. Those molecules in a dumbbell configuration [Fig. 5.17(a)] extended faster than folded molecules, and the residence time at which significant stretching occurs was highly variable. The actual probability distribution of extension is close to, but more complex than the Keunings L-distribution (Figs 5.12 and 5.13) and Perkins *et al.* express doubts that a simple dumbbell model will give accurate predictions of stress. Also the mean spring law was found to be roughly of the form

$$\frac{F^{(c)}L}{3NkT} = \frac{2}{3}\left[0.25/(1 - R/L)^2 - 0.25 + R/L\right] \tag{5.132}$$

Harrison *et al.* (1998) also studied very dilute solutions of polystyrene in an extension-dominated flow. They inferred from birefringence measurements that many of their molecules reached 90 per cent of full extension and that the Chilcott–Rallison model (5.108) fitted the data well. However, they also preferred a spring of a form different from the Warner spring to describe their results. They suggested replacing the $f(R)$ on the right-hand side of (5.108) by

$$f_1(R) = \frac{1 + 2.3(R^2)/L^2}{(1 - R^2/L^2)^{0.1}}. \tag{5.133}$$

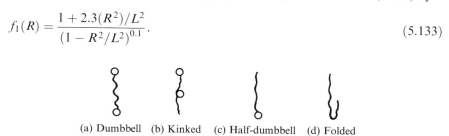

(a) Dumbbell (b) Kinked   (c) Half-dumbbell (d) Folded

Fig. 5.17 Sketches of DNA molecules after elongation straining. Most of the molecules are in one or other of the above configurations. After a photograph by Perkins *et al.* (1997).

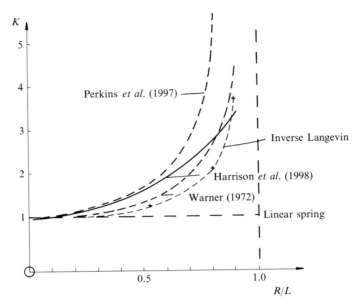

Fig. 5.18   Experimental and theoretical spring 'constants' for molecules. K is the ratio of the spring 'constant', as a function of (R/L), normalized to the Gaussian spring constant [eqn (5.48)]; R/L is the fraction of extension related to maximum length of the molecules. Perkins *et al.* (1997)—[eqn (5.132)]; Harrison *et al.* (1998)—eqn (5.133). Warner and Inverse Langevin, see eqns (5.22) and (5.21).

The spring laws [(5.20) (Inverse Langevin), (5.22) (Warner 1972), (5.133) (Harrison *et al.* 1988), and (5.132) (Perkins *et al.* 1997)] are compared in Fig. 5.18, all normalized to 1.0 for small extensions. The differences between the two experimentally fitted laws (Perkins *et al.* and Harrison *et al.*) are small for $R/L < 0.6$, with wide divergence thereafter, in accordance to observed differences in extensions.

The considerable differences between molecules' experimental behaviours and the theoretical Warner and inverse Langevin laws are also clear.

5.6   Theories for molten polymers and concentrated solutions

Section 5.5, dealing with dilute solutions, shows how difficult this simpler case is, and few successful attempts have been made to deal with rheological problems in the intermediate case of medium concentrations. For the high concentrations of greater interest in polymer engineering, other paths are needed. The most successful basic ideas are the Yamamoto–Lodge network theory and the developments by Doi and Edwards. In both cases the concentration is so high that gross interference between molecules is inevitable, and one is well beyond the 'knee' point of the zero-shear viscosity vs. molecular weight curve where the zero-shear rate viscosity begins to rise proportional to the 3.4-power of the molecular weight

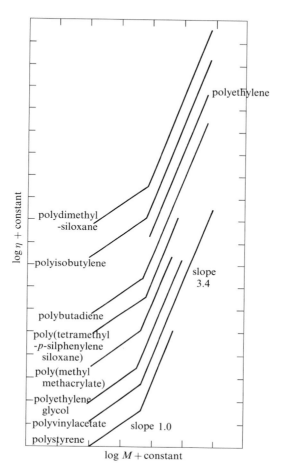

Fig. 5.19 Relationship between zero-shear viscosity and molecular weight for several melts of narrow polydispersity from Berry and Fox (1968).

(Berry and Fox 1968) due to entanglements. See Fig. 5.19. In the following sections we will present some of these theories.

5.6.1 Theory of rubber elasticity

The theory of macromolecular substances developed in the 1930s via the work of H. Staudinger, H. Mark, and W. Kuhn (Tanner and Walters 1998). Application to rubber elasticity followed; the general concept of an entropic force (see Section 5.1.3) in long-chain molecules was used to construct a theory based on cross-linked molecules. If in Fig. 5.20(a) one imagines physical links at points A, B, and other crossings, then one can develop a theory connecting stress and deformation for polymers. The book by Treloar (1958) gives a clear picture of earlier work; the vast scope of the statistical theory of rubber elasticity can be seen

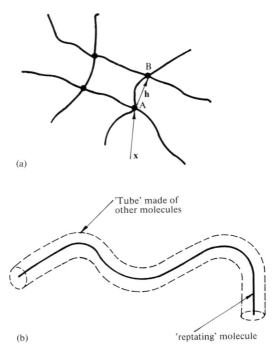

(a)

'Tube' made of
other molecules

(b)   'reptating' molecule

Fig. 5.20 Concepts for concentrated polymer solutions. (a) Yamamoto–Lodge network. (b) Doi–Edwards virtual tube.

in the book by Erman and Mark (1997). However, given the concept of stored energy and elasticity, then the equilibrium stress-deformation relations must fall within the continuum elasticity theory of Section 4.7. One is left to consider non-equilibrium effects in non-ideal rubberlike bodies from a microstructural standpoint, but this would require too large an excursion from the main purpose of this book. We now consider how the theories of rubber elasticity were adapted to non-crosslinked materials or fluids.

5.6.2   The Lodge–Yamamoto network theory

These theories of networks were formulated following the concepts given by the pioneering work of Green and Tobolsky (1946). Both Lodge (1956) and Yamamoto (1956, 1957, 1958) produced similar theories at the same time; the publication of the book by Lodge (1964) undoubtedly drew great attention to these contributions.

Suppose that in the polymer liquid one has a 'network' of molecules with temporary junctions [A, B in Fig. 5.20(a)]. We shall not be much concerned here with the exact nature of the junctions, which has been disputed; the junctions[†] are

---

[†]Often called entanglements; we shall not use this term.

simply supposed to appear and disappear, and the strands connecting them are able to transmit force. The network is not permanent, but constantly changes its topology. Lodge (1968) has carefully discussed the physical picture, including the Brownian agitation of individual strands. We neglect completely the contribution to the stress from the solvent; it will usually be negligible as may be seen by comparing solvent and solution viscosities. Thus the theory is useful for both molten polymers and solutions, and both will be regarded as incompressible.

Let us call the element AB of the network a chain. Each chain is, from Section 5.1, properly regarded as a non-linear spring. However, it is often permissible to consider chains as linear springs, and this will be initially adhered to. Each chain is supposed to be made up of a certain number N of molecules or 'submolecules' and we can describe each chain as being of type $(\mathbf{h}, N)$, where $\mathbf{h}$ is the end-to-end vector [Fig. 5.20(a)].

Consider a position $\mathbf{x}$ in the fluid, relative to a laboratory frame of reference. There will be a distribution function ϕ which describes the distribution of chains of type $(\mathbf{h}, N)$ around the point $\mathbf{x}$ at time t, so that $\phi(\mathbf{h}, \mathbf{x}, N, t)\,\mathrm{d}\mathbf{h}^3\,\mathrm{d}\mathbf{x}^3$ is the number of chains in a small element of physical space $(\mathbf{x})$ and chain space $(\mathbf{h})$ around the point $\mathbf{x}$ at time t. To begin with, we will suppose that N is constant for all chains; eventually this will lead to a single relaxation time, and one can see that each group of chains with different N will contribute on a different time-relaxation scale. We can write a balance equation for $\phi(\mathbf{h}, N, \mathbf{x}, t)$:

$$\frac{\mathrm{d}}{\mathrm{d}t}(\phi\,\mathrm{d}\mathbf{h}^3\,\mathrm{d}\mathbf{x}^3) = \begin{bmatrix} \text{rate of chain creation} \\ -\text{rate of destruction} \end{bmatrix} \mathrm{d}\mathbf{h}^3\,\mathrm{d}\mathbf{x}^3, \tag{5.134}$$

where the rates are for chains of type $(\mathbf{h}, N)$. Henceforward, we shall drop the N, it being taken equal for all chains at present. We shall write the rate of chain creation as g, and the rate of destruction as $\beta\phi$, it being natural to write the destruction rate as being proportional to the number of chains present. Thus, eqn (5.134) can be written as:

$$\frac{\mathrm{d}}{\mathrm{d}t}(\phi\,\mathrm{d}\mathbf{h}^3\,\mathrm{d}\mathbf{x}^3) = [g(\mathbf{h}) - \beta(\mathbf{h})\phi]\,\mathrm{d}\mathbf{h}^3\,\mathrm{d}\mathbf{x}^3. \tag{5.135}$$

We now introduce the affine deformation hypothesis which relates an $\mathbf{h}$ vector for a chain before deformation to the same quantity for the same chain after deformation. From our previous discussion an element of the $\mathbf{x}$ vector, $\Delta\mathbf{x}$ at time t, was represented by $\Delta\mathbf{r}$ at time t', and hence

$$\Delta\mathbf{r}(t') = \frac{\partial\mathbf{r}}{\partial\mathbf{x}}\Delta\mathbf{x}(t) = \mathbf{F}\Delta\mathbf{x}. \tag{5.136}$$

The affine deformation hypothesis is that **h** vectors also transform by (5.136) so that

$$\mathbf{h}(t') = \mathbf{Fh}(t). \tag{5.137}$$

In both the h-space and the x-space small volume elements are conserved ('incompressible'), and so

$$\frac{\mathrm{d}}{\mathrm{d}t}(\,\mathrm{d}\mathbf{h}^3) = \frac{\mathrm{d}}{\mathrm{d}t}(\mathrm{d}\mathbf{x}^3) = 0. \tag{5.138}$$

Hence (5.135) becomes

$$\frac{\partial \phi}{\partial t} + \dot{x}_i \frac{\partial \phi}{\partial x_i} + \dot{h}_i \frac{\partial \phi}{\partial h_i} = g - \beta \phi. \tag{5.139}$$

From (5.137) we find

$$\frac{\mathrm{d}\mathbf{h}}{\mathrm{d}t'} = \frac{\partial}{\partial \mathbf{x}} \left(\frac{\mathrm{d}\mathbf{r}}{\mathrm{d}t'} \right) \mathbf{h}(t), \tag{5.140}$$

and if we evaluate this at time t we find

$$\dot{\mathbf{h}} = \mathbf{L}(\mathbf{x}, t)\mathbf{h}(t). \tag{5.141}$$

If we concentrate on homogeneous stress states, then the second term in (5.139) is zero, and using (5.141) we get

$$\frac{\partial \phi}{\partial t} + \frac{\partial v_i}{\partial x_j} h_j \frac{\partial \phi}{\partial h_i} = g - \beta \phi. \tag{5.142}$$

Solution of this equation yields $\phi(\mathbf{h}, t)$ when g, β and suitable initial and boundary conditions are given; for example, ϕ may be given at some initial time and we may require $\phi \to 0$ as $|\mathbf{h}| \to \infty$. Given ϕ, then the stress tensor may be computed, using arguments similar to those yielding (5.46), from

$$\tau_{ij} = \sigma_{ij} + p\delta_{ij} = \iiint_{\mathbf{h}} \phi \, \mathrm{d}^3 \mathbf{h} \mathbf{F}^{(c)} \mathbf{h} = \langle \mathbf{F}^{(c)} \mathbf{h} \rangle, \tag{5.143}$$

where the integration is over all **h**-configuration space, and $\mathbf{F}^{(c)}$ is the force in the chain. Clearly, (5.143) is similar to (5.46), but (5.142) is not the same as the diffusion eqn (5.41) for ψ. If we take the hypotheses of Lodge (1964) so that

$$\mathbf{F}^{(c)} = K\mathbf{h}, \tag{5.144}$$

where K is a constant, and also assume β is constant, and that g is a function only of the magnitude of **h**, denoted by h, then one can derive a constitutive model.

First, we note that the solution of (5.142) in this case is

$$\phi = \int_{-\infty}^{t} dt' g[h(t')] \exp -\beta(t - t'), \tag{5.145}$$

where any initial conditions were imposed at $t = -\infty$. From (5.143)

$$\tau_{ij}(\mathbf{x}, t) = K \int\!\!\int\!\!\int_{\mathbf{h}(t)} d^3\mathbf{h} h_i(t) h_j(t) \int_{-\infty}^{t} dt' g([h(t')]) \exp -\beta(t - t'). \tag{5.146}$$

But by using (5.137) we can change to integrating over $\mathbf{h}(t')$, and also we can invert the time–space order of integration to obtain

$$\tau_{ij} = K \int_{-\infty}^{t} dt' \exp -\beta(t - t') \int\!\!\int\!\!\int_{\mathbf{h}(t')} d^3\mathbf{h}(t')(\mathbf{F}^{-1})_{ik} h_k h_m (\mathbf{F}^{-1})_{jm} g(h). \tag{5.147}$$

Now $\mathbf{F}(t')$ is not a function of $\mathbf{h}$, and so it can be moved outside the triple integration. The remaining integral, $\int\!\int\!\int g h_k h_m d^3\mathbf{h}$, is spherically symmetric, and is a multiple of the unit tensor δ_{km}. Hence we can rewrite (5.147) as

$$\tau_{ij} = K' \int_{-\infty}^{t} \exp -\beta(t - t') F_{ik}^{-1} F_{jk}^{-1} \, dt'. \tag{5.148}$$

Now $F_{ik}^{-1} F_{jk}^{-1} = (\mathbf{C}^{-1})_{ij}$, the inverse of the strain tensor, and hence (5.148) becomes a constitutive equation identical in form to (5.58).

$$\tau = \sigma + p\mathbf{I} = K' \int_{-\infty}^{t} \exp -\beta(t - t') \mathbf{C}^{-1} \, dt'. \tag{5.149}$$

If we let $\beta = 1/\lambda$, $K' = \eta_0/\lambda^2$, then we have the convected Maxwell model (5.58). By allowing each chain category (N) to have its independent relaxation time, and summing over these contributions, we obtain the Lodge (1964) model

$$\sigma + p\mathbf{I} = \int_{-\infty}^{t} \mu(t - t') \mathbf{C}^{-1} \, dt'. \tag{5.150}$$

Clearly, this form has the same virtues and vices as the convected Maxwell form for dilute solutions, and the response is identical for identical relaxation spectra. However, in concentrated solutions the Zimm spectrum discussed in Section 5.3.2 is not recommended, and measurements should be made wherever possible.

It is remarkable that two such diverse microstructural ideas arrive at the same constitutive model and we begin to see that the constitutive equation, in broad form, is relatively insensitive to details of the microstructural picture. This is fortunate, and gives confidence in the microstructural procedure being adopted.

5.6.3   Variable rates of network creation and destruction

Lodge assumed that the junction creation function g [eqn (5.135)] was Gaussian; one does not alter the results significantly by changing this assumption; the response remains Maxwellian.

By contrast, the fluid response depends critically on the destruction coefficient β for which various forms have been suggested. As written in eqn (5.135), β is given as a function of **h**. Takano (1974) has investigated various forms of β numerically; specifically she takes $\beta = \beta_0 h^2$, so that longer junctions tend to disappear more readily. The computed results show decreasing viscosity and many other realistic features of real solutions. Approached in this way the theory is, of course, unwieldy but the following approximate approach, following the dilute solution pattern (Section 5.5) yields a constitutive equation.

In eqn (5.143), suppose that the network has a linear force–chain length characteristic:

$$\mathbf{F}^{(c)} = K\mathbf{h}. \tag{5.151}$$

Then, (5.143) becomes, with a constant K

$$\tau_{ij} = K\langle h_i h_j \rangle, \tag{5.152}$$

or, taking the trace, we find

$$\langle h^2 \rangle = \tau_{ii}/K. \tag{5.153}$$

If we now assume that most of the chains are oriented in the same direction, then we have the equivalent of the delta-function approximation used for dilute solutions. In addition, if a group of highly-extended chains dominates the mean-square value $\langle h^2 \rangle$, we can write as a (closure) approximation

$$\langle h^2 \rangle \approx h^2.$$

We may then replace the function $\beta(h)$ by $\beta(\mathrm{tr}\,\boldsymbol{\tau})$ in (5.142). Since $\boldsymbol{\tau}$ is not now regarded as a function of **h**, we can complete the analysis to obtain a constitutive equation of the Kaye (1966) family:

$$\boldsymbol{\tau} = \int_{-\infty}^{t} \mu(t - s,\ \mathrm{tr}\,\boldsymbol{\tau}(s))\mathbf{C}^{-1}(s)\,\mathrm{d}s. \tag{5.154}$$

For the case when $\beta = \mathrm{const} \times h^2$ (Takano 1974) the kernel μ takes the form (for one set of uniform chains)

$$\mu = \mathrm{const} \times \exp -\frac{1}{\lambda K}\int_{t'}^{t} \mathrm{tr}\,\boldsymbol{\tau}(s)\,\mathrm{d}s, \tag{5.155}$$

where λ and K are constant.

This model shows many realistic features but has an implicit form, requiring computation. If the functions g and β in eqn (5.142) are not functions of $\mathbf{h}$, but are functions of the invariants of τ (or another related stress) or the local rate of deformation $\mathbf{d}$, then one can still multiply (5.142) by $h_k h_m$ and integrate over the space of $\mathbf{h}$. Use of the divergence theorem yields

$$\frac{\Delta \tau}{\Delta t} + \frac{\tau}{\lambda} = g_0 \mathbf{I}, \tag{5.156}$$

where $g = g_0 \psi_0$, with ψ_0 an equilibrium distribution, a function of $\mathbf{h}$. Also $\lambda = 1/\beta$; g_0 and β are supposed independent of $\mathbf{h}$. Larson (1988) has termed (5.156) the General Network Model; he also contemplates replacing $\Delta/\Delta t$ by a more general derivative in (5.156). Equation (5.156) can be put in an equivalent integral form (Phan-Thien and Tanner 1977), for multiple relaxation times:

$$\tau = \sum_i \int_{-\infty}^t \frac{g_0(t')}{\lambda_i(t')} \exp\left[- \int_{t'}^t \frac{dt''}{\lambda_i(t'')} \right] \mathbf{E}(t, t') \, dt' \tag{5.157}$$

where the strain tensor $\mathbf{E}$ satisfies

$$\frac{D\mathbf{E}}{Dt} - \mathbf{w}^T \cdot \mathbf{E} - \mathbf{Ew} - a(\mathbf{d} \cdot \mathbf{E} + \mathbf{Ed}) = \mathbf{0} \tag{5.158}$$

Here $\mathbf{w}$ is the vorticity tensor $1/2\,(\mathbf{L}^T - \mathbf{L})$ (Chapter 2) and a is a constant; if $a = 1$, we have the upper convected derivative; $a = 0$ gives the corotational derivative, and $a = -1$ the lower convected derivative. $\mathbf{E}$ can be a combination of $\mathbf{C}$ and $\mathbf{C}^{-1}$; for example, a useful form is to set

$$\mathbf{E} = \left(1 + \frac{\varepsilon}{2}\right)\mathbf{C}^{-1} + \frac{\varepsilon}{2}\mathbf{C} \tag{5.159}$$

which can generate a second normal stress difference such that $N_2/N_1 = \frac{1}{2}\varepsilon$, where ε is a constant.

Modifications made to the Lodge model include 'network rupture', the replacement of tr $\tau(s)$ in (5.155) by some function of the rate of deformation invariants, and the KBKZ theory (see Section 4.8) where strain invariants are used in the kernel. Another possibility is to replace $\mathbf{C}^{-1}$ in all these forms by an optimum strain measure $\mathbf{H}$ of the form

$$\mathbf{H} \approx f_1(\mathrm{I}_{c^{-1}}, \mathrm{II}_{c^{-1}})\mathbf{C}^{-1} + f_2(\mathrm{I}_{c^{-1}}, \mathrm{II}_{c^{-1}})\mathbf{C}, \tag{4.64}$$

as suggested in Section 4.8. Then we recover a KBKZ form, but without the U-function. Most of these forms may be expressed as differential forms by differentiating as in Section 5.3.1; an exception is the KBKZ model. Larson (1988) has given an approximate way of finding differential equivalents for this case.

Despite the many attempts, the results of these efforts are not on the whole better than the KBKZ theory, and in some cases they are worse. For example, the use of the strain-rate invariants (Carreau 1972) yields the same type of infinite Trouton viscosity at finite elongation rates as does the Lodge model; it also may be objected to in ultra-high frequency small oscillation tests because then the rate of deformation may not be small although the strain is. Experiments by Gross and Maxwell (1972) (see Section 4.11.5 and Fig. 4.9) and many others support the idea that the departure from linear viscoelastic behaviour depends on strain magnitude, not on strain-rate magnitude, and hence the Carreau (1972) forms are not thought to be fundamentally sound.

Even excluding this class of behaviour, then one still has to face the problem of recoil and double-step relaxation results described in Section 4.8. and 4.8.2. Since most of the models discussed here behave poorly in these respects, we shall return to this point yet again in Section 5.6.6. In summary, although the assumption of variable rates of creation and destruction of junctions yields good agreement with most experiments, there are still other ideas that need exploration.

5.6.4   The relaxation of the Gaussian spring assumption

Lodge (1968) showed that one could permit the network 'springs' to be non-linear, as in the dilute solution case, but that the results were not very useful. Specifically, the shear viscosity is now a rising function of shear rate and the ratio N_2/N_1 is positive. Neither of these results is realistic, and since it seems, from the microscopic point of view, that spring stiffness must increase with extension, no further work will be done with this hypothesis.

An extreme case of this model is the network-rupture theory of Tanner and Simmons (1967) where the network is assumed to break when tr $\mathbf{C}^{-1}$ reaches a critical value; this equation shows possible behaviour, but needs multiple relaxation times to be at all realistic.

5.6.5   The non-affine deformation assumption—PTT model

Phan-Thien (1978) has considered relaxing the hypothesis of affine deformation [eqn (5.137)]. In particular, if $\mathbf{h}$ is the end-to-end vector of a network segment, he assumed that

$$\dot{\mathbf{h}} = \mathbf{L}\mathbf{h} - \xi\mathbf{d}\mathbf{h}, \tag{5.160}$$

where $\mathbf{L}$ is the macroscopic velocity gradient in the fluid, $\mathbf{d} = 1/2(\mathbf{L} + \mathbf{L}^T)$, and ξ is a constant. (In the original paper a further term in (5.160) of the form $-\sigma\mathbf{h}$ was used, but Bird et al. (1987b) have shown that this term must be zero for stability.)

From this assumption, a constitutive equation for the network contribution to the stress tensor $\boldsymbol{\tau}$ is found (for a material with a single relaxation time λ)

$$\lambda\left[\frac{D}{Dt}\boldsymbol{\tau} - \mathscr{L}\boldsymbol{\tau} - \boldsymbol{\tau}\mathscr{L}^T\right] + Y\boldsymbol{\tau} = 2\lambda G\mathbf{d}, \tag{5.161}$$

where $\mathscr{L}$ is the effective velocity gradient tensor, defined by

$$\mathscr{L} = \mathbf{L} - \xi\mathbf{d}, \tag{5.162}$$

where Y is a function of tr $\boldsymbol{\tau}$, and G is a constant. This form can easily be generalized for multiple relaxation times λ_i as follows. Let

$$\boldsymbol{\tau} = \sum_i \boldsymbol{\tau}^{(i)},$$

then, for the isothermal case,

$$\lambda_i \frac{\delta\boldsymbol{\tau}^{(i)}}{\delta t} + Y\boldsymbol{\tau}^{(i)} = 2\lambda_i G_i\mathbf{d}, \tag{5.163}$$

where Y is a function of $tr\boldsymbol{\tau}^{(i)}$, G_i is the usual linear-viscoelastic shear modulus, and

$$\frac{\delta\boldsymbol{\tau}^{(i)}}{\delta t} \equiv \frac{D\boldsymbol{\tau}^{(i)}}{Dt} - \mathscr{L}\boldsymbol{\tau}^{(i)} - \boldsymbol{\tau}^{(i)}\mathscr{L}^T. \tag{5.164}$$

This form of constitutive equation gives the following results:

(i) Linear viscoelastic behaviour at small strains.

(ii) Good fit to viscosity and first normal stress difference data for low-density polyethylene for both steady and transient shearing.

(iii) A non-zero second normal stress difference, where $N_2 = -(\xi/2)N_1$, [eqn (5.163)].

(iv) Reasonable elongational behaviour at all elongation rates.

Two forms for Y have been extensively used:

(a) $Y = 1 + \dfrac{\varepsilon}{G_i}$ tr $\boldsymbol{\tau}^{(i)}$ (linear form), $\qquad\qquad$ (5.165 a)

and

(b) $Y = \exp\left\{\dfrac{\varepsilon}{G_i}\text{ tr }\boldsymbol{\tau}^{(i)}\right\}$ (exponential form). $\qquad\qquad$ (5.165 b)

Here, G_i is a rigidity, and ε is a constant.

Apart from the linear viscoelastic data λ_i and G_i, eqns (5.162)–(5.165) contain only two constants, ε and ξ. It was found that ξ may be estimated in several ways from linear-viscoelastic and viscometric data, while ε (principally) governs the extensional flow response. When $\varepsilon = 0$, the constitutive equations are identical to those of Johnson and Segalman (1981) which were deduced by an alternative (continuum) method; in this case the model becomes one of the Oldroyd family, eqn (4.13). The main advantage of the PTT model over both the Johnson–Segalman model and the model of Acierno *et al.* (1976) seems to be in elongational flows; the former does not show a limiting viscosity with elongation rate, while the latter shows an overshoot in transient elongation which is hard to reconcile with a network-type theory. It should be noted, however, that such an overshoot has been seen by Wagner *et al.* (1979).

Thus the model behaves well in many tests, and we now examine it in a single-step test. If a sudden (single or multiple) shear strain is applied to a sample, and the resulting stress relaxation is studied, then this stress is a function of the shear strain (γ) and the time (t). In this test the results (for $\varepsilon = 0$, $\xi \neq 0$) are equivalent to an integral model, but with a specific kernel. The problems previously found for double-step relaxation persist with the model and complicated remedies for this behaviour may be needed. For the single-step relaxation, if we divide the shear stress τ by $\gamma G(t)$, where $G(t)$ is the linear shear stress relaxation function, then we can define a function f, where

$$f(\gamma) = \tau(t)/\gamma G(t). \tag{5.166}$$

For linear viscoelastic response $f = 1$; typical experimental results for f as a function of γ are shown in Fig. 5.21 and predictions for the present model are also plotted. The predictions are considerably in error as the change with γ is initially too slow and is then too fast. The form found for f for this model is

$$f = \left(\sin \gamma \sqrt{\xi(2 - \xi)} \right) / \gamma \sqrt{\xi(2 - \xi)}. \tag{5.167}$$

For $\xi = 0$ (Lodge's model) $f = 1$. Hence, the present model is an improvement on this model in this flow. It is clear that if γ exceeds $\pi\sqrt{\xi(2 - \xi)}$, f will become negative. This is not an acceptable result and thus predictions for large γ are unrealistic. If $\xi = 0.1$, then the critical value of γ for the onset of negative f is 7.21.

This model embraces the extremes of the Lodge model and the corotational models of Bird *et al.* (1977, 1987a) as ξ goes from 0 to 1. It should also be noted that in a simple shear flow the shear stress goes through a maximum as the shear rate is increased. This is not usually acceptable. One can add a 'solvent viscosity' to overcome this problem if required, but other ideas have usually been preferred, such as using multiple relaxation times.

5.6.6 *Affine PTT models* ($\xi = 0$)

In this case the motion is again affine, and the response of the model to step strains is always reasonable. The solution of the response for simple shearing and steady elongation for the linear model (5.165a) is given below for two values of ε, for a single relaxation time $\lambda \equiv \eta_0/G$.

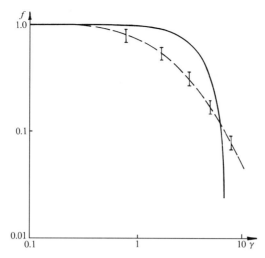

Fig. 5.21 Non-linear behaviour in single-step relaxation; f is the ratio of shear stress to $\gamma G(t)$; where γ is the step of strain and $G(t)$ is the linear relaxation function. The solid line is the Phan-Thien prediction eqn (5.161) with $\xi = 0.1$ and the dashed line is experimental data.

(a) Steady shearing

In this case let $\mathbf{v} = \dot{\gamma} y \mathbf{i}$ where $\dot{\gamma}$ is the shear rate and $\mathbf{i}$ is the unit vector in the x-direction. Then the stress components τ_{yy}, τ_{zz}, τ_{zy} and τ_{zx} are seen to be zero and the remaining equations for τ_{xx} and τ_{xy} are

$$-2\lambda\tau_{xy}\dot{\gamma} + \tau_{xx}f = 0 \tag{5.168}$$

$$\tau_{xy}f = \eta_0\dot{\gamma} \tag{5.169}$$

Here

$$f = 1 + \frac{\lambda\varepsilon}{\eta_0}\tau_{xx} \quad \text{or} \quad \exp\left[\frac{\lambda\varepsilon\tau_{xx}}{\eta_0}\right] \tag{5.169a}$$

$$\frac{\tau_{xy}}{\eta_0\dot{\gamma}} \equiv \frac{\eta}{\eta_0} = f^{-1}$$

so that f^{-1} is the viscosity ratio. From (5.168), after eliminating τ_{xy} using (5.169) we find

$$N_1 = \tau_{xx} = 2\eta_0\lambda\dot{\gamma}^2 f^{-2} \tag{5.170}$$

which is a single equation for τ_{xx} (or N_1). We can rearrange it in terms of f and the Weissenberg number (W_i) $\lambda\dot{\gamma}$ to find

$$Xf^2(X) = 2\varepsilon(Wi)^2$$

where $X = \varepsilon\lambda\tau_{xx}/\eta_0$.

Thus, for any given f, ϵ we can find the corresponding Weissenberg number. The second normal stress difference is zero with this model, and N_1 is given by eqn (5.170).

In practice, the two forms of f given in eqn (5.169a) have both been used; the following asymptotic behaviour as $(Wi) \to \infty$ can be verified:

(i) $f(X) = 1 + X$ gives $\dfrac{\tau_{xy}}{\eta_0 \dot{\gamma}} \sim (2\varepsilon)^{-1/3} (Wi)^{-2/3}$

$$\frac{\varepsilon \lambda \tau_{xx}}{\eta_0} \sim (2\varepsilon)^{1/3} (Wi)^{2/3} \tag{5.171}$$

(ii) $f(X) = e^X$ gives $\dfrac{\tau_{xy}}{\eta_0 \dot{\gamma}} \sim \dfrac{1}{(Wi)}$

$$\frac{\varepsilon \lambda \tau_{xx}}{\eta_0} \sim \ln(Wi)$$

The behaviour (5.171) is consistent with some dilute solution theories discussed above (Section 5.5).

(b) *Steady elongation*
Here the equations to be solved are ($\dot{\varepsilon} = \partial u / \partial x$ and $\tau_{yy} = \tau_{zz}$)

$$-2\lambda \dot{\varepsilon} \tau_{xx} + \tau_{xx} f\{Y^*\} = 2\eta_0 \dot{\varepsilon}$$

$$\lambda \dot{\varepsilon} \tau_{yy} + \tau_{yy} f\{Y^*\} = -\eta_0 \dot{\varepsilon}$$

where $Y^* = \varepsilon \lambda (\tau_{xx} + 2\tau_{yy})/\eta_0$.
The asymptotic behaviour in these cases ($\lambda \dot{\varepsilon} \gg 1$) is

(i) $f(y) = 1 + Y^* : \dfrac{\tau_{xx} - \tau_{yy}}{\eta_0 \dot{\varepsilon}} = \dfrac{2}{\varepsilon}$ \hfill (5.172a)

(ii) $f(y) = \exp(Y^*) : \dfrac{\tau_{xx} - \tau_{yy}}{\eta_0 \dot{\varepsilon}} = \dfrac{1}{\varepsilon} \left[\dfrac{\ln(Wi)}{(Wi)} \right]$ \hfill (5.172b)

For molten polymers (5.172b) is preferred (Larson 1988).

To illustrate solutions, we take the linear form $f = 1 + (\varepsilon \lambda / \eta_0) \mathrm{tr}\, \tau$; two values of ε, 0.02 and 0.25 are used. The former reflects dilute solution behaviour, while $\varepsilon = 0.25$ is closer to the response of some HDPE melts (Larson 1988).

Figure 5.22 shows these responses ($\dot{\varepsilon} > 0$ only: the figures for $0 > \dot{\varepsilon}$ are slightly different.)

To compute the curves in Fig. 5.22, eqns (5.170) and (5.171) were used and the equations were rearranged by defining

$$X = \varepsilon \lambda \tau_{xx}/\eta_0$$
$$Y = \varepsilon \lambda \tau_{yy}/\eta_0$$

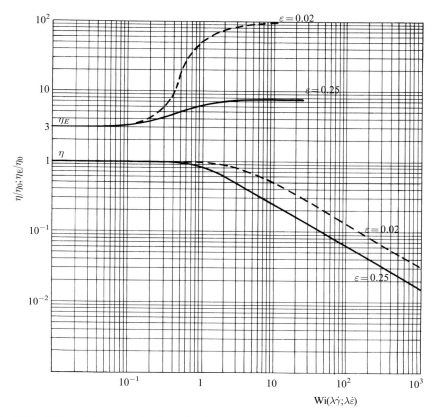

Fig. 5.22 PTT model response in elongation (top) and shear (below) for two values of ε; λ is the (single) relaxation time and $\xi = 0$ (affine motion).

and hence

$$Y = -\varepsilon X/(3X + 2\varepsilon)$$

$$\lambda\dot{\varepsilon} = X(1 + X + 2Y)/2(X + \varepsilon)$$

Thus, given a choice of X, Y and ε, $\lambda\dot{\varepsilon}$ can be found. Then the ratio of η_E/η_0, which is equal to

$$\frac{\eta_E}{\eta_0} = \frac{\tau_{xx} - \tau_{yy}}{\eta_0\dot{\varepsilon}} = (X - Y)/\varepsilon(\lambda\dot{\varepsilon}),$$

can also be found.

Example: Poiseuille flow in a channel—PTT model
Let the channel be of width $2h$, and the wall shear stress magnitude be τ_w. Then

$$\tau_{xz} = -\tau_w y/h = \eta\frac{\mathrm{d}u}{\mathrm{d}y}$$

From (5.169) and (5.170) we find

$$f = 1 + \frac{\varepsilon \lambda \tau_{xx}}{\eta_0} \equiv 1 + X = \frac{\eta_0}{\eta},$$

and from (5.170)

$$2\varepsilon \lambda^2 \left[\frac{du}{dy}\right]^2 = X(1+X)^2 = \left[\frac{\eta_0}{\eta} - 1\right]\left[\frac{\eta_0}{\eta}\right]^2.$$

Rearranging, one has

$$\frac{\eta_0}{\eta} = \frac{2\varepsilon \lambda^2}{\eta_0^2}\left[\frac{\tau_w^2 y^2}{h^2}\right] + 1,$$

and

$$\frac{du}{dy} = -\frac{\tau_w y}{\eta h} = -\frac{\tau_w y}{\eta_0 h}\left[1 + \frac{2\varepsilon \lambda^2}{\eta_0^2}\frac{\tau_w^2 y^2}{h^2}\right].$$

Integrating,

$$u = u_0 - \frac{\tau_w y^2}{2\eta_0 h} - \varepsilon \lambda^2 \frac{\tau_w^3 y^4}{2\eta_0^3 h^3},$$

when $y = h$, $u = 0$ (no-slip condition), hence

$$u_0 = \frac{\tau_w h}{2\eta_0} + \frac{\varepsilon \lambda^2 \tau_w^3 h}{2\eta_0^3},$$

and

$$u = \frac{\tau_w h}{2\eta_0}\left[\left[1 - \frac{y^2}{h^2}\right] + \frac{\varepsilon \lambda^2 \tau_w^2}{\eta_0^2}\left[1 - \frac{y^4}{h^4}\right]\right].$$

Also, if the mean velocity in the channel is $\bar{u}$, one finds

$$\bar{u} = \frac{\tau_w h}{3\eta_0}\left[1 + \frac{6}{5}\frac{\varepsilon \lambda^2 \tau_w^2}{\eta_0^2}\right].$$

Hence, by choosing τ_w, $\bar{u}$ can be found; and the equation can be put in the form

$$\frac{\lambda \bar{u}}{h} = \frac{1}{3}\left[\frac{\lambda \tau_w}{\eta_0}\right]\left[1 + \frac{6}{5}\varepsilon\left[\frac{\lambda \tau_w}{\eta_0}\right]^2\right].$$

By solving for $\lambda\tau_w/\eta_0$ as a function of $\lambda\bar{u}/h$ the velocity profile can be expressed in terms of $\bar{u}$ and the other parameters.

5.6.7 Fitting experimental data with PTT models

Considerable work has been done in fitting the PTT-type models[†] to experimental data (Larson 1988; Byars *et al.* 1997), and provided enough modes are used, the fits obtained are reasonable. For example, Byars *et al.* (1997) fitted data for the standard sample S1 and compared the Giesekus (1982) model, the PTT model, and an integral model. The latter is a White–Tokita model like eqns (4.63–4.64) where the strain function $h_1 = 0$ and

$$h_{-1} = \alpha[\alpha - 3 + \beta I_1 - (1 - \beta)I_2]^{-1} \tag{5.173}$$

Here α and β are constants, $I_1 = \mathrm{tr}\,\mathbf{C}^{-1}$ and I_2 is the second invariant of $\mathbf{C}^{-1}$ also equal in magnitude to $\mathrm{tr}\,\mathbf{C}$. A four-mode model with a solvent viscosity was fitted. The linear viscoelastic parameters used are those in Table 3.8, and the non-linear model parameters are shown in Table 5.1. The Giesekus model (5.111) has the extra parameter α; $G_i = \eta_i/\lambda_i$ for each mode. The PTT model used the exponential form (5.165b) in (5.163), and the slip parameter ξ was also allowed to be non-zero.

All the models fit the measured linear viscoelastic, shear viscosity and normal stress data well, hence only the fits of the steady elongational viscosity (Fig. 5.23) are displayed. Baaijens *et al.* (1997) also fitted the exponential PTT and the Giesekus model to low-density polyethylene data with some success for both models, but there were areas (that is, wake behind a cylinder) where the models appeared to be in some disagreement with the experiments. That it seems essential to use several modes in these models is clear from the results. This point was made by Azaiez *et al.* (1996) where 4-mode Giesekus and PTT models were compared with a one-mode PTT model.

Table 5.1   Coefficients for the four-mode Giesekus, PTT, and White–Tokita models

Mode No.		1	2	3	4
Giesekus	α	0.27	0.29 0.001 0.04	0.03	0.5
PTT	ξ	0.072	0.072	0.072	0.072
	ε	0.2	4.9×10^{-4}	0.2	0.2
White–Tokita	α	25 for all modes			
	β	1.1×10^{-3} for all modes			

[†] Here the PTT models are assumed to be of differential forms (5.161) and (5.156); via eqn (5.157) one can also write these models in integral form.

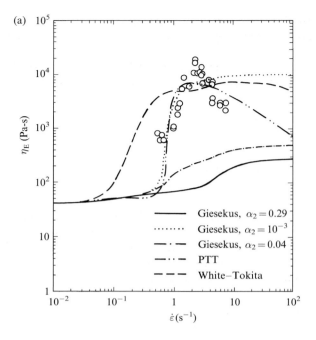

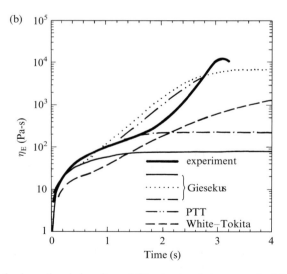

Fig. 5.23   (a) Steady elongational viscosity and (b) extensional stress growth at $\dot{\varepsilon} = 2.2\,\text{s}^{-1}$ (21 °C). The parameters are those of Table 5.1. From Byars *et al.* (1997).

Langouche and Debbaut (1999) have also fitted a 9-mode PTT model and have shown, remarkably, that it gives fairly good results, even for recoil, for High-Density Polyethylene.

To alleviate errors with the one-mode model, Phan-Thien (1984) has modified the PTT model so that the viscosity $\eta(=\lambda G)$ on the right hand side of eqn (5.163) has the form

$$\eta = \eta_0 \frac{1 + \xi(2 - \xi)\lambda^2\dot{\gamma}^2}{(1 + \Gamma^2\dot{\gamma}^2)^{(1-n)/2}}, \tag{5.174}$$

where Γ is another time-constant.

This yields a Carreau form for the viscosity function, $N_1 = 2\lambda\dot{\gamma}^2\eta$, and $N_2 = -(\xi/2)N_1$.

While these forms are useful in shear-like flows, these modifications are difficult to regard as microstructurally inspired. Suggestions that (5.174) be expanded to include a stress-invariant form have also been made (Tanner 1965); one could also consider a Reiner–Rivlin form (eqn 4.1) in order to model N_2, but this has not, to date, been much explored.

5.6.8 The Kaye and Leonov models

Kaye (1966) made the assumption that the destruction of junctions function β in (5.154) was a function of the invariants of the deviator of the stress (τ_d). By carrying out an analysis similar to that of eqn (5.6.3) one obtains a result in terms of the deviator τ_d

$$\tau_\mathrm{d} = \tau - \tfrac{1}{3}\mathbf{I}\,\mathrm{tr}\,\tau,$$

so that

$$\tau = \sigma + p\mathbf{I} = \sum_j \int_{-\infty}^{t} \frac{\eta_j}{\lambda_j^2} g_j(t') \exp\left\{-\int_{t'}^{t} \frac{g_j(t'')}{\lambda_j}\,\mathrm{d}t''\right\} \mathbf{C}^{-1}\,\mathrm{d}t', \tag{5.175}$$

where the $g_j(x) = g_j(\mathrm{II}_{\tau_d}(x), \mathrm{III}_{\tau_d}(x))$ are functions of the two independent invariants of τ_d; (5.175) is a model of the form (5.148).

By differentiating (5.175) with respect to time t, we find, for a single relaxation process j, if $\tau = \sum_j \tau_j$, the result

$$g_j^{-1}(t)\lambda_j \frac{\Delta\tau_j}{\Delta t} + \tau_j = \frac{\eta_j}{\lambda_j}\mathbf{I}, \quad \text{(no sum on } j) \tag{5.176}$$

which is a Maxwell element of the form (5.55) but with a time 'constant' $\lambda_j g_j^{-1}$ which varies with the current stress level.

Leonov (1976) introduced a constitutive model which appears to be most easily visualized as a rubberlike response from the present configuration to a 'relaxed' or stress-free state. This 'relaxed' state may not ever be occupied by the fluid, but it can be properly defined. The notation used here differs from that of Leonov; Leonov's $\mathbf{C}$ is here called $\mathbf{C}^{-1}$ to conform with previous sections. If an infinitesimal element of length $d\mathbf{x}$ in the current state corresponds to an element $d\mathbf{r}$ in the relaxed state, then, using the definitions (2.31) and (2.32) the relevant strain measure is $\mathbf{C}$. To track the evolution of the $\mathbf{r}$-state, we find

$$dv_i = \overline{dx_i} = L_{ij}\,dx_j. \tag{2.21}$$

By differentiating (2.31) with respect to time one finds

$$\overline{dr^2} = d\mathbf{x}^T \left\{ \frac{D\mathbf{C}}{Dt} + \mathbf{C}\mathbf{L} + \mathbf{L}^T\mathbf{C} \right\} d\mathbf{x} \equiv d\mathbf{x}^T \frac{\mathscr{D}\mathbf{C}}{\mathscr{D}t}\,d\mathbf{x} \tag{5.177}$$

which gives a convected derivative rule for $\mathbf{C}$; note that this rate rule is the same as that used in the Oldroyd equation (4.13). We shall denote the bracketted term in (5.177) by $\mathscr{D}\mathbf{C}/\mathscr{D}t$. It is now assumed that the stress in the fluid is developed from an isotropic elastic relation similar to (4.53), plus a viscous component of stress so the model represents a Maxwell model with a parallel viscous dashpot, see Fig. 1.16. The stress is assumed to be of the form of a sum over N relaxation processes

$$\boldsymbol{\sigma} + p\mathbf{I} = \eta_0 s\mathbf{A}_1 + \sum_{K=1}^{N} \boldsymbol{\tau}_K, \tag{5.178}$$

where in the viscous part s is a number between 0 and 1, and the elastic stress components $\boldsymbol{\tau}_K$ are each of the form (4.53):

$$\boldsymbol{\tau}_K = -2\left\{ \frac{\partial W}{\partial \mathbf{I}_c}\mathbf{C} - \frac{\partial W}{\partial \mathbf{I}_{c^{-1}}}\mathbf{C}^{-1} \right\}_K. \tag{5.179}$$

We shall set $K = 1$ for initial exposition here corresponding to a single Maxwell element. Each component has its own relaxed state, its own strain tensor and its own free (or strain) energy function W. Although one can envisage strain tensors which are not incompressible, we shall assume that $\mathbf{C}$ is. An evolution-rate equation for the rate of change of $\mathbf{C}$ is also written

$$\frac{\mathscr{D}\mathbf{C}}{\mathscr{D}t} = \mathbf{f}(\mathbf{C}), \tag{5.180}$$

where $\mathbf{f}$ is an isotropic function of $\mathbf{C}$.
   We denote

$$\mathbf{e} = -\frac{1}{2\mu\lambda}\left\{ \frac{\partial W^s}{\partial \mathbf{I}_c}(\mathbf{C} - \tfrac{1}{3}\mathbf{I}_c\mathbf{I}) - \frac{\partial W^s}{\partial \mathbf{I}_{c^{-1}}}(\mathbf{C}^{-1} - \tfrac{1}{3}\mathbf{I}_{c^{-1}}\mathbf{I}) \right\}, \tag{5.181}$$

where λ is a time-constant for the relaxation process being considered; $\mathbf{e}$ thus contains the deviatoric part of the stress, and is derived from the potential function W^s which is symmetric in $\mathbf{I}_c$ and $\mathbf{I}_{c^{-1}}$.

$$2W^s = W(\mathbf{I}_c, \mathbf{I}_{c^{-1}}) + W(\mathbf{I}_{c^{-1}}, \mathbf{I}_c). \tag{5.182}$$

The neo-Hookean potential W is the simplest choice:

$$W = \mu(\mathbf{I}_{c^{-1}} - 3). \tag{5.183}$$

This completes a simple set of relations; more complex versions can be written down. In the Leonov (1976) paper $\mathbf{f}(\mathbf{C})$ is set equal to

$$\mathbf{f}(\mathbf{C}) = \mathbf{e}\mathbf{C} + \mathbf{C}\mathbf{e}. \tag{5.184}$$

These are the set of equations as used by Leonov *et al.* (1976); other functions $\mathbf{f}(\mathbf{C})$ and much more generality are possible at the cost of simplicity; see Larson (1988).

The resulting equations are fairly easy to solve analytically for steady flows. They appear to give adequate description in weak flows. In strong flows it can be shown that the elongational viscosity is only slightly increased at high shear rates (Fig. 4.15) contrary to experiment.

The connection of the above simplest model with other continuum models can be established. From (5.179) and (5.183) we find for a single relaxation mode

$$\boldsymbol{\tau} = 2\mu\mathbf{C}^{-1}. \tag{5.185}$$

Using a formula of Leonov (1976) we can replace (5.180) by an evolution equation for $\mathbf{C}^{-1}$:

$$\frac{\Delta\mathbf{C}^{-1}}{\Delta t} \equiv \overline{\dot{\mathbf{C}^{-1}}} - \mathbf{C}^{-1}\mathbf{L}^{\mathrm{T}} - \mathbf{L}\mathbf{C}^{-1} = -\mathbf{C}^{-1}\mathbf{e} - \mathbf{e}\mathbf{C}^{-1}, \tag{5.186}$$

where (5.184) has been used for the right-hand side of (5.186). Multiplication of (5.186) by 2μ converts this equation into an evolution equation for $\boldsymbol{\tau}$

$$\frac{\Delta\boldsymbol{\tau}}{\Delta t} = -(\boldsymbol{\tau}\mathbf{e} + \mathbf{e}\boldsymbol{\tau}). \tag{5.187}$$

From (5.181) and (5.182) we have

$$4\lambda\mathbf{e} = \mathbf{C}^{-1} - \mathbf{C} + \tfrac{1}{3}\mathbf{I}(I_c - I_{c^{-1}}). \tag{5.188}$$

In terms of $\boldsymbol{\tau}$ we find (5.187) to be

$$\mathbf{0} = \lambda\frac{\Delta\boldsymbol{\tau}}{\Delta t} + \frac{1}{2}\boldsymbol{\tau}\left\{\frac{\boldsymbol{\tau}}{2\mu} - 2\mu\boldsymbol{\tau}^{-1} + \frac{1}{3}\mathbf{I}\left[2\mu\,\mathrm{tr}\,(\boldsymbol{\tau}^{-1}) - \frac{\mathrm{tr}\,\boldsymbol{\tau}}{2\mu}\right]\right\}, \tag{5.189}$$

or, rewriting the bracketted term as $\mathbf{K}(\tau)$

$$\lambda \frac{\Delta \tau}{\Delta t} + \mathbf{K}(\tau) = \mathbf{0},\tag{5.190}$$

which is a non-linear convected Maxwell model.

The similarity between the Kaye result (5.176) and this result should be noted; the Kaye result is a special case of (5.190), corresponding to the right-hand side of (5.186) being a function of the invariants of $\mathbf{C}$ multiplied by $\mathbf{C}^{-1}$. We shall therefore not investigate the Kaye model separately. Leonov *et al.* (1976) have shown that the model describes steady and unsteady shearing well; see also Upadhyay *et al.* (1981) for transient shear flow behaviour.

The response of the Leonov model in double-step shearing is compared with the KBKZ model in Fig. 4.13 (a, b) using the spectrum of eqn (4.77); the dashpot term in (5.178) is set equal to zero ($s = 0$). There is some inaccuracy in the *single-step* response. With the elongational viscosity (Fig. 4.16) we note that the response is not as good as the KBKZ model; here the relaxation spectrum of Fig. 4.14 was used. (See also Table 2.1.) Since the number of parameters that have been chosen is minimal for the Leonov model, it is attractive. The step response might also be improved by using a different *W*-function in conjunction with another $\mathbf{f}(\mathbf{C})$ function.

5.7   Doi–Edwards theory

Although the Green–Tobolsky concept of a network structure has proved durable and useful, there has always been some obscurity about the junction concept. An alternative view of the structure of entangled polymer fluids has been given by Doi and Edwards (1986). The theory is based on the reptating chain model of Edwards (1967) and de Gennes (1979). [See Fig. 5.20(b).]

An essential idea in the Doi–Edwards theory is that the transverse segmental motions of each chain are impeded by the meshwork of strands from other chains in its neighbourhood. Large-scale configurational rearrangement and diffusion are assumed to proceed mainly by reptation; that is, by random snakelike motions of each chain along its own length. Each chain continually disengages itself from its current cage of strands, creating new cages and eventually a new configuration as its emerging end worms randomly through the mesh. Since the strands of the mesh are parts of chains which diffuse similarly, the lifetime of each cage constraint is itself comparable to the disengagement time for the entire chain. Thus, for chains which are long compared to the mesh size, the frequency of transverse jumps is small enough to make plausible the idea of a semi-permanent cage and predominantly snakelike motion.

If the system is deformed the cages are distorted and the chains are carried into new configurations. Stress relaxation proceeds first by a relatively rapid equilibration of chain configurations within the distorted cages, then by a relatively slow diffusion of chains out of the distorted cages into random configurations.

Doi and Edwards represent the cage by a tube of diameter a and contour length L enclosing each chain. The chains are random coils with molecular weight M and mean-square end-to-end distance R^2. There are ν chains per unit volume. The tube diameter is assumed to be independent of M and the tube length directly proportional to M. Each chain and the path of its associated tube have random walk configurations with the same end-to-end vector. The distance a represents the mesh size and is assumed to correspond also to the step-length of the tube path. Therefore, we have

$$aL = R^2. \tag{5.191}$$

Alternatively, the tube path is a random walk of N steps, each of length a, and

$$Na^2 = R^2. \tag{5.192}$$

The self-diffusion coefficient for the chains is given by (Graessley 1980)

$$D = kT/3N\zeta_r, \tag{5.193}$$

where k is the Boltzmann constant, T is the absolute temperature, and ζ_r is the molecular friction coefficient of a Rouse chain ($D_r = kT/\zeta_r; \zeta_r \propto M$).

Doi and Edwards calculate the stress following an instantaneous deformation. Immediately after deformation the stress is that for an affinely deformed Gaussian network with νN strands/volume:

$$\tau_{ij}(0) = 3\nu NkT \langle (\mathbf{F} \cdot \mathbf{u})_i (\mathbf{F} \cdot \mathbf{u})_j \rangle, \tag{5.194}$$

in which τ is the extra stress tensor, $\mathbf{F}$ is the deformation gradient tensor, and $\langle \rangle$ denotes an average taken over all directions of the unit vector $\mathbf{u}$. Equilibration within the distorted tubes requires a time of the order of the Rouse relaxation time [$\lambda_r = (1/6\pi^2)\zeta_r R^2/kT; \lambda_r \propto M^2$]. After this process is complete the stress is

$$\tau_{ij}(\lambda_r) = 3\nu NkT \left\langle \frac{(\mathbf{F} \cdot \mathbf{u})_i (\mathbf{F} \cdot \mathbf{u})_j}{|\mathbf{F} \cdot \mathbf{u}|} \right\rangle \frac{1}{\langle |\mathbf{F} \cdot \mathbf{u}| \rangle}, \tag{5.195}$$

where the first $|\mathbf{F} \cdot \mathbf{u}|$ accounts for the equilibration of local stretches within the tube and the second for the assumed retraction of the tube path to its equilibrium length L. For longer times ($t \geq \lambda_r$) the stress decays according to

$$\tau_{ij}(t) = \tau_{ij}(\lambda_r) \frac{8}{\pi^2} \sum_{\text{odd } n} \frac{1}{n^2} \exp\left(-\frac{n^2 t}{\lambda_d}\right), \tag{5.196}$$

in which λ_d is the tube disengagement time given by

$$\lambda_d = L^2 \zeta_r / \pi^2 kT. \tag{5.197}$$

The long-time process [eqn (5.197)] corresponds to diffusion of chains out of their original cages. Since $\lambda_d \propto M^3$, the equilibration and disengagement processes will be widely separated in time-scale if the chains are long enough.

Results have been obtained for several of the flows mentioned above and it is possible to obtain (Currie 1982) from the theory a KBKZ constitutive model of the form (4.63) see also Larson (1988). Use of the Doi–Edwards spectrum (5.196) and associated potential gives only moderate agreement with experiments (Saab *et al.* 1982; Currie 1982). Doi and Edwards (1986) have shown that the 'irreversible' network idea of Wagner (1979) arises naturally in the theory; in physical terms the behaviour of a caged molecule is not independent of the previous strain pattern. Curtiss and Bird (1980) and Bird *et al.* (1987*b*) have presented an alternative derivation of this type of theory showing its relation to the bead-spring (Rouse) theory. They have also extended the theory and made comparisons with experiments (Saab *et al.* 1982). Currie (1982) has discussed the Curtiss–Bird theory; it often shows better agreement with experiment than the Doi–Edwards version, but it cannot describe step-strains. While these newer theories are attractive intuitive pictures of polymer response, they are at least as complex as the KBKZ theory in computations.

It is perhaps useful to contrast here the Doi–Edwards theory with the earlier network theories. The network models make no attempt to connect the memory function $\mu(s)$ with molecular parameters directly; the Doi–Edwards theory actually predicts the relaxation spectrum. Similarly, the rest of the Doi–Edwards (D–E) theory contains few adjustable parameters, and is therefore a more basic theory. The D–E theory gives (Doi and Edwards 1986; Doi 1995) the following positive results:

   (i) The diffusion of polymeric molecules is successfully predicted;

  (ii) The stress-optical law is predicted;

 (iii) General agreement with polymer melt behaviour is predicted.

Note that (i) is not addressed in network theories and (iii) needs considerable experimental input with the network theories.

However, there are problems with the D–E theory—for example, the viscosity function is atypical, and a maximum in shear is predicted; the D–E viscosity function decreases too rapidly. Also, the theoretical relaxation spectrum is close to a single line, and lacks breadth. The general form of the D–E constitutive model is again the separable single-integral KBKZ model, and so recoil remains a problem. A critical review of the description of stretching experiments on HDPE by D–E theory is that of Wagner *et al.* (1998). They conclude that the damping functions arising in the D–E theory underestimate the stresses arising in HDPE considerably and so new concepts are needed; see also Larson (1988) and Kasehagen and Macosko (1998). One example is the 'double reptation concept' of Graessley (Wasserman and Graessley 1996); see also Ianniruberto and Marrucci (1998) for some improved D–E-type theories.

5.8 Rigid particle microstructures

So far we have mainly discussed fluids with very flexible microstructures. In the present Section the particles are essentially rigid, and interactions can only be avoided at low concentrations. We will be concerned with suspensions, colloids and briefly, liquid crystals. In all cases the size of the particles is assumed to be several orders of magnitude smaller than the geometrical features of the flow.

The forces on the particles may include (viscoelastic) hydrodynamics, Brownian motion, gravity, electric forces between charged particles, inertia and other attractive forces. This wide field has been discussed by Russel *et al.* (1995) in their book *Colloidal dispersions*. In the present treatment, we will not consider electric and other attractive forces. The Reynolds number is generally assumed to be so small that inertia effects can be neglected, and gravity forces will also be ignored, despite their importance and interest [see, for example Davis and Acrivos (1985)] in settling problems.

Particle sizes range widely, from colloidal particles of order $10^{-2}-10^{-3}$ μm in diameter, up to sandlike particles, of order 0.1 mm in diameter. Particle shapes also vary, from near-spheres to rods; even more complex bead-rod shapes have also been discussed. (Bird *et al.* 1987*b*.)

5.8.1 Suspensions of spheres

To begin we suppose the suspending fluid is Newtonian with viscosity η_s. One is studying the interaction of Brownian, viscous, gravity, and inertia forces. These give rise, in a particle of radius a, to three dimensionless numbers characterizing behaviour. The first is the volume concentration (ϕ), the second the Reynolds number. To form the Reynolds number (Re) for a particle one uses the local velocity (U) and finds $(Re) = \rho U a / \eta_s$, which we will often ignore. The other dimensionless group is the ratio of viscous to Brownian forces. Here the local velocity U is of no significance, and the viscous forces will be written as $\eta_s \dot{\gamma} a^2$, where $\dot{\gamma}$ is a characteristic (generalized) shear rate. Brownian forces are, from (5.26), of order kT/a. The ratio of these forces determines a Peclet number (Pe) defined as

$$(Pe) = \dot{\gamma}\eta_s a^3 / kT \tag{5.198}$$

Hence, a large Peclet number is associated with rapid shearing, or very viscous solvent and/or large particles.

Alternatively, one can define a diffusivity D_s of a sphere in the suspending fluid as $D_s = kT/6\pi\eta_s a$, so obtaining a Peclet number $\dot{\gamma}a^2/D_s$ which is 6π times the value defined in (5.198). To put matters in concrete terms, consider a 1000 Pa-s solvent at 300 °K. Then since $k = 1.3807 \times 10^{-23}$ J/K, we have from (5.198), $(Pe) = 2.5 \times 10^{23} a^3 \dot{\gamma}$, or for $\dot{\gamma} = 0.1$ s^{-1}, $(Pe) = 2.5 \times 10^{22} a^3$. If $(Pe) > 1000$, we need $a > 0.3$ microns.

To avoid settling out, the Brownian 'forces' must counterbalance gravity; for a particle of radius a, we require $a^4 \Delta\rho g / kT < 1$, where $\Delta\rho$ is the density difference between sphere and suspending fluid, and g is the gravitational acceleration. Thus for a neutrally buoyant, inertialess suspension of identical spheres dimensional analysis tells us that the suspension viscosity (η) relative to that of the suspending fluid (η_s) is a function of (Pe) and the volume fraction (ϕ) only:

$$\eta/\eta_s = f[(Pe), \phi]. \tag{5.199}$$

When $(Pe) \rightarrow \infty$, then the (Pe) dependence disappears from (5.199).

The beginning of suspension theory is the result of Einstein (1905) for very dilute suspensions of spheres (volume fraction of spheres less than about 0.03 of the total volume) in any flow field:

$$\frac{\eta}{\eta_s} = 1 + 2.5\phi, \tag{5.200}$$

where η/η_s is the average relative viscosity of the suspension.

Einstein's (1905) derivation of (5.200) was arrived at via a dissipation argument, balancing micro- and macro-scales; Huilgol and Phan-Thien (1997) give a derivation starting purely from viscous, inertia-free hydrodynamics, and hence Brownian motion is not a factor in (5.200). It was a considerable time until Batchelor and Green (1972) extended (5.200) in shearing flow to add the term $5.2\phi^2$. If Brownian motion is considered, this term was found to be $6.2\phi^2$; for elongational flow the coefficient was found to be 7.6 (Batchelor 1974). Thus the flow is already 'anisotropic' in the sense that shear and elongation have different effective viscosities.

There are many proposed forms for the viscosity function (5.199) for every large Peclet numbers (Metzner 1985; Utracki 1988). One of the simplest ideas has been discussed by Phan-Thien and Pham (1987). They consider a suspension with volume fraction ϕ and viscosity η. Addition of a small amount of particles $d\phi$ to the $1 - \phi$ fraction of fluid remaining raises the volume fraction by $d\phi/(1 - \phi)$. Using the viscosity η as an average viscosity of the fluid, and using the Einstein relation (5.200) the increment in viscosity $d\eta$ is

$$d\eta = \frac{5}{2} \frac{d\phi}{(1 - \phi)} \eta. \tag{5.200a}$$

Integrating, one finds the 'self-consistent' result

$$\frac{\eta}{\eta_s} = (1 - \phi)^{-5/2}. \tag{5.201}$$

For ϕ values near $\phi_{\max}$, the maximum packing fraction (~ 0.64), the expression in (5.201) gives too low a value. The result can be extended to several sphere sizes and it has an expansion at low ϕ which agrees with the Einstein result (5.200). However, the next term in the expansion of (5.201) gives $4.375\phi^2$, which is too low.

Ball and Richmond (1980) had a more refined argument which pointed out that the increment $d\phi$ was really added to less than $1 - \phi$, due to a 'crowding' effect. They replaced $1 - \phi$ by $1 - \phi/\phi_m$ in (5.200a) and so obtained

$$\eta/\eta_s = (1 - \phi/\phi_m)^{-2.5\phi_m} \tag{5.201a}$$

which agrees with the Krieger–Dougherty formula. The results are shown in the fourth column of Table 5.2, assuming $\phi_m = 0.64$. This approach can also be extended to several particle sizes (Barnes *et al.* 1989).

The approach of Lundgren (1971) yields

$$\frac{\eta}{\eta_s} = (1 - 2.5\phi)^{-1}, \tag{5.202}$$

which becomes infinite at $\phi = 0.4$.

According to the excellent survey of Metzner (1985) the best theoretical equation is that of Frankel and Acrivos (1967):

$$\frac{\eta}{\eta_s} = \frac{9(\phi/\phi_m)^{1/3}}{8\{1 - (\phi/\phi_m)^{1/3}\}} \tag{5.203}$$

where ϕ_m is the maximum packing fraction ($\sim 0.62 - 0.64$ for uniform spheres). Frankel and Acrivos recognized that in dense suspensions the 'lubrication' (see Chapter 6) forces between neighbouring particles dominated, and that sphere–sphere distance was an important parameter. By considering a simple cubic lattice, side b, with spheres of radius a, we find $\phi_{max} = \pi/6 \approx 0.52$, and the

Table 5.2 Relative viscosity formulae for smooth sphere suspensions. This table gives the relative viscosity (η/η_s) as a function of the volume concentration ϕ, for various formulae, identified below

ϕ	Einstein (5.200)	Batchelor[1] and Green	Eqn (5.201a)	Frankel[2]	Eqn[3] (5.204)	Experiment[4]
0.01	1.025	1.026	1.026	0.375	1.030	1.029
0.05	1.125	1.141	1.139	0.840	1.165	1.156
0.1	1.250	1.312	1.312	1.313	1.375	1.365
0.2	1.500	1.748	1.821	2.375	2.007	1.978
0.3	1.750	2.308	2.751	3.916	3.202	3.052
0.4	2.000	2.992	4.803	6.633	5.898	5.697
0.5	2.250	3.800	11.38	13.12	14.27	15.75
0.6	2.50	4.73	84.45	51.73	72.25	63.89
0.64	2.60	5.14	∞	∞	289.0	118.9

[1]Batchelor and Green (1972): $\eta/\eta_s = 1 + 2.5\phi + 6.2\phi^2$
[2]In this formula $\phi_m = 0.64$
[3]Here $A = 0.68$ for smooth spheres.
[4]An average experimental fit gives

$$\eta/\eta_s = 1 + 2.5\phi + 10.05\phi^2 + 0.00273\exp(16.6\phi)$$

(dimensionless) gap between spheres is $b/a - 1$, or $(\phi_m/\phi)^{1/3} - 1$. Thus the form (5.203) appears to have the correct asymptotic value as $\phi \to \phi_{max}$. This equation does not agree with Einstein's result as $\phi \to 0$, and Metzner (1985) recommends

$$\eta/\eta_s = [1 - \phi/A]^{-2} \qquad (5.204)$$

where A is to be found empirically. For uniform smooth spheres $A = 0.680$. Again, some disagreement with the Einstein relation as $\phi \to 0$ is evident. Comparison of the various formulas (5.200) to (5.204) with the excellent experimental data of Thomas (1965) is shown in Table 5.2 and Fig. 5.24. An empirical fit was made to the data ($\pm$ around 5 per cent experimental scatter) and this is plotted as the full line in Fig. 5.24. One sees:

(i) Up to about $\phi = 0.02$ the Einstein formula is adequate.

(ii) Up to about $\phi = 0.04$ the Batchelor formula is adequate (see Table 5.2).

(iii) For $\phi > 0.1$ up to $\phi = 0.5$ (5.201a) or (5.204) may be used, but beyond $\phi = 0.6$ results are not very accurate.

(iv) The empirical equation of Thomas (1965) is given in Table 5.2.

One should note that more experimental scatter than indicated by Thomas (1965) is usual (Rutgers 1962).

While the above results are for monodisperse sphere sizes, it is possible to reduce the viscosity of suspensions at a given ϕ, by having a wide spectrum of sphere sizes (Metzner 1985; Utracki 1988; Barnes *et al.* 1989).

Variable viscosity and normal stress effects

So far non-linear effects in suspensions have not been examined. Some non-Newtonian effects are noticeable with Newtonian solvents at low Peclet numbers; with non-Newtonian solvents shear-thinning behaviour is no surprise. When Brownian motion is important, shear-thinning can occur. Curiously, and contrary to the ideas set out in Chapter 3, since the entire fluid/solid system is linear, so are the normal stresses linear in shear rate. However, reversal of shear does not reverse the normal stresses. One may refer to the paper of Nunan and Keller (1984) to see that in an ideal cubic-lattice system the viscosity is constant and normal stress differences are zero, as one would expect in a linear creeping flow. However, in real suspensions the configuration changes and particle migration occurs. Some non-Newtonian effects in monodisperse spherical particle/Newtonian suspending fluid have been found (Gadala-Maria and Acrivos 1980). They showed a decline of relative viscosity with *time* at a constant shear rate, slightly non-Newtonian behaviour at higher ϕ values (Fig. 5.25) and some 'elastic' effects. The Peclet numbers [eqn. (5.190)] were in excess of 10^5 in these experiments, and so the effects were attributed to structural rearrangements as shearing continued. For spherical particles normal stress differences are low and not easy to measure. Phan-Thien (1995) gives some data for $N_1 - N_2$ for the fluid

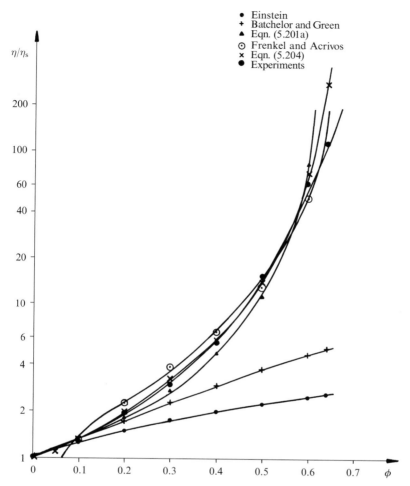

• Einstein
+ Batchelor and Green
▲ Eqn. (5.201a)
⊙ Frenkel and Acrivos
× Eqn. (5.204)
● Experiments

Fig. 5.24   Relative viscosity versus volume fraction (ϕ) for smooth sphere suspensions. ● Experimental average; • Einstein formula; + Batchelor and Green; ▲ eqn (5.201a); ⊙ Frenkel and Acrivos; × eqn (5.204).

tested by Gadala-Maria and Acrivos (1980). Jomha and Reynolds (1993) found shear-thinning, -thickening (Fig. 5.26), and normal stresses in several suspensions. The Peclet number was smaller in these experiments, $O(10^3)$. Barnes (1989) has surveyed some shear-thickening behaviour.

For smaller Peclet numbers Brownian motion is clearly important, and shear-thinning and normal stress effects have been predicted by Brady and Vicic (1995) and Brady and Morris (1997). They find $N_2/N_1 = -0.89$ for very low

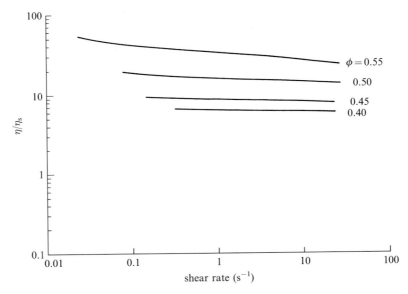

Fig. 5.25 Dependence of the asymptotic relative shear viscosity on the shear rate for suspensions of several concentrations. Polystyrene spheres, 40–50 μm in diameter in a mixture of silicone oils.

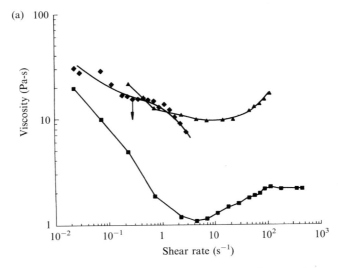

Fig. 5.26(a) Viscosity–shear rate data for (■) Superclay, (◆) polystyrene latex and (▲) PVC suspensions. The arrow indicates the point at which polystyrene latex fractured. Lines are drawn through the data to guide the eye.

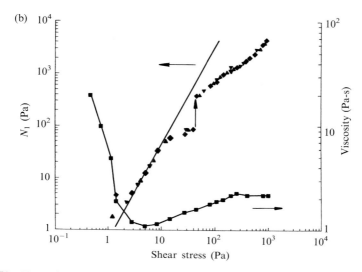

Fig. 5.26(b) $N_1 - \tau$ data for Superclay/water suspension at ($\blacklozenge$) 20 °C, ($\blacktriangle$) 27 °C and ($\blacktriangledown$) 36 °C. The straight line has a slope of 2. The viscosity data ($\blacksquare$) at 20 °C are shown.

Peclet numbers, $N_1 \sim \phi^2(Pe)$, and the viscosity ratio $\eta/\eta_s = 1 + 2.5\phi + 5.91\phi^2 + O[(Pe^2\phi^2)]$. Thus the Newtonian constitutive equation is not always appropriate even for hard-sphere suspensions, although it is often assumed to be.

The evolution of the microstructure can only be derived from a full solution of the multibodied interaction problem. In the absence of this, we need to postulate a reasonable evolution equation for $\mathbf{R}$, the centre-to-centre vector between two generic spheres. Based on some limited numerical data, Phan-Thien (1995) has proposed the simple model

$$\dot{\mathbf{R}} = \mathbf{L} \cdot \mathbf{R} + \mathbf{B}(t), \tag{5.205}$$

where $B(t)$ is the fluctuating component of the motion (in a Lagrangian sense); it is not Brownian motion, but reflects the spatial fluctuations being experienced by the two generic particles. Here, $\mathbf{L}$ is the velocity gradient tensor. The simplest statistical model of this fluctuation component is white noise; however, this would require a vanishing time scale for the fluctuations compared with the time scale of the average motion, and Phan-Thien (1995) replaced this unrealistic assumption with a diffusive process having a diffusivity d_R, such that

$$d_R = Ka^2\dot{\gamma} \tag{5.206}$$

where K is a constant [O(1)] and $\dot{\gamma}$ is the generalized strain-rate. Phan-Thien finds the particle stress τ^P to be

$$\tau^P = \eta(\phi)\{\mathbf{d} : \langle \mathbf{uuuu} \rangle + \tfrac{1}{2}K\dot{\gamma}\langle \mathbf{uu} \rangle\} \tag{5.207}$$

where **d** is the rate-of-deformation tensor, and the unit vector **u** obeys the evolution equation

$$\lambda \left\{ \frac{\Delta \mathbf{A}}{\Delta t} + 2\mathbf{d} : \langle \mathbf{uuuu} \rangle \right\} + \mathbf{A} = \frac{1}{3}\mathbf{I} \tag{5.208}$$

where $\Delta/\Delta t$ is the upper convected derivative (4.24) and **A** is the average of **uu**, $\langle \mathbf{uu} \rangle$. Here the time constant $\lambda = 2/3K\dot{\gamma}$. The need for a closure approximation is evident. Feng *et al.* (1998) have surveyed several possibilities for closure schemes, but Phan-Thien used the Hinch and Leal approximation (1976):

$$\mathbf{d} : \langle \mathbf{uuuu} \rangle = \frac{1}{5} \left(6\langle \mathbf{uu} \rangle \cdot \mathbf{d} \cdot \langle \mathbf{uu} \rangle - \mathbf{d} : \langle \mathbf{uu} \rangle \langle \mathbf{uu} \rangle - 2\langle \mathbf{uu} \rangle^2 : \mathbf{dI} \right.$$

$$\left. + 2\langle \mathbf{uu} \rangle : \mathbf{dI} \right), \tag{5.209}$$

which is valid in both weak and strong flows.

If the front factor on the right-hand side of (5.207) is chosen empirically as

$$\eta(\phi) \simeq 8\eta_s \left(1 - \frac{\phi}{\phi_m} \right)^{-\alpha},$$

then the viscosity form (5.204) results in a simple shear flow.

The constitutive model has several features, some of which have been observed in concentrated systems:

(i)   An instantaneous response at the inception of the flow.

(ii) The stresses instantaneously reduce to zero when the flow is stopped.

(iii) If the flow is restarted in the same direction, then the stresses will recover their previous values instantaneously, with the period of rest being of no consequence.

(iv) If the shear rate changes from the previous value, the stresses instantaneously attain the steady state values corresponding to this new shear rate.

(v)   If the flow is restarted but in the opposite direction, then the stresses only recover partially, and then relax to their steady-state values.

(vi) The stresses are linear in the strain rate, which leads to a Newtonian viscosity, and normal stress differences which are linear in the magnitude of the shear rate, and the stress is anisotropic with respect to the strain rate tensor so that the flow resistance will depend on the nature of the flow field.

(vii) A universal transient response is found when the stresses (reduced by $\eta(\phi)\dot{\gamma}$) are plotted against $\dot{\gamma}t$; and also in oscillatory flow.

(viii) the dependence of the stresses on the volume fraction is the same in all flows.

The predictions of the model in some simple flow fields have been given by Phan-Thien (1995).

5.8.2 *Effects of particle geometry*

So far only spherical particles have been discussed. For the case when the particles are non-spherical but reasonably compact it has been found (Metzner 1985), for modest aspect ratios (<30) that the relative viscosity can again be described by (5.204) provided A is chosen appropriately. For uniform spheres $A = 0.68$, as discussed above. For rough crystals A is around 0.44. For aspect ratios of short fibres varying from 6 to 27, one finds, approximately (Kitano *et al.* 1981) in the range of aspect ratio (AR) of $5 < AR < 30$, that

$$A = 0.55 - 0.013(AR) \tag{5.210}$$

When $AR = 1$, this expression lies between the smooth sphere and rough crystal values. Nevertheless, the expression needs to be used cautiously.

Once one moves away from spherical particles there is an expectation of non-Newtonian behaviour. Nguyen and Boger (1992) have measured typical viscoelastic behaviour in a fibre suspension.

Given the complexity of the result for spheres, it is not surprising that results for non-spherical particles are mostly for dilute suspensions; in this case one can also permit individual particles to deform in the flow (Schowalter 1978, p. 271). However, Batchelor (1971) has computed the elongational flow characteristics of a solution of parallel rods which interact significantly with one another, but the corresponding problem for shear flow is very difficult and hence no constitutive equation is available yet for such a suspension.

Leal and Hinch (1973) have made a valuable contribution, beginning from suspension theory, that bears some resemblance to the dilute solution theories we have studied above. Since it can in some cases produce a constitutive equation, we reproduce the conclusions here.

For a dilute solution of small rigid, axially symmetric ellipsoidal particles, the unit vector directed along the longest axis may be denoted by $\mathbf{u}$, and it obeys an evolution equation, in the absence of Brownian motion, of the Jeffrey (1922) form

$$\dot{\mathbf{u}} = \mathbf{w} \cdot \mathbf{u} + \frac{(AR)^2 - 1}{(AR)^2 + 1}(\mathbf{d} \cdot \mathbf{u} - \mathbf{d} : \mathbf{uuu}), \tag{5.211}$$

where AR is the aspect ratio of the particle (length to diameter ratio), $\mathbf{w}$ is the vorticity tensor, and $\mathbf{d}$ is the rate of deformation tensor. The particle-contributed stress may be shown to be

$$\boldsymbol{\sigma}^{(p)} = 2\eta_s \phi \{A\mathbf{d} : \langle \mathbf{uuuu} \rangle + B(\mathbf{d} \cdot \langle \mathbf{uu} \rangle + \langle \mathbf{uu} \rangle \cdot \mathbf{d}) + C\mathbf{d} + d_R F \langle \mathbf{uu} \rangle \}, \tag{5.212}$$

where the angular brackets denote the ensemble average with respect to the distribution function of $\mathbf{u}$; A, B, C, and F are some shape factors, and d_R is the rotational diffusivity. If the particles are large enough so that Brownian motion can be ignored, then the last term, as well as the angular brackets, can be omitted in the previous expression. The asymptotic values of the shape factors are given in Table 5.3.

The rheological predictions of this constitutive equation have also been considered by Hinch and Leal (1972); the shear viscosity is shear-thinning, the first normal stress difference is positive while the second normal stress difference is negative, but of a smaller magnitude. The precise values depend on the aspect ratio and the strength of the Brownian motion.

When Brownian motion is absent, the complication due to the ensemble averages disappears because of the large size of the particles (Huilgol and Phan-Thien 1997). One finds:

1. The reduced viscosity (where $\langle \tau_{12} \rangle$ is the average shear stress; see Huilgol and Phan-Thien 1997):

$$\frac{\langle \tau_{12} \rangle - \eta_s \dot{\gamma}}{\eta_s \dot{\gamma} \phi} = 2A u_1^2 u_2^2 + B(u_1^2 + u_2^2) + C. \tag{5.213}$$

2. The reduced first normal stress difference:

$$\frac{N_1}{\eta_s \dot{\gamma} \phi} = 2A u_1 u_2 (u_1^2 - u_2^2), \tag{5.214}$$

and

3. The reduced second normal stress difference:

$$\frac{N_2}{\eta_s \dot{\gamma} \phi} = 2A u_1 u_2 (A u_2^2 + B). \tag{5.215}$$

The particles tumble in the flow and the viscosity is periodic in time; most of the time they are aligned with the flow; N_1 and N_2 are also periodic, with the same period as the viscosity.

Table 5.3 Asymptotic values of the shape factors

Asymptotic limit	$AR \to \infty$ (rod-like)	$AR = 1 + \delta, \delta \ll 1$ (near-sphere)	$AR \to 0$ (disc-like)
A	$\dfrac{(AR)^2}{2(\ln 2AR - 1.5)}$	$\dfrac{395}{147}\delta^2$	$\dfrac{10}{3\pi AR} + \dfrac{208}{9\pi^2} - 2$
B	$\dfrac{6\ln 2AR - 11}{(AR)^2}$	$\dfrac{15}{14}\delta - \dfrac{395}{588}\delta^2$	$-\dfrac{8}{3\pi AR} + 1 - \dfrac{128}{9\pi^2}$
C	2	$\dfrac{5}{2}\left(1 - \dfrac{2}{7}\delta + \dfrac{1}{3}\delta^2\right)$	$\dfrac{8}{3\pi AR}$
F	$\dfrac{3(AR)^2}{\ln 2AR - 1/2}$	9δ	$-\dfrac{12}{\pi AR}$

In the start-up of elongation the particle is quickly aligned with the flow in a time scale $O(\dot\gamma^{-1})$. At a steady state, the reduced elongational viscosity (elongation rate $\dot\varepsilon$) is

$$\frac{N_1 - 3\eta_s\dot\varepsilon}{\eta_s\dot\varepsilon\phi} = 2(A + 2B + C) \approx \frac{(AR)^2}{\ln 2(AR) - 1.5} \qquad (5.216)$$

The dilute solution assumption used here means that the volume fraction is low enough, so that a particle can rotate freely without any hindrance from its nearby neighbours. The distance Δ between any two particles must therefore satisfy $l < \Delta$, so that a volume of l^3 contains only one particle, where l is the length of the particle and d is its diameter. The volume fraction therefore satisfies

$$\phi \sim \frac{d^2 l}{\Delta^3}, \qquad \text{or} \qquad \phi(AR)^2 < 1.$$

Thus, the reduced elongational viscosity is only $O(1)$ in the dilute limit, not $O(AR^2)$ as suggested by the formula. As the concentration increases, we get subsequently into the semi-dilute regime, the isotropic concentrated solution, and the liquid crystalline regime. The reader is referred to Doi and Edwards (1986) for more details. Here, we simply note that the concentration region $1 < \phi(AR)^2 < AR$ is called semi-concentrated. Finally, suspensions with $\phi AR > 1$ are called concentrated, where the average distance between fibres is less than a fibre diameter and therefore fibres cannot rotate independently except around their symmetry axes. Any motion of the fibre must necessarily involve a co-operative motion of surrounding fibres.

5.8.3 Fibre suspensions

As the volume concentration rises, the freedom of an individual fibre to move is curtailed, until eventually only co-operative motion is allowed in concentrated solutions. Microstructural theories developed by various workers (see Huilgol and Phan-Thien 1997) for the dilute case have forms similar to the early work of Ericksen (1960) and Hand (1962). These theories have an evolution equation and a stress calculator, similar to the work described in Section 5.5.2.

For non-dilute solutions, fibre–fibre interactions are paramount. Folgar and Tucker (1984) have developed an evolution equation for concentrated fibre suspensions, where the fibre–fibre interactions are taken into account by adding a diffusion term to Jeffery's equation. Dinh and Armstrong (1984) discuss the dynamics of non-Brownian particles and derive a constitutive equation for semi-dilute suspensions; the model takes into account the fibre–fibre interaction and uses a distribution function to describe the orientation state.

In the Folgar and Tucker (1984) model, the diffusivity is assumed to be of the form $C_i\dot\gamma$, where $\dot\gamma = (2\,\mathrm{tr}\,\mathbf{d}^2)^{1/2}$ is the generalized strain rate, and the aspect ratio of the fibres is assumed infinite. The parameter C_i is known as the *interaction coefficient*, which has been experimentally determined to lie in the range of $10^{-2} - 10^{-3}$. Yamane *et al.* (1994) obtained the interaction coefficient in a numerical simulation of semi-dilute suspensions of rod-like particles in a shear flow. However, the predicted values of C_i are about two orders of magnitude smaller than those suggested by experiment.

Fan *et al.* (1998) have computed the C_i coefficient by direct numerical simulation and have found much larger values than previous analyses gave. It seems that this phenomenological constant must be a function of the volume fraction of the fibres, and its aspect ratio; it may even be a tensorial quantity, reflecting the anisotropy of the fluid. A closure approximation relating $\langle \mathbf{uuuu} \rangle$ to $\langle \mathbf{uu} \rangle$ is again needed; see Feng *et al.* (1998).

Various other models have been proposed; see, Dinh and Armstrong (1984) and Phan-Thien and Graham (1991). See also Huilgol and Phan-Thien (1997); they discuss particle migration in suspensions.

5.9 Anisotropic fluids

Constitutive modelling for isotropic non-Newtonian fluids was dominant until the mid-1960s, but was also complemented by the study of anisotropic fluids. The works of Ericksen (1961) and Leslie (1968) are important milestones in this study.

Work on anisotropic fluid theory was given a boost by applications to liquid crystal displays and similar devices, and a further impetus was provided by the growth of interest in liquid crystal polymers (LCPs) (for example, Kevlar).

Continuum theory for anisotropic fluids requires an additional variable over and above those employed in the isotropic theory. A unit vector field, here denoted by u_i, is commonly referred to as the director vector. Another added complication is that the question of the symmetry of the stress tensor needed to be examined closely in these systems.

In view of these complications, it is not surprising that initial developments of the theory were quite special and nothing like as general as, say, the simple fluid theory of Section 4.4, for isotropic fluids. We quote the immensely influential Leslie (1968)/Ericksen (1961) theory, which is even in u_i and linear in the rate of strain d_{ik} and u_i, where

$$N_i = \frac{\mathrm{D}u_i}{\mathrm{D}t} - \omega_{ip}u_p, \tag{5.217}$$

ω_{ik} being the vorticity tensor component and $\mathrm{D}/\mathrm{D}t$ denoting the material time derivative.

Under these restrictive conditions, the extra stress tensor τ_{ik} can be written in the form:

$$\begin{aligned} \tau_{ik} = {} & \alpha_1 d_{jl}u_j u_l u_i u_k + \alpha_2 N_i u_k + \alpha_3 N_k u_i + \alpha_4 d_{ik} + \alpha_5 d_{ij}u_j u_k \\ & + \alpha_6 d_{kj}u_j u_i, \end{aligned} \tag{5.218}$$

where here the αs are the so-called Leslie coefficients. Parodi (1970), applying Onsager relationships, showed that the constants can be related through $\alpha_6 - \alpha_5 = \alpha_3 + \alpha_2$. Such a special theory is still very much in vogue. There is no doubt that the initial motivation was pragmatic, given the obvious difficulties of attempting any level of generality. However, some of the implications of the theory, like the scaling laws for simple shear and Poiseuille flow (see, for example, Leslie 1987), were shown to be in surprisingly good agreement with experiment, and any further complication to the theory would be likely to destroy that agreement. With hindsight, therefore, there was little merit in attempting a more sophisticated theory. For this reason, the Leslie–Ericksen theory has been and continues to be a focal point in modern developments in liquid-crystal theory. For example, between 1968 and 1974, P. G. de Gennes and his group applied the Leslie–Ericksen theory to a number of experimental situations (see, for example, de Gennes 1974). We refer to the extensive literature on liquid crystals for applications and further considerations (for example, Chandrasekhar 1980; Larson 1999).

5.10 Use of constitutive models for engineering problems

Except in the simplest continuum theories (elasticity, Newtonian fluids) it is normal and necessary to utilize various stages of approximation to real material behaviour depending on the case in hand. For example, in plasticity theory one might use elastic-plastic behaviour, work-hardening theory, or simply rigid-plastic analysis depending on the problem being considered. The author believes that this must also be done in viscoelastic and non-Newtonian case studies, and that it is wasteful and counterproductive to always insist on using the most complex constitutive model for all purposes; it merely needs to be adequate for the purpose in hand. When assessing adequacy, the experimental errors mentioned in Section 3.8.1 should be firmly in mind.

One also needs to have, whenever possible, a model relevant to the range of generalized shear rates in the flow process being investigated. Rates of deformation vary enormously; in the low range, from 10^{-6} s^{-1} to 10^{-4} s^{-1} for the sedimentation of fine powders, and from 10^{-3} to 10 s^{-1} in gravity-driven flows. In the medium range, 1 s^{-1} to 10^3 s^{-1}, lie many common mixing, extrusion and pipe flows. For the higher ranges, $10^4 - 10^7$ s^{-1}, found in brushing, lubrication, and coating flows, data may be hard to find. Generally it is not difficult to find an average shear rate for a given process. Fluid particles may also experience rates of shear for several decades on either side of this average, and several relaxation times will usually be necessary to cover the expected range properly. Attempts to use a single relaxation time can give inaccurate results (Aziez et al. 1996).

There are a number of classes of flow problems which arise in applications:

1. Small-strain behaviour.

2. Steady weak flows, of which simple shearing may be taken as an example.

3. Steady strong flows, including simple elongation as a prime example.

Table 5.4 Various constitutive relations for polymer fluids and their performance

Constitutive model	Small-strain	Steady viscometric			Steady elongation	Start/stop in shearing flow	Elongational start/recoil	Single shear step	Double-step shear	Remark
		η	N_1	N_2						
Newtonian, eqn (1.7)	P	M	U	U	M	P	U	U	U	Infinite stresses in step strains.
Generalized Newtonian, e.g. eqns (4.4, 4.11)	P	E	U	U	M	P	U	U	U	Infinite stresses in step strains.
Reiner–Rivlin, eqn (4.1)	P	E	U	E	E	P	U	U	U	As above.
Second-order, eqn (4.30c), η, Ψ_1, Ψ_2 constant	P	M	M	M	U	U	U	U	U	At high elongation rates η_E becomes negative.
Higher-order fluids	P	M	M	M	U-P	U	U	U	U	See Schowalter (1978) for forms of equations.
Criminale, Ericksen, Filbey, eqn (4.32)	P	E	E	E	U-P	U	U	U	U	Useful for viscometric flows.
Linear viscoelastic, eqn (2.81)	E	M	U	U	M	M	M	M	M	No non-linear effects.
Oldroyd, eqn (4.13)	M	M	M	M	P	M	M	M	M	Limited viscosity variation.
Green–Rivlin, eqn (4.42)	E	M	M	M	P	M	M	M	M	Double integral is hard to use.

									Comments
Lodge–Maxwell, (UCM) eqns (4.25, 5.150)	E	M	M	U	P	M	M	M	Useful for illustrative purposes.
White–Metzner	M	E	E	U	P	M	M	P	See footnote 1.
Co-rotational	E	E	M	M	P	P	P	U	Oscillation in start of shearing. See footnote 2.
Bird–Carreau, Carreau	E	E	E	G	P	M	M	P	See Bird et al. (1987a, b) and Problem 5.9.
Phan–Thien–Tanner (PTT) eqn (5.161)	E	E	E	G	G	G	G	M	Unsuitable for very large step strains, unless $N_2=0$
Giesekus eqn (5.111)	E	E	G-E	G	G	G	G	M	
Acierno et al., eqns (5.119–123)	E	E	G	G	G	G	M	M	Poor step strain response.
KBKZ, eqn (4.54)	E	E	G	G	G	G	E	G-M	
KBKZ (Wagner), eqn (4.75)	E	E	G	G	G	G	E	G	See footnote 3.
Eqn (5.103)	G	G	G	U	G	G	M	U	Dilute solution theory: large step strains, cannot be applied.
Leonov	E	E	G	M	M	G	M	M	
Doi–Edwards	M	M	M	M	M	M	M	M	

[1] The White–Metzner model is of the form

$$\lambda(\mathrm{II_D})\frac{\Delta \boldsymbol{\tau}}{\Delta t} + \boldsymbol{\tau} = 2\eta(\mathrm{II}_D)\mathbf{d};$$

that is, it is a convected Maxwell model where the time constant and viscosity are allowed to vary with the rate of shearing.

[2] An extensive discussion is given by Bird et al. (1977) in Chapter 8.

[3] There is some lack of fit with this theory, but the qualitative response is very good overall.

4. Weak flows with discontinuous velocity histories; this class includes stopping and starting shearing flows.

5. Strong flows with discontinuous velocity histories.

6. Flows with single strain steps. (It is not necessary to distinguish between the weak and strong cases here because a step-strain is necessarily finite).

7. Flows with multiple-strain jumps, especially jumps of varying sign.

These seven categories form a heirarchy of hurdles of increasing difficulty on which constitutive equations can be tested. Results of calculations will show either

(a) No result or a physically impossible result. (U)

(b) A possible result but not in good agreement with typical experimental data. (P)

(c) Moderate agreement with typical behaviour. (M)

(d) Good agreement with experiments. (G)

(e) Exact agreement due to use of fitted data. (E)

These categories will be denoted by the letters U (unsuitable or useless), P (poor or possible), M (modest agreement meaning the sign and trend of the results are correct), G (good, meaning that the data are fitted within (say) ± 20 per cent in most places) and E (exact or excellent, meaning that the data are, or can be, fitted within a few per cent). Table 5.4 shows the results of this exercise for polymer melts and concentrated solutions. Other models are given in the books by Bird *et al.* (1987*a*), Larson (1988), and Huilgol and Phan-Thien (1997).

From Table 5.4 it will be clear that one should always avoid the cases where a U is indicated for the class of response under consideration; generally one should also avoid the P cases. The equations are roughly in order of increasing complexity as one goes down the table, and the overall results get, on average, better as one moves down. Nevertheless, one should not automatically choose the most complex type of equation, (for example, KBKZ) if it is not required for accuracy.

Clearly, if step-strains occur, then the choice is limited if a moderate or better description is needed. None of the theories except Wagner's KBKZ variant, perhaps, handles double-step strains very accurately. For unsteady strong flows the choice is also limited, but otherwise a reasonable choice is available, especially if only illustrative problems are to be worked out. When the flow is steady but not nearly-viscometric, one should beware of problems near points where large stresses occur (see Chapter 8) and it may still be necessary to control the large tensile stresses that occur with the Lodge and related Maxwell models. Finally, when the flow is nearly viscometric, then one can use the methods developed in Chapter 4 for this class of constitutive model. For illustrative cases, one often requires a simple model with a single time constant; in that case it may be possible to use one of the Oldroyd models or the Maxwell model.

For realistic modelling, multiple relaxation times are generally required. In these cases the PTT model and its variants, and the Giesekus model, are useful.

The older models of White–Metzner and Tanner (1965) are not recommended except for illustrative purposes. They show atypical behaviour in all flows except nearly-viscometric cases. When the flow is viscometric, we have no real need of a constitutive model as explained in Chapter 3.

For some materials (for example, polyvinylchloride) an elastic response (Section 4.7) is sometimes useful. Phan-Thien *et al.* (1997) have used an elastic plus viscoelastic non-linear model for bread dough.

For dilute solutions equations (5.106) and the Chilcott–Rallison model (5.110) avoid most of the obvious problems in this type of system. For suspensions and like systems there are relatively few constitutive equations; see Section 5.8.1 for an equation of Phan-Thien (Huilgol and Phan-Thien 1997) as an example. The Leslie–Ericksen equations continue to be used for liquid crystals, and the Folgar–Tucker (Section 5.8.3) equation and its variants for fibre-filled melts. Pastes and suspensions (Benbow and Bridgwater 1993) are usually treated as a Bingham-type inelastic model, and concentrated suspensions are often described by a generalized Newtonian model.

Clearly, some judgement must be exercised in the choice of model for a particular situation, and examples of this philosophy will be given in the following chapters.

References

Acierno, D., La Mantia, F. P., Marrucci, G., and Titomanlio, G. (1976). *J. Non-Newtonian Fluid Mech.*, **1**, 125.

Azaiez, J., Guénette, R., and Ait-Kadi, A. (1996). *J. Non-Newtonian Fluid Mech.*, **66**, 271.

Baaijens, F. P. T., Selen, S. H. A., Baaijens, H. P. W., Peters, G. W. M., and Meijer, H. E. H. (1997). *J. Non-Newtonian Fluid Mech.*, **68**, 173.

Ball, R. and Richmond, P. (1980). *J. Phys. Chem. Liquids.*, **9**, 99.

Barnes, H. A. (1989). *J. Rheol.*, **33**, 329.

Barnes, H. A., Hutton, J. F., and Walters, K. (1989). *An introduction to rheology.* Elsevier, Amsterdam.

Batchelor, G. K. (1971). *J. Fluid Mech.*, **6**, 227.

Batchelor, G. K. (1974). *Ann. Rev. Fluid Mech.*, **6**, 227.

Batchelor, G. K. and Green, J. T. (1972). *J. Fluid Mech.*, **56**, 401

Bellman, R. E. (1960). *Introduction to matrix analysis.* McGraw-Hill, New York.

Benbow, J. J. and Bridgewater, J. (1993). *Paste flow and extrusion*, Oxford University Press

Berry, G. C. and Fox, T. G. (1968). *Adv. in Polymer Science*, **5**, 261.

Billmeyer, F. W. (1964). *Textbook of polymer science*, 2nd edn. McGraw-Hill, New York.

Bird, R. B. (1982). *Chem. Engn. Commun.*, **16**, 175.

Bird, R. B. and DeAguiar, J. R. (1983). *J. Non-Newtonian Fluid Mech.*, **13**, 149.

Bird, R. B., Armstrong, R. C., and Hassager, O. (1977). *Dynamics of polymeric liquids*, Vol. 1. *Fluid mechanics.* Wiley, New York.

Bird, R. B., Armstrong, R. C., and Hassager, O. (1987a). *Dynamics of polymeric liquids*, Vol. 1. *Fluid mechanics.* Wiley, New York. 2nd edn.

Bird, R. B., Hassager, O., Armstrong, R. C. and Curtiss, C. F. (1987*b*). *Dynamics of polymeric liquids*, Vol. 2, *Kinetic theory*. Wiley, New York.

Boger, D. V. (1977). *J. Non-Newtonian Fluid Mech.*, **3**, 87.

Brady, J. F. and Morris, J. F. (1997). *J. Fluid Mech.*, **348**, 103.

Brady, J. F. and Vicic, M. (1995). *J. Rheol.*, **39**, 545.

Byars, J. A., Binnington, R. J., and Boger, D. V. (1997). *J. Non-Newtonian Fluid Mech.*, **27**, 219.

Carreau, P. J. (1972). *Trans. Soc. Rheol.*, **16**, 99.

Chandrasekhar, S. (1980). *Liquid crystals*. Cambridge University Press.

Chilcott, M. D. and Rallison, J. M. (1988). *J. Non-Newtonian Fluid Mech.*, **29**, 381.

Currie, P. K. (1982). *J. Non-Newtonian Fluid Mech.*, **11**, 53.

Curtiss, C. F. and Bird, R. B. (1980). *Rheology Research Centre Rept. No. 65*, University of Wisconsin.

Davis, K. E. and Acrivos, A. (1985). *Ann. Rev. Fluid Mech.*, **17**, 91.

DeGennes, P. G. (1974). *The physics of liquid crystals*, Oxford University Press.

DeAguiar, J. R. (1983). *J. Non-Newtonian Fluid Mech.*, **13**, 161.

DeGennes, P. G. (1979). *Scaling concepts in polymer physics*. Cornell University Press, London.

Dinh, S. H. and Armstrong, R. C. (1984). *J. Rheol.*, **28**, 207.

Doi, M. (1995). *Introduction to polymer physics*, Clarendon Press, Oxford.

Doi, M. and Edwards, S. F. (1986). *The theory of polymer dynamics*. Oxford University Press.

Doyle, P. S., Shaqfeh, E. S. G., and McKinley, G. (1998). *J. Non-Newtonian Fluid Mech.*, **76**, 79.

Edwards, S. F. (1967). *Proc. Phys. Soc.*, **92**, 9.

Einaga, Y., Osaki, K., Kurata, M., Kimura, S., and Tamura, M. (1971). *Polymer J. (Japan)*, **2**, 550.

Einstein, A. (1905). *Ann. Physik*, **17**, 549.

Ericksen, J. L. (1960). *Arch. Rational Mech. Anal*, **4**, 231.

Ericksen, J. L. (1961). *Trans. Soc. Rheol.*, **5**, 22.

Erman, B. and Mark, J. E. (1997). *Structure and properties of rubberlike networks*. Oxford University Press, New York.

Fan, X.-J. (1984). *Rheol. Res. Centre Rept. No. 91*, University of Wisconsin.

Fan, X.-J. and Bird, R. B. (1984). *Rheol. Res. Centre Rept.* No. 96, University of Wisconsin.

Fan, X.-J., Phan-Thien, N., and Zheng, R. (1998). *J. Non-Newtonian Fluid Mech.*, **74**, 113.

Feng, J., Chaubal, C. V., and Leal, L. G. (1998). *J. Rheol.*, **42**, 1095.

Flory, P. J. (1969). *Statistical mechanics of chain molecules*. Wiley, New York.

Folgar, F. P. and Tucker, C. L. (1984). *J. Reinforced Plastics and Composites*, **3**, 98.

Frankel, N. A. and Acrivos, A. (1967). *Chem. Eng. Sci.*, **22**, 847.

Gadala-Maria, F. and Acrivos, A. (1980). *J. Rheol.*, **24**, 799.

Giesekus, H. (1966). *Rheol. Acta*, **5**, 29.

Giesekus, H. (1982). *J. Non-Newtonian Fluid Mech.*, **11**, 69.

Graessley, W. W. (1980). *J. Polymer. Sci. (Polymer. Phys. edn)*, **18**, 27.

Green, M. S. and Tobolsky, A. V. (1946). *J. Chem. Phys.*, **14**, 80.

Gross, L. H. and Maxwell, B. (1972). *Trans. Soc. Rheol.*, **16**, 577.

Hand, G. L. (1962). *J. Fluid Mech.*, **13**, 33.

Harrison, G. M., Remmelgas, J., and Leal, L. G. (1998). *J. Rheol.*, **42**, 1039.

Hinch, E. J. and Leal, L. G. (1972). *J. Fluid Mech.*, **52**, 683.

Hinch, E. J. and Leal, L. G. (1976). *J. Fluid Mech.*, **76**, 187.

Huilgol, R. R. and Phan-Thien, N. (1997). *Fluid Mechanics of viscoelasticity*. Elsevier Amsterdam.

Ianniruberto, G. and Marrucci, G. (1998). *J. Non-Newtonian Fluid Mech.*, **79**, 225.
James, D. F. and Sridhar, T. (1995). *J. Rheol.*, **39**, 713.
Jeffery, G. B. (1922). *Proc. Roy. Soc. London*, **A102**, 161.
Johma, A. I. and Reynolds, P. A. (1993). *Rheol. Acta*, **32**, 457.
Johnson, M. W. and Segalman, D. (1981). *J. Non-Newtonian Fluid Mech.*, **9**, 481.
Kasehagen, L. G. and Macosko, C. W. (1998). *J. Rheol.*, **42**, 1303.
Kaye, A. (1966). *Brt. J. appl. Phys.*, **17**, 803.
Keunings, R. (1997). *J. Non-Newtonian Fluid Mech.*, **68**, 85.
Kitano, T., Kataoka, T., and Shirota, T. (1981). *Rheol. Acta*, **20**, 207.
Langouche, F. and Debbaut, B. (1999). *Rheol. Acta*, **38**, 48.
Larson, R. G. (1988). *Constitutive equations for polymer melts and solutions*. Butterworths, Boston.
Larson, R. G. (1999). *The structure and rheology of complex fluids*. Oxford University Press, New York.
Leal, L. G. and Hinch, E. J. (1973). *Rheol. Acta*, **12**, 127.
Leonov, A. I. (1976). *Rheol. Acta*, **15**, 85.
Leonov, A. I., Lipkina, E. H., Pashkin, E. D., and Prokunin, A. W. (1976). *Rheol. Acta*, **15**, 411.
Leslie, F. M. (1968). *Arch. Rat. Mech. Anal.*, **28**, 265.
Leslie, F. M. (1987). In *Theory and applications of liquid crystals*. (ed. J. L. Ericksen and D. Kinderlehrer), p. 235. Springer, New York.
Lielens, G., Halin, P., Jaumain, I., Keunings, R., and Legat, V. (1998). *J. Non-Newtonian Fluid Mech.*, **76**, 79.
Lodge, A. S. (1956). *Trans. Faraday Soc.*, **52**, 120.
Lodge, A. S. (1964). *Elastic liquids*. Academic Press, London.
Lodge, A. S. (1968). *Rheol. Acta*, **7**, 379.
Lodge, A. S. (1974). *Body tensor fields in continuum mechanics*. Academic Press, New York.
Lundgren, T. S. (1971). *J. Fluid Mech.*, **51**, 273.
Metzner, A. B. (1985). *J. Rheol.*, **29**, 739.
Nguyen, Q. D. and Boger, D. V. (1992). *Ann. Rev. Fluid Mech.*, **24**, 47.
Nunan, K. C. and Keller, J. B. (1984). *J. Fluid Mech.*, **142**, 269.
Parodi, O. (1970). *J. de Physique*, **31**, 581.
Perkins, T. T., Smith, D. E., and Chu, S. (1997). *Science*, **276**, 2016.
Phan-Thien, N. (1978). *J. Rheol.* **22**, 259.
Phan-Thien, N. (1984). *J. Non-Newtonian Fluid Mech.*, **16**, 329.
Phan-Thien, N. (1995). *J. Rheol.*, **39**, 679.
Phan-Thien, N. and Graham, A. L. (1991). *J. Rheol.*, **30**, 44.
Phan-Thien, N. and Pham, D. C. (1997). *J. Non-Newtonian. Fluid Mech.*, **72**, 305.
Phan-Thien, N., Safari-Ardi, M., and Morales-Patino, A. (1997). *Rheol. Acta*, **36**, 38.
Phan-Thien, N. and Tanner, R. I. (1977). *J. Non-Newtonian Fluid Mech.*, **2**, 353.
Phan-Thien, N. and Tanner, R. I. (1978). *Rheol. Acta.*, **17**, 568.
Rallison, J. M. and Hinch, E. J. (1988). *J. Non-Newtonian Fluid Mech.*, **29**, 37.
Russel, W. B., Saville, D. A. and Schowalter, W. R. (1995). *Colloidal dispersions*. Cambridge University Press.
Rutgers, R. (1962). *Rheol. Acta*, **2**, 305.
Saab, H. H., Bird, R. B., and Curtiss, C. F. (1982). *J. Chem. Phys.*, **77**, 4758.
Schowalter, W. R. (1978). *Mechanics of non-Newtonian fluids*. Pergamon Press, Oxford.
Takano, Y. (1974). *Polymer. J.*, **6**, 61.
Tanner, R. I. (1965). *Trans. Am. Soc. Lubric. Engrs.*, **8**, 179.
Tanner, R. I. (1975). *Trans. Soc. Rheol.*, **19**, 557.
Tanner, R. I. and Simmons, J. M. (1967). *Chem. Eng. Sci.*, **22**, 1803.

Tanner, R. I. and Walters, K. (1998). *Rheology: an historical perspective*, Elsevier, Amsterdam.

Thomas, D. G. (1965). *J. Colloid Sci.*, **20**, 267.

Treloar, L. R. G. (1958). *The physics of rubber elasticity*, (2nd edn). Oxford University Press.

Upadhyay, R. K., Isayev, A. I., and Shen, S. F. (1981). *Rheol. Acta*, **20**, 443.

Utracki, L. A. (1988). in Collyer, A. A., and Clegg, D. W. (ed.), *Rheological measurements*, p. 479, Elsevier, London.

Verhoef, M. R. J., Van den Brule, B.H.A.A., and Hulsen, M. A. (1999). *J. Non-Newtonian Fluid Mech.*, **80**, 155.

Wagner, M. H. (1979). *Rheol. Acta*, **18**, 681.

Wagner, M. H., Ehrecke, P., Hachmann, P., and Meissner, J. (1998). *J. Rheol.*, **42**, 621.

Wagner, M. H., Raible, T., and Meissner, J. (1979). *Rheol. Acta.*, **18**, 427.

Warner, H. R. (1972). *Ind. Eng. Chem. Fund.*, **11**, 379.

Wasserman, S. H. and Graessley, W. W. (1996). *Polymer Eng. and Sci.*, **36**, 852.

Wiest, J. M. and Tanner, R. I. (1989). *J. Rheol.* **33**, 281.

Yamakawa, H. (1971). *Modern theory of polymer solutions*. Harper & Row, New York.

Yamamoto, M. (1956). *J. phys. Soc. Japan*, **11**, 413.

Yamamoto, M. (1957). *J. phys. Soc. Japan*, **12**, 1148.

Yamamoto, M. (1958). *J. phys. Soc. Japan*, **13**, 1200.

Yamane, Y., Kaneda, Y., and Doi, M. (1994). *J. Non-Newtonian Fluid Mech.*, **54**, 405.

Zimm, B. H. (1956). *J. chem. Phys.*, **24**, 269.

Problems

1. Using the Warner spring [eqn (5.22)] derive its potential χ and hence find the distribution function ψ for this molecular model in a fluid at rest.

2. Repeat Problem 1 when the fluid is undergoing a flow $\mathbf{v} = (\dot{\varepsilon}x, -\dot{\varepsilon}y, 0)$. Compute the stress difference $\tau_{xx} - \tau_{yy}$ for large values of $\lambda\dot{\varepsilon}$, where λ is an appropriate time-constant.

3. Consider a constitutive equation of the form

$$\tau_{ij}(\mathbf{x}, t) = K \int_{-\infty}^{t} \exp(-(t - t')/\lambda) f[A(t')] \frac{\partial x_i}{\partial r_m} \frac{\partial x_j}{\partial r_m} \, \mathrm{d}t'$$

which is a generalization of eqn (5.58). The quantity A may be stress-dependent (Kaye model), strain-rate dependent (Carreau model) or strain-dependent (KBKZ model) via the invariants of $\boldsymbol{\tau}(t')$, $\mathbf{d}(t')$ or $\mathbf{C}$ respectively. Shown that the first two cases may be reduced to a differential type of constitutive equation but that the KBKZ cannot be so reduced.

4. Derive eqn. (5.102) and use it to show that N_2 is zero for a dilute solution of dumbbells in a steady shear flow, no matter what the spring law is.

5. Show that N_1 relaxes slower than τ for a Lodge model with two relaxation times.

6. Find the response of the convected Maxwell model to a suddenly started elongational flow.

7. Consider the response of the constitutive equation (5.106) at high shear-rates ($\lambda\dot{\gamma} \gg 1$). Show that the response for the viscosity is $\eta \sim \dot{\gamma}^{-2/3}$, that for N_1 is $\sim \dot{\gamma}^{2/3}$ and that $N_2 = 0$.

8. Demonstrate that the flow in an eccentric disc viscometer [eqn (4.123)] is a weak flow. Find where it fits on the Pipkin diagram (Fig. 3.9).

9. Consider the Carreau model where (Problem 3) the non-linearity of the kernel is present via the second invariant of the rate of deformation tensor (d_{ij}). (Consider only a single time constant.) What is the response of this model in:

(i) a steady elongational flow

(ii) a sinusoidal shear strain of small amplitude ε and frequency ω, where ω is permitted to become large, so that $\omega\varepsilon$ is not negligible?

10. You are asked to choose a constitutive model for some computations in processing. What would you choose for the following processes:

(a) Modelling flow through a bed of spheres.

(b) Flow in a screw extruder.

(c) A high-frequency damper (~ 1 kHz) consisting of a moving blade between two fixed surfaces parallel to the blade.

(d) Spinning of a fibre.

(e) Expansion of a balloon by internal air pressure.

(f) An eccentric disc rheometer (Fig. 4.19) where one plate is suddenly started up resulting in an initial difference in speed between the two plates.

(g) Turbulent flows of dilute polymer solutions.

11. For a dilute solution of rigid dumbbells, Section 5.5.5, compute the steady-state elongational viscosity and verify that the polymer contribution is $6\eta_0$, where η_0 is evaluated at very small rates of deformation.

6

LUBRICATION, CALENDARING, AND RELATED FLOWS

6.1 Nearly viscometric flows. The lubrication approximation

THE processes described in this chapter are examples of nearly-viscometric flows. A typical (two-dimensional) problem is shown in Fig. 6.1. where the space between two surfaces in relative motion is filled with fluid. In such problems the characteristic length L in the x-direction (and y-direction if relevant) is much greater than the characteristic length h in the z-direction, and the relative slope of the surfaces $(\mathrm{d}h/\mathrm{d}x)$ is also assumed to be small, of order (h/L). We shall consequently assume that the x-component of velocity (u) is much greater than the z-component. Formally, considering the mass conservation equation and the order of magnitude of the terms, we have

$$\frac{\partial u}{\partial x} + \frac{\partial w}{\partial z} = 0$$

$$\left(\frac{u}{L}\right)\left(\frac{w}{h}\right).$$

(6.1)

But u is $\mathrm{O}(U)$ where U is a characteristic speed on the boundary, and hence

$$w = \mathrm{O}\left(\frac{h}{L}U\right).$$

(6.2)

Now consider the equations of motion (neglecting body forces)

$$-\frac{\partial p}{\partial x} + \frac{\partial \tau_{xx}}{\partial x} + \frac{\partial \tau_{xz}}{\partial z} = \rho\frac{\mathrm{D}u}{\mathrm{D}t},$$

(6.3)

$$-\frac{\partial p}{\partial z} + \frac{\partial \tau_{xz}}{\partial x} + \frac{\partial \tau_{zz}}{\partial z} = \rho\frac{\mathrm{D}w}{\mathrm{D}t},$$

(6.4)

where τ_{ij} is the extra-stress tensor.

To begin with we shall neglect inertia forces (negligible Reynolds numbers). This is usual in lubrication studies; when we consider large inertia forces we have boundary-layer flows, to be considered later (Section 6.8). We suppose for the moment that the stresses τ_{xx}, τ_{xz}, and τ_{zz} are comparable in magnitude. Then order of magnitude arguments reduce (6.3) to

$$-\frac{\partial p}{\partial x} + \frac{\partial \tau_{xz}}{\partial z}\left\{1 + \mathrm{O}\left(\frac{h}{L}\right)\right\} = 0.$$

(6.5)

If, contrary to our hypothesis, $(h/L)\tau_{xx}$ happens to be comparable to τ_{xz}, then this is not expected to be a valid approximation. By a similar argument, the

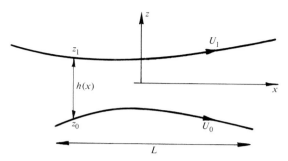

Fig. 6.1 Axes for general two-dimensional lubrication problem. Fluid is contained between the z_0 and z_1 moving surfaces.

pressure p is large $[\mathrm{O}(L/h)]$ compared with τ_{xz}, and (6.4) reduces to the simple form

$$\frac{\partial p}{\partial z} = 0, \quad \text{or} \quad p = p(x). \tag{6.6}$$

This latter reduction will also be invalid if τ_{zz} is of order $(L/h)\tau_{xz}$, hence we require that both of the normal stresses τ_{xx} and τ_{zz} be much less than $\mathrm{O}[(L/h)\tau_{xz}]$ in magnitude for validity of the theory; with this requirement the *differences* of normal stresses are also of the same order of magnitude relative to the shear stress.

Accepting these restrictions on normal stress magnitudes for the moment, we can integrate (6.5) to give

$$\tau_{xz} = z\frac{\mathrm{d}p}{\mathrm{d}x} + C(x). \tag{6.7}$$

Also, from mass conservation, the mass flux through the gap is constant at all x-values, and hence, ignoring terms of order $(h/L)^2$

$$\int_{z_0}^{z_1} u\,\mathrm{d}z = q \text{ (constant)}. \tag{6.8}$$

The equation (6.7) thus holds, within the given restrictions on normal stress magnitudes and (h/L) ratios, for both Newtonian and viscoelastic systems.

6.2 Newtonian and generalized Newtonian lubrication theory

In the Newtonian case we have from the constitutive law

$$\tau_{xz} = \eta\frac{\partial u}{\partial z}\left(1 + \mathrm{O}\left(\frac{h^2}{L^2}\right)\right) \tag{6.9}$$

and the direct stresses are of order (h/L) times the shear stress, as required above.

Substituting (6.9) in (6.7) and integrating again with respect to z, we have

$$u = \frac{z^2}{2\eta}\frac{\mathrm{d}p}{\mathrm{d}x} + \frac{C}{\eta}z + D. \tag{6.10}$$

It is now usual to use as reference one of the surfaces (the 0-subscript surface in Fig. 6.1), so that this is the surface $z = 0$, and the other surface (1-subscript) is then the surface $z = h$. If, following Fig. 6.1, the speeds of the 0 and 1 surfaces are U_0 and U_1 respectively, then C and D can be determined.

We have $D = U_0$, $C = \eta(U_1 - U_0)/h - \frac{1}{2}h(\mathrm{d}p/\mathrm{d}x)$, and hence the complete velocity field is

$$u = U_0 + (U_1 - U_0)\frac{z}{h} - \left(\frac{1}{2\eta}\right)\frac{\mathrm{d}p}{\mathrm{d}x}(h - z)z. \tag{6.11}$$

The Reynolds equation for the pressure is found by using (6.8)

$$q = \frac{1}{2}(U_0 + U_1)h - \frac{h^3}{12\eta}\frac{\mathrm{d}p}{\mathrm{d}x}. \tag{6.12}$$

Since q is an (unknown) constant, and integration of (6.12) to find p introduces another constant, it is necessary to know p at two places to complete the problem and evaluate p explicitly. In the present approximation, $p \sim -\sigma_{xx}$ at the ends of the fluid film, and hence we require knowledge of the component σ_{xx} at the film ends.

6.2.1   The plane slider bearing case

Consider the thrust bearing sketched in Fig. 6.2. Its invention by A.G.M. Michell enabled the propeller thrust of large ships to be taken up by a much smaller bearing area than was possible previously. For the present we shall assume the flow is in the x–z plane. In this case the gap geometry is

$$h = h_{\mathrm{o}} + (h_{\mathrm{i}} - h_{\mathrm{o}})\frac{x}{L} \tag{6.13}$$

and, recalling $\sigma_{xx} \sim -p$, we set $p = 0$ at $x = 0$, L. The Newtonian solution is well known.

We find, defining $h^* = -2q/U$

$$p = \frac{6\eta UL}{(h_{\mathrm{i}}^2 - h_{\mathrm{o}}^2)}\left[-1 + \frac{(h_{\mathrm{i}} + h_{\mathrm{o}})}{h} - \frac{h_{\mathrm{i}}h_{\mathrm{o}}}{h^2}\right] \tag{6.14}$$

and

$$h^* = 2h_{\mathrm{i}}h_{\mathrm{o}}/(h_{\mathrm{i}} + h_{\mathrm{o}}). \tag{6.14a}$$

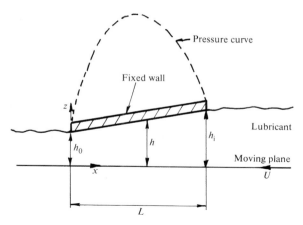

Fig. 6.2 Two-dimensional thrust bearing.

From (6.12) and (6.14), noting that $h^* = -2q/U$, we see that when $h = h^*$ $(\mathrm{d}p/\mathrm{d}x)$ is zero; this shows that h^* is the point of maximum pressure. This result concerning the significance of h^* is general.

The pressure is a humped distribution (Fig. 6.2). All other quantities can be found when p is known. The total load capacity W $(W = \int_0^L p\,\mathrm{d}x)$ and friction force F $\{F = \int_0^L \tau_{xz}|_{z=0}\,\mathrm{d}x\}$ can be evaluated as (all forces per unit width)

$$W = \frac{6\eta UL^2}{(h_i^2 - h_o^2)}\left[\frac{(h_i + h_o)}{(h_i - h_o)}\ln\left(\frac{h_i}{h_o}\right) - 2\right] \qquad (6.15)$$

and

$$F = \frac{4\eta UL}{(h_i - h_o)}\left[\ln\left(\frac{h_i}{h_o}\right) - \frac{3(h_i - h_o)}{2(h_i + h_o)}\right]. \qquad (6.16)$$

Thus, the friction coefficient F/W is of order h_o/L. The shear stress varies linearly across the gap:

$$\tau_{xz} = \frac{\eta U}{h} - \frac{1}{2}\frac{\mathrm{d}p}{\mathrm{d}x}(h - 2z). \qquad (6.17)$$

See Table 6.1 for the factors $Wh_o^2/\eta UL^2$ and $Fh_o/\eta UL$ for this geometry. We shall now examine some non-Newtonian cases.

6.2.2 Inelastic fluids

It is not difficult to see how to generalize the Newtonian solution to inelastic fluid models. Returning to (6.7) if we can express $\partial u/\partial z(\approx \partial u/\partial z + \partial w/\partial x)$ as a function of τ_{xz}, then we can find $\partial u/\partial z$ as a (complicated) function of z. In other words, we can often make progress if the fluidity function $\partial u/\partial z = \tau\phi(\tau)$ is

Table 6.1 Dimensionless factors for the Newtonian plane slider

h_i/h_o	$K_p(\equiv Wh_o^2/\eta UL^2)$	$K_f(\equiv Fh_o/\eta UL)$
1.2	0.0755	0.9192
1.5	0.1312	0.8437
2.0	0.1589	0.7726
3.0	0.1479	0.6972
4.0	0.1242	0.6484

known; see Chapter 3. The integration can then sometimes be completed analytically as in the Newtonian case. For example, in the case of the power-law fluid [eqn (1.10)] we will have approximately

$$\tau_{xz} = k\left|\frac{\partial u}{\partial z}\right|^{n-1}\left(\frac{\partial u}{\partial z}\right). \tag{6.18}$$

From eqn (6.7) we find an equation for $\partial u/\partial z$:

$$k\left|\frac{\partial u}{\partial z}\right|^{n-1}\left(\frac{\partial u}{\partial z}\right) = z\frac{dp}{dx} + c. \tag{6.19}$$

Care must be taken when taking the roots of the left-hand side to get the sign of $\partial u/\partial z$ correct:

$$\frac{\partial u}{\partial z} = \pm\left|\frac{z}{k}\frac{dp}{dx} + \frac{c}{k}\right|^{1/n}. \tag{6.20}$$

Integrating eqn (6.20), one finds

$$u = \pm\frac{k}{(1+1/n)(dp/dx)}\left[\frac{z}{k}\frac{dp}{dx} + \frac{c}{k}\right]^{1/n+1} + D. \tag{6.21}$$

Insertion of the boundary velocities U_0, U_1 at $z=0$, h and elimination of the constant D gives a complex connection between c, (dp/dx) and the boundary speeds which makes it difficult to continue the general analysis.

Exceptionally, when we know the shear stress on a plane $z=$ constant the analysis becomes tractable because the constant c can be eliminated. Fortunately, the squeeze film, roller lubrication and some calendaring problems fall into this category. These will be discussed later (Section 6.4).

For the easier case of a slightly non-Newtonian fluid where

$$\eta = \eta_0(1 - \alpha\dot{\gamma}^2) \tag{6.22}$$

and the term $\alpha\dot{\gamma}^2$ is small compared with unity, we can recast (6.22) in the form of a stress-dependent fluidity

$$\frac{1}{\eta} = \frac{1}{\eta_0}[1 + \beta\tau_{xz}^2] + O(\beta^2\tau_{xz}^4), \tag{6.23}$$

where $\beta = \alpha/\eta_0^2$ and the error is supposed small in the whole flow field.

There are many plane perturbation analyses for various lubrication problems, often beginning from elaborate constitutive equations, but after noting that the flow is steady, and after neglecting the normal stress contributions as discussed above, they all must reduce to the constitutive assumption (6.23) above. Hence there is no loss in generality in using that constitutive model for illustrative calculations.

A perturbation calculation will now be made for the plane slider (Fig. 6.2). We take

$$\frac{\partial u}{\partial z} = \frac{1}{\eta_0}\{1 + \beta\tau_{xz}^2\}\tau_{xz}, \tag{6.24}$$

where $\beta > 0$ for a shear-thinning fluid.

If we treat the case $\beta\tau_{xz}^2 \ll 1$, then we can replace this term by the Newtonian value $\beta\tau_{xz}^{02}$ in the perturbation analysis.

The results are cumbersome, and for the case when $h_i = 2h_o$, we find

$$\frac{h^*}{h_o} = \frac{4}{3} + \frac{56}{300}\beta\left(\frac{\eta_0 U}{h_o}\right)^2 \tag{6.25}$$

$$\frac{Wh_o^2}{\eta_0 UL^2} = 0.159\left[1 - 1.76\beta\left(\frac{\eta_0 U}{h_o}\right)^2\right] \tag{6.26}$$

$$\frac{Fh_o}{\eta_0 UL} = 0.773\left[1 - 0.828\beta\left(\frac{\eta_0 U}{h_o}\right)^2\right]. \tag{6.27}$$

Hence the friction coefficient for this case is

$$\mu = \frac{F}{W} = 4.86\left(\frac{h_o}{L}\right)\left[1 + 0.93\beta\left(\frac{\eta_0 U}{h_o}\right)^2\right]. \tag{6.28}$$

Clearly, the friction coefficient is higher for the non-Newtonian case. See also the work of Milne (1957) who considered the slider bearing problem for a Bingham body.

The conclusion from many calculations using a small perturbation of the Newtonian viscosity in the form (6.22) seems to be that both the load capacity and the friction are reduced for shear-thinning fluids, but the former more so than the latter, so that an increase in friction coefficient occurs with these non-Newtonian lubricants, at least in the isothermal case.

For the case where the angle of the slider is so small that the deviation from a parallel shear flow can be neglected, the friction force is just that due to shear flow, so $F = \eta U L/h$, where the viscosity η is evaluated at the shear rate U/h ($\equiv \dot{\gamma}_0$). It is straightforward to show that the load for this case, for any inelastic fluid with a viscosity function $\eta(\dot{\gamma})$, is given by

$$W = W_N \left(1 + \frac{\mathrm{d} \ln \eta}{\mathrm{d} \ln \dot{\gamma}} \bigg|_{\dot{\gamma}_0} \right) \tag{6.29}$$

where W_N is the Newtonian load. This result follows from the work of Davies and Walters (1973). Thus, for shear-thinning fluids, $W < W_N$, and the friction is essentially unchanged, so the friction coefficient is increased.

6.3   Lubrication with viscoelastic fluids

Most of the difficulties in the previous section were due to integration problems with the variable viscosity cases. Clearly any attempt to include normal-stress effects as well as variable viscosity effects will certainly restrict us to perturbation calculations or force us to use numerical methods. We first look at the second-order fluid as a lubricant, thereby avoiding variable viscosity.

6.3.1   *The second-order fluid*

Because of difficulties with boundary conditions (see Section 6.3.3) which tend to obscure the significance of many of the perturbation results in the literature we begin with an exact treatment based on the second-order fluid theory. Since lubrication flows have been shown above to be nearly viscometric, we believe the second-order model is a useful one for this type of problem. We suppose that the flow is plane and inertia-less. To make the boundary conditions definite, we also suppose that the fluid before and after the slider (Fig. 6.2) is very deep, so that velocity boundary conditions may properly be applied. Using the theorem discussed in Section 4.11.1, we can assert that the velocity field will be the same as that for the Newtonian case. From this, using eqns (4.87) and (4.95) we can compute the difference between the stress on the moving plane with and without the non-Newtonian terms.

The result for σ_{zz} on the moving plane $z = 0$ may be shown to be, from eqn (4.87)

$$\sigma_{zz} = -p^0 - \frac{\Psi_1 U}{2\eta} \frac{\mathrm{d}p^0}{\mathrm{d}x} - \frac{\Psi_1}{4} \left(\frac{\partial u}{\partial z} \right)^2 \bigg|_{z=0}, \tag{6.30}$$

where p^0 is the Newtonian pressure and Ψ_1 is the first normal stress coefficient. Computing the normal load per unit width, $-\int_{-\infty}^{\infty} \sigma_{zz}\, dx$, and noting that the Newtonian pressure and the shear rate $\partial u/\partial z$ are zero outside the bearing length in the lubrication approximation, we have

$$W = -\int_{-\infty}^{\infty} \sigma_{zz}\, dx = -\int_{0}^{L} \sigma_{zz}\, dx = \int_{0}^{L} p^0\, dx + \frac{\Psi_1}{4}\int_{0}^{L}\left(\frac{\partial u}{\partial z}\right)^2 dx. \qquad (6.31)$$

The dp^0/dx term does not contribute to the load, since p^0 is zero at the bearing ends; the first term on the right-hand side of (6.30) is the Newtonian load (per unit width), W_N say.
Thus

$$W - W_N = \frac{\Psi_1}{4}\int_{0}^{L}\left(\frac{\partial u}{\partial z}\right)^2\Bigg|_{0}\, dx \qquad (6.32)$$

and $W > W_N$, if $\Psi_1 > 0$. The expression for u, (6.18) may be used to evaluate the integral in (6.41), giving the result (Tanner 1969) for a straight slider,

$$\frac{W}{W_N} = 1 + \frac{\Psi_1 U}{\eta L K_p}\frac{(1 + m + m^2)}{(1 + m)(2 + m)^2}, \qquad (6.33)$$

where $m = (h_i - h_o)/h_o$ and K_p values $(= Wh_o^2/\eta U L^2)$ are given in Table 6.1. When $m = 1$ (inlet film thickness h_i is twice outlet thickness h_o) (6.33) becomes

$$\frac{W}{W_N} = 1 + 1.05\left(\frac{\Psi_1 U}{\eta L}\right). \qquad (6.34)$$

These results are valid for arbitrarily large normal stress differences. The fractional increase in load in (6.34) is seen to be of order $(h/L)N_1/\tau$ where N_1 and τ are representative normal and shear stress magnitudes. Due to the small size of the factor (h/L) it is clear that the extra positive lift due to the normal stresses is small, and occurs by direct normal force action. This is in contrast to the wedge action of the shear forces, which is much more effective. Because the velocity field is unchanged, it is easy to show that the friction force is unchanged from the viscous case. Unlike the wedge action, the normal force action occurs even with a parallel slider. In this case the shear rate under the slider is constant, with a value U/h_o, and the load capacity is just $0.25\Psi_1 U^2 L/h_o^2$, or $N_1 L/4$.

The general conclusion from the previous sections is that the combined effects of perturbations in variable viscosity and normal stresses give rise to load variations of the form

$$W = W_N\left[1 - a\beta\left(\frac{\eta_0 U}{h_o}\right)^2 + b\frac{h_o}{L}\left(\frac{\Psi_1 U}{\eta_0 h_o}\right)\right], \qquad (6.35)$$

where all constants are positive and a, b are of order unity. If we wish to apply the formula (6.35) to practical cases then one can set $\beta = \eta_0 \lambda^2$ and also $b\Psi_1/\eta_0 \approx b'\lambda$, where λ is a time constant. Then (6.35) becomes

$$W/W_N = 1 - a(\lambda U/h_o)^2 + b'(h_o/L)(\lambda U/h_o) + \cdots. \tag{6.36}$$

Now b' is also of order unity, and hence there is an initial very small rise in capacity up to a small Weissenberg number of order (h_o/L) and then the shear thinning effect becomes dominant and reduces W.

6.3.2   Plane journal bearings

The geometry of the journal bearing is shown in Fig. 6.3. The difference between the radii of the shaft and journal is c, and the inner surface turns at a speed Ω. The offset along the y-axis is εc where ε is the eccentricity ratio. The quantities of interest are the torque to turn the inner cylinder and the x and y components of the resultant load generated by the motion.

Davies and Walters (1973) have analysed the problem for the second- and third-order fluids for arbitrary ε values when no cavitation occurs. It should be emphasized in passing that this is generally an unrealistic assumption, and consequently applications of this and similar analyses should be restricted to cases where the negative pressures are small compared with atmospheric pressure. It is supposed that $c/r \ll 1$, the usual case. Their results in the second-order fluid case, for the torque M and the load components respectively, are

$$M = \int_0^{2\pi} \tau r^2 \, d\theta = -\frac{4\eta r^3 \Omega \pi (1 + 2\varepsilon^2)}{c(1 - \varepsilon^2)^{1/2}(2 + \varepsilon^2)}, \tag{6.37}$$

$$W_x = -\int_0^{2\pi} pr \cos\theta \, d\theta = -\frac{12r^3 \Omega \eta \pi \varepsilon}{c^2(1 - \varepsilon^2)^{1/2}(2 + \varepsilon^2)}, \tag{6.38}$$

$$W_y = -\int_0^{2\pi} pr \sin\theta \, d\theta = \frac{\Psi_1 r^3 \Omega^2 \varepsilon \pi (8 - \varepsilon^2 + 2\varepsilon^4)}{c^2(1 - \varepsilon^2)^{9/2}(2 + \varepsilon^2)^2}, \tag{6.39}$$

where η is the (constant) fluid viscosity and Ψ_1 is the first normal-stress difference coefficient [eqn (3.18)].

Note that the x-component of load is solely due to viscous action and the y-component is due to normal-stress effects. The couple M has the Newtonian value. The results are therefore similar to the slider-bearing results. They also analysed the third-order fluid, where the viscosity function has the form $\eta_0(1 - \lambda^2\dot{\gamma}^2)$, but the normal stress differences are still quadratic in shear rate, as in the second-order fluid. Results were given for a range of eccentricity ratios 0–0.7. Basically, the load in the x-direction is still viscous, and that in the

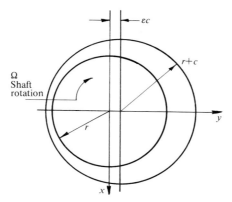

Fig. 6.3 Bearing geometry.

y-direction is elastic. For the case of a purely viscous fluid, and $\varepsilon \to 0$, they gave the result

$$M = \frac{2\pi r^3}{c} \Omega\eta\{1 + \mathrm{O}(\varepsilon)\}, \tag{6.40}$$

$$W_x = \frac{6\pi r^3}{c^2} \Omega\varepsilon\left(\eta + \dot{\gamma}\frac{\mathrm{d}\eta}{\mathrm{d}\dot{\gamma}}\right) + \mathrm{O}(\varepsilon^2) \tag{6.41}$$

$$W_y = 0, \tag{6.42}$$

where the viscosity function $\eta(\dot{\gamma})$ is evaluated at $\dot{\gamma}_0 = \Omega r/c$. Combining W_x above and W_y from (6.39) gives a rough estimate of load for small eccentricity. Davies and Walters also presented a numerical analysis for a special Oldroyd fluid and $\varepsilon \to 0$ and large Weissenberg numbers $(\lambda\dot{\gamma})$ up to 1000.

For the case of small eccentricity, it is interesting to solve the uncavitated journal problem for the general viscometric equation (4.32). The main results have been discussed by Phan-Thien and Tanner (1981). A straight-forward extension of Davies and Walters's (1973) work shows that

$$\frac{Wc}{\varepsilon r^2 \tau(1 + n)} = f\left\{\frac{N_1 c}{\tau r(1 + n)}, m\right\} \equiv f\{\Phi, m\}, \tag{6.43}$$

where τ is the shear stress at the shear rate $\dot{\gamma}_0$, (where $\dot{\gamma}_0 = \Omega r/c$), n is the viscosity slope function $\mathrm{d}\ln\eta/\mathrm{d}\ln\dot{\gamma}$, evaluated at $\dot{\gamma}_0$, N_1 is the first normal stress difference, and m is the normal stress slope function $\mathrm{d}\log\Psi_1/\mathrm{d}\log\dot{\gamma}$, evaluated at $\dot{\gamma}_0$. The previous results given in this section are special cases of (6.43).

The results of this analysis may be compared with those of Davies and Walters (1973) for the Oldroyd fluid; they give good results only when the transit time of a particle around the bearing is less than the fluid characteristic time. Results for several constitutive models are given by Beris *et al.* (1983); the point about the transit time/characteristic time is emphasized there.

6.3.3 Normal stresses in lubrication

In the previous section, it was shown that the increase in load (per unit width) due to 'normal stress' effects is proportional to LN_1; the load due to viscous effects is of order $L^2\tau/h$, and hence improvements in load capacity are of order $(h/L)(N_1/\tau)$. Usually h/L is of order 10^{-3} for typical machine bearings, so N_1 has to be of order 10^3 times the shear stress for this mechanism to be effective. Measurements (Fig. 6.4) available up to a shear rate of $5 \times 10^6\,\mathrm{s}^{-1}$ show N_1/τ of order 10 (Williamson *et al.* 1997). Nevertheless these authors invoke normal stresses to explain the behaviour of journal bearings at very high eccentricity (~ 0.99). In this careful set of experiments Newtonian and non-Newtonian lubricants are subjected to very high shear rates (about $10^6\,\mathrm{s}^{-1}$ or greater) in very narrow gaps (1 micron) and some improved load carrying capacity was noticed with the polymeric (non-Newtonian) oils but only at very high ε (>0.95). This is hard to explain using the classical viscoelastic theory of hydrodynamic lubrication discussed here. Since the nominal gap and the surface roughness of the lubricated surfaces are now of the same order of magnitude, local shear rates can be expected which will be much larger than the nominal rates (10^6) quoted above. Also, the usual hydrodynamic (continuum) theory of lubrication is no longer expected to be valid, because the thinnest gaps will only be of molecular dimensions. It is possible that in these extremely thin, rapidly sheared lubrication situations the explanation of enhanced load capacity lies in the molecular domain (Jabbarzadeh *et al.* (1998); Granick *et al.* (1994); Gunsel *et al.* (1998)).

See also Chapter 8.

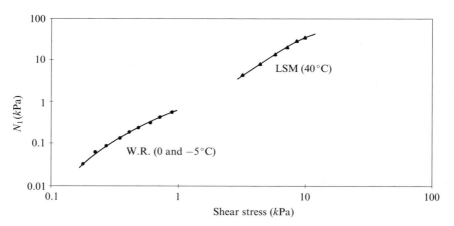

Fig. 6.4   Representative viscometric data for an oil using Weissenberg Rheogoniometer (W.R.) and Lodge stress meter (LSM) (from Williamson *et al.* 1997).

One can also consider the 'strength' of lubrication flows (see Section 5.4). From the flow field (6.11) and the continuity eqn (6.1) one can find that for the plane slider bearing the flow will be strong along the mid-plane $z = h/2$ if

$$\frac{3}{2} \lambda U h' \frac{h^*}{h^2} > 1. \tag{6.44}$$

On the solid surface the flow is always weak.

For the straight slider, $h' = mh_o/L$, and the least value of (6.44) is at the inlet plane, where $h = h_o(1 + m)$, thus the flow is strong all along the plane $z = h/2$ if

$$\frac{3\lambda U m}{2L(1 + m)^2} \left(\frac{h^*}{h_o}\right) > 1. \tag{6.45}$$

Typically, $m \approx 1$ ($h_i/h_o \sim 2$), and h^*/h_o is given in (6.14a). Using $m = 1$, we find the flow is strong on the plane $y = h/2$ if

$$\frac{\lambda U}{L} > 2 \tag{6.46}$$

but that the flow is always weak at the solid boundaries.

The Weissenberg number is clearly of order $\lambda U/h_o$, and the Deborah number of order $\lambda U/L$.

Now one can assess the flow dynamics. In an elongational flow particles separate exponentially, and a large extension of the structure in a simple elongation will only occur if the product of $\partial u/\partial x$ and the residence time t_r exceeds unity. To secure a relative elongation of (say) 100, it is necessary that $t_r(\partial u/\partial x) \sim \log_e 100 = 4.6$. It seems impossible to achieve this with the bearing analysed because the product of effective elongation rate and t_r, from the arguments just given, is about unity on the plane $z = h/2$ and less near the solid boundaries. Thus it is not expected that the structure will be much disturbed from the equilibrium configuration at least in most of the flow field. This, in turn, implies that the non-viscometric terms will not be very important in this flow. This conclusion does not seem to be very sensitive to the exact flow field postulated; the mean residence time is not expected to change much from the magnitude calculated above, and, similarly, the rate of stretching is always of order U/L. Hence, no great structural changes are to be expected. Consequently, non-viscometric effects will generally not be pronounced in the slider bearing. For a journal bearing where fluid particles can recirculate, we can turn to the analysis of Beris et al. (1983). These authors consider a two-dimensional journal bearing with small eccentricity. A lubricant model of the second-order kind used above does not show memory, and so Beris et al. (1983) used the UCM model (see Section 4.3.1) which can 'remember' stress states at a particle as it cycles through the bearing. The results were presented for various comparatively large gaps. If the inner shaft radius is r, and the outer radius is $r + c$, then the ratio c/r was taken as 0.1 (compared to the usual value of 10^{-3} for practical bearings). Figure 6.5 shows a comparison

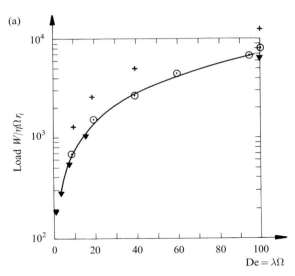

Fig. 6.5(a)   Load–Deborah number curves for a wide-gap ($c/r_i = 0.1$) low eccentricity ($\varepsilon = 0.1$) plane-flow bearing. Here $W = \sqrt{W_x^2 + W_y^2}$: + Second-order model; ⊙ Huang *et al.* (Computation), UCM model; — Beris *et al.* (1983) Computation; ▼ Beris *et al.* (1983) Perturbation solution.

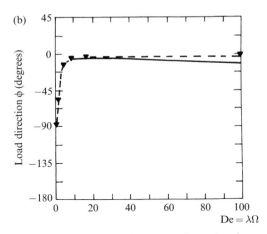

Fig. 6.5(b)   Phase-angle vs Deborah number curves. Same situations as in Fig. 6.5(a).

of the second-order theory (6.38–6.39) where W is the total load, and the results of a perturbation theory for the UCM model and two sets of computations for this model, one by Beris *et al.* (1983) and another by Huang *et al.* (1996). The Deborah number (*De*) here is $\lambda\Omega$, where Ω is the shaft speed in radians/second and the Weissenberg number is $\lambda\Omega r/c$. If we take $\lambda = \Psi_1/2\eta$ for the second-order model, then both models behave identically in a shear flow, with a constant viscosity.

The second normal stress difference is irrelevant in this plane flow. Results in Fig. 6.5(a) show considerable differences in load carrying behaviour between the two models, although the phase angles [Fig. 6.5(b)] ($\phi = \arctan(W_y/W_x)$) are similar. Normal stress effects can be expected to be significant if $(N_1/\tau)(c/r)$ is greater than about one. If $c/r = 10$, then $N_1/\tau = 2(De)$ for both models, for $(De) \geq 2$ one expects normal stress effects to be important here (see Section 6.3.1). However, the second-order model gives load values W about twice that for the UCM model. To explain this one can follow a particle on the rotating shaft and it is clearly subject to a pulsating flow of frequency Ω. Fluid particles on the stationary bearing do not undergo pulsation. The second-order model does not consider these effects, but the UCM model does. If the entire UCM lubricant film were pulsated at Ω, then one might expect the load capacity to be reduced, by a factor $|G^*|/\eta\Omega$, approximately. For a Maxwell model [eqn (2.94)] shows

$$|G^*|/\eta\Omega = \{1 + (\Omega\lambda)^2\}^{-1/2}. \tag{6.47}$$

As a simple approach one can expect an effective 'softening' of the film by this amount for particles near the shaft, without a corresponding softening at the bearing. Very roughly, one can correct the second-order results by taking the average of these two extremes, multiplying (6.38 and 6.39) by $\frac{1}{2}$ [1+ $(1 + De^2)^{-1/2}$]. The results correct the load fairly well, and the angle ϕ is unchanged (Fig. 6.5).

The UCM model also shows a thin stress boundary layer (Beris et al. 1983) for $\tau_{\theta\theta}$, in contrast to the second-order case (Huang et al. 1996 did not find this layer due to the coarse mesh they used; however, reference to Fig. 6.5 shows that their load results were not significantly affected by this omission).

6.3.4 Boundary conditions in lubrication

One of the most difficult problems to discuss in this area is the question of proper boundary conditions in viscoelastic lubrication flows. Consider the situation in Fig. 6.2. What should be the correct boundary conditions at inlet and outlet to the bearing? With the Newtonian case the lubrication flow will not be fully developed near the ends; roughly we expect disturbances to penetrate about one film thickness (h_o) into the bearing. Nevertheless, the relative error in neglecting this zone is not serious with the Newtonian case; it will be some multiple of h_o/L in fact. For very high N_1/τ values we can expect the disturbance to be of order $h_o N_1/\tau L$, which is larger. Unless one is prepared to solve the entire flow external to the bearing as part of the problem, it is not clear that in highly viscoelastic cases [that is, when $(hN_1)/(L\tau)$ is of order unity or more] that an especially simple lubrication theory exists.

To clarify this point consider again the parallel slider ($m = 0$) of Section 6.3.1, where the normal stresses contributed a load (per unit width) of $N_1 L/4$ when the inlet and outlet regions were deeply immersed in fluid. A plausible boundary condition when the ends are not deeply immersed would be to set the average axial stress to zero at inlet and outlet, i.e., to set

$$\int_0^h \sigma_{xx}\,\mathrm{d}z = 0 \quad \text{at} \quad x = 0, L. \tag{6.48}$$

In the case of a parallel slider, we have a simple shearing flow so that $\sigma_{zz} = \sigma_{xx} - N_1$, and hence $\sigma_{zz} = -N_1$ everywhere. Thus in this calculation the total lift is $N_1 L$, four times the previous result. Because of this disagreement the following calculations have been made with more realistic boundary conditions.

The parallel slider geometry is shown in Fig. 6.6, only the trailing edge or exit region is considered, and it is clear that on the scale shown the non-parallelism of the planes is inconsequential. The inclination does, however, affect the pressure gradient and the velocity profile of the fluid reaching the trailing edge. In Fig. 6.6 we show the case where there is no upstream pressure gradient, and the upstream velocity profile is linear (shear case). The curved meniscus shape is shown; it was computed using a finite element program for both Newtonian and second-order fluids ignoring surface tension and inertia. Two other cases were also treated; one corresponding to the Newtonian case where $h_i/h_o = 2$ ($m = 1$) the upstream velocity profile having the dimensionless form $u = 1 - (z/h)^2$; and in the other only the fluid moves; both planes are stationary and the inlet velocity profile is parabolic; *both* planes end at the exit here (Poiseuille case).

In all cases computed the free-surface shape changed little up to a value of N_1/τ (evaluated at the upper surface) of about 1. The pressure increase due to exit effects far inside a given bearing therefore depends only on the ratio $(N_1/\tau)_w$. Figure 6.7 shows a plot for the three cases considered; here Δp is the increase in 'pressure' $(-\sigma_{zz})$ at the wall far upstream due to exit effects. Note that normal stress effects tend to increase the load capacity by an amount roughly proportional to $N_1 L$, as found above. The increases (above the mentioned case) are, respectively $0.78 N_1 L$, $0.39 N_1 L$, and $0.25 N_1 L$ for the shear, intermediate, and Poiseuille cases. These increases are larger than eqn (6.32) but not as large as the result from (6.48). Upstream effects have not been considered here, but it is clear that there are substantial problems in extending the classical theory of lubrication to find the terms of order $(h_o/L)(\lambda U/h_o)$ occurring in (6.36). Simply

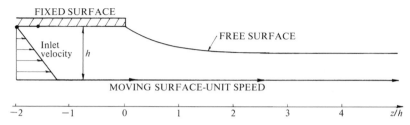

Fig. 6.6 Boundary conditions in lubrication at film exit. The exit profile shown is for the shearing case. The free surface shape was computed using a finite element program. For $(N_1/\tau)_w < 1.0$ all surface profiles are nearly the same.

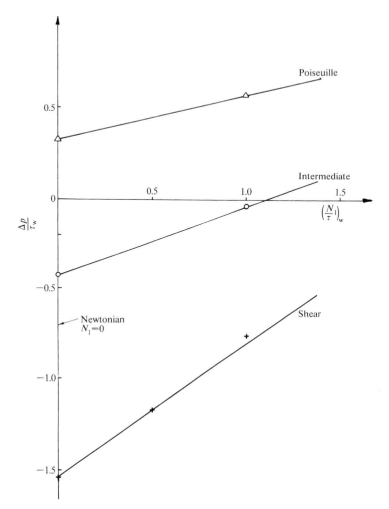

Fig. 6.7 Boundary conditions in lubrication. Plotted is increased load capacity Δp due to non-Newtonian effects at exit. The labels Poiseuille, intermediate and shear refer to the exit velocity profiles shown in Fig. 6.6. $(N_1/\tau)_w = 0$ is the Newtonian case. τ_w is the shear stress on the lower plane far upstream.

perturbing the Newtonian flow within the bearing without careful consideration of the boundary conditions will not be exact, but it can be a guide to trends; when normal stress effects are very large, so that $(N_1 h)/(\tau L) \gg 1$, it will be necessary to solve the entire flow field, including that external to the bearing, without any special simplification.

See Phan-Thien *et al.* (1989) for further discussion of the effects of inlet and exit flows on load and friction.

The problems are more difficult when there is no well-defined separation point. This is the case in calendaring (Fig. 6.8) and roller and journal bearing lubrication.

The separation problem here falls into two classes:

(a) Cases such as calendaring where the sheet fails to adhere to the rolls at a certain point [Fig. 6.8(a)]. Soon after the separation point the stresses decay to zero and the sheet moves as a rigid body.

Here we usually assume that $\sigma_{xx} = 0$ at the separation point; for all inelastic models this also implies $\sigma_{zz} = 0$ and $p = 0$ at this point, since $\eta \partial u / \partial x$ and $\eta \partial w / \partial z$ are neglected in the theory. The shear stress σ_{xz} will also vanish if we make $\partial p / \partial x = p = 0$, at the separation point. Hence the Swift condition, $\mathrm{d}p/\mathrm{d}x = p = 0$ at separation, is found. To get some idea of the error we note that in a Newtonian creeping flow it takes about one film thickness for an initial stress field (between parallel plates) with a mean value of zero to decay to less than 1 per cent of its initial amplitude; thus the error due to this approximation is often small.

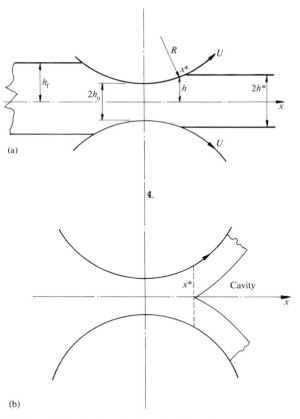

(a)

(b)

Fig. 6.8   Ideal exit fluid behaviour in (a) calendaring; (b) lubrication and coating.

(b) In bearing lubrication the film could split as in Fig. 6.8(b). In this case surface tension forces will not permit σ_{xx} to vanish at the end of the cavity. Alternatively, a wavy instability may set in, with 'fingering' of the lubricant in a three-dimensional pattern. Savage (1977) discusses the manifold possibilities occurring with lightly loaded bearings. Ruschak (1982) shows that $\partial p/\partial x$ is not exactly zero at the interface in the Newtonian case, but depends on the capillary number $\eta U/\sigma$; here σ is the surface tension coefficient. Carvalho and Scriven (1996) have given a survey of boundary conditions for lightly loaded deformable rollers with Newtonian fluids, with applications to printing press problems.

For heavily-loaded bearings, where maximum pressures are many atmospheres, the situation is simpler. Floberg (1957) has shown that it is adequate in this case to let $p = -p_m$, $\partial p/\partial x = 0$ at the separation point. Here the quantity p_m is the amount of subatmospheric pressure which is needed to vaporize the lubricant (or to cause air to burst out of the lubricant). Hence in this case the Swift condition is also a reasonable approximation if p_m is negligible compared to other pressures of interest in the system. Some numerical results for similar Newtonian problems have been given by Kistler and Scriven (1983).

In the case of inelastic non-Newtonian flows the situation is similar, but with viscoelastic flows the correct conditions are more complex and, due to memory effects, caution should be observed. For the second-order and other inelastic models, application of the above arguments again finds the Swift condition.

We now consider the case shown in Fig. 6.8(b), assuming separation occurs at the point when $p = \mathrm{d}p/\mathrm{d}x = 0$ (Swift condition). When this is done, solution of the problem in terms of a perturbation series about the Newtonian case is possible (Tanner 1960). The reduction in load capacity shown in this solution is mainly due to shear-thinning (see Section 6.2). The small shift in the separation point is also due to this cause. Since shear-thinning is so important, we now consider this effect in detail for some specific cases. We shall not consider the plane journal bearing with the Swift condition because it neglects side leakage. For details of the Newtonian calculations see Pinkus and Sternlicht (1961).

See also Section 9.2.1 for a further discussion on cavitation.

6.4 Calendaring and related problems

Calendaring is a process whereby a molten material is formed into a sheet by being passed through a pair of rollers [Fig. 6.8(a)]. Control of the exit thickness and finish of the sheet are important. In the present investigation we shall not be able to connect the variables in the problem to the surface finish, but we shall be able to discuss the control of thickness through the nip force, which pushes the rollers together, the speed, and the material properties.

The problem will be approached as a lubrication analysis problem. Assuming that both rollers have the same speed and radius, then if we take advantage of the symmetry of the problem by placing the x-axis at the centre of the nip, we find that the shear stress τ_{xz} in the fluid has the simplified form

$$\tau_{xz} = z \frac{\mathrm{d}p}{\mathrm{d}x}. \tag{6.49}$$

(Note that the film thickness is $2h$ in this problem.) We can conveniently analyse the power-law fluid in this case; returning to eqn (6.20) and setting $c = 0$, we have, for $z > 0$,

$$\frac{\partial u}{\partial z} = \left[\frac{z}{k} \frac{\mathrm{d}p}{\mathrm{d}x} \right]^{1/n} \mathrm{sgn} \left(\frac{\mathrm{d}p}{\mathrm{d}x} \right) \tag{6.50}$$

and the regions where $\mathrm{d}p/\mathrm{d}x$ is greater and less than zero have to be treated separately. Integrating, we find, after using the velocity boundary condition that $u = U$ at $z = h$, that

$$u = U + \frac{1}{\mu} \left[\frac{1}{k} \frac{\mathrm{d}p}{\mathrm{d}x} \right]^{1/n} \mathrm{sgn} \left(\frac{\mathrm{d}p}{\mathrm{d}x} \right) \{ z^{\mu} - h^{\mu} \} \tag{6.51}$$

where $\mu = (1 + n)/n$.

In the Newtonian case, $n = 1$, this reduces to the known form

$$u = U + \frac{1}{2\eta} \frac{\mathrm{d}p}{\mathrm{d}x} (z^2 - h^2). \tag{6.52}$$

Integrating from $z = 0$ to $z = h$, and letting the total nip flow be q we have

$$q = 2 \int_0^h u \, \mathrm{d}z = 2Uh + \frac{2}{\mu} \left| \frac{1}{k} \frac{\mathrm{d}p}{\mathrm{d}x} \right|^{1/n} \mathrm{sgn} \left(\frac{\mathrm{d}p}{\mathrm{d}x} \right) \left[\frac{h^{\mu+1}}{\mu + 1} - h^{\mu+1} \right] \tag{6.53}$$

from which an equation for the pressure can be found. If we define $q = 2Uh^*$; then

$$\left[\frac{1}{k} \frac{\mathrm{d}p}{\mathrm{d}x} \right]^{1/n} \mathrm{sgn} \left(\frac{\mathrm{d}p}{\mathrm{d}x} \right) = -(\mu + 1)(h^* - h) Uh^{-(\mu+1)}. \tag{6.54}$$

When the gap exceeds h^*, $\mathrm{d}p/\mathrm{d}x$ is positive; as $x \to \infty$, $\mathrm{d}p/\mathrm{d}x \to 0$. It is now necessary to specify h as a function of x. It is usual to make a parabolic approximation to the circular roll form and set

$$h \approx h_o \left(1 + \frac{x^2}{2h_o R} \right). \tag{6.55}$$

Before proceeding with the case of the general power-law fluid it is useful to study the Newtonian case, $n = 1$.

It is convenient to define dimensionless variables

$$x' = \frac{x}{\sqrt{2Rh_o}}; \quad u' = \frac{u}{U}$$

$$z' = \frac{z}{h_0}; \quad p' = \frac{ph_o}{\eta U}.$$

(6.56)

Then eqn (6.51) may be written as

$$u' = 1 + \sqrt{\frac{h_o}{8R}}[z'^2 - (1+x'^2)^2]\frac{dp'}{dx'}.$$

(6.57)

We may find an expression for the pressure gradient by using a mass balance

$$q = 2\int_0^h u\,dz = 2hU - \left[\frac{2h^3}{3\eta}\frac{dp}{dx}\right].$$

(6.58)

We may solve for dp/dx and find, in dimensionless form,

$$\frac{dp'}{dx'} = \sqrt{\frac{18R}{h_o}}\frac{x'^2 - x^{*2}}{(1+x'^2)^3},$$

(6.59)

where x^* is the (dimensionless) coordinate at which in our previous terminology $h = h^*$ and $dp/dx = 0$.

From the definition of x^*, we see that $(h^*/h_o) = 1 + x^{*2}$ where x^* is also dimensionless. We assume that right at separation, the Swift condition $p = (dp/dx) = 0$ holds. Thus, the separation point is at x^* (Fig. 6.8). From the boundary condition and (6.55) and (6.58) we find that the sheet leaves the roll at speed U and that $q = 2Uh^*$. It also follows that

$$x^* = \sqrt{\frac{h^*}{h_o} - 1}.$$

(6.60)

Thus, if h^* is measured, x^* may be found.

Equation (6.59) may be integrated with respect to x', and if the boundary conditions are imposed the solution for $p'(x')$ is found to be

$$p' = \sqrt{\frac{9R}{32h_o}}\left[\frac{x'^2(1 - 3x^{*2}) - 1 - 5x^{*2}}{(1+x'^2)^2}x'\right.$$

$$\left. +(1 - 3x^{*2})(\tan^{-1}x' - \tan^{-1}x^*) + \frac{1 + 3x^{*2}}{1 + x^{*2}}x^*\right].$$

(6.61)

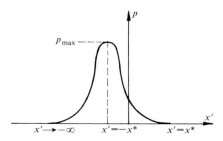

Fig. 6.9 Pressure distribution in calendaring and roll lubrication.

Figure 6.9 shows the shape of the pressure distribution. The maximum pressure occurs just upstream of the nip at $x' = -x^*$ and has the value

$$p'_{max} = \frac{3C}{2}\sqrt{\frac{R}{2h_o}} \tag{6.62}$$

where

$$C = \frac{1 + 3x^{*2}}{1 + x^{*2}} x^* - (1 - 3x^{*2}) \tan^{-1} x^*. \tag{6.63}$$

Thus, there are several extrema: At $x' = x^*$ the vanishing gradient occurs at $p' = 0$; at $x' = -x^*$ the maximum pressure occurs. The third value of x' at which the gradient vanishes is seen [from eqn (6.59)] to be at $x' = -\infty$. The pressure there is obtained upon setting $x' = -\infty$ in eqn (6.61). One finds that

$$p'(x' \to -\infty) = \sqrt{\frac{9R}{32h_o}} \left[\frac{1 + 3x^{*2}}{1 + x^{*2}} x^* - (1 - 3x^{*2}) \left(\frac{\pi}{2} + \tan^{-1} x^* \right) \right] \tag{6.64}$$

If an external pressure were imposed far upstream of the nip, then one could obtain x^* in terms of that pressure from eqn (6.64). One does not normally impose such a pressure on a calendar. The most reasonable assumption would be that $p'(x' \to -\infty) = 0$, from which it follows that x^* has a specific value, namely, $x^* = 0.475$.

From eqn (6.55), knowing x^*, we see that

$$\frac{h^*}{h_o} = 1 + x^{*2} = 1.226, \tag{6.65}$$

and so the sheet thickness depends only on h_o.

The roll-separating force can be calculated from

$$W = \int_{-\infty}^{x^*} (-\sigma_{zz}) \, dx. \tag{6.66}$$

Within the lubrication approximation we ignore the viscous contribution to $-\sigma_{zz}$ and simply equate $-\sigma_{zz}$ to p. Then we calculate the force from

$$W = \int_{-\infty}^{x^*} p(x)\,dx = \eta U \sqrt{\frac{2R}{h_o}} \int_{-\infty}^{x^*} p'\,dx' \tag{6.67}$$

with the result

$$W = \frac{\eta U R}{h_o} G(x^*) \tag{6.68}$$

where $G(x^*)$ is a complicated function whose value, for the expected case of $x^* = 0.475$, is 1.22. Thus we have

$$W = 1.22\frac{\eta U R}{h_o}. \tag{6.69}$$

The drag force on one roll is calculated from

$$F = \int_{-\infty}^{x^*} \sigma_{xz}|_{z=h(x)}\,dx. \tag{6.70}$$

The result for $x^* = 0.475$ is

$$F = 3U\eta_0\sqrt{\frac{2R}{h_o}}M(x^*), \tag{6.71}$$

where

$$M(x^*) = (1 - x^{*2})\left(\tan^{-1} x^* + \frac{\pi}{2}\right) - x^*. \tag{6.72}$$

For $x^* = 0.475$, M is 1.08 and

$$F = 4.58U\eta_0\sqrt{\frac{R}{h_o}}. \tag{6.73}$$

The friction coefficient $F/W = 3.75\sqrt{h_o/R}$ from which the power dissipation $2UF$ can be found. This completes the Newtonian case.

Returning to the power-law fluid, we may use the preceding dimensionless variables provided we generalize the definition of p' to read $p' = p(h_o/U)^n/k$. It is then straightforward to show that

$$p' = \left(\frac{2n+1}{n}\right)^n \sqrt{\frac{2R}{h_o}} \int_{x'}^{x^*} \frac{|x^{*2} - x'^2|^{n-1}(x^{*2} - x'^2)\,dx'}{(1 + x'^2)^{2n+1}}, \tag{6.74}$$

where the Swift boundary condition at $x = x^*$ has been used. Numerical solution may be used to complete the problem.

The complete results as a function of n for the loads and friction are given in the form

$$W = k\left(\frac{U}{h_o}\right)^n RW^*(n), \tag{6.75}$$

$$F = k\left(\frac{U}{h_o}\right)^n \sqrt{Rh_o}\,F^*(n),$$

where W^* and F^* are given in Fig. 6.10. The thickness function h^*/h_o is given in Fig. 6.11(a). We see that a shear-thinning fluid gives rise to a thicker sheet. In the models considered above it was assumed that the calendar was fed with a mass of fluid so large that an infinite reservoir of fluid existed up-stream from the nip. It is possible, of course, to feed the calendar with a sheet of fluid of known width, as suggested in Fig. 6.8(a). In this case a new boundary condition $p' = 0$ at $x' = -x'_f$ must be introduced, and the lower limit in the problem becomes $-x'_f$.

The thickness of the feedstock enters through the definition

$$x'_f = \left(\frac{h_f}{h_o} - 1\right)^{1/2}. \tag{6.76}$$

Equation (6.74) may be solved for x^* as a function of n and h_f/h_o, and the results are shown in Fig. 6.11(b). Several points are of interest.

First, we see, as expected, that the calendared thickness is reduced somewhat if h_f/h_o is finite, for all values of n. Next, we see that while $h_f/h_o = 20$ is large enough that the separation point is practically equal to x^* for the case of the Newtonian

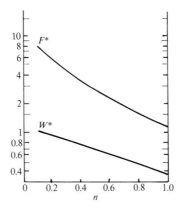

Fig. 6.10   Calendaring of power-law fluids; n is the index and F^* and W^* are the dimensionless friction and load forces respectively (from Middleman 1977).

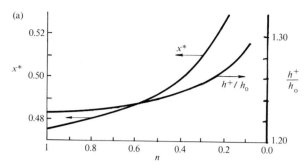

Fig. 6.11(a) Calendared thickness in terms of x^* or h^*/h_o as a function of the power-law parameter n (from Middleman 1977).

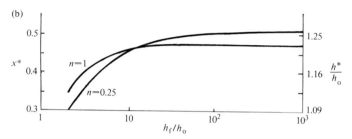

Fig. 6.11(b) Calendared thickness in terms of x^* or h^*/h_o as a function of h_f/h_o for the power-law parameter $n = 1.0$ and 0.25. Note $h^*/h_o = 1 + x^{*2}$ (from Middleman 1977).

fluid, a much larger value, say $h_f/h_o = 200$, is required for a fluid with $n = 0.25$ to be near its infinite thickness separation point.

Overall, it is apparent that as far as sheet thickness is concerned, there is not much variability with changes in n, or with reasonable variations in h_f/h_o. There are, however, considerable variations in the forces, so that calendaring with a fixed nip force, roll speed, and roll radius, will yield sheets of variable thickness as the material changes.

It is clear that the present analytical technique is difficult to apply to more complex cases and numerical methods are necessary for practical applications. As an example, Kiparissides and Vlachopoulos (1976) have shown how to use finite element methods for the analysis of inelastic non-Newtonian calendaring problems.

Zheng and Tanner (1988) used perturbation and numerical methods to study calendaring. They used the modified PTT model (see Section 5.6.7) of the affine form

$$\lambda \frac{\Delta \boldsymbol{\tau}}{\Delta t} + \left(1 + \varepsilon \frac{\lambda}{\eta_0} \operatorname{tr} \boldsymbol{\tau}\right) \boldsymbol{\tau} = 2\eta_0 (1 + 2\varepsilon \lambda^2 \dot{\gamma}^2)\mathbf{d}, \qquad (5.163a)$$

with $\varepsilon = 0.01$. Perturbation solutions were used for $Wi(\sim \lambda U/h)$ less than one. Numerical simulations up to a Weissenberg number (Wi) of 3 were made (h_o is half the minimum gap size), and in all cases the upstream feed sheet was 20 times the nip size. The separation criterion used was that of zero tangential traction. In these calculations a trial and error procedure was used to locate the separation point, where the tangential traction just upstream of the separation point is zero. In all cases the separation point (x^*) was found to be about 0.47, as in the Newtonian case. The normal traction just at the separation point was found to be tensile, so that the sheet was 'peeled' off the rolls; no attempt to set the normal stresses at peel-off was made, although that would have been possible.

Unlike the inelastic case, the sheet was found to thicken after leaving the nip (similar to die-swell) so that at $(Wi) = 3$ the ultimate thickness of the sheet was about 5 per cent greater than the Newtonian case. Some changes of up to 10 per cent in the roll-separating force were also found.

6.4.1 Connection with roller and ball lubrication

Because of the small-slope approximation used previously, the calendaring analysis is applicable to the lubrication problem between two cylinders of any radius provided the two surfaces move at the same speed U. If the cylinders have radius R_0 and R_1 respectively, then the (total) gap $2h$ is

$$2h = 2h_o + \frac{x^2}{2}\left\{\frac{1}{R_1} + \frac{1}{R_0}\right\}. \tag{6.77}$$

Hence, by placing the surface $z = 0$ at the centre of the gap, the problem is reduced to the previous equal cylinder case calendaring by using the equivalent radius $2/(1/R_1 + 1/R_0)$ instead of R. In particular, if R_0 is infinite, then the problem of a cylinder rolling over a plane is solved.

The separation at the exit from the nip is not envisaged as shown in Fig. 6.8(a), but rather as in Fig. 6.8(b), where the emergent film is split between the rollers, often with a ribbed cavitation region between. As mentioned previously, the Swift condition is not exact but it is the most often used condition in lubrication problems. When it is used, the results are identical to the calendaring problem.

For a further discussion of high-pressure rheology and ball and roller lubrication, see Jacobson (1991).

6.5   Coating flows

Coating is a process in which a liquid is applied continuously to a moving sheet in order to produce a uniform application of the fluid onto and/or within the sheet.

Figure 6.12(a) shows, schematically, a roll coater. The lower roll picks up liquid from a bath and delivers it to a second roll, or directly to the moving web. The web is 'squeezed' between two rolls, and the amount of coating applied depends upon fluid properties, roll spacing and speed.

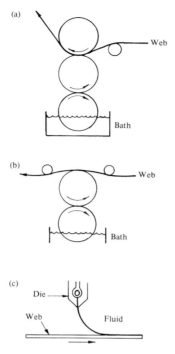

Fig. 6.12   Three types of coating flows. (a) Roll coater. (b) Kiss coater. (c) Curtain coater.

In some applications, the nip separation is controlled; in others the nip pressure is controlled.

Similar to the roll coater is the 'kiss' coater, shown in Fig. 6.12(b) in which the web is run over the roll without any backup roll on the other side. The amount of wrap around the roll, the tension in the web, and fluid properties control the amount of coating applied to the web.

Various other arrangements have been discussed by Middleman (1977). They are weak or lubrication type flows which can be analysed by the methods given above. Wirecoating is discussed by Binding *et al.* (1976).

6.5.1   Roll coating

A roll-coating process is similar to calendaring, but with a split film in the exit region as shown in Fig. 6.8(b).

The boundary condition at the nip-exit must be considered. Generally in coating flows the pressures are low and hence cavitation may not occur abruptly as in lubrication where the (potentially) very low minimum pressure ensures cavitation. Middleman (1977) uses the (approximate) condition in the cavitation region that the speed and the pressure are zero simultaneously. Again, we reiterate that the true separation condition is complex, see Savage (1977), Ruschak (1982) and Carvalho and Scriven (1996).

In the case where the sheet has the same velocity as the peripheral roll-speed and moves in the same direction, the solution is very similar to that for calendaring and lubrication discussed previously. See Middleman (1977) for details. Differential speeds, or reverse coating (where the roll motion is opposite that of the sheet), are other commonly encountered cases in coating.

6.5.2   Free-surface coating

In this case the surface to be coated is withdrawn from a bath of fluid. Gravity and surface tension are both expected to be important in these low-stress conditions. Gravity causes the fluid to tend to drain off the sheet.

Two of the simplest free-coating problems arise when the object to be coated is either a continuous web or belt of material (such as in film coating) or a cylindrical filament (such as in fibre coating). The principal complicating feature of these problems is the strong role played by the free surface with its attendant surface tension forces in controlling the coating dynamics in the region where the object leaves the surface of the fluid. The Newtonian case is complex (Scriven and Kistler 1982; Kistler and Scriven 1983) and there does not appear to be any simple non-Newtonian model for these processes (see Middleman 1977). These flows are not nearly-viscometric flows; in the bath region there is considerable elongation and this has to be taken into account in a proper theory. We shall not therefore discuss this type of problem here.

6.6   Three-dimensional lubrication theory

In most practical bearings the two-dimensional cases described previously give correct qualitative predictions, but are not adequate for quantitative predictions. This is because in a bearing of finite width (y-direction) there is a large pressure gradient $\partial p/\partial y$ at right angles to the main motion in the x-direction, since the pressure falls to zero at the ends of the bearing. This pressure gradient gives rise to a leakage flow and a consequent loss of pressure (and hence load capacity) in the bearing. Thus we now turn to three-dimensional lubrication theory.

It is not difficult to generalize the inelastic lubrication theory to the general case when the flow is three-dimensional and is confined between two rigid surfaces in relative motion. One obtains the results

$$p = p(x, y), \tag{6.78}$$

$$\tau_{xz} = z\frac{\partial p}{\partial x} + C_1(x, y), \tag{6.79}$$

$$\tau_{yz} = z\frac{\partial p}{\partial y} + C_2(x, y), \tag{6.80}$$

and in the Newtonian case one finds

$$u = U_0 + (U_1 - U_0)\frac{z}{h} - \frac{z}{2\eta}\frac{\partial p}{\partial x}(h - z) \tag{6.81}$$

$$v = V_0 + (V_1 - V_0)\frac{z}{h} - \frac{z}{2\eta}\frac{\partial p}{\partial y}(h - z). \tag{6.82}$$

In place of (6.8) the averaged continuity equation becomes

$$\frac{\partial h}{\partial t} + \frac{\partial}{\partial x}(\bar{u}h) + \frac{\partial}{\partial x}(\bar{v}h) = 0, \tag{6.83}$$

where time variations in the film thickness are permitted, and the mean velocities $\bar{u}, \bar{v}$, are given by

$$(\bar{u}, \bar{v}) = \frac{1}{h}\int_0^h (u, v)\,\mathrm{d}z. \tag{6.84}$$

After forming $(\bar{u}, \bar{v})$ and substituting in (6.83) we obtain the Reynolds equation for the pressure

$$\frac{\partial}{\partial x}\left(\frac{h^3}{\eta}\frac{\partial p}{\partial x}\right) + \frac{\partial}{\partial y}\left(\frac{h^3}{\eta}\frac{\partial p}{\partial y}\right) = 6\frac{\partial}{\partial x}(h(U_1 + U_0)) + 12\frac{\partial h}{\partial t}. \tag{6.85}$$

Boundary conditions on the pressure field complete the specification of problem.

The solution of the Reynolds equation is non-trivial in most cases (Pinkus and Sternlicht 1961) even for the steady-flow Newtonian case.

For the non-Newtonian case, the flow is still regarded as nearly-viscometric; locally it is a skew flow (Section 3.2.2). For the generalized Newtonian fluid, we now have the coupling between the x- and y-motions through the viscosity function, since in this case the shear rate is given by

$$\dot{\gamma}^2 = \left(\frac{\partial u}{\partial z}\right)^2 + \left(\frac{\partial v}{\partial z}\right)^2. \tag{6.86}$$

In most cases it is necessary to resort to numerical methods to solve the non-Newtonian case. A useful exception is discussed next.

6.6.1 Newtonian short-bearing theory

We consider an approximate theory valid for complete journal bearings whose length/diameter ratio is small (Fig 6.13). In such cases the fluid 'leaks' rapidly out of the sides of the bearing and the pressure developed in the fluid is fairly small compared with the two-dimensional case. Since $D \gg L$, by hypothesis, we expect,

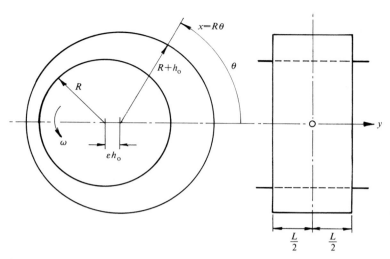

Fig. 6.13 Short-bearing geometry. Clearance is greatly exaggerated; usually h_0/R is $O(10^{-3})$. Radius $R = \frac{1}{2}D$.

on average, that

$$\frac{\partial p}{\partial x} \ll \frac{\partial p}{\partial y}. \tag{6.87}$$

We now make the assumption that $\partial p/\partial x$ may be completely neglected. Then, at least in the steady Newtonian case, the circumferential flow is everywhere a steady shearing, locally $\omega R/h$ in magnitude. If we let $U_1 = 0$, $U_0 = \omega R$, $V_1 = V_0 = 0$, then

$$u = \omega R\left(1 - \frac{z}{h}\right); \quad \bar{u} = \tfrac{1}{2}\omega R \tag{6.88}$$

$$v = -\frac{z}{2\eta}(h - z)\frac{\partial p}{\partial y}; \quad \bar{v} = -\frac{h^2}{12\eta}\frac{\partial p}{\partial y}. \tag{6.89}$$

Substituting in (6.83) we find

$$\frac{\partial}{\partial y}\left\{h^3\frac{\partial p}{\partial y}\right\} = 6\eta\omega R\frac{\partial h}{\partial x}. \tag{6.90}$$

If we now 'unwrap' the bearing, so that

$$x = R\theta,$$

where θ is the angular position, (Fig. 6.13), and we consider the clearance h of the form (Pinkus and Sternlicht 1961)

$$h = h_o(1 + \varepsilon \cos \theta),$$ (6.91)

where ε is the eccentricity ratio (e/h_0), then we find a simple equation for $\partial^2 p/\partial y^2$:

$$\frac{\partial^2 p}{\partial y^2} = -6\eta \omega h_o^{-2} \varepsilon \sin \theta / (1 + \varepsilon \cos \theta)^3.$$ (6.92)

Hence, using the boundary condition $p = 0$ at $y = \pm L/2$ (Fig. 6.13),

$$p = 3\eta \omega h_o^{-2} \varepsilon \left(\frac{L^2}{4} - y^2 \right) \sin \theta / (1 + \varepsilon \cos \theta)^3.$$ (6.93)

Further integration gives the total load components as

$$W_h = \int_{-L/2}^{L/2} \int_0^\pi pR \cos \theta \, d\theta \, dy = \left(\frac{\eta \omega R L^3}{h_o^2} \right) \frac{\varepsilon^2}{(1 - \varepsilon^2)^2},$$

$$W_v = \int_{-L/2}^{L/2} \int_0^\pi pR \sin \theta \, d\theta \, dy = \left(\frac{\eta \omega R L^3}{h_o^2} \right) \frac{\pi}{4} \frac{\varepsilon}{(1 - \varepsilon^2)^{3/2}},$$ (6.94)

and

$$W = \sqrt{W_h^2 + W_v^2}$$

and the friction moment as (including the drag of the cavitated region $2\pi > \theta > \pi$)

$$M = \int_{-L/2}^{L/2} \int_0^{2\pi} \tau R^2 \, d\theta \, dy = \frac{2\pi \eta \omega R^3}{h_o} (1 - \varepsilon^2)^{-1/2}.$$ (6.95)

Here we assume that the pressure is zero, due to cavitation, over the segment $\pi < \theta < 2\pi$. This boundary condition is compatible with the short bearing theory. It is found that this formula represents quite well the behaviour of bearings even when $L/D \simeq 1$, which is close to common cases of practical interest and Barwell (1956) made the solutions (6.94) and (6.95) the basis of a design method. Subsequently, Raimondi and Boyd (1958) solved the full Reynolds equation by computer methods to produce design charts which are an improvement over the short-bearing design method for the Newtonian case. However, it is not clear that this is a feasible strategy for every non-Newtonian case, and thus it is of interest to try and generalize the simple short-bearing solution to inelastic non-Newtonian fluids.

6.6.2   Inelastic short-bearing theory

From (6.89) and (6.90) we can see that the shear rate $\partial v/\partial z$ is of order $\omega L/h$, whereas the shear rate $\partial u/\partial z$ is of order $\omega R/h$. This suggests that when $R^2 \gg L^2$ we may assume that the circumferential shearing is dominant, and then we have

$$\dot{\gamma} \sim \frac{\partial u}{\partial z} = \omega R/h. \tag{6.96}$$

The inelastic problem is then straightforward, since the viscosity is constant across the film at any angular position, (but varies with θ) and so the sideways flow has a parabolic profile. Hence the solution (6.93) for p still applies, it being understood that η is now a function of θ, through $\dot{\gamma}$. The load capacity and frictional torque may be computed. The results for a family of power-law fluids are given in Fig. 6.14. One sees that for any given eccentricity, speed, load and bearing geometry the power-law fluid displays a higher friction coefficient than a Newtonian fluid, as in the simpler cases discussed above. More details are given by Tanner (1963).

6.6.3   Viscoelastic short-bearing theory

If one is prepared to accept the above restrictions on the analysis, then it is possible to extend the above short-bearing theory to unsteady loads with visco-elastic fluids, at least when the eccentricity is low, that is when the circumferential flow remains an almost constant simple shear flow with a shear rate $\omega R/h_0$.

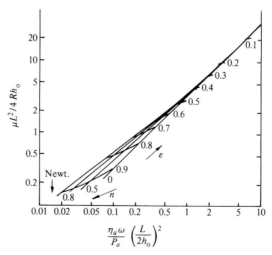

Fig. 6.14 Friction and Sommerfeld number for power-law fluid is short-bearing theory. In this dimensionless plot $P_a = W/2RL$; η_a is the viscosity at the mean shear rate $\omega R/h_0$, and $\mu =$ turning moment M/WR.

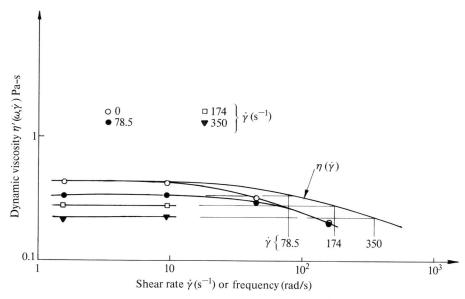

Fig. 6.15 Dynamic viscosity of 4.4 per cent polyisobutylene ($M_\mathrm{w} \sim 10^6$) in cetane solution at 25 °C. Sample is subjected to sinusoidal shearing at frequency ω and steady shearing at a rate $\dot\gamma$ simultaneously; axes of shearing are orthogonal. Note that $\eta'(0, \dot\gamma) = \eta(\dot\gamma)$.

The basic problem in this unsteady skew flow is to find the response in the y-direction when the flow is disturbed so as to produce a (steady or unsteady) cross-shear $\partial v/\partial z$. There are limited experimental data which are relevant (Tanner and Williams 1971) to this problem. In these experiments a simple shearing flow ($\dot\gamma$) was disturbed by a small sinusoidal (in time) cross-flow and the linear response functions $G'(\dot\gamma, \omega)$, $G''(\dot\gamma, \omega)$ were measured. The results of Williams are shown in Fig. 6.15. We see that the fluid response is essentially Newtonian at a reduced viscosity $\eta(\dot\gamma)$ when $\omega \ll \dot\gamma$, and the response is the same as the unsheared fluid when $\omega \gg \dot\gamma$. Thus for low-frequency disturbances the response is Newtonian with a viscosity $\eta(\dot\gamma)$ and the theory of (6.6.1) is applicable. For high-frequency disturbances the problem is reduced to a linear viscoelastic problem with a modified spectrum. However, the latter limit is somewhat academic because in order to reach this high-frequency range the frequency of application must be greater than $\omega R/h_0$, or of order 10^3 times the rotation frequency. Hence it will not be explored further here. Further problems of the type discussed in Section 6.3.3 (stress boundary layers) may also arise with some models.

The solutions discussed here are clearly only relevant in certain ranges (short bearings, low eccentricity and frequency) and more remains to be done.

6.6.4 Viscosity a known function of position

When the viscosity function η is a known function of x, y, and z, then the lubrication problem may be formally solved in the following way. Define

$$F_0(z) = \int_0^z \frac{dz}{\eta}; \quad F_1(z) = \int_0^z \frac{z\, dz}{\eta} \quad \text{and} \quad F_2 = \int_0^z \frac{z^2\, dz}{\eta}. \tag{6.97}$$

Then if we take the equations governing lubrication and write them as

$$\frac{\partial u}{\partial z} = \frac{z}{\eta}\frac{\partial p}{\partial x} + \frac{c_1}{\eta} \tag{6.98}$$

$$\frac{\partial v}{\partial z} = \frac{z}{\eta}\frac{\partial p}{\partial y} + \frac{c_2}{\eta}, \tag{6.99}$$

where η is a known function, then we can integrate to find

$$u = U_0 + F_1(z)\frac{\partial p}{\partial x} + \left(U_1 - U_0 - F_1(h)\frac{\partial p}{\partial x}\right)\frac{F_0(z)}{F_0(h)}$$

$$v = V_0 + F_1(z)\frac{\partial p}{\partial y} + \left(V_1 - V_0 - F_1(h)\frac{\partial p}{\partial y}\right)\frac{F_0(z)}{F_0(h)}. \tag{6.100}$$

Since the viscosity is supposed to be a known function of position, the functions F_0, F_1, and F_2 can be found numerically.

By integrating (6.100) we find

$$h\bar{u} = \int_0^h u\, dz = U_0 h + \frac{\partial p}{\partial x}(hF_1(h) - F_2(h))$$

$$+ \left[(U_1 - U_0) - F_1(h)\frac{\partial p}{\partial x}\right]\left[\frac{hF_0(h) - F_1(h)}{F_0(h)}\right] \tag{6.101}$$

and a similar expression for $h\bar{v}$. These results can be substituted into the continuity equation (6.85) to find a generalized Reynolds equation

$$\frac{\partial h}{\partial t} + \nabla \cdot \left\{\left(h - \frac{F_1}{F_0}\right)\mathbf{V}_1 + \frac{F_1}{F_0}\mathbf{V}_0 + \left(\frac{F_1^2}{F_0} - F_2\right)\nabla p\right\} = 0, \tag{6.102}$$

where

$$\mathbf{V}_1 = U_1\mathbf{i} + V_1\mathbf{j},$$

and

$$\mathbf{V}_0 = U_0\mathbf{i} + V_0\mathbf{j}, \tag{6.103}$$

and the *F*-functions are evaluated at $z = h$. Solution of (6.102) can be carried out numerically as a routine operation.

Some analytical progress can sometimes be made if η can be expressed in terms of stresses; see Section 6.6.5.

These results provide a method of solving non-Newtonian lubrication problems by an iterative process. By solving (6.102) using an initial guess for the viscosity distribution and appropriate boundary conditions we can find the pressure, and hence the velocity gradients, from (6.100). These can be used to recalculate the viscosity function. Hence a new, more accurate, distribution of viscosity can be found. Then the iteration cycle can be repeated until convergence occurs. Note that the temperature variation of viscosity can also be included in this scheme. Dowson (1962) gives an even more general formulation for the case when the density of the fluid is allowed to vary. The advantage of the proposed method of integration is that it reduces a complex three-dimensional flow to a simple two-dimensional problem, with advantages in computing speed. Such a numerical approach makes the solution of inelastic non-Newtonian lubrication-type flows possible. The necessary boundary conditions, as always, need careful consideration.

6.6.5 Hele–Shaw flow and die filling

Many die-filling flows use narrow channels where the flow is approximately two-dimensional. For stationary walls one can use eqn (6.85) with U_1, U_0, and $\partial h / \partial t$ equal to zero, in the Newtonian case. This is then a two-dimensional flow, and is readily solved. In the non-Newtonian case the problem is more complex, but the approach of Section 6.6.4 is applicable. Generally, it is necessary to also solve for the temperature field, and this usually will need a fully three-dimensional temperature computation (Chapter 9).

When a fluidity ϕ ($\equiv 1/\eta$, a function of the second invariant of stress) can be written, considerable progress can be made, as seen in the example following. For further applications to injection moulding see Kennedy (1995).

Example: Power-law Hele–Shaw flow

If one rewrites the power-law fluid (3.63) in terms of a fluidity ϕ, one finds that, in this form of flow,

$$\phi = \frac{1}{\eta} = k^{-1/n}(\tau_{yz}^2 + \tau_{xz}^2)^{(1-n)/2n}.$$

But in the Hele–Shaw flow where z is measured from the midplane, we have, from (6.79) and (6.80)

$$\tau_{yz} = z\frac{\partial p}{\partial y}, \quad \tau_{xz} = z\frac{\partial p}{\partial x}.$$

Hence ($z > 0$)

$$\phi \equiv \frac{1}{\eta} = k^{-1/n} z^{(1-n)/n} \left(\left(\frac{\partial p}{\partial x} \right)^2 + \left(\frac{\partial p}{\partial y} \right)^2 \right)^{(1-n)/2n}.$$

From (6.102), when the walls are stationary and the flow is steady, we have the equation

$$\nabla \cdot \left\{ \left(\frac{F_1^2}{F_0} - F_2 \right) \nabla p \right\} = 0. \tag{6.102a}$$

But in this case one finds, from (6.97)

$$F_0 = \int_0^h \frac{dz}{\eta} = \int_0^h \phi\, dz = -nk^{-1/n} |\nabla p|^{(1-n)/2n} h^{2+1/n}$$

and similar expressions for $F_1(h)$ and $F_2(h)$. Hence the equation for the pressure becomes

$$\nabla \cdot \{h^{2+1/n} |\nabla p|^{(1-n)/n} \nabla p\} = 0 \tag{6.104}$$

which may be solved with appropriate boundary conditions on pressure.

6.7   Unsteady flows: the squeeze film problem

In the previous section we have mainly investigated Eulerian steady flows where normal stress effects were overshadowed by shear-thinning, so that the classical wedge mechanism of load generation dominated over the direct or normal-stress mechanism of load generation. Consider now the squeezing of fluid between two approaching parallel plates (Fig. 6.16). These are shown as circular, radius R, but a two-dimensional version of this problem is possible. The fluid is being forced from the space between the plates and its resistance generates the squeeze-film load W. 'Normal stress effects' are also possible, as will be demonstrated. The problem is to predict W given R, $h(t)$ and the fluid properties, and the interest lies in the fact that the fluid flow is an unsteady Eulerian flow and one therefore expects viscoelasticity to be important.

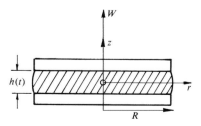

Fig. 6.16   Squeeze-film geometry. Hatched area is fluid.

6.7.1 The Newtonian solution

We can with advantage take the origin of coordinates at the centre of the gap so that the problem is symmetrical and each plate approaches the origin at a speed $-\dot{h}/2$ while supporting a load W. The conservation of mass equation is

$$\frac{1}{r}\frac{\partial(ru)}{\partial r} + \frac{\partial w}{\partial z} = 0. \tag{6.105}$$

We can integrate from $z = -h/2$ to $h/2$ to find

$$\frac{dh}{dt} = -\frac{1}{r}\frac{d}{dr}(r\bar{u}h), \tag{6.106}$$

where the average radial velocity $\bar{u}$ is defined by

$$\bar{u}h = 2\int_0^{h/2} u(z,r)\,dz. \tag{6.107}$$

From (6.105) it is clear that $\bar{u}$ is much greater in magnitude than the speed dh/dt; in fact $\bar{u} = O(R\dot{h}/h)$.

Also, since it is supposed $R \gg h$, then we may again make the lubrication approximation. When inertia, body forces, and normal stresses are ignored, we find

$$\frac{dp}{dr}(r,t) = \frac{\partial}{\partial z}(\sigma_{rz}) \approx \eta\frac{\partial^2 u}{\partial z^2}, \tag{6.108}$$

where u is a function of r, z, and t.

Integrating and using the boundary conditions $u=0$ on $z=\pm h/2$, finds

$$u = \frac{1}{2\eta}\left\{z^2 - \frac{h^2}{4}\right\}\frac{dp}{dr}. \tag{6.109}$$

To find the pressure, we average (6.109) over the gap and use (6.107), hence

$$\frac{1}{r}\frac{d}{dr}\left\{r\frac{dp}{dr}\right\} = \frac{12\eta}{h^3}\frac{dh}{dt}. \tag{6.110}$$

Integrating, noting that dp/dr is zero at $r=0$ and that $p=0$ at $r=R$, we find

$$p = \frac{3\eta}{h^3}\frac{dh}{dt}\{r^2 - R^2\}, \tag{6.111}$$

and hence the force W required to maintain the plate motion is

$$W = - \int_0^R 2\pi r p \, dr = \frac{3\pi}{2} \eta R^4 \dot{h}/h^3.$$ (6.112)

When the plates approach, $\dot{h}$ is negative, and so is W.

The solution for W can be found in a similar manner taking a small amount of inertia into account; if ρ is the fluid density, one finds to the first order in ρ (Bird *et al.* 1977):

$$W = \frac{3\pi\eta R^4 \dot{h}}{2h^3} \left[1 - 0.1786 \frac{\rho h \dot{h}}{\eta} \{1 - 0.56 h \ddot{h}/\dot{h}^2 \} \right].$$ (6.112a)

Henceforth we shall usually ignore inertia in our calculations while noting that it must often be used when comparing experimental results with theory.

6.7.2   Inelastic fluids

Under the lubrication approximation, for an inelastic fluid we have $\tau_{rz} \sim \eta(|\partial u/\partial z|)(\partial u/\partial z)$, and also $\tau_{rz} = z(dp/dr)$ for all fluids. When it is possible to find a fluidity $\phi(\tau)$ and solve for $\partial u/\partial z$ one can integrate to find u, apply the boundary conditions as given above and solve for the pressure and load by integration. Clearly shear-thinning will reduce the load capacity relative to the Newtonian case; for the constitutive model (6.22) we find

$$W/W_N = 1 - 3.6\alpha^2 \dot{h}^2 R^2/h^4,$$ (6.113)

where W_N is the Newtonian value (6.112).

For the power-law fluid, a similar analysis leads to Scott's result

$$W = \frac{2\pi(2 + 1/n)^n}{(n + 3)} k \, \text{sgn}(\dot{h})|\dot{h}|^n h^{-(2n+1)} R^{n+3}$$ (6.114)

which has been carefully compared with experiment (Grimm 1978).

For materials with a yield stress, where the possibility of unyielded regions exists, see Sections 4.2.1 and 4.2.2 for a discussion of squeeze flows.

6.7.3   Viscoelastic effects

The squeeze-film action generates pressures of order (R/h) times the shear stresses, while the normal stresses are, as for the second order fluid, of the order of the square of the shear stresses, and they act directly, so that the situation is similar to the steady-flow bearing problems discussed previously. Therefore, the normal-stress effects connected with the predominantly shearing flow will be ignored; an approximate treatment of normal-stress effects has been given by McClelland and Finlayson (1983).

An exact treatment has been presented (Phan-Thien and Tanner 1983), and a stress-dependent viscosity ignoring normal stresses was used by Tanner (1965).

The squeeze-film flow is not a steady flow and this new element in the analysis must be investigated. We consider a linear viscoelastic problem where the plates are initially at rest and then begin to move towards each other at a constant speed V, so that

$$\dot{h} = -VH(t), \quad \text{and} \quad h(0) = h_0, \tag{6.115}$$

where $H(t)$ is the Heaviside step function. It is convenient to use the linear viscoelastic law in the form (2.70) so that

$$\tau_{rz} = \int_{-\infty}^{t} G(t - t') \frac{\partial u(t')}{\partial z} dt'. \tag{6.116}$$

Laplace transforming this equation and the equation $\tau_{rz} = zp'$, we find

$$\frac{\partial \tilde{u}}{\partial z} = z\tilde{p}'/\tilde{G}, \tag{6.117}$$

where $(\tilde{\ })$ denotes 'Laplace transform of ( )'. Inverting, integrating and then putting in the spatial boundary conditions on u, we find

$$u = \left(\frac{z^2}{2} - \frac{h^2}{8}\right) L^{-1}\{\tilde{p}'/\tilde{G}\}, \tag{6.118}$$

where $L^{-1}\{\ \}$ denotes 'inverse transform of $\{\ \}$'.

Integration with respect to z then gives

$$\bar{u}h = -\frac{h^3}{12} L^{-1}\{\tilde{p}'/\tilde{G}\}, \tag{6.119}$$

where $p' = dp/dr$.

This cumbersome procedure is used because h is a function of t.

One has, by considering overall mass conservation out to a radius r, $\bar{u}h = -r\dot{h}/2$, for any h. This can be used to eliminate $\bar{u}h$ from (6.119), finding, after integrating with respect to r and inserting the boundary condition $p = 0$ at $r = R$,

$$3(r^2 - R^2)\dot{h}/h^3 = L^{-1}\{\tilde{p}/\tilde{G}\}, \tag{6.120}$$

Taking Laplace transforms again, and integrating to find the load, one finds

$$W = \frac{3\pi R^4}{2} \int_{0}^{t} G(t - t')\left\{h^{-3}\frac{dh}{dt}\right\}_{t'} dt'. \tag{6.121}$$

Alternatively, one can assume u is of the form $rg(t)\,(z^2 - h^2/4)$, find $g = 3\dot{h}/h^3$ from mass-conservation, then use (6.116) to produce a result for p'; integration with respect to r then finds (6.120–6.121).

If one uses a Maxwell form (2.94) for $G(t)$, it is easy to show that the initial force is zero; only if a Newtonian component $\eta_\infty \delta(t)$ is present in G will the force be non-zero. Thus one concludes that, compared to a Newtonian film of viscosity η_0, the viscoelastic film is 'softer' or less resistant to deformation.

These conclusions also arise from the 'exact' solution for a non-linear Maxwell model (Phan-Thien and Tanner 1983), in which normal stress effects are included.

6.7.4 Experimental evidence

Tichy and Winer (1978) have studied the squeeze film problem experimentally and have shown that the load capacity, corrected for inertia, sometimes is less than and sometimes exceeds the Newtonian value. Fig. 6.17 shows their results: the error ranges should be noted. Grimm (1978) did some careful experiments with various fluids using a constant load technique. He showed that some of the viscoelastic fluids were 'stiffer' than an inelastic fluid with the same viscosity function. Note that the Deborah number (De) is in his case defined as the ratio $\lambda/t_{\frac{1}{2}}$, where $t_{\frac{1}{2}}$ is the time required to reduce the gap to half its initial value, whereas the Deborah number of Tichy and Winer, with a uniform speed of gap closure V, is defined as $\lambda V/h$. Thus there appears to be no question that increased resistance to deformation can arise from viscoelastic effects, and we now look at several possible explanations.

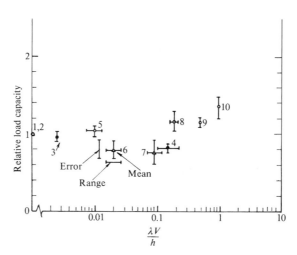

Fig. 6.17 Load capacity of squeeze film relative to Newtonian case showing load enhancement for polymer solutions: □ Newtonian petroleum oil, fluids 1,2 $0.5 < (Re) < 25$; ● Silicones $0.5 < (Re) < 25$, fluids 3,4; △ Oil-polymer solutions $0.5 < (Re) < 25$, fluids 5–7; ○ Water-glycerol-polymer solutions $(Re) \sim 20$, fluids 8–10. The vertical bars are error bars on the loads. The horizontal bars show the range of Weissenberg number $\lambda V/h$. The Reynolds number $(Re) = \rho V h/\eta_0$ (from Tichy and Winer 1977).

6.7.5   Pure squeezing action

It has been argued that the elongational aspects of the squeeze-film should not be ignored. As a simplified model of this action we can suppose that a shear-less flow occurs between the platens, so that slip at the boundaries must occur. The flow is then easily shown to be a simple elongational flow with a stretching rate $(\partial w/\partial z)$ of $\dot{h}/h$, which is usually negative. The particles spend a variable time under the squeezing action, depending on their initial location in the gap. If we denote the original radial location of a particle by r_0, when the gap is h_0, then one can show by considering mass conservation that the gap thickness when the particle initially at r_0 reaches the exit or platen radius R is just $h_0(r_0/R)^2$, and thereafter the particle is 'free'. Therefore, if the platens approach one another at a constant speed $-\dot{h}$, the residence time of a particle initially at r_0 is just $h_0\{1-(r_0/R)^2\}/|\dot{h}|$, and the product of residence time and rate of stretching is $(h_0/h)\{1-(r_0/R^2)\}$. This can be of any magnitude which contrasts with the slider bearing problem where it is difficult to keep fluid in the bearing zone long enough to stretch the structure much.

In the case where the plates approach at a uniform speed the rate of squeezing increases with time, and the load at any given time depends on the history of the squeeze rate. To an order of magnitude, if we are given the Trouton viscosity η_E (Section 3.7) at the current squeeze rate $|\dot{h}|/h$, then one expects that this will give an overestimate of the stress. Then the squeeze load capacity W_S will be bounded by

$$W_S \lesssim \pi R^2 \eta_E(|\dot{h}|/h)\frac{|\dot{h}|}{h}, \tag{6.122}$$

where the equality sign will hold for inelastic fluids. The ratio of W_S to W_N, the Newtonian load capacity [eqn (6.112)] is then

$$\frac{W_S}{W_N} \lesssim \frac{2}{3}\left(\frac{\eta_E}{\eta}\right)\left(\frac{h}{R}\right)^2, \tag{6.123}$$

so that the elongational contribution will normally be insignificant compared with the viscous component when (h/R) is of order 10^{-2} or less, even if allowance is made for the reduction of η with shear rate. Therefore we shall ignore this component in our future calculations.

6.7.6   Unsteady overshoot action

The above discussion does not explain the several observations that viscoelastic squeeze films are stiffer than equivalent inelastic films. In numerical calculations with a squeeze-film flow of a non-linear convected Maxwell fluid [eqn (5.57)] a decrease in load capacity relative to the Newtonian case was found (Phan-Thien and Tanner 1983). All of the effects mentioned above are modelled in this calculation, including normal-stress effects and boundary condition changes.

We now consider the remaining factor in the problem, which is that the shearing action in the fluid is unsteady and therefore the 'overshoot phenomenon' [Fig. 3.8(b)] must be considered.

To begin, consider the kinematics of the flow field. It will be similar to the Newtonian field, and we will use this flow to illustrate the main points. The radial velocity (u) is much greater than the axial (w) velocity and the field can be written as

$$u = 3\frac{\dot{h}r}{h}\left[\left(\frac{z}{h}\right)^2 - \frac{1}{4}\right] \tag{6.124}$$

$$w = -2\dot{h}\left(\frac{z}{h}\right)\left[\left(\frac{z}{h}\right)^2 - \frac{3}{4}\right]. \tag{6.125}$$

Now it appears that a sharp peak occurs in a suddenly started shear flow only when the product of the characteristic time λ and the imposed shear rate ($\dot{\gamma}$) exceeds unity by a considerable margin. Examples of solutions given by Bird *et al.* (1977) show that overshoots of 50 per cent or so in stress occur when $\lambda\dot{\gamma}$ is about 10–100, and for $\lambda\dot{\gamma}$ of about 1000 overshoots up to 200 per cent have been observed. Low density polyethylene (Fig. 3.18) shows a similar behaviour. The maximum shear rate in the gap is, from 6.124, of order $3RV/h^2$ and thus we need

$$3\lambda RV/h^2 \gg 1 \tag{6.126}$$

to begin observing overshoot effects. In terms of the Deborah number $(De) = \lambda V/h$, we expect $(De) \gg h/R$ before effects are visible. In the experiments quoted above, Tichy and Winer (1978) seem to have observed viscoelastic effects when their Deborah number was in the range 0.5–1.0 and their R/h ratios were in the range 17–100. Grimm (1978) used (slightly differently defined) Deborah numbers from less than 0.5 up to 1000, with $R/h > 50$ and found increased stiffness only when his Deborah number exceeded unity. Since his Deborah number is about twice that of Tichy and Winer, the results are consistent, and if we take the truly relevant speed VR/h, then the criterion for increased stiffness seems to be approximately

$$\lambda VR/h^2 \gtrsim 100. \tag{6.127}$$

Thus the 'overshoot' explanation of increased load capacity presented here is plausible.

To illustrate this hypothesis, Phan-Thien *et al.* (1987) considered the response of a modified PTT model [see eqn (5.174)] in a squeezing flow. In order to obtain an overshoot it is necessary to set the parameter ξ (eqn 5.162) to be non-zero. The first peak in the shear stress following a suddenly started shear flow occurs (roughly) at $t_p = \pi/2\dot{\gamma}\sqrt{\xi}\,(2 - \xi)$. As $\dot{\gamma}$ and ξ increase ($\xi < 1$) the time to the first maximum decreases; the magnitude of the overshoot, defined as $\tau(t_p)/\tau(\infty)$

is given by

$$\tau(t_p)/\tau(\infty) = 1 + \exp(-2t_p/\lambda). \tag{6.128}$$

In this modified PTT model the effects of overshoot and shear thinning can therefore be studied separately. Results were calculated for $n = 0.56$, $\xi = 0.4$, $\varepsilon = 0$, and $\Gamma = 11.1\lambda$ [eqns (5.174) and (5.162)] for several Weissenberg numbers and are compared with inelastic models ($\lambda = 0$) of the Carreau and power-law types in Fig. 6.18; here $(Wi) = -\lambda\dot{h}_0/h_0$.

From the results we conclude:

1. There is a load enhancement which increases with (Wi) when $\xi \neq 0$, over and above the load of an inelastic fluid of comparable consistency.

2. There is an overshoot in the load capacity which increases with (Wi) and ξ.

3. At a fixed (Wi) and at long times where the initial dynamic effects are no longer important, the load appears to be independent of ξ, but there is still some load enhancement.

Finally, we compare the results for a *constant load* squeezing with the experiments of Leider (1974). In Fig. 6.19 we plot the dimensionless film thickness versus dimensionless time. In this figure the dimensionless quantities are $t_{\frac{1}{2}}/n\Gamma$ (abscissa) and $k(\pi R^2 m/WR^n)^{1/n}$ $(R/h_0)^{(n+1)/n}$ (ordinate), to compare with Leider (1974). The agreement is seen to be quite good, especially as one expects $N_2/N_1 = -\xi/2 = -0.1$ and their computed results are for $\xi = 0.2$.

The overshoot mechanism seems to be the only one among those considered which is a plausible explanation of observed load enhancement. Considering a constant rate squeeze situation, we have $t_{\frac{1}{2}} = -h_0/2\dot{h}_0$, and one expects significant overshoot if $\lambda > t_{\frac{1}{2}}$, or $(Wi) > 0.5$, so that there is a persistence of memory during the squeeze. This agrees with the data of Leider (1984) shown above, although the squeeze rate varied in those experiments.

For further computations on combined squeezing and sliding, see Phan-Thien *et al.* (1989), and for the question of film compressibility in ultrafast impacts, see Hirst and Lewis (1973).

6.8 Boundary-layer flows

In classical fluid mechanics where inertia forces in the fluid are, on average, large compared with viscous forces (high Reynolds number), the flow near solid surfaces often forms a (velocity) boundary layer, in which the fluid speed rapidly changes from the free stream speed to that of a solid surface. In the boundary layer the flow is nearly parallel to the wall, and has much in common with the lubrication flows discussed above.

In viscoelastic problems it is also possible to have stress boundary layers near surfaces, as in the bearing solution of Beris *et al.* (1983) described in Section 6.6.3, without a velocity boundary layer developing. See also Chapter 8 for another instance in connection with singularities in the flow of the upper convected

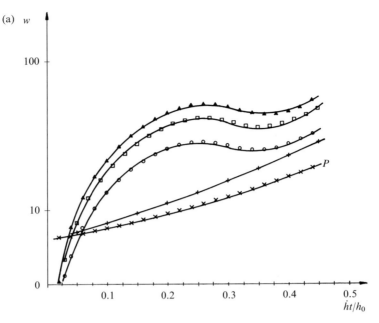

Fig. 6.18(a) Dimensionless load w versus dimensionless time for the MPTT model where $n = 0.56$, $\xi = 0.4$, $\varepsilon = 0$, $\Gamma = 11.1\lambda$. The Weissenberg number is defined as $-\lambda \dot{h}_0/h_0$. $\times$ Carreau model where $(Wi) = 1$; $+$ Newtonian model; $\bigcirc (Wi) = 1$; $\square (Wi) = 2$; $\triangle (Wi) = 3$. The solid line marked P is the power-law prediction [eqn (6.114)]. The dimensionless load is $W/(\pi\eta|\dot{h}|R^4/4h^3)$.

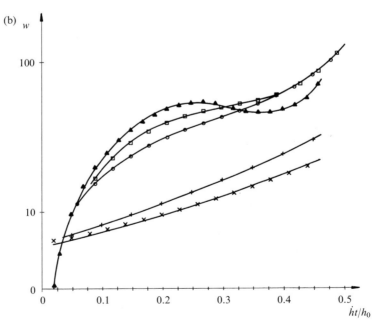

Fig. 6.18(b) Same as Fig. 6.18(a) but for $(Wi) = 3$ and different values of ξ. $\times$ Carreau model where $(Wi) = 1$; $+$ Newtonian model; $\bigcirc \xi = 0.2$; $\square \xi = 0.3$; $\triangle \xi = 0.4$.

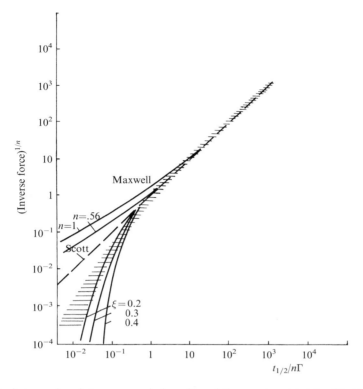

Fig. 6.19 A comparison between numerical results and the experimental data of Leider (1974). These data were collected with four different fluids and are displayed as the hatched region in the figure. The inverse force is normalized to agree with the Scott result (6.114), and Γ is the time constant defined in (5.174). With this scaling the Scott model plots as the line [ordinate $= t_{1/2}/n\Gamma$].

Maxwell model. For these stress boundary layers to develop, it is not necessary to have any inertia contributions.

In problems where elasticity and inertia are both important, one also has the possibility of generating and propagating waves (Joseph 1990). We now consider variations on classical boundary layer theory, where inertia is important.

Non-Newtonian materials often exhibit very high viscosities and hence the effective Reynolds number is frequently low. Thus the importance of boundary layer flows is not quite as great as in the Newtonian case. We begin with the usual steady two-dimensional equations of motion, in the absence of gravitational effects, and the continuity equation

$$\rho\left(u\frac{\partial u}{\partial x} + w\frac{\partial u}{\partial z}\right) = -\frac{\partial p}{\partial x} + \frac{\partial \tau_{xx}}{\partial x} + \frac{\partial \tau_{xz}}{\partial z}, \tag{6.129}$$

$$\rho\left(u\frac{\partial w}{\partial x} + w\frac{\partial w}{\partial z}\right) = -\frac{\partial p}{\partial z} + \frac{\partial \tau_{xz}}{\partial x} + \frac{\partial \tau_{zz}}{\partial z}. \tag{6.130}$$

Fig. 6.20 Co-ordinates for two-dimensional boundary layers.

The same order of magnitude arguments applied in lubrication flows will be valid here, so that $u \gg w$.

We assume that far from the surface of the body in question, the flow in the x-direction is described by the equation of motion for an inviscid fluid. This leads to the conclusion that the streamwise pressure gradient in the boundary layer can be found from solution of the inviscid equation of motion along the body surface. Thus if $U(x)$ denotes the x-component of velocity found from the inviscid equation of motion for $x \geq 0$, $z = 0$, we have

$$U\frac{\mathrm{d}U}{\mathrm{d}x} = -\frac{1}{\rho}\frac{\mathrm{d}p}{\mathrm{d}x} \qquad (6.131)$$

with respect to the coordinate system shown in Fig. 6.20. Since the boundary layer is assumed thin relative to the radius of curvature of the body, the coordinate system of Fig. 6.20 can be treated as if it were rectangular Cartesian. Reduction to dimensionless form and application of an order-of-magnitude argument lead to the appropriate non-Newtonian Reynolds number (Re) and the form of the two-dimensional boundary-layer equations.

6.8.1 The inelastic case

For the generalized Newtonian fluid, results can be obtained in some simple cases. In particular, for the power-law fluid, in practically all the cases where a similarity solution exists for the Newtonian case one can construct the corresponding non-Newtonian solution. By a similarity solution we mean here one in which the equations of motion and continuity are reduced to a single ordinary differential equation. Schowalter (1978) gives details for the general case in which there is a pressure gradient ($\partial p/\partial x$) in the free stream outside the boundary layer. To illustrate the method we shall consider only the case of a flat plate at zero incidence where the external stream has a constant speed U and $\partial p/\partial x$ is zero [eqn (6.131)]. Using the order of magnitude arguments used in lubrication theory (with the small boundary-layer thickness δ in place of h) we can reduce eqns (6.129) and (6.130) to their approximate boundary layer forms

$$\frac{\partial p}{\partial z} = 0 \qquad (6.132)$$

$$\rho\left(u\frac{\partial u}{\partial x} + w\frac{\partial u}{\partial z}\right) = -\frac{\partial p}{\partial x} + \frac{\partial \tau_{xz}}{\partial z}. \qquad (6.133)$$

In the case of the flat plate where the pressure is constant, we are left with (6.133) and the continuity equation to satisfy. If we introduce the stream function ψ so that

$$u = \frac{\partial \psi}{\partial z}, \tag{6.134}$$

$$w = -\frac{\partial \psi}{\partial x}, \tag{6.135}$$

then the equation of continuity (6.1) is satisfied.

For the power-law fluid in a unidirectional shear flow $(\partial u/\partial z > 0)$ we have

$$\tau_{xz} = k(\partial^2 \psi/\partial z^2)^n. \tag{6.136}$$

We now follow the Newtonian theory (Schlichting 1977) and let a dimensionless dependent variable η be defined as

$$\eta = z/ax^N, \tag{6.137}$$

where $a = ((n+1)k/U^{2-n}\rho)^N$ and $N = 1/(1+n)$.

In the Newtonian case $k/\rho = \nu$, the kinematic viscosity, $n = 1$ and $N = \frac{1}{2}$ we have the familiar result (Schlichting 1977) $\eta = z/\sqrt{2\nu x/U}$. We also define the stream function ψ in terms of η as

$$\psi = U\left(\frac{kx(n+1)}{\rho U^{2-n}}\right)^N F(\eta). \tag{6.138}$$

Substituting (6.136) into (6.133), setting $\partial p/\partial x = 0$, and using the definitions (6.137) and (6.138) we arrive at a single ordinary differential equation

$$nF''' + F(F'')^{2-n} = 0. \tag{6.139}$$

Boundary conditions for this equation are $u = w = 0$ on the plane $z = 0$ $(x > 0)$ and $u \to U$ as $z \to \infty$. In terms of the variables F and η, we have

$$u = UF'(\eta), \quad w = NU((n+1)k/\rho U^{2-n})^N x^{N-1}(\eta F' - F),$$

and therefore the boundary conditions for (6.139) are

$$F'(0) = 0, \quad F'(\infty) = 1, \quad F(0) = 0. \tag{6.140}$$

The solution of (6.139) must be found numerically. Solutions for velocity profiles and drag coefficients have been tabulated by Acrivos and co-workers (1960) over a range of n between 0.1 and 5.0. The details are similar to the Newtonian case and will not be exhibited; the solutions show the diffusion of vorticity into the fluid from the fixed surfaces. At the leading edge of the plate, the analysis shows that the boundary-layer thickness $\delta \to 0$, and that very high rates of strain occur. In such a region one cannot expect that the inelastic theory presented here is a useful approximation for viscoelastic fluids.

Care must be taken when attempting to compute a perturbation series solution for small elasticity. Serth (1973) discusses these problems. It is not clear that this approach yields a realistic solution near the leading edge of a flat plate.

Various other problems may be solved (Schowalter 1978). A problem which has attracted the continuing interest of workers in Newtonian fluid mechanics is the manner in which a laminar flow velocity profile is developed as the fluid enters a conduit from a reservoir. There has been a parallel development of the same class of problems for entry flow of non-Newtonian fluids, beginning with boundary-layer analyses of power-law fluids in two-dimensional channels and circular conduits. Experimental data are available which indicate that if the Reynolds number is sufficiently high so that there is negligible upstream diffusion of vorticity, and if the fluids are not appreciably elastic, then the inelastic analysis is useful. In Fig. 6.21 the theoretical predictions (Schowalter 1978) are shown for entry lengths of pseudo-plastic power-law fluids in tubes along with experimental results. Entry length x_e is defined as the length required for the centreline velocity to reach 98 per cent of its fully developed value, and Reynolds number is based on the pipe diameter.

Analyses of turbulent boundary layers are necessarily more empirical than those of the laminar counterpart. The situation with non-Newtonian fluids is similar. Wilkinson (1960) has developed correlations for turbulent boundary layers of power-law fluids flowing over flat plates. Adaption for flow through tubes shows good agreement with data. (See Chapter 10.) We now consider viscoelastic effects.

6.8.2   Results for viscoelastic fluids

Elastic effects can have a profound influence on fluid response in a region where a material element is undergoing a rapid change in deformation with time, such as

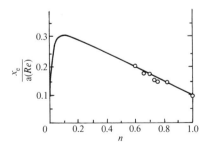

Fig. 6.21   Entrance length x_e as a function of flow behaviour index for power-law fluids. Solid line is theory, open circles are data. (Re) is based on the diameter $2a$ and is defined as $(Re) = (2a)^n \rho \bar{u}^{(2-n)}/k$ where $\bar{u}$ is the mean speed in the tube and k is the consistency (from Schowalter 1978).

near the leading edge of a flat plate or other body. Hermes and Fredrickson (1967) have studied the flow of several aqueous solutions of carboxymethylcellulose (CMC) in an approximation to uniform flow past a semi-infinite flat plate. Their results show that fluid elasticity does have a substantial effect on the velocity profile in the boundary layer (Fig. 6.22). Their work also casts doubt upon the validity of analyses which do not allow for important memory effects near the leading edge of the plate. Unfortunately, the data of Hermes and Fredrickson (1967) cannot be compared directly with theories because the elastic parameters of the CMC solutions were not measured.

Joseph (1990) has discussed these experiments in the light of his discussion on waves in viscoelastic fluids. He shows that there is a 'change of type' in the set of equations, from elliptic to hyperbolic, when the fluid speed exceeds a certain wave speed (Mach number > 1). When inertia effects are omitted, or a 'solvent' viscosity acts as well as the Maxwell component, there is, strictly speaking, no change of type.

Because of the difficult analysis required, the pipe-entry problem, well-treated for Newtonian fluids, has not been subjected to a comparable amount of analysis for viscoelastic fluids; see Eggleton et al. (1996).

Most of the interest in turbulent boundary layers of viscoelastic fluids has centred on the phenomenon of drag reduction. We treat this topic separately in Chapter 10. Some insight into the behaviour of viscoelastic fluids can be obtained by studying a simpler problem, the Rayleigh problem, which we now consider.

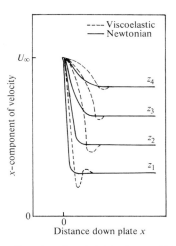

Fig. 6.22 Sketch of velocity profiles at various heights above a flat plate z; here $z_4 > z_3 > z_2 > z_1$. In the Newtonian profiles the decay from the free-stream speed U_∞ is monotonic as a particle moves away from the leading edge of the plate. Over- and undershoots are observed in the viscoelastic case.

6.8.3   The Rayleigh problem

Viscoelastic boundary-layer theory contains problems similar to those already encountered in lubrication. For the flat plate one can introduce an inertia factor, or an Oseen factor, $\rho c U_\infty (\partial u / \partial x)$, instead of the full term shown in eqn (6.133). Here c is a constant chosen so that the average speed $c U_\infty$ represents the inertia terms as well as possible; $c \sim 0.4$ is found to be appropriate in the Newtonian case. Then the non-linear Newtonian boundary layer problem for the flat plate can be replaced by the linear problem

$$\rho c U_\infty \frac{\partial u}{\partial x} = \frac{\partial \tau_{xz}}{\partial z}, \tag{6.141}$$

at least when the normal stress effects are neglected, as we do here.

If we consider, after Rayleigh, the one-dimensional problem where $u = u(z, t)$, $v = w = 0$, then the only non-trivial equation of motion is

$$\rho \frac{\partial u}{\partial t} = \frac{\partial \tau_{xz}}{\partial z}. \tag{6.142}$$

Hence, the steady boundary-layer dynamics may be studied approximately via this problem by replacing $x / c U_\infty$ by t, as originally suggested by Rayleigh for the Newtonian case. A similarity solution for the power-law case is possible (Bird *et al.* 1977), but we shall only consider the viscoelastic case. In solving (6.142) we shall need a constitutive model; specifically the non-linear convected Maxwell model (5.57) will be used. One finds from the constitutive model that provided $\tau_{yy} = \tau_{zz} = \tau_{zy} = \tau_{zx} = 0$ at some time then all these variables remain zero as time progresses. An equation containing τ_{xz} only emerges:

$$\lambda \frac{\partial \tau_{xz}}{\partial t} + \tau_{xz} = \eta \frac{\partial u}{\partial z}. \tag{6.143}$$

Elimination of τ_{xz} from (6.142) and (6.143) yields the equation

$$\frac{\partial u}{\partial t} + \lambda \frac{\partial^2 u}{\partial t^2} = \frac{\eta}{\rho} \frac{\partial^2 u}{\partial z^2}. \tag{6.144}$$

This equation exhibits damped wave solutions. If one has a boundary plane $z = 0$ suddenly set in motion with a speed U at time zero, the fluid in the half-space $z > 0$ being otherwise at rest, then it can be shown (Tanner 1962) that the solution for u is

$$\frac{u}{U} = \exp(-r/2) H(t^* - r) + \frac{r}{2} \int_0^{t^*} \frac{e^{-t/2}}{\sqrt{t^2 - r^2}} I_1(\tfrac{1}{2} \sqrt{t^2 - r^2})$$
$$\times H(t - r) \mathrm{d}t, \tag{6.145}$$

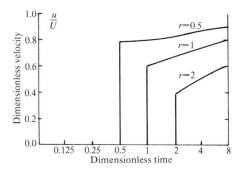

Fig. 6.23 Rayleigh problem for Maxwell fluid showing shear wave propagation. Dimensionless time is t/λ, dimensionless distance r is $z(\rho/\eta\lambda)^{1/2}$; U is the plate speed.

where

$$r = z\sqrt{(\rho/\eta\lambda)} \quad \text{and} \quad t^* = t/\lambda$$

are dimensionless variables and I_1 is the modified Bessel function of order one.

The solution is plotted in Fig. 6.23. The damped transverse wave progresses into the static fluid at a speed $\sqrt{\eta/\rho\lambda}$. Evidence of such waves were found in the solution of the squeeze-film problem by Phan-Thien and Tanner (1983), and one expects that the solution of the boundary layer problem will also show wave phenomena as in the experimental data of Fig. 6.22. Tanner (1962) also considered the case where the constitutive equation was of the Oldroyd type (eqn 4.13). These results amount to increasing the shear stress in (6.143) by the addition of a viscous term $\eta_\infty(\partial u/\partial z)$. Let the shear stress resulting from the Maxwell model (6.143) be $\tau_{xz} = \tau_m$. Then the Oldroyd shear stress is $\tau_0 = \tau_m + \eta_\infty(\partial u/\partial z)$. Substituting $\tau_m = \tau_0 - \eta_\infty(\partial u/\partial z)$ into (6.143) one finds an equation for τ_0:

$$\lambda\frac{\partial\tau_0}{\partial t} + \tau_0 = \eta_0\left\{\frac{\partial u}{\partial z} + \lambda'\frac{\partial^2 u}{\partial z\partial t}\right\}, \tag{6.146}$$

where $\eta_0 = \eta + \eta_\infty$ and $\lambda' = \lambda\eta_\infty/\eta_0$.

Elimination of the shear stress using (6.142) gives a diffusion equation and steep-fronted waves are not propagated. For details of the solution see Tanner (1962).

In summary, study of the Rayleigh problem gives useful insight into the development of viscoelastic boundary layers. However, it is not a complete model, since the boundary layer equations are non-linear; also, the lack of stress overshoot in the Maxwell and Oldroyd models must not be forgotten. See Joseph (1990) for further examples. For a discussion of stress boundary layers, see Chapter 8.

References

Acrivos, A., Shah, M. J. and Petersen, E. E. (1960). *Am. Inst. chem. Engrs J.*, **6**, 312.
Barwell, F. T. (1956). *Lubrication of bearings*. Butterworths, London.
Beris, A. N., Armstrong, R. C. and Brown, R. A. (1983). *J. Non-Newtonian Fluid Mech.*, **13**, 109.
Binding, D. M., Blythe, A. R., Gunter, S., Mosquera, A. A., Townsend P., and Webster, M. F. (1996). *J. Non-Newtonian Fluid Mech.*, **64**, 191.
Bird, R. B., Armstrong, R. C., and Hassager, O. (1977). *Dynamics of polymeric liquids*, Vol. 1, *Fluid mechanics*. Wiley, New York.
Carvalho, M. S. and Scriven, L. E. (1996). *J. Tribology*, **118**, 872.
Davies, M. J. and Walters, K. (1973). In T. C. Davenport (ed.) *Rheology of lubricants*. Applied Science Publications, London.
Dowson, D. (1962). *Int. J. mech. Sci.*, **4**, 159.
Eggleton, C. D., Pulliam, T. H., and Ferziger, J. H. (1996). *J. Non-Newtonian Fluid Mech.*, **64**, 269.
Floberg, L. (1957). *Trans. Chalmers Univ. Tech.* No. 189.
Granick, S., We Hu, H., and Carson, A. (1994) *Langmuir*, **10**, 3857.
Grimm, R. J. (1978). *Am. Inst. chem. Engrs J.*, **24**, 427.
Gunsel, S., Smeeth, M., and Spikes, H. (1998). *S.A.E. Technical Paper 982579*, Warrendale, Pennsylvania.
Hermes, R. A. and Fredrickson, A. G. (1967). *Am. Inst. chem. Engrs J.*, **13**, 253.
Hirst, W. and Lewis, M. G. (1973). *Proc. Roy. Soc.*, **A334**, 1.
Huang, X., Phan-Thien, N., and Tanner R. I. (1996). *J. Non-Newtonian Fluid Mech.*, **64**, 71.
Jabbarzadeh, A., Atkinson, J. D., and Tanner R. I. (1998). *J. Non-Newtonian Fluid Mech.*, **77**, 53.
Jacobson, B. O. (1991). *Rheology and elastohydrodynamic lubrication*. Elsevier, Amsterdam.
Joseph, D. D. (1990). *Fluid dynamics of viscoelastic liquids*, Springer-Verlag, New York.
Kennedy, P. (1995). *Flow analysis of injection molds*, Hanser, Munich.
Kiparissides, C. and Vlachopoulos, J. (1976). *Polymer Eng. Sci.*, **16**, 712.
Kistler, S. F. and Scriven, L. E. (1983). In *Computational analysis of polymer processing* (ed. J. R. A. Pearson and S. M. Richardson), p. 243. Applied Science Publications, London.
Leider, P. M. (1974). *I. and E.C. Fundamentals*, **13**, 342.
McClelland, M. A. and Finlayson, B. A. (1983). *J. Non-Newtonian Fluid Mech.*, **13**, 181.
Middleman, S. (1977). *Fundamentals of polymer processing*. McGraw-Hill, New York.
Milne, A. A. (1957). *Proc. Conf. Lubric. and Wear*. Inst. Mech. Engrs (Lond.), paper 102.
Phan-Thien, N. and Tanner, R. I. (1981). *J. Non-Newtonian Fluid Mech.*, **9**, 107.
Phan-Thien, N. and Tanner, R. I. (1983). *J. Fluid Mech.*, **129**, 265.
Phan-Thien, N., Jin, H., and Tanner, R. I. (1989). *Wear*, **133**, 323.
Phan-Thien, N., Sugeng, F., and Tanner, R. I. (1987). *J. Non-Newtonian Fluid Mech.*, **24**, 97.
Pinkus, O. and Sternlicht, B. (1961). *Theory of hydrodynamic lubrication*. McGraw-Hill, New York.
Raimondi, A. A. and Boyd, J. (1958). *Trans. Am. Soc. Lubric. Engrs*, **1**, 159.
Ruschak, K. J. (1982). *J. Fluid Mech.*, **119**, 107.
Savage, M. D. (1977). *J. Fluid Mech.*, **80**, 743.
Schlichting, H. (1977). *Boundary-layer theory*, 7th edn. McGraw-Hill, New York.
Schowalter, W. R. (1978). *Mechanics of non-Newtonian fluids*. Pergamon Press, Oxford.

Scriven, L. E. and Kistler, S. F. (1982). In *Finite element flow analysis* (ed. T. Kawai), p. 503, University of Tokyo Press.
Serth, R. W. (1973). *Am. Inst. chem. Engrs. J.*, **19**, 1089, 1275.
Tanner, R. I. (1960). *Int. J. mech. Sci.*, **1**, 206.
Tanner, R. I. (1962). *Z. angew. Math. Phys.*, **13**, 573.
Tanner, R. I. (1963). *Aust. J. Applied Sci.*, **14**, 129.
Tanner, R. I. (1965). *ASLE Trans.*, **8**, 179.
Tanner, R. I. (1969). *J appl. Mech.*, **36**, 634.
Tanner, R. I. and Williams G. (1971). *Rheol. Acta*, **10**, 528.
Tichy, J. A. and Winer, W. O. (1978). *J. Lubric. Technol.* (Am. Soc. Mech. Engrs), **100**, 56.
Wilkinson, W. L. (1960). *Non-Newtonian fluids*. Pergamon Press, Oxford.
Williamson, B. P., Walters, K., Bates, T. W., Coy, R. C., and Milton, A. L. (1997). *J. Non-Newtonian Fluid Mech.*, **73**, 115.
Zheng, R. and Tanner, R. I. (1988). *J Non-Newtonian Fluid Mech.*, **28**, 149.

Problems

1. Use lubrication theory to solve the stepped slider problem (Fig. 6.24) for a Newtonian fluid when $h_i/h_o = 2$ and $\varkappa = \frac{1}{2}$.

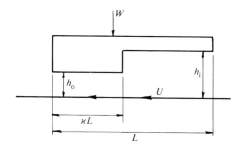

Fig. 6.24 Stepped slider. Problems 1–2.

Note: the pressure and the flux of fluid must be continuous at the step. Find the load and the friction forces.

2. Rayleigh showed that among all film shapes $h(x)$ the stepped bearing gave the highest load capacity for a fixed speed, length and outlet film thickness. Choose h_i/h_o and κ for optimal load capacity in the Newtonian case.

3. Show from dimensional considerations that for a journal bearing one can write a relationship of the form

$$\frac{h_{min}}{C} = f(S, L/D)$$

where

$$h_{min} = C(1 - \varepsilon).$$

C is the clearance, and S is the dimensionless Sommerfeld number $\eta(N/P)(R/C)^2$ where N is the speed of the bearing (in revs/s) and P is the mean load, (W/LD). This relationship can be conveniently presented on a single graph.

4. Repeat Question 3 for a power-law fluid and for a Maxwell fluid.

5. Consider the flow of a slightly compressible flow in a circular tube. Let $\rho = \rho_0(1+p/K)$, where p is the pressure. Find an expression for the pressure loss in a tube of length L, radius R when the total mass flow rate is $\dot{m}$. Suppose the viscosity η is constant.

6. Solve the Rayleigh (suddenly started plate) problem for a Newtonian fluid by introducing a dimensionless variable $\zeta = z\sqrt{(\rho/4\eta t)}$ and hence reducing the partial differential equation for u to an ordinary differential equation for $\phi = u/U$, finding

$$\phi'' + 2\zeta\phi' = 0.$$

The solution for u is then $u = U(1 - \mathrm{erf}\,\zeta)$.

7. Generalize this result to a power-law fluid and find a corresponding ordinary differential equation. Can this be integrated analytically?

8. Find an expression for the velocity in a film of a power-law fluid flowing down a plane surface making an angle β with the horizontal.

9. Two plates, very long and wide in the x and y directions are separated by a gap of width $2h$ in the z-direction. A small sinusoidal pressure gradient $-p'e^{i\omega t}$ acts in the x-direction.

Find an expression for the motion when the fluid is a Maxwell material. Show that the flow rate is given by (at low frequency) the real part of

$$Q = \left[\frac{2p'h^3 e^{i\omega t}}{3\eta}\left(1 + i\frac{\eta\omega}{G} + \cdots \mathrm{O}(\omega^2)\right)\right]$$

where ω is the pulsation frequency and G, η are fluid parameters.

10. Fluid is fed into the centre of the space between two parallel discs. Assume the radial velocity $u = f(z)/r$ (is this exact?) and the other velocity components are zero.

(a) Find the radial location of a fluid particle at time t' that will be at location r at time t. Calculate the C_{rz} component of the strain tensor.

(b) Solve the problem for a Newtonian fluid.

11. Consider the calendaring of a Newtonian sheet fed with stock of thickness H_f. Let $\lambda^2 = (H/h_0) - 1$. Show that the solution for the calendared thickness H obeys the relation

$$2(1+\lambda^2)\beta - (1-3\lambda^2)\left[\left(\frac{H_f}{h_0}\right)\beta + \tan^{-1}\left(\left(\frac{H_f}{h_0}\right)^2\beta\right)\right]$$

$$= -\frac{2\lambda}{1+\lambda^2} + (1-3\lambda^2)\left(\frac{\lambda}{1+\lambda^2} + \tan^{-1}\lambda\right)$$

where $\beta = \sqrt{(H_f/h_0 - 1)}/(H_f/h_0)^2$. Tabulate solutions of H_f/h_0 as a function of H/h_0.

12. A 1 per cent aqueous solution of carboxymethylcellulose (CMC) flows past a flat plate at zero incidence. The plate is 6 mm long (that is, along the direction of flow), and the total force measured per unit width of one side of the plate is $1.26\ N/m$. Approach velocity U_∞ of the CMC ($\rho \sim 1000\ kg/m^3$) is $1.1\ m/s$. A rheogram of the CMC solution is approximated by the power-law constants $n = 0.5$ and $k = 2.5(kg/m\,(s)^{2-n})$ (data adapted from Hermes and Fredrickson 1967).

For a power-law fluid, assume the following expression for the drag force per unit area τ_w at a distance x from the leading edge of the plate:

$$\frac{\tau_w}{\rho U_\infty^2} = c(n)|(Re)|^{-1/1+n}$$

where $c(0.5) = 0.58$ and $(Re) = \rho U_\infty^{2-n} x^n/(k)$. Compare the measured drag with the inelastic prediction. What is your conclusion?

7
FIBRE SPINNING AND FILM BLOWING

7.1 Almost-elongational flows

THE processes in this chapter are basically shear-free processes, in direct contrast to those discussed previously. We discuss fibre spinning and film blowing as examples. Constant strain rate elongational flows were discussed in Chapter 3. Much useful background information on these elongational flows can be found in Petrie's (1979) monograph. The steady, uniform rate almost-elongational theory corresponding to the almost-viscometric theory of Criminale, Ericksen, and Filbey is the Reiner–Rivlin theory (see Section 4.1). Many nearly-elongational flows are unsteady from the Lagrangian or particle-following viewpoint, however, and so the Reiner–Rivlin model is not very useful. A more elaborate theory has been discussed by Huilgol (Section 4.10) which contains five unknown kernels. These are difficult to find experimentally, and thus the use of some other definite constitutive model is usually necessary. The basic difference in the philosophy of approach to strong and weak flows lies in kinematics: it is easy to arrange that a particle in a viscometric flow spends a long time in the same kinematic environment, because the slip planes can be replaced by unstretched solid surfaces (see Chapter 3) and so the steady Criminale, Ericksen, and Filbey equation (4.32) is useful. By contrast, simple elongational flows generally need to be bounded by free surfaces to avoid shearing. It is practically difficult to devise a very long period of dwell in such a flow. Hence it is more important to consider (Lagrangian) unsteadiness in this case, and so the Reiner–Rivlin theory is not very useful. Before beginning on the case of fibre spinning, we first discuss some exact solutions which are useful for illustrating boundary condition problems.

7.1.1 Source and sink flows

We cite these simple flows (Fig. 7.1) to show that the development of infinite stresses does not necessarily follow the application of large strain rates, even with the Maxwell fluid model. We can consider either a point source or a line source. Detailed calculations will be made for the line source or sink, where the symmetrical velocity field has a radial component $u = q/2\pi r$, and the other components are zero. For sink (inward) flow, q is negative. The single momentum equation has the form (no body forces or accelerations are assumed)

$$\frac{\mathrm{d}\sigma_{rr}}{\mathrm{d}r} + \frac{\sigma_{rr} - \sigma_{\theta\theta}}{r} = 0. \tag{7.1}$$

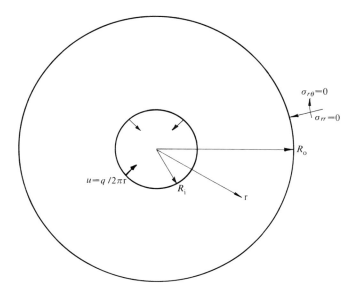

Fig. 7.1   Sink flow inwards to origin. A zero-traction boundary condition is applied at $r = R_o$.

For the second-order fluid [eqn (4.30c)] the response is easily found to be

$$\sigma_{rr} = \frac{\eta q}{\pi} \left\{ \frac{1}{R_o^2} - \frac{1}{r^2} \right\} + \Psi_1 \left(\frac{q}{2\pi} \right)^2 \left\{ \frac{1}{R_o^4} - \frac{1}{r^4} \right\}, \tag{7.2}$$

where the radial stress σ_{rr} is set equal to zero on $r = R_o$; this shows a sharp increase as r diminishes. It follows directly that

$$\sigma_{\theta\theta} = \frac{\eta q}{\pi} \left[\frac{1}{R_o^2} + \frac{1}{r^2} \right] + \Psi_1 \left(\frac{q}{2\pi} \right)^2 \left[\frac{1}{R_o^4} + \frac{3}{r^4} \right], \tag{7.3}$$

and $\sigma_{zz} = -p$, $\sigma_{rz} = \sigma_{r\theta} = \sigma_{\theta z} = 0$.
   The pressure is given by

$$p = -\frac{\eta q}{\pi R_o^2} - \frac{\Psi_1}{R_o^4} \left(\frac{q}{2\pi} \right)^2 + (\Psi_1 + 4\Psi_2) \left(\frac{q}{2\pi} \right)^2 r^{-4}. \tag{7.4}$$

The flow is plane (and also potential) so the pressure modification theorems (4.86) and (4.91) are applicable here. No extra boundary conditions beyond the Newtonian case (velocity components set on $r = R_i$, the inner radius, tractions set equal to zero on R_o) are needed for the complete solution, and no stresses approach infinite values unless R_i approaches zero.
   For the convected Maxwell fluid, eqn (5.57), we note the absence of shearing and set all shear stresses to zero. The velocity gradient tensor **L** is diagonal, with

components $(u', -u', 0)$ where $u' = -q/2\pi r^2$. Thus, the constitutive equations (5.57) reduce to

$$u\frac{d\tau_{rr}}{dr} - 2\tau_{rr}u' + \frac{\tau_{rr}}{\lambda} = \frac{2\eta}{\lambda}u'$$

$$u\frac{d\tau_{\theta\theta}}{dr} + 2\tau_{\theta\theta}u' + \frac{\tau_{\theta\theta}}{\lambda} = -\frac{2\eta}{\lambda}u' \qquad (7.5)$$

$$u\frac{d\tau_{zz}}{dr} + \frac{\tau_{zz}}{\lambda} = 0.$$

The solutions of (7.5) are found to be

$$\tau_{rr} = -\frac{\eta q}{\pi r^2} + \frac{C_r}{r^2}\exp\left(-\frac{\pi r^2}{\lambda q}\right) \qquad (7.6)$$

$$\tau_{\theta\theta} = \frac{2\eta}{\lambda}r^2\exp\left(-\frac{\pi r^2}{\lambda q}\right)\int\frac{dr}{r^3}\exp\left(\frac{\pi r^2}{\lambda q}\right) + C_\theta r^2\exp\left(-\frac{\pi r^2}{\lambda q}\right) \qquad (7.7)$$

$$\tau_{zz} = C_z\exp\left(-\frac{\pi r^2}{\lambda q}\right). \qquad (7.8)$$

For a sink flow, where q is negative, let $y = -\pi r^2/\lambda q$, so y is positive. Then (7.7) can be written as

$$\frac{\lambda}{\eta}\tau_{\theta\theta} = C'_\theta y e^y - [1 - y e^y E_1(y)], \qquad (7.9)$$

where $E_1(y) = \int_y^\infty (e^{-t}/t)\,dt$ (see Abramowitz and Stegun 1965, p. 228) and C'_θ is a dimensionless constant.

In terms of this variable (7.6) and (7.8) become

$$\frac{\lambda}{\eta}\tau_{rr} = C'_r e^y/y + y^{-1} \qquad (7.10)$$

$$\frac{\lambda}{\eta}\tau_{zz} = C'_z e^y. \qquad (7.11)$$

The pressure can also be found from (7.1):

$$p\frac{\lambda}{\eta} = \frac{C'_r}{2}\left(\frac{e^y}{y} + Ei(y)\right) - \frac{C'_\theta}{2}e^y + \frac{1}{2y} - \frac{e^y}{2}E_1(y) + C'_p, \qquad (7.12)$$

where $Ei(y)$ is the principal value of $\int_{-\infty}^y dt\,e^t/t$ (Abramowitz and Stegun 1965, p. 228) and C'_p is a constant. When we set in boundary conditions at $r \to \infty$ ($y \to \infty$) such that the fluid is 'relaxed' there, meaning that all stresses τ_{ij} are zero, we find that all the constants in (7.9)–(7.12) are zero.

Figure 7.2 shows a graph of τ_{rr} and $\tau_{\theta\theta}$ together with corresponding results for the second-order fluid. At large values of y (small Weissenberg numbers) the expressions are identical up to terms in y^{-2}; they differ markedly for small y, except for τ_{rr}, where they are identical (and also equal to the Newtonian result). Note that when $y < 2$ that $\tau_{\theta\theta}$ is of opposite sign to the strain-rate (u/r) for the second-order fluid; when $y < 1$ the local rate of dissipation of energy for this fluid is negative. By contrast, the more realistic Maxwell model is always well-behaved. The pressure curves (not shown) can be found in both cases by dividing the results for $\tau_{\theta\theta}$ by $-2y$.

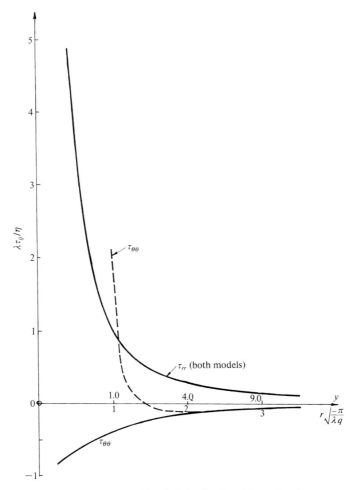

Fig. 7.2 Results for sink flow stresses. Flow is fully developed here; that is, stresses are zero at a very large value of r. Solid line, Maxwell model. Dashed line, second-order flow; in this case $\Psi_2 = 0$,
$$\lambda = \Psi_1/2\eta_0.$$

This flow could be useful for calibrating computer programmes. For this purpose, and also to show the effects of boundary conditions, we show the results in Fig. 7.3 for the case where the fluid is 'relaxed' (that is, $\tau_{ij} = 0$) at $r = 5$, with $q = -2\pi$, $\eta = 1$ and λ taking values of 0.1, 1, 10, and 100 respectively. Flow continues down to $r = 1$. The condition $\sigma_{rr} = 0$ at $r = 5$ is also imposed. Only the results for τ_{rr} are shown; the other variables show a similar behaviour. Note that for low Weissenberg numbers (small λ) the inlet 'boundary-layer' is small; it becomes dominant for $\lambda \gg 1$. For large λ the results are nearly

$$\tau_{rr} \sim \frac{\eta}{\lambda} \left\{ \left(\frac{R_0}{r} \right)^2 - 1 \right\} \tag{7.13}$$

provided $\pi r^2 / \lambda |q| \ll 1$.

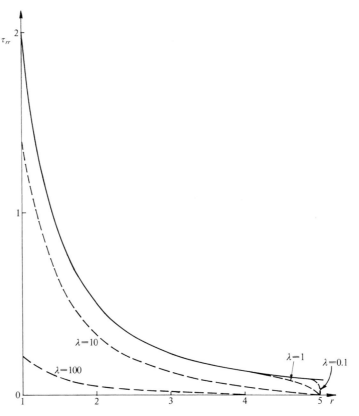

Fig. 7.3   Inlet boundary layers in sink flow problem for Maxwell fluid; the stress component τ_{rr} is shown. The solid line is fully-developed as in Fig. 7.2. The dashed lines correspond to the various time constants λ shown with fluid relaxed at $r = 5$. Here $u = -1$ for $r = 1$ and $\eta = 1$.

The results presented here show the need for extra boundary conditions (above the Newtonian and second-order cases) for elastic liquids. Note that the use of the integral form of the Maxwell model [eqn (5.62)] does not avoid this problem; one still has to put into the problem the value of the particle stress (or integral of strain) at entry to the flow region.

For the half-space ($x > 0$) with a wall ($x = 0$) and a line sink at the origin, Hull (1981) has solved the problem of the flow of a Maxwell fluid exactly.

7.2   Fibre spinning

The continuous stretching of viscous liquids to form fibres is a primary manu-facturing process for textiles and glass fibres (Ziabicki 1976). The melt spinning process for the manufacture of fibres is shown schematically in Fig. 7.4. Molten material is extruded through a small hole into cross-flowing ambient air at a temperature below the solidification temperature of the material. The solidified polymer or glass is wound up (taken up) on a reel moving at a higher speed than the mean extrusion velocity, resulting in thinning or drawing of the filament. The steady-state ratio of extrusion to take-up area is known as the draw ratio (D_R): the draw ratio for incompressible fluids is also equal to the steady-state ratio of take-up to extrusion velocity. Typical processing variables for the manufacture of poly(ethylene terephthalate) (PET) fibre are given by Denn (1980). From a spinneret (die) with holes of diameter 0.25–0.5 mm the fluid undergoes a draw ratio of about 200 before being taken up at 15–60 m/s in a solid amorphous

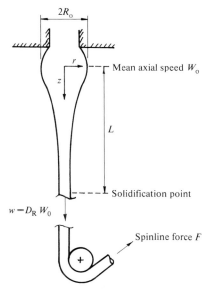

Fig. 7.4   Sketch of spinning process.

state. The solidified filament is typically subjected to further downstream processing for property development. A single spinneret, or spinning head, contains many holes, and the individual solidified filaments from each head are taken up together to form a yarn.

The primary goals of a rheological analysis are to be able to compute the stress and to define regions of instability where large diameter and property fluctuations might occur.

There is no mass transfer between the filament and the surroundings in the melt drawing process. Other fibre formation processes (wet spinning, dry spinning) require mass transfer to remove a solvent or to enable a chemical reaction to occur. We shall not pursue these problems here.

For other aspects of fibre formation, see the review of Denn (1980) and the volume on high-speed spinning of Ziabicki and Kawai (1985).

7.2.1 Experimental observations

Certain general qualitative experimental observations about continuous drawing of filaments can be made. The most significant is that the stress and rate of diameter attenuation depend strongly on the rheological properties of the liquid being drawn. Figure 7.5 shows results of laboratory experiments using a Newtonian corn syrup ($\eta = 25$ Pa-s) and a dilute solution of polyacrylamide in corn syrup (Denn 1980). The polymer solution has a nearly constant viscosity equal to

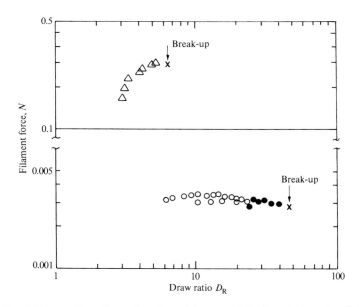

Fig. 7.5 Total filament force F as a function of draw ratio D_R for a Newtonian fluid (circles, *bottom*) and a non-Newtonian fluid (triangles, *top*) of almost equal viscosity. Filled circles indicate oscillations are taking place.

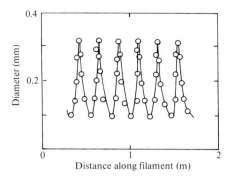

Fig. 7.6   Diameter variations in a drawn filament of poly(ethylene terephthalate).

that of the corn syrup, but it has measurable viscoelastic properties. The two liquids were drawn isothermally at the same extrusion velocity and filament or spinline lengths. The force required to draw the viscoelastic liquid is nearly two orders of magnitude greater than the force to draw the Newtonian liquid of equal viscosity at the same draw ratio. (The draw ratio, D_R, is defined in terms of spinneret area by some authors, and in terms of the maximum area by others. We use the latter definition throughout this chapter; see Chapter 8 for a discussion of swelling of extrudates). Similar large takeup forces have been recorded for laboratory experiments on a variety of polymer melts of commercial interest (Petrie 1979), but direct comparison with a comparable Newtonian liquid is not possible because of the more complex melt rheology. We note that most laboratory spinning experiments have been carried out at take-up speeds that are well below commercial conditions.

Sustained oscillations in drawn filament diameter and take-up force are sometimes observed when the drawing zone is maintained close to the extrusion temperature, followed by rapid cooling if solidification is to take place. Diameter fluctuations observed by Ishihara and Kase (1976) in a PET filament are shown in Fig. 7.6. This phenomenon, known as draw resonance, is not exclusively associated with viscoelasticity; sustained oscillations are observed in the continuous drawing of Newtonian fluids (Denn 1980), corresponding to the filled data points in Fig. 7.5. This phenomenon is considered later (Chapter 10).

7.2.2   Steady-state isothermal theory for inelastic fluids

We have already discussed elongational flows in Section 3.7, and some insight into fibre-spinning can be obtained from these studies. Figure 3.14 shows the increasing stiffness as elongation increases with time, and the high stress in the ultimate steady state. These features are important in spinning.

It is usual to base spinning analyses on average values of the velocity w and stresses $(\sigma_{zz}, \sigma_{xx})$ over the fibre cross-section, which varies in the spinning (z) direction. The resulting steady-state momentum equation is (referring to Fig. 7.4) easily shown to be

$$\frac{d}{dz}[\pi R^2(\sigma_{zz})] = \rho\pi R^2 w \frac{dw}{dz} + 2\pi R\tau_f - \pi R^2 \rho g - 2\pi\sigma\frac{dR}{dz}, \tag{7.14}$$

where z is the vertical drawing direction and R is the local filament radius; w is understood to be the average value of velocity over the filament section. The last three terms in eqn (7.14) represent air drag on the surface, gravity and surface tension, the first is the inertia term. We assume here and in the calculation of stresses from the constitutive equation that we can ignore variations in axial velocity over the cross-section (and hence the velocity components u and v will also be assumed to be small); the 'smallness' of dR/dz has not been made explicit. Air drag can be important but we shall not include it to begin with. The surface-tension and gravity terms are often not important in polymer melt drawing and will be ignored; thermal considerations will be postponed until Chapter 9. Inertia effects can be important; see Problem 7.3 and also Middleman (1977).

Equation (7.14) together with the steady-state equation of conservation of mass, $\rho\pi R^2 w = \text{constant}$ and an appropriate constitutive model, define the spinning problem. In most situations the flow rate will be specified at the beginning of the spinline and the take-up velocity at the end; the take-up force cannot be specified independently unless the take-up velocity or filament length is left unspecified.

When all of the terms on the right-hand side of eqn (7.14) are neglected, one finds the result

$$F = \pi\sigma_{zz}R^2 = \text{constant}, \tag{7.15}$$

and if the constitutive model is Newtonian:

$$\sigma_{zz} = -p + 2\eta\frac{\partial w}{\partial z}, \tag{7.16}$$

$$\sigma_{rr} = -p + 2\eta\frac{\partial u}{\partial r}. \tag{7.17}$$

We assume that the local state of flow is a simple elongation, so that

$$\frac{\partial u}{\partial r} = \frac{u}{r} = -\frac{1}{2}\frac{\partial w}{\partial z}. \tag{7.18}$$

This is compatible with the assumption of a uniform stress state in the cross-section. Ignoring the slope of the surface, on the free surface we have $\sigma_{rr} \approx 0$, and hence

$$p \approx 2\eta\frac{\partial u}{\partial r} \approx -\eta\frac{\partial w}{\partial z}. \tag{7.19}$$

From (7.19), (7.16), and (7.15) we find

$$\sigma_{zz} = 3\eta \frac{dw}{dz} = \frac{F}{\pi R^2},$$

(7.20)

where w, the average velocity, is not a function of r. Mass conservation is imposed on average, so that we have

$$q = \pi R^2 w.$$

(7.21)

Eliminating R between (7.20) and (7.21) gives

$$\frac{dw}{dz} = \frac{Fw}{3q\eta}.$$

(7.22)

Integrating and applying the boundary condition that $w = W_0$ at $z = 0$ gives

$$\frac{w}{W_0} = \exp(zF/3\eta q),$$

(7.23)

when $z = L$, w/W_0 equals D_R, the draw ratio. Hence $FL/3\eta q = \ln D_R$ and (7.23) can be written

$$\frac{w}{W_0} = \exp\left(\frac{z}{L}\ln D_R\right),$$

(7.24)

and

$$\frac{R}{R_0} = \exp\left(-\frac{z}{2L}\ln D_R\right).$$

(7.25)

This completes the isothermal Newtonian solution. Inelastic solutions can easily be generated, and will not be detailed. For the power-law fluid we find (Petrie 1979)

$$\frac{w}{W_0} = \left[1 + (D_R^m - 1)\frac{z}{L}\right]^{1/m}$$

(7.26)

where $m = 1 - 1/n$. Observations on spinning lines show that an extensional-thickening index ($n > 1$) would be needed to make theory and experiment agree even in materials which thin in shear ($n < 1$).

The proper initial conditions for the problem have not been definitely established, since the averaged one-dimensional equations given above are not applicable at the spinneret, but it has been assumed that the asymptotic equations apply from that point. A detailed study of the approach to the asymptotic equations starting from the spinneret has been made by Fisher et al. (1980) for isothermal drawing of a Newtonian liquid under conditions in which inertia, air drag, gravity, and surface tension can be neglected. The solution to the averaged equations in that case is given by eqn (7.24). The mean velocity (flow rate/area) obtained by a finite-element solution of the Navier–Stokes equations is shown in Fig. 7.7 for $d_0 F/3\eta Q$ ranging from 0.22 to 1.0. The point of maximum extrudate swell is always located less than one radius from the spinneret, and the asymptotic solution is reached within slightly more than one radius. Negligible stress variation over the cross-section also occurs at this distance from the die. Extrudate swell ranges from zero to six per cent for these calculations, compared with a free jet swell in the absence of drawing of 13 per cent (see Chapter 8) and the amount of swell decreases with applied force. It is likely that use of spinneret conditions as initial conditions for the asymptotic equations will cause little error under conditions of processing interest, where the length of the melt zone is very long relative to the spinneret diameter.

Agreement between theory and experiment is good in the Newtonian case: see Denn (1980).

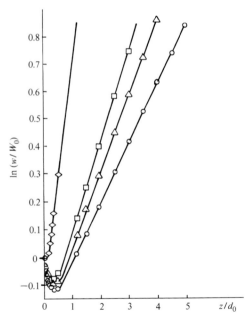

Fig. 7.7   Mean velocity profile development for low-speed isothermal spinning of a Newtonian liquid. The points are a finite-element solution and the straight lines are asymptotic solutions valid for large z/d_0. Values of parameter $d_0 F/3\eta Q$ for the various curves are ○ 0.22 △ 0.28 □ 0.33 ◇ 1.00.

7.2.3 Spinning of viscoelastic fluids

The failure of the inelastic model to be realistic in describing spinning suggests that more complex rheological constitutive models must be used in this essentially strong flow. As a particle traverses the fibre, it undergoes increasing stress and elongation rate, and hence the flow is unsteady in the Lagrangian sense. In the Newtonian velocity field eqn (7.24) the rate of elongation dw/dz is $w \ln D_R/L$, and a Weissenberg number of the form $\lambda\, dw/dz$ will often be very large.

A particle is in the fibre for a time $L(1 - D_R^{-1})/W_0 \ln D_R$ and hence the average Deborah number is, for $D_R \gg 1$, about $(\lambda W_0/L) \ln D_R$.

The strain of a small fibre lying parallel to the axis of stretching is just D_R, which is usually large. Hence we can expect considerable elongation of the macroscopic network structure and relevant constitutive models must behave correctly under these conditions.

One of the first non-linear viscoelastic models to be used was the convected Maxwell model, eqn (5.57) (see Denn 1980), and we now examine its response.

The calculation of the average stresses in terms of the gradient of mean velocity dw/dz requires the averaging of non-linear terms for any non-Newtonian fluid. The assumption has been made by most investigators that averages of products can be replaced by products of averages and the order of the errors so introduced is small. The Maxwell model, eqn (5.57) then becomes

$$\tau_{zz} + \lambda\left[w\frac{d\tau_{zz}}{dz} - 2\tau_{zz}\frac{dw}{dz}\right] = 2\eta\frac{dw}{dz}, \tag{7.27}$$

$$\tau_{rr} + \lambda\left[w\frac{d\tau_{rr}}{dz} + \tau_{rr}\frac{dw}{dz}\right] = -\eta\frac{dw}{dz}. \tag{7.28}$$

The equilibrium condition (7.15) may be rewritten

$$\frac{F}{\pi R^2} = \sigma_{zz} = -p + \tau_{zz}. \tag{7.29}$$

Similarly, from radial equilibrium σ_{rr} is zero, and hence

$$p = \tau_{rr}, \tag{7.30}$$

which may be used to eliminate p from (7.29). It is convenient to make the problem dimensionless in terms of the variables

$$v = w/W_0, \quad y = z/L, \quad T = \pi\tau_{zz}\,R_0^2/F, \quad P = \pi\tau_{rr}\,R_0^2/F, \quad \text{obtaining}$$

$$T - P = v, \tag{7.31}$$

$$T + (Wi)\left[v\frac{dT}{dy} - 2T\frac{dv}{dy}\right] = 2\varepsilon\frac{dv}{dy}, \tag{7.32}$$

$$P + (Wi)\left[v\frac{dP}{dy} + P\frac{dv}{dy}\right] = -\varepsilon\frac{dv}{dy}, \tag{7.33}$$

where the Weissenberg number $(Wi) = \lambda W_0/L$, and ε is the force ratio $\eta q/FL$. The dimensionless upstream boundary condition is $v(0) = 1$, and other boundary conditions will be needed. By using (7.31) in (7.33) we can eliminate P finding, with the help of (7.31)

$$v + v'(Wi)(2v - 3T) = 3\varepsilon v', \tag{7.34}$$

where $v' = dv/dy$.

Eliminating T from (7.32) by using (7.34) gives a single equation for v:

$$vv'' + 2(Wi)v'^3 - v'^2 - v'/(Wi) + 3\varepsilon v'^2/v(Wi) = 0, \tag{7.35}$$

which evidently needs two boundary conditions. In the limit $(Wi) \to 0$, the equation reduces to a first-order equation and only one boundary condition is needed; the solution is dimensionless terms is the Newtonian solution $v = C\exp(y/3\varepsilon)$. In the Maxwell case it will be necessary to specify $v'(0)$ [or, perhaps, $T(0)$].

Evidently, when $(Wi) \to 0$ one has a singular perturbation problem, as the v'' term vanishes.

To deal with this problem one can use the method of matched asymptotic expansions (Nayfeh 1973). First we rewrite (7.35) in terms of $\rho \equiv v'$ as dependent variable, and v as independent variable, so that the phase-plane trajectory equation is (if $\rho \neq 0$)

$$(Wi)v\frac{d\rho}{dv} + 3\varepsilon\rho/v - (Wi)\rho + 2(Wi)^2\rho^2 = 1. \tag{7.36}$$

The 'outer' solution is generated by expanding ρ in a series in (Wi):

$$\rho = \rho_0 + (Wi)\rho_1 + (Wi)^2\rho_2 + O(Wi^3). \tag{7.37}$$

The first term gives the Newtonian result $\rho_0 = v/3\varepsilon$; ρ_1 is zero, and ρ_2 is easily shown to be $-2v^3/27\varepsilon^3$. Integrating ρ to find $v(y)$ gives

$$v = A_0\exp(y/3\varepsilon) - \frac{(Wi)^2 A_0^3}{9\varepsilon^2}\exp(y/\varepsilon) + O(Wi)^3. \tag{7.38}$$

From (7.32) it follows that

$$T = \tfrac{2}{3}A_0\exp(y/3\varepsilon) + \frac{2(Wi)}{9\varepsilon}A_0^2\exp(2y/3\varepsilon) + O(Wi^2). \tag{7.39}$$

The outer solutions (7.38) and (7.39) are valid as $(Wi) \to 0$ with a fixed $y > 0$. To generate the 'inner' solution valid near an end we must involve the derivative term in (7.36). To do this we can use $Y = (\ln v)/(Wi)$ as a stretched variable. Then (7.36) becomes

$$\frac{\mathrm{d}\rho}{\mathrm{d}Y} + 3\varepsilon\rho\exp[-Y(Wi)] = 1 + (Wi)\rho + \mathrm{O}(Wi)^2. \tag{7.40}$$

Letting $(Wi) \to 0$ we find

$$\rho = \frac{1}{3\varepsilon} + \frac{B_0}{3\varepsilon}\exp(-3\varepsilon Y), \tag{7.41}$$

where B_0 is a constant of integration. In terms of the original variables,

$$\frac{\mathrm{d}v}{\mathrm{d}y} = \frac{1}{3\varepsilon} + \frac{B_0}{3\varepsilon}v^{-(3\varepsilon/Wi)}. \tag{7.42}$$

At the beginning of the fibre, $v = 1$, and thereafter it increases. Hence $(1 + B_0)/3\varepsilon$ is the initial value of $\mathrm{d}v/\mathrm{d}y$ for the system which may be prescribed as $v'(0)$. To the lowest order, we can replace v in (7.42) by the Newtonian value, integrate and find

$$v = 1 + y/3\varepsilon + (Wi)\{v'(0) - 1/3\varepsilon\}\{1 - \exp[-y/(Wi)]\}. \tag{7.43}$$

This solution has the properties that $v(0) = 1$ ($v'(0)$ is a free parameter) and that it is part of the Newtonian expansion when $(Wi) = 0$. As y increases, the exponential term dies away rapidly and the outer solution becomes relevant. A composite expansion can be formed which contains both (7.38) and (7.43) in the appropriate regions.

$$v = \left[1 + (Wi)\left(v'(0) - \frac{1}{3\varepsilon}\right) + \frac{(Wi^2)}{9\varepsilon^2}\right]e^{y/3\varepsilon}$$
$$- \frac{(Wi^2)}{9\varepsilon^2}e^{y/\varepsilon} - (Wi)\left(v'(0) - \frac{1}{3\varepsilon}\right)e^{-y/Wi}. \tag{7.44}$$

When y is of moderate size and $(Wi) \to 0$, we recover (7.38); when $y/\varepsilon \to 0$ and $(Wi) \to 0$ we find (7.43). This composite expansion is identical to that of Petrie (1979). To verify this note that the initial value of $T(\equiv T_0)$ appears in Petrie's work and we have the connection

$$T_0 = \tfrac{2}{3} - \frac{\varepsilon}{(Wi)}(3\varepsilon v'(0) - 1) + \mathrm{O}(Wi). \tag{7.45}$$

Therefore the order of the equations in (Wi) appears to change depending on the variables used; in terms of T_0 (7.44) is complete to $\mathrm{O}(Wi^2)$.

There seems to be no obvious way of completing the one-dimensional calculation without taking some more or less arbitrary value for T_0 or $v'(0)$. Considering the discussion of Section 7.1.1 (sink flow) one notes that the situation there is very similar to the present case; there we found that at low Weissenberg numbers the boundary layer at inlet died away very quickly and that the stress component $\tau_{\theta\theta}$ was always much smaller than τ_{rr} at high Weissenberg numbers. Hence at low (Wi) the choice of boundary condition is not critical, while at high (Wi) the stress orthogonal to the main stretching is small.

The value of $v'(0)$, or, equivalently T_0, that should be taken is therefore not immediately clear. One suggestion is to take the radial stress $P(0)$ to be zero; then, from (7.31) one has $T_0 = 1$. However, (7.33) is not satisfied in this case. Beris and Liu (1988) have suggested taking

$$\left(\frac{P}{T}\right)_0 = \frac{(Wi)(D_R - 1) - 1}{2 + (Wi)(D_R - 1)} \tag{7.46}$$

In this expression, for (Wi) equal to zero, the correct Newtonian value of the ratio (-0.5) is found, while when $(Wi) \to \infty$, since $D_R \to 1 + 1/(Wi)$ in this case, $P \to 0$ is the relevant boundary condition. (We discuss the high Weissenberg number limit in Section 7.2.4). Given the value of $(P/T)_0$, one can then complete the calculation. For example, if one takes $P = 0$ and $y = 0$, then $T_0 = 1$, and, from (7.34), $v'(0) = [(Wi) + 3\varepsilon]^{-1}$.

Inserting in (7.44) we get, to $O(Wi)^2$

$$v = e^{y/3\varepsilon} + \frac{(Wi)^2}{9\varepsilon^2}\{\exp[-y/(Wi)] - \exp(y/\varepsilon)\} \tag{7.47}$$

which completes the solution to $O(Wi)^2$.

7.2.4　Results for large Weissenberg numbers

The previous results were for small viscoelastic parameter values. When the force is large, so that $\varepsilon \to 0$, and (Wi) is large we look for asymptotic results. In this case, assuming $\varepsilon/(Wi) \to 0$ as (Wi) becomes very large, we see from (7.35) that $v \to$ constant. If we retain the next terms in the expansion, so that (7.35) becomes

$$vv'' + 2(Wi)v'^3 - v'^2 - \frac{v'}{(Wi)} = 0, \tag{7.48}$$

we may integrate to find the general solution. The solution is

$$y = c + v(Wi) + \frac{(Wi)}{b}\left[\frac{1}{2}\ln\left\{\frac{(x-1)^2}{x^2 + x + 1}\right\} - \sqrt{3}\tan^{-1}\left(\frac{2x+1}{\sqrt{3}}\right)\right], \tag{7.49}$$

where $x = bv$ and the constant b in the expression

$$b^3 = 1 + (Wi)/\varepsilon, \tag{7.50}$$

is found from the condition on $v'(0)$, eqn (7.46), (corresponding to $P=0$ at $y=0$) and the constant c is to be found by setting $v(0)=1$. Using these results and expanding in terms of $\varepsilon/(Wi)$, we find, as $(Wi) \to \infty$,

$$v = 1 + y(Wi)^{-1}[1 - 3\varepsilon/(Wi)] + O(Wi)^{-2}. \tag{7.51}$$

Thus for very large Weissenberg numbers the filament speed changes linearly along the threadline (Denn *et al.* 1975). We may also note from (7.51) that for large forces ($\varepsilon \to 0$) the draw ratio approaches a limiting value of $1 + (Wi)^{-1}$.

7.2.5 *Numerical results and comparison with experiments*

Equation (7.35) can be integrated by numerical methods, using as given initial conditions $v(0)$ and $v'(0)$, as a given boundary condition $v(1)$, and either (Wi) or ε being given. Figure 7.8(a) shows some results for a draw ratio of 20. In this case one has boundary conditions on velocity at both ends of the spinline, and an iterative search using a Runge–Kutta–Gill integration method was applied (Denn *et al.* 1975). Figure 7.8(b) shows the relation found between ε and (Wi) for this particular draw ratio. (Note: $T(0)$ is 1 for all curves in Fig. 7.8(a) except for the Newtonian curve where it necessarily has the value $2/3$, as demonstrated above). The asymptotic solutions for small (Wi) and small $\varepsilon/(Wi)$ are shown in Fig. 7.8(b).

To compare with experiment we consider the work of Phan-Thien (1978) who solved the problem for a Maxwell model and the PTT model with multiple relaxation times [see eqns (5.163–4)]. In this case the constitutive equations were fitted to data for low-density polyethylene (LDPE) and polystyrene and compared with spinning experiments. The results are shown in Fig. 7.9 and demonstrate that the PTT model is adequate for this flow; the major omission is the effect of temperature (Chapter 9).

It should perhaps be emphasized that the problem of initial conditions in the threadline was solved here by setting the radial stress for each relaxation time (eight were used) to zero. Care must be taken because the differential equations can become 'stiff' when a wide range of relaxation time are present.

Further discussion of the mathematical formulation is given by Beris and Liu (1988). This paper also surveys the state of approximation in the one-dimensional theory, and concludes that this simplification is useful and valid. They investigate the mathematical nature of the one-dimensional equations in the case of the convected Maxwell model. The problems of boundary conditions and the nature of time-dependent spinning problems are also discussed. Note that this analysis can also include temperature variations (Chapter 9).

The question of the stability of the drawing process will be considered in Chapter 10. Finally, we note the use of spinning-type experiments to characterize polymers (Wagner *et al.* 1988).

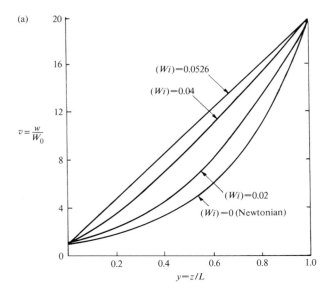

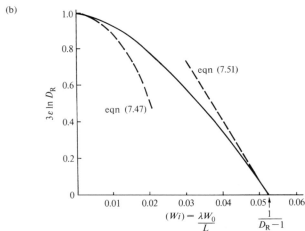

Fig. 7.8 Spinning of a Maxwell fluid. (a) Dimensionless velocity as a function of distance for various Weissenberg numbers $(Wi)(= \lambda W_0/L)$. $D_R = 20$, $T(0) = 1.0$. (b) Force parameter ε as function of Weissenberg number (Wi) for $D_R = 20$, $T(0) = 1.0$.

7.3 Film blowing and biaxial stretching

Spinning is but one example of a nearly shear-free process; certain parts of coating processes also involve free webs of material that are stretched without appreciable shear (Middleman 1977) and film blowing (Fig. 7.10) is also an important shear-free motion.

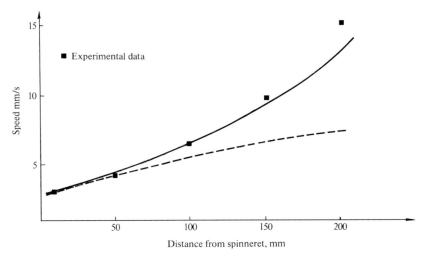

Fig. 7.9 Comparison of spinning theories and experimental data. Full line, PTT model, $\xi = 0.1$, $\varepsilon = 0.015$, [eqn (5.163)]; Dashed line, Maxwell model, eqn (5.57). Experimental data are for polystyrene at 170 °C. Eight time-constants were used in the model.

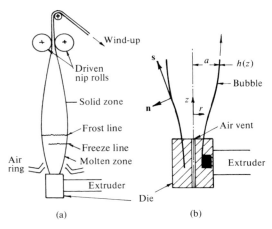

Fig. 7.10 Sketch of film-blowing process. (a) Section of whole proecss. (b) Enlarged section showing flow in die and in molten zone of bubble, with the intrinsic co-ordinate system (**n**,**s**). The air vent is used to regulate the internal pressure. The frost line is an optical effect due to material property changes with cooling; material is solid above freeze line.

A plane web-stretching operation is very similar to spinning; instead of an axisymmetric stretching motion the stretch is along one axis, the contraction is mainly at right angles to the plane of the sheet, but the third direction may also show contraction. Thus a theory may be built up along the lines indicated for spinning (see also Problem 7.8). It will not be detailed here.

Film blowing is a process whereby a thin tube of molten polymer is extruded through an annular die and is extended in both the radial and axial directions by a differential pressure Δp between the inside and outside of the tube and by mechanical tension in the axial direction. The tension is provided by a set of driven 'nip'-rolls, which also seal off the gas within the 'bubble'. A diagram of the arrangement is shown in Fig. 7.10; the air ring is a device sometimes used to 'stabilize' the bubble and to cause a rapid increase of heat transfer at the point where the turbulent stream of air from the ring strikes the bubble. Thicknesses from die to finished film are usually in the range of a 20–300 reduction ratio.

Inertia, surface tension, air drag and gravity forces will be neglected; due to the thinness of the membrane these are often realistic assumptions, but gravity and air drag are often significant in typical processes.

Considering Fig. 7.10, we now make the further assumption that the bubble is axisymmetric, at least until freezing sets in. Then the problem is conveniently referred to intrinsic coordinates described by the unit vectors i_s, i_t, and i_n respectively in the meridional, tangential, and normal directions [Fig. 7.11(a)] at the centreline of the sheet.

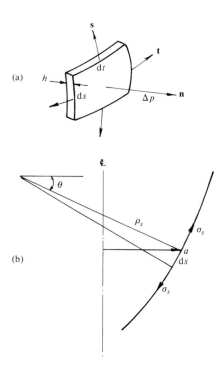

Fig. 7.11   Film geometry. (a) Element of film. (b) Bubble film radii.

Mass conservation demands that

$$q = 2\pi a h v_s = \text{constant,} \tag{7.52}$$

where v_s is the meridional velocity component, q is the total flow rate through the die, a is the local bubble radius and h is the local film thickness. Since the problem is axisymmetric, v_t is zero; v_n is not exactly zero since the film is of changing thickness, but it is negligible, in the same way that it is negligible in spinning and lubrication flows.

If we take the derivative of (7.52) with respect to s, the distance along the film, then we find

$$\frac{dv_s}{ds} = -\frac{1}{h} v_s \frac{dh}{ds} - \frac{1}{a} v_s \frac{da}{ds}. \tag{7.53}$$

The left-hand side of (7.53) is the rate of stretching along the film, and the two terms on the right-hand side of (7.53) are, respectively, the negatives of the stretch rates in the thickness (**n**) and the tangential directions (**t**).

The equilibrium of the material can now be considered. As in the kinematic section above, all quantities are averages over the thickness, and shears are zero. A small element of material (Fig. 7.11) is in equilibrium under a set of membrane forces. Let σ_t be the tangential stress and σ_s the meridional stress; shears are ignored. Then the equilibrium equation in the normal direction is found to be

$$\frac{\Delta p}{h} = \frac{\sigma_s}{\rho_s} + \frac{\sigma_t}{\rho_t} \tag{7.54}$$

where ρ_s and ρ_t are the principal curvatures in the two directions s and t and Δp is the internal pressure measured relative to the external (atmospheric) pressure. Explicitly,

$$\left.\begin{array}{l} \rho_s = -\{1 + (da/dz)^2\}^{3/2} / \dfrac{d^2 a}{dz^2} \\[2mm] \rho_t = a(1 + (da/dz)^2)^{1/2} \end{array}\right\} \tag{7.55}$$

where $a(z)$ is the local bubble radius.

In order to find a further equilibrium statement we may consider a force balance on a plane z. Then

$$-\pi a^2 \Delta p + 2\pi a h \sigma_s \cos\theta = T, \tag{7.56}$$

where $\cos\theta = \{1 + (da/dx)^2\}^{-1/2}$, and T is the (constant) total force on the shell. When T is known, for a given shell, then σ_s is determined from (7.56), σ_t can be found from (7.54), and the stresses are statically determined, irrespective of the constitutive equation. Equation (7.56) ignores inertia, air drag, gravity, and surface tension.

7.3.1 Newtonian solution

In this problem the stresses in the **s**, **t**, and **n** directions are the principal stresses. In general, the film stresses will be much larger than the pressure difference Δp; it is easy to see from (7.54) that the stresses are of order $(a\Delta p/h)$, so that $\sigma_s \gg p$, since $a \gg h$. Hence, to this order of accuracy, the through-thickness principal stress is zero. We use this to evaluate the pressure in the film

$$p = 2\eta \frac{v_s}{h} \frac{dh}{ds} = 2\eta \frac{v_s}{h} \frac{dh}{dz} \cos\theta, \tag{7.57}$$

since

$$dz = ds \cos\theta. \tag{7.58}$$

One can now substitute (7.57) into the Newtonian constitutive law for the stresses, thereby eliminating the pressure as a variable. We find, after eliminating v_s by using (7.52)

$$\sigma_s = -\frac{\eta q \cos\theta}{\pi a h} \left\{ \frac{1}{a} \frac{da}{dz} + \frac{2}{h} \frac{dh}{dz} \right\} \tag{7.59}$$

$$\sigma_t = \frac{\eta q \cos\theta}{\pi a h} \left\{ \frac{1}{a} \frac{da}{dz} - \frac{1}{h} \frac{dh}{dz} \right\} \tag{7.60}$$

These relations can be substituted into eqns (7.54) and (7.56) to yield a pair of (non-linear) ordinary differential equations for h and a. We now suppose that all quantities are made dimensionless. The mean inlet bubble radius a_0 is chosen as a unit of length, so that $r = a/a_0$, $x = z/a_0$. h is left dimensional, it can also be scaled with respect to h_0, the thickness at $z = 0$. We also define the dimensionless quantities $B = \pi a_0^3 \Delta p/\eta q$ and $T^* = Ta_0/\eta q$. Then (7.56) reduces

$$\frac{h'}{h} = -\frac{1}{2} \frac{r'}{r} - \frac{1}{4}(T^* + r^2 B) \sec^2\theta, \tag{7.61}$$

where $h' = dh/dx$, etc. By using this relationship in the equilibrium equation (7.54) to eliminate h'/h, we can obtain an equation containing only r:

$$2r^2(T^* + r^2 B)r'' = 6r' + r(T^* - 3r^2 B)(1 + r'^2). \tag{7.62}$$

This equation may be integrated to find the bubble shape. For boundary conditions we have $r = 1$ at $x = 0$ and provided the film 'freezes' at $x = x_f$, then we can set $r' = 0$ there: the tube will be parallel for $x > x_f$. Pearson and Petrie (1970) have discussed this boundary condition more fully. They also solved the problem numerically by integrating (7.62) backwards from an origin at the freeze-line.

In this process the value of r_f is guessed and used with the condition $r' = 0$ at the freeze-line. The freeze-line radius r_f is then adjusted until the desired radius ($r = 1$) is reached at $x = 0$. The dimensionless pressure B, the dimensionless force T^* and the freeze-line distance x_f then specify the problem completely.

It is not possible to have totally unrestricted values of the parameters; in particular one assumes $B > 0$ (positive internal pressure). Inspection of (7.62) shows that if $T^* + r^2 B$ becomes zero at any x then the curvature r'' must become very large: a 'kink' develops in the film. To avoid this we suppose that the tension applied to the bubble is sufficient to avoid this problem. This may be guaranteed by setting the dimensionless total force T^* to be greater than $-B$; if the force F applied to the film is used, then $T = F - \pi a^2 \Delta p$, and in terms of F we must have

$$F > \pi \Delta p a^2 (r_f^2 - 1). \tag{7.63}$$

When the value of T is zero, a rapid blow-up is obtained which is uncharacteristic of the practical process, so only the case $T^* > 1$ is to be considered. In the solution one can choose either r_f, B or T^* arbitrarily: once fixed, the solutions to the problem will be curves in the $(r_f, h_0/h)$ plane along which the other two parameters will be constrained. Figure 7.12 shows the results for $x_f = 10$. In this solution $B = 0.2$, $T^* = 2.0$; here the final blow-up ratio is 2.7. Further details of the solution procedure are given by Pearson and Petrie (1970). The long-necked bubble shapes produced by this analysis are not close to typical experimental shapes for polymeric fluids. Therefore we now consider viscoelastic effects.

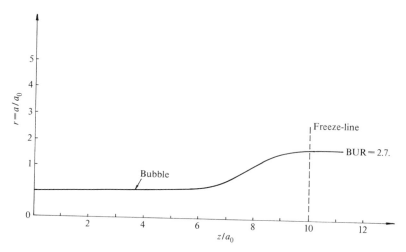

Fig. 7.12 Typical long-necked isothermal Newtonian bubble. Freeze-line is at $z/a_0 = 10$, $B = 0.2$, $T^* = 2.0$. Blow-up ratio (BUR) is 2.7.

7.3.2   Non-Newtonian solution

It is not difficult to include the effect of gravity in the Newtonian film-blowing analysis (see Problem 7.7). Major omissions from the previous section are the viscoelastic behaviour and thermal effects. We now consider the solution for the convected Maxwell model [eqn (5.57)].

Let the dimensionless stresses in the **s** and **t** directions be S and C respectively, where

$$S = a_0 \sigma_{ss}/\eta_0 V_0, \quad C = a_0 \sigma_{tt}/\eta_0 V_0. \tag{7.64}$$

Here the subscript zero refers to the beginning of the bubble; in terms of the volumetric flow rate q we have

$$V_0 = q/2\pi a_0 h_0, \tag{7.65}$$

where h_0 is the inlet film thickness. The linear distances are to be made dimensionless with respect to a_0; exceptionally the dimensionless thickness (h) is in terms of h_0; here $h = \text{thickness}/h_0$. The parameters B and T^* (dimensionless pressure difference and film tension respectively) are as defined in the Newtonian case (Section 7.3.1).

The equilibrium equation (7.56) reduces to

$$S = (T^* + Br^2)\sqrt{1 + r'^2}/rh, \tag{7.66}$$

and (7.54) reduces to

$$(T^* + Br^2)r'' = hC\sqrt{1 + r'^2} - 2Br(1 + r'^2), \tag{7.67}$$

where (7.66) has been used to eliminate S. The through-thickness stress σ_{nn} is nearly zero, and hence the extra stress $\tau_{nn} = \sigma_{nn} + p^*$ is nearly equal to the pressure p^*. We assume there are no shear stresses, hence the dimensionless extra stresses τ_{ss} and τ_{tt} are connected to S and C by

$$\begin{aligned} \tau_{ss} &= S + p \\ \tau_{tt} &= C + p, \end{aligned} \tag{7.68}$$

where $p = a_0 p^*/\eta V_0$ is the dimensionless pressure.

If the speed is made dimensionless with respect to V_0, and we define a Weissenberg number $(Wi) = \lambda V_0/a_0$, then the dimensionless constitutive equations can be written as

$$(Wi)p' = 2\frac{h'}{h}[(Wi)p + \mu] - prh\sqrt{1 + r'^2}, \tag{7.69}$$

which follows directly from the τ_{nn} equation,

$$(Wi)C' = 2(Wi)C\frac{r'}{r} - rhC\sqrt{1+r'^2} + 2\left(\frac{r'}{r} - \frac{h'}{h}\right)(\mu + (Wi)p), \qquad (7.70)$$

which is obtained from the hoop stress equation and uses (7.69), and, finally by using (7.69), (7.66), and (7.67) one can reduce the streamwise equation to

$$\frac{h'}{h}\{(Wi)S + 4[\mu + (Wi)p]\} = -rhS\sqrt{1+r'^2}$$
$$-\frac{r'}{r}[(Wi)(S+C) + 2(Wi)p + 2\mu]. \qquad (7.71)$$

In these equations $\mu = \eta/\eta_0$; this ratio is unity for isothermal flow. These equations can then be arranged in the form of five coupled equations

$$y_i' = f_i(y_i, x) \quad (i = 1, 5),$$

to solve for the five y_i variables r, r', C, p, and h. Note that S may be found from (7.66) without integration. In addition to the Newtonian boundary conditions $r = 1$, $h = 1$ at $x = 0$, $r' = 0$ at $x = x_f$, we need two additional conditions on p and C. Following the discussion on the spinline problem, we shall assume that $p = C = 0$ at $x = 0$.

Integration from $x = x_f$ backwards towards $x = 0$ is highly unstable and hence integration from $x = 0$ forwards is preferred. Luo and Tanner (1985) have used a fourth-order Runge–Kutta program to integrate in this manner using an initial guess for $r'(0)$. Even so, the process is unstable for comparatively low Weissenberg numbers. Results are shown in Fig. 7.13. The freeze-line occurs

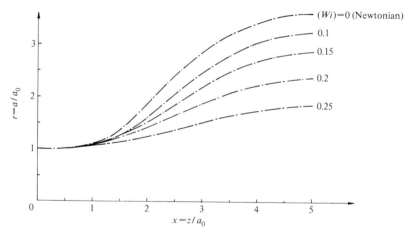

Fig. 7.13 Effect of Weissenberg number $(Wi) = \lambda V_0/a_0$ on bubble shape for Maxwell fluid; $T^* = 2.9$, $B = 0.2$, Freezeline at $z = 5a_0$.

at $x_f = 5a_0$ in these solutions and the values of T^* and B were 2.9 and 0.2 respectively. The effect of viscoelasticity is to *decrease* the ultimate bubble radius. This is expected due to the extra stiffness of the Maxwell fluid in stretching motions. Hence the resulting bubbles are not close to experimentally observed shapes (Petrie 1975).

These computations have been confirmed by André *et al.* (1998) and by Beaulne and Mitsoulis (1999) using a finite element method. These authors also survey the literature usefully.

Petrie (1975) has considered an elastic membrane theory of bubble shapes; this was regarded as the limit as $(Wi) \to \infty$ of the Maxwell model. He also points out the very great importance of temperature in this problem. We shall consider temperature effects in general in Chapter 10; here we shall consider the application to the blown film.

7.3.3 *Maxwell fluid with temperature variation*

A model of the heat-loss process for a film element is shown in Fig. 7.14. The loss from the film is via a heat transfer coefficient h_t which occurs over an area $2\pi a\,ds$ and over a temperature difference $T - T_a$, where T is the local film temperature and T_a is the ambient temperature; h_t can vary from position to position, depending on temperature, air speed over the outside and other factors. We shall assume it is constant to begin with. The heat transfer coefficient as defined is assumed to take care of losses by convection and radiation from both sides of the film; we shall consider $T(s)$ to be the average film temperature; conduction along the film is also neglected. A heat balance on the element then gives the relation

$$\rho C_p q \frac{dT}{ds} + 2\pi a h_t (T - T_a) = 2\pi a h \left\{ \sigma_{ss} \frac{dv}{ds} + \sigma_u \frac{v\,da}{a\,ds} \right\}. \tag{7.72}$$

The dissipation terms on the r.h.s. of (7.72) may often be neglected. We define a dimensionless temperature θ,

$$\theta = \frac{T - T_a}{T_0 - T_a}, \tag{7.73}$$

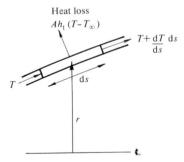

Fig. 7.14   Heat balance for film element.

and dimensionless coefficients C_H and C_E where

$$C_H = 2\pi a_0^2 h_t / \rho C_p q$$

$$C_E = \eta_0 q / 2\pi a_0^2 h_0 \rho C_p (T_0 - T_a)$$

(7.74)

Here C_p is the specific heat at constant pressure. In dimensionless form

$$\theta' = -C_H r \theta \sqrt{1 + r'^2} + C_E \left[C \frac{r'}{r} - S \left(\frac{h'}{h} + \frac{r'}{r} \right) \right].$$

(7.75)

With an initial condition $\theta(0) = 1$, (7.75) can be included in the set of equations for the film. It is supposed that the viscosity ratio $\eta/\eta_0(\equiv \mu)$ is a function of temperature:

$$\frac{\eta}{\eta_0} = \exp A \left\{ \frac{1}{T} - \frac{1}{T_0} \right\}$$

(7.76)

and also we let

$$\lambda = \frac{\eta(T)}{G(T)} = \lambda(T),$$

(7.77)

where

$$G = G_0 + \alpha(T - T_0).$$

(7.78)

After reducing (7.76)–(7.78) to dimensionless terms one can integrate to find the bubble shapes. Luo and Tanner (1985) have done this for the data of Gupta (1980). The values for Styron 666 that were used are $\rho = 1050 \text{ kg/m}^3$, $C_p = 1710 \text{ J/kg K}$, $T_0 = 443 \text{ K}$, $\eta_0 = 8.8 \times 10^4 \text{ Pa-s}$, $A = 18\,900 \text{ K}$, $G_0 = 200 \text{ kPa}$, $\alpha = -2500 \text{ Pa/K}$. The average heat-transfer coefficient h_t was found by matching temperatures at the freeze-line, and was found to be about $4 \text{ W/m}^2 \text{ K}$. Comparison of bubble shapes for run 18 (Gupta 1980) are shown in Fig. 7.15(a). The temperature profiles are shown in Fig. 7.15(b).

Although the agreement is not exact, we believe that the inclusion of temperature modelling is a great improvement over the isothermal case. In many other cases this is also true and we devote Chapter 9 to thermal and pressure effects.

7.3.4 Further analyses

In the above analyses the 'freeze' line is imposed by assuming a constant diameter bubble from a certain point on. Cao and Campbell (1990) added instead a yielding process; the 'yield strength' was assumed to be a function of temperature and deformation rate; essentially a strain-hardening mechanism is assumed. They

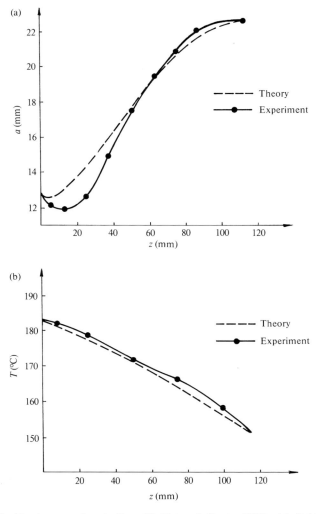

Fig. 7.15 Film-blowing experiment, Run 18 (data of Gupta 1980). (a) Bubble shape. (b) Temperature profile.

claim improved agreement with the experiments of Gupta (1980). It seems that this emphasis on the physics of the film near the freeze point is highly appropriate and needs to be continued. These authors (Campbell *et al.* 1992) also looked at the effect of aerodynamic drag on the bubble shape; the flow over the bubble can result in a variable differential pressure, and consequent changes in bubble shape.

7.4 Other extensional processes

Other processes which are nearly shearfree include extrusion blow moulding (Hensen 1988), where internal pressure in an extruded tube is made to fill a die; stretch/blow moulding (Schmidt *et al.* 1996) and film casting, which is similar to a two-dimensional version of fibre-spinning kinematically, but is more complex due to edge effects (Iyengar and Co 1993).

References

Abramowitz, M. and Stegun, I. A. (1965). *Handbook of mathematical functions.* Dover, New York.
André, J.-M., Agassant, J.-F., Demay, Y., Haudin, J.-M., and Monasse, B. (1998). *Int. J. Forming Processes,* **1**, 187.
Beaulne, M. and Mitsoulis, E. (1999). *Int. J. Forming Processes,* **2**, 41.
Beris, A. N. and Liu, B. (1988). *J. Non-Newtonian Fluid Mech.,* **26**, 341.
Campbell, G. A., Obot, N. T., and Cao, B. (1992). *Polymer Eng. and Sci.,* **32**, 751.
Cao, B. and Campbell, G. A. (1990). *AIChE J.,* **36**, 420.
Denn, M. M. (1980). *Ann. Rev. Fluid Mech.,* **12**, 365.
Denn, M. M., Petrie, C. J. S., and Avenas, P. (1975). *Am. Inst. chem. Engrs J.,* **21**, 791.
Fisher, R. J., Denn, M. M., and Tanner, R. I. (1980). *Ind. Eng. Chem. Fund.,* **19**, 195.
Gupta, R. K. (1980). Ph.D. Thesis, University of Delaware.
Hensen, F. (ed.) (1988). *Plastics extrusion technology.* Hanser, Munich.
Hull, A. M. (1981). *J. Non-Newtonian Fluid Mech.,* **8**, 327.
Ishihara, U. and Kase, S. (1976). *J. appl. Polymer Sci.,* **20**, 169.
Iyengar, V. R. and Co, A. (1993). *J. Non-Newtonian Fluid Mech.,* **48**, 1.
Luo, X.-L. and Tanner, R. I. (1985). *Polymer Engr. Sci.,* **25**, 620.
Middleman, S. (1977). *Fundamentals of polymer processing.* McGraw-Hill, New York.
Nayfeh, A. (1973). *Perturbation methods.* Wiley, New York.
Pearson, J. R. A. and Petrie, C. J. S. (1970). *J. Fluid Mech.,* **92**, 609.
Petrie, C. J. S. (1975). *Am. Inst. chem. Engrs. J.,* **21**, 275.
Petrie, C. J. S. (1979). *Elongational flows.* Pitman, London.
Phan-Thien, N. (1978). *J. Rheol.,* **22**, 259.
Schmidt, F. M., Agassant, J. F., Bellet, M., and Desoutter, L. (1996). *J. Non-Newtonian Fluid Mech.,* **64**, 19.
Wagner, M. H., Bernnat, A., and Schulze, V. (1998). *J. Rheol.,* **42**, 917.
Ziabicki, A. (1976). *Fundamentals of fibre formation.* Wiley, London.
Ziabicki, A. and Kawai, H. (1985). *High-speed Fiber Spinning.* Wiley, New York.

Problems

1. Solve the source/sink flows in the following cases.
(a) Line sink—Newtonian plus inertia.
(b) Point sink—Inertia-less second-order flow.
(c) Point sink—Inertia-less Maxwell flow.
(d) Line source, Maxwell fluid.
Comment on boundary conditions and boundary layers as appropriate.

2. Find the steady-state spinning solution for an isothermal power-law fluid [eqn (7.26)].

3. Consider the Newtonian spinning problem with inertia added and find an approximate solution in this case. Detail your assumptions carefully.

4. Set up the inertia-less spinning equation for an Oldroyd model where the deviatoric stress τ_{ij} is given by

$$\tau_{ij} = \tau_{ij}^{(M)} + 2\eta_\infty d_{ij}$$

where $\tau_{ij}^{(M)}$ is the Maxwell stress, eqn (5.57) and η_∞ is a constant viscosity. Use a computer to produce the solution for this case. What boundary conditions do you need?

5. In Problem 4 find the result when the appropriate Weissenberg number $(\lambda V_0/L)$ becomes very large.

6. Derive eqn (7.49).

7. Set up the Newtonian film-blowing equations with gravity present. Solve them by a suitable numerical scheme.

8. Consider a series of stretching motions in a Maxwell fluid [eqn (5.57)], as follows:

(a) Equal biaxial stretching along two axes at right-angles. Show that this is equivalent to uniaxial compression and find the Trouton viscosity.
(b) Stretching along one axis is inhibited. Find the effective viscosity for stretching.

9. Set up a computer program to integrate (7.62) and (7.61). Investigate integrating in the forward and backward ways along the film. Which is more economical?

8
COMPUTATIONAL RHEOLOGY AND APPLICATIONS

8.1 Computational rheology

The mathematical complexity of rheological problems forces one to use computation for the effective solution of many problems arising in practice. That this is the case will be clear from studying the previous chapters, where simple flow fields were studied. For example, in fibre spinning one is led to numerical methods to determine the flow at exit from the spinneret (Fig. 7.7) even for a Newtonian fluid; the power-law analysis of the Hele–Shaw flow [Section 6.6.5, eqn (6.104)] is also generally beyond simple analytical techniques.

The relatively new field of computational rheology (see Tanner and Walters 1998) has developed two branches:

(a) Given the mass and momentum balances (Chapter 2), a choice of constitutive equation (Chapters 4 and 5) and boundary conditions appropriate to the flow of interest, find a numerical solution of the problem;

(b) Beginning with a microstructural concept, use computation to avoid the use of a constitutive relation, while still observing mass and momentum conservation.

Subfield (a) has been used in rheology since 1964, but (b) is much younger, and has mainly developed since 1990. We shall address (b) relatively briefly, concentrating the bulk of this chapter on solutions to boundary problems with a chosen constitutive relation. Only an overview of computational methods is given here, and the reader will have to refer to the literature for many details of computer program construction and implementation. We will instead concentrate on providing significant, successful examples of various solution techniques. Two useful sources for non-Newtonian numerical fluid mechanics are Crochet et al. (1984) and Huilgol and Phan-Thien (1997).

8.2 Computational problems for non-Newtonian incompressible flows

Here we assume that a constitutive relation and boundary conditions have been chosen. The choice (Chapter 5, Table 5.4) between differential and integral models has a profound effect on the computing strategy adopted, and for viscoelastic materials the question of boundary conditions also needs careful scrutiny.

Let us begin by assuming an inelastic model (Section 4.1). Then

$$\tau_{ik} = 2\eta(\dot{\gamma})d_{ik} \tag{8.1}$$

where $\dot{\gamma} = \sqrt{2d_{ik}d_{ik}}$, the equivalent shear rate. This is by far the most common sort of practical problem, and will serve as an introduction. The Newtonian case, $\eta =$ constant, is a subset of (8.1). Boundary conditions are well-known for this (elliptic) set of equations: (Chapter 1; Table 1.4). We require, on each portion of the boundary, that the velocities or tractions are given; sometimes the direction of the flow is also given, which gives a relation between the velocity components, and sometimes one traction and one velocity component are given; in cases where slip occurs, a relation between traction and slip speed could be given (see Section 3.9).

When free surfaces are present, and the flow is steady, then the velocity boundary condition $\mathbf{v} \cdot \mathbf{n} = 0$ must also be enforced, as well as the traction boundary conditions on the free surface; the position of the free surface becomes another unknown.

For unsteady free surface flows, if $h(\mathbf{x}, t)$ is the free surface location, then one has (Stoker 1957)

$$\frac{\partial h}{\partial t} + \mathbf{v} \cdot \boldsymbol{\nabla} h = 0 \tag{8.2}$$

plus appropriate traction conditions. When surface tension is important, then the normal traction is related to the two principal curvatures $1/R_1$ and $1/R_2$, so that $\sigma(1/R_1 + 1/R_2)$ is the jump in normal traction on the surface. In cases where the surface tension varies on the free surface, then an effective shear traction proportional to $\nabla\sigma$ is applied to the surface (Levich 1962; Leal 1992).

Additional boundary conditions needed for viscoelastic flows will be discussed later (Section 8.4); initial conditions are also required for unsteady flow problems. An example of a steady flow problem with a free surface (extrusion) is given in Fig. 1.7.

Following the setting up of the problem, there is a considerable number of choices for discretization of the partial differential equations and their solution.

8.3   Discretization schemes

Numerical solution entails the discretization of the fields for $\mathbf{v}$ and p, and any other required variables. In all cases the object of the discretization is to reduce the partial differential equations of the problem to a set of (generally non-linear) simultaneous equations for a known finite number of nodal variables. There are many schemes: only a selection of popular methods is given here.

Generally, a mesh is set in the domain of solution, although some research on spectral and other meshless numerical schemes has been done. We now survey the principal methods of discretization. With scarcely an exception, they can be regarded as variants of the methods of weighted residuals (Finlayson 1972; Ames

1977). In these methods the field equations are written as

$$\mathbf{Lu} - \mathbf{f} = \mathbf{0}, \tag{8.3}$$

where $\mathbf{L}$ is an operator, $\mathbf{u}$ is the solution vector, and $\mathbf{f}$ is a driving function. $\mathbf{u}$ is then expressed as a sum of N modes and (8.3) is weighted over the space Ω of interest with respect to each mode g_n:

$$\int_\Omega (\mathbf{Lu} - \mathbf{f}) g_n \, dV = \mathbf{0} \quad (n = 1, N) \tag{8.4}$$

thereby producing a weak-form solution with a given number of unknowns. The g_n depend on the method chosen (Finlayson 1972).

8.3.1 Finite differences

The oldest discretization method is that of finite differences (FD), which has been used since (at least) 1908 when Runge (1908) and Richardson (1910) investigated some elasticity problems. Southwell (1946) discusses the pre-computer use of finite differences in continuum mechanics. The first use of finite differences for non-Newtonian flow appears to be the work of Young and Wheeler (1964) on the flow of a power-law fluid in a square duct.

Since there are many books on finite differences and a comparatively small use of them in rheology, we shall be brief. An extended discussion is given by Richtmyer and Morton (1967), Roache (1976) and Crochet *et al.* (1984).

In this technique the fluid field is covered with a (usually) uniform grid of size h (Fig. 8.1).

The nodes are numbered in an (i,j) space and the variables ψ, v, p, etc., are discretized at the nodal points so that one has, for example, $p_{i,j}$ as the value of the pressure p at the node (i,j). By utilizing a Taylor expansion,

$$p_{i+1,j} = p_{i,j} + h \frac{\partial p}{\partial x}\bigg|_{i,j} + \frac{h^2}{2} \frac{\partial^2 p}{\partial x^2}\bigg|_{i,j} + \mathrm{O}(h^3), \tag{8.5}$$

and a similar expression for $p_{i-1,j}$ found by putting $-h$ for h in (8.5). One can find values for $\partial p/\partial x|_{i,j}$ and $\partial^2 p/\partial x^2|_{i,j}$ by adding and subtracting equations:

$$\frac{\partial p}{\partial x}\bigg|_{i,j} = \frac{p_{i+1,j} - p_{i-1,j}}{2h} + \mathrm{O}(h^2), \tag{8.6}$$

$$\frac{\partial^2 p}{\partial x^2}\bigg|_{i,j} = \frac{p_{i+1,j} - 2p_{i,j} + p_{i-1,j}}{h^2} + \mathrm{O}(h^2). \tag{8.7}$$

These are the central difference formulas.

Alternatively, one can define a backward difference or a forward difference for the first-order derivatives: Respectively, these are

$$\frac{\partial p}{\partial x} \sim (p_{i,j} - p_{i-1,j})/h; \quad \frac{\partial p}{\partial x} \sim (p_{i+1,j} - p_{i,j})/h. \tag{8.8}$$

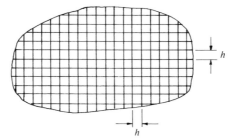

(a) Mesh pattern – Finite Differences

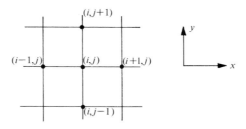

(b) Computational Molecule

Fig. 8.1   Finite difference grid.

In contrast to (8.5–8.7), which are of $O(h^2)$ accuracy, (8.8) is only of $O(h)$ accuracy. One can easily rewrite (8.5–8.7) for $\partial p/\partial y$ and other derivatives. The scheme can be extended to three spatial dimensions.

For time-dependent problems, for example the one-dimensional diffusion equation for temperature T, where α is the thermal diffusivity:

$$\frac{\partial T}{\partial t} = \alpha \frac{\partial^2 T}{\partial x^2},$$

(8.9)

one can discretize the left-hand side using backward, forward or central differences. Nearly all computations discretize time-dependent terms using finite differences, in contrast to the variety of spatial discretization schemes (Richtmyer and Morton 1967).

Let T_i^n be the value of the temperature at the ith x-node and n denote the time scale, in increments of Δt.

Then the simplest (Euler) scheme corresponding to (8.9) is the explicit rule, using a forward time difference:

$$(T_i^{n+1} - T_i^n)/\Delta t = (\alpha/h^2)(T_{i+1}^n - 2T_i^n + T_{n-1}^n),$$

which can be rearranged as

$$T_i^{n+1} = T_i^n + (\alpha \Delta t/h^2)(T_{i+1}^n - 2T_i^n + T_{i-1}^n). \tag{8.10}$$

This enables very fast computations, since no simultaneous equations need be solved, but unless $\Delta t < h^2/2\alpha$ the scheme is unstable (Richtmyer and Morton 1967); for two- and three-dimensional problems $\Delta t < h^2/4\alpha$ and $h^2/6\alpha$ respectively for stability. On the other hand, if a backward time difference is used, then the bracketed terms in (8.8) have a superscript $n+1$, and the scheme is implicit, yielding the sets of equations

$$-T_{i-1}^n + (2 + M)T_i^n - T_{i+1}^n = MT_i^{n-1} \quad (i = 1, N), \tag{8.11}$$

where $M = h^2/\alpha \Delta t$.

The tri-diagonal matrix arising from the left-hand side can be solved using a very fast tridiagonal (Thomas) algorithm, and the scheme is stable for all values of Δt. However, very large values of Δt may lead to stable, but inaccurate, results.

For non-linear problems the coefficients of the various terms depend on the solution and iteration is needed. This process is discussed in Section 8.5 below.

By using the estimates (8.6–8.8) an entire set of partial differential equations can be reduced to a set of algebraic equations in the nodal variables. The main problem with the method lies at the edges of the region in irregular domains, since the mesh will not fit the body exactly, and accuracy is lost.

Finite difference methods have been used for axisymmetric potential flow free jet problems and results for final jet size were not accurate (Hunt 1968), probably due to difficulties in enforcing the normal force boundary condition on the free surface which does not fit the mesh neatly, in general.

A stream-function-vorticity finite-difference formulation (see Roache 1976) was used for a viscous jet and produced a swelling result of about half of that observed. It seems likely that this failure was again mainly due to problems in trying to enforce the normal stress boundary condition where the mesh fits worst. Hill *et al.* (1981) used a marker-and-cell-type finite difference program and produced results that show too little swelling (about 60 per cent of observed swelling). Apart from the difficulty in the free surface fit, they used too short a field to accomodate the proper boundary conditions shown in Fig. 1.7 and most likely this affected their swelling ratios. Since this problem arises with all computer methods we shall discuss it now. Looking at Fig. 1.7 one can regard the transition from a Poiseuille flow inside the tube to the rigid body motion downstream in terms of a disturbance superposed on a Poiseuille flow or on a rigid cylinder motion. In both cases the disturbance flow has a zero net flux along the z-axis, and it is known (Tanner 1963) that in creeping flows these zero-flux disturbances in a tube or channel attenuate exponentially away from the source of disturbance (in this case, the exit plane). The rate of attenuation (upstream) inside the tube (the smallest eigenvalue in the problem) is about $\exp + 4.466z/R$ in axisymmetric creeping flow and for plane flow about $\exp + 3.749x/R$. (Note x and z are negative here). In the exterior free jet the rates are about $\exp - 2.811z/R_f$ and $\exp - 2.106x/R_f$ respectively. Thus for axisymmetric creeping flows an upstream and downstream field of about $2R$ gives a minimum attenuation of 250. For plane flows it is safer to use $2.5–3R$ as a field length. With viscoelastic and/or inertial effects longer fields will often be needed, especially downstream, and failure to provide these will render the results questionable; the jet form is sensitive to small constraints. If analytical results, such as those quoted above for creeping flow, are available, they should be used, otherwise careful numerical experiments are necessary.

In another finite difference study Ryan and Dutta (1981) avoided the lack of fit between the free surface and the mesh by mapping the jet and tube wall on to a rectangular space and performing difference calculations in the rectangular space. They also used a stream function/vorticity scheme in the mapped space and satisfied the normal stress boundary condition on the transformed free surface. Results for a circular die are better than the previous efforts described above (but somewhat low, only 12 per cent swelling for a circular tube instead of the nearly 13 per cent expected; see Table 8.8, p. 424). Thus, although the finite difference scheme is very easy to set up, it needs considerable care, especially with irregularly shaped spaces and free surfaces.

The following finite difference Example shows an investigation of numerical stability.

Example
We use a one-dimensional model of the inelastic second-order type [eqn (4.30c)] to show how a finite range of convergence in Weissenberg number space can occur due to employment of a particular numerical scheme.

Let us study the one-dimensional model for the stress τ

$$\tau = \eta \frac{\mathrm{d}V}{\mathrm{d}x} + \Psi_2 \left(\frac{\mathrm{d}V}{\mathrm{d}x} \right)^2 - \frac{1}{2} \Psi_1 V \frac{\mathrm{d}^2 V}{\mathrm{d}x^2}. \tag{8.12}$$

The one-dimensional equation of motion is $\mathrm{d}\tau/\mathrm{d}x = 0$, and no incompressibility constraint is used. The equation of motion is then

$$\eta \frac{\mathrm{d}^2 V}{\mathrm{d}x^2} + \left(2\Psi_2 - \frac{1}{2}\Psi_1 \right) \left(\frac{\mathrm{d}V}{\mathrm{d}x} \right) \frac{\mathrm{d}^2 V}{\mathrm{d}x^2} - \frac{1}{2} \Psi_1 V \frac{\mathrm{d}^3 V}{\mathrm{d}x^3} = 0, \tag{8.12a}$$

where V is the velocity.

To study the behaviour of eqn (8.12a) numerically we use a finite difference method. Replacing (8.12a) by its finite difference equivalent with a grid length h, we get, centred at node j, the typical equation

$$\frac{\eta}{h^2}(V_{j+1} - 2V_j + V_{j-1}) + \left(2\Psi_2 - \frac{1}{2}\Psi_1\right)\frac{(V_{j+1} - V_{j-1})(V_{j+1} - 2V_j + V_{j-1})}{2h^3}$$

$$-\frac{\Psi_1}{4h^3}V_j(V_{j+2} - 2V_{j+1} + 2V_{j-1} - V_{j-2}) = 0, \qquad (8.13)$$

where V_i is the velocity at node i.

To consider the stability of (8.13) we need to postulate an iterative process. To begin with, suppose that the velocities at all nodes can be written $V_j = U + u_j$ where U is constant, with a small perturbation u_j at node j. Then (8.13) may be linearized

$$u_{j+1} - 2u_j + u_{j-1} = \frac{\Psi_1 U}{4h\eta}(u_{j+2} - 2u_{j+1} + 2u_{j-1} - u_{j-2}). \qquad (8.14)$$

Let us now represent the current state (after n iterations) at the jth node by u_j, while the new velocity to be found at the next iteration is W_j. Boundary conditions are imposed at the ends of the line where the u_j are zero ($j \to \pm\infty$). Under these circumstances, the appropriate solution of the linearized problem is that $V_j = U$ everywhere, so $u_j \to 0$ for all j; any solution scheme which does not produce this behaviour after a large number of iterations is unstable. We can represent the solution as a Fourier series with an amplification factor depending on iteration number; each component is represented as ($i = \sqrt{-1}$ here)

$$\left.\begin{array}{l} u_j = \sum_m a_m e^{2\pi i m j} \\[2mm] W_j = \sum_m b_m e^{2\pi i m j} \end{array}\right\},$$

where m is the mode number. We shall now study the growth of a single Fourier component; since (8.14) is linear, the modes do not interact.

In particular, suppose each new estimate W_j is computed by putting the current estimate of u_j in the right-hand side of (8.14). For the mth component, we get

$$b_m[\cos 2\pi m - 1] = a_m \frac{i}{4}\frac{\Psi_1 U}{h\eta}[\sin 4\pi m - 2\sin 2\pi m].$$

Simplifying

$$\frac{b_m}{a_m} = \frac{i\Psi_1 U}{2h\eta}\sin 2\pi m.$$

Now if $|b_m/a_m| > 1$, the m-mode is growing. The fastest growing mode is the one which makes $|b_m/a_m|$ largest. This occurs when m is any odd multiple of $1/4$ and thus

$$\left|\frac{b_m}{a_m}\right| = \frac{\Psi_1 U}{2h\eta}.$$

Alternatively stated, for stability we must have

$$(Wi)_c = \frac{\Psi_1 U}{h\eta} < 2. \tag{8.15}$$

In terms of $\lambda = \Psi_1/2\eta$, (8.15) becomes $\lambda U/h = 1$; we interpret $\Psi_1 U/h\eta$ (or N_1/τ) as a Weissenberg number based on mesh size and a typical velocity.

A similar analysis can be carried out for plane creeping flow of a second-order fluid (Tanner 1982).

While many of the schemes proposed in the literature are not based on the simple iterative approach given above, the replacement of the left-hand side in (8.13) by any similar linear operator involving one iteration step only will not affect the lack of convergence of the scheme, which appears to be due to the higher order terms on the right-hand side. One of the most disquieting aspects of the stability criterion (8.15) is that any attempt to refine the solution by using a finer mesh will often cause lack of convergence, and in the limit $h \to 0$, no solution is ever possible with these simple schemes. There is evidence that the above criterion is relevant (Tanner 1982), but such instabilities if observed must not be confused with physical instabilities built into the constitutive equations themselves (Chapter 10).

8.3.2 Finite volume methods

Two such methods have been adapted to non-Newtonian flows. In the first, Chorin (1967) proposed an artificial compressibility method (plus a time-marching scheme) which was used for viscoelastic flow by Jin *et al.* (1994); see Huilgol and Phan-Thien (1997) for a further description.

The second method is that described by Patankar (1980) which has been widely used in Newtonian fluid mechanics because of its speed and low memory demands. We illustrate the discretization scheme by considering the compressible continuity equation

$$\frac{\partial \rho}{\partial t} + \nabla \cdot (\rho \mathbf{v}) = 0. \tag{2.52}$$

Consider a fixed volume Ω of unit depth in the z-direction, on a square area BCDE (of side h) whose area is A. The centre of the area is the node P. B, C, D, and E can be considered as the centres of other areas, as sketched in Fig. 8.2. Integrating over A and applying the divergence theorem to (2.52) one finds

$$\frac{\mathrm{d}}{\mathrm{d}t} \int_\Omega \rho \, \mathrm{d}V + \int_s \rho \mathbf{v} \cdot \mathbf{n} \, \mathrm{d}s = 0, \tag{8.16}$$

where s is the contour BCDEB.

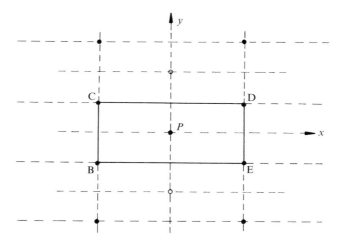

Fig. 8.2 A finite volume element BCDE. Nodal values are attached to points PBCDE in the element shown.

Now suppose that the velocity vector field is $(u, v, 0)$ so that u_B is the x-component of velocity at B, and so on. Then the discretized form of (8.16) is

$$A \frac{\mathrm{d}\rho_p}{\mathrm{d}t} + \frac{h}{2} (\rho_E u_E + \rho_D u_D - \rho_B u_B - \rho_C u_C + \rho_D v_D + \rho_C v_C - \rho_B v_B - \rho_E v_E) = 0.$$

$$(8.17)$$

The advantage of this formulation is that mass is always conserved precisely.

We cast the conservation equation and constitutive relations in the form (Patankar 1980; Xue *et al.* 1998)

$$\frac{\partial}{\partial t} (\Lambda \Phi) + \frac{\partial}{\partial x_k} (\Lambda u_k \Phi) = \frac{\partial}{\partial x_k} \left(\Gamma \frac{\partial \Phi}{\partial x_k} \right) + S_\phi$$

$$(8.18)$$

where Φ is a solution vector composed of the velocity components u_k, the pressure, the extra stress components, and possibly the density, temperature and other variables. The four terms of (8.18) are the unsteady term, the convection term, the diffusion term, and a source term S_ϕ. The functions Λ, Γ and S_ϕ depend on the unknown variable being considered; the source term sweeps up all terms not reducible to the other three forms.

The field is discretized into non-overlapping volumes (which can be irregular in shape, but are often rectangular boxes). Values of the dependent variables Φ are stored at the centroid of the volumes, as in Fig. 8.2.

Equation (8.18) is integrated over each control volume (ΔV) at each time interval δt, to find, using the divergence theorem:

$$\int_{\delta t}\int_{\Omega}\Lambda\frac{\partial\Phi}{\partial t}\,\mathrm{d}V\,\mathrm{d}t + \int_{\partial t}\int_{S}J_k n_k\,\mathrm{d}\Omega\,\mathrm{d}t = \int_{\delta t}\int_{\Omega}S_\phi\,\mathrm{d}V\,\mathrm{d}t \tag{8.19}$$

where n_k is a component of the outward normal unit vector and the flux J_k is given by

$$J_k = \Lambda u_k\Phi - \Gamma\frac{\partial\Phi}{\partial x_k}; \tag{8.20}$$

that is, a convective plus diffusive flux.

By assuming that the value at the central node P of the control volume prevails throughout the control volume for the terms which are not explicitly a function of position, such as the time difference term $\partial\Phi/\partial t$ and the source term S_Φ, and adopting the first-order backward Euler implicit formula over the time interval δt in view of its simplicity and unconditional stability for numerical calculations, one has (Xue *et al.* 1999)

$$a_p^0(\Phi_p - \Phi_p^n) + \int_{\Omega_{cv}}J_k n_k\,\mathrm{d}\Omega = \bar{S}_c + \bar{S}_p\Phi_p, \tag{8.21}$$

with

$$a_p^0 = \frac{\Lambda}{\delta t}\Delta V,$$

where the superscript n denotes the values taken from the previous time level, and the volume integral of the source term S_Φ, which is generally dependent on the dependent variable itself, has been linearized as a function of the Φ_p with $\bar{S}_c$ being the part that does not explicitly depend on Φ_p and $\bar{S}_p$, the coefficient of Φ_p. An overbar means the variable values are evaluated using the known field at time level n (or from the previous iteration).

Now the only integral term left is that involving the total flux J_k. Once one knows the values of the dependent variable at the control volume faces by using an appropriate spatial variation approximation scheme, that is, the control volume-based interpolation function, which relates the total flux J_k on each control volume face to the values of the dependent variable at the neighbouring nodes, then this integral can be worked out using the values multiplied by the respective areas of the face. Thus the final form of the integrated equation— namely, an algebraic approximation equation which establishes the relation of the value Φ_p at the centred node P to its neighbouring nodal values is determined and can be symbolically expressed in a generalized (nominally linear) form for each of the control volumes in the domain

$$a_p \Phi_p = \sum_{nb} a_{nb} \Phi_{nb} + \bar{S}_c + a_p^0 \Phi_p^n, \tag{8.22}$$

with

$$a_p = \sum_{nb} a_{nb} + a_p^0 - \bar{S}_p,$$

where the summation is to be taken over all of the neighbouring nodes nb of the centred node P. The coefficients a_{nb} are the functions of the dependent variables, and their contents depend on the relation of the total flux J_k to the values of the dependent variable at the neighbouring nodes, thus depending on both the form of the control volume used and the approximation scheme adopted. It is these coefficients that determine the spatial accuracy of the final solution.

After implementing the above mentioned steps for each of the control volumes in the domain, then each of the nodes in the flow domain has its own discretized equation like eqn (8.22). The expressions for $\bar{S}_c$ and $\bar{S}_p$ will play a crucial role for computational success in many cases as emphasized by Patankar (1980). One of the basic rules required to enhance the numerical stability of the discretized equation system and prevent a physically unrealistic solution is that the sign of $\bar{S}_c$ should be the same as that of a_{nb} and a_p (say, positive), and the sign of $\bar{S}_p$ should be opposite to that of a_{nb} and a_p so that $a_p \geq \sum_{nb} a_{nb}$ and $\bar{S}_p$ has an under-relaxation function for calculations.

In developing the approximation scheme, that is, in choosing the form of the total flux J_k, avoidance of false diffusion and instability of the resultant discretized system associated with high convection rates or negative coefficients is the main issue of concern.

To date, various schemes suitable for structured mesh systems have been developed, such as the PL (*Power-Law*) scheme of Patankar (1980), which is based on a curve fit to the exact exponential solution of the one-dimensional convection-diffusion equation without any source, the SUD (*Skew Upwind Differencing*) scheme in which the flow direction was taken into account in determining the values of the dependent variables at the faces by applying the upwind method in a vectorial rather than a componential sense, and the QUICK scheme proposed by Leonard (1979) with the values of the dependent variables on the control volume face being obtained by fitting a parabola to the values of the dependent variable at three consecutive nodes (for a one-dimensional problem) with the two nodes located on either side of the face plus the next node on the upstream side. All of these schemes involve a balance between accuracy (less false diffusion) and stability, but none of the schemes covers all of the following three desirable features:

(a) Complete elimination of the spatial oscillations due to lack of diagonal dominance (negative coefficient problem) in the discretized system of equations;

(b) Reasonable accounting for the coupling between streamwise and cross-stream gradients as well as the influence of the source terms so that false diffusion can be reduced or even eliminated; and

(c) Ease of implementation in 3-D flows at low computational expense.

Numerical evaluations show that none of the schemes seems superior to the others in multidimensional flow situations, and the performance of a particular scheme can vary for different physical problems. In the work of Xue *et al.* (1998), the Power-law (PL) scheme was chosen for the discretization of the convective-diffusive (momentum) equation mainly because of the simplicity of its implementation and low computational expense as well as excellent conservation properties.

With these schemes, the coefficients a_{nb} take the form:

$$a_{nb} = D_{nb}f(|P_{nb}|) + \lceil \text{sign}(nb)F_{nb}, 0 \rceil, \tag{8.23}$$

where the 'sign(nb)' is $+$ for the upstream faces and $-$ for the downstream faces. The function $f(|P_{nb}|)$ has the form

$$f(|P_{nb}|) = \lceil 0, (1 - 0.1|P_{nb}|)^5 \rceil, \tag{8.24}$$

where

$$F_{nb} = (\Lambda u_k \Omega)_{nb}; \quad D_{nb} = \left(\frac{\Gamma\Omega}{\delta x_k}\right)_{nb},$$

can be thought of as being the strength of the convection (or flow) through the face nb; and the diffusion conductance, respectively. The ratio of the two strengths is called the local Peclet number:

$$P_{nb} = \left(\frac{F}{D}\right)_{nb} = \left(\frac{\Lambda u_k \delta x_k}{\Gamma}\right)_{nb},$$

with δx_k the distance between the central node P and its neighbouring node in the k direction.

With the discretized form given in eqn (8.22) for the general transport equation (Eqn 8.18) the discretized form for a specific equation can be readily written out by imposing different coefficients Λ, Γ, and the discretized source terms. The discretized equation obtained in this manner expresses the conservation principle for Φ in the finite volume. The most attractive feature of the control-volume formulation is that the resulting solution would imply that the integral conservation of quantities such as mass, momentum, and energy is exactly satisfied over any group of control volumes and, of course, over the whole computational domain. This characteristic exists for any number of grid points, not just in a limiting sense. Thus, even the coarse-grid solution exhibits the exact integral balances.

For further details the reader is referred to Patankar (1980) and Huilgol and Phan-Thien (1997).

8.3.3 The boundary-element method

In the finite-volume and finite-difference schemes, one finds the complete solution field in all the fluid-filled region whether or not this is of interest. In some problems only the domain shape and/or the pressure losses may be needed, and it is therefore interesting to look at the boundary-element method and its derivatives which avoid evaluating all of the internal variables, at least for linear creeping flow problems. Essentially, the techniques are derived from reciprocal theorems familiar in linear elasticity, combined with a knowledge of the relevant Green's function. The effective adaptation of these methods for computer use in elasticity is due to Rizzo and Cruse (Cruse and Rizzo 1968). Here we begin by discussing viscous incompressible flows; the problem follows closely the discussion for linear elasticity (Brebbia 1980), but incompressibility needs to be considered.

Consider an arbitrary set of fields $v_i^*, p^*, \sigma_{ij}^*$. We multiply the equation of motion (2.60) and the mass conservation eqn (2.53) by v_i^*, and p^*, respectively, and integrate over the body; this is the Galerkin procedure (Finlayson 1972), but we also add on a surface term over the part S_v of the surface where the velocities are given. Thus we have (where Ω denotes the entire body)

$$0 = \int_\Omega \left[\frac{\partial \sigma_{ij}}{\partial x_i} + \rho(f_i - a_i) \right] v_i^* \, dv + \int_\Omega p^* \frac{\partial v_i}{\partial x_i} \, dv + \int_{S_v} (v_i - \hat{v}_i) t_i^* \, dS, \qquad (8.25)$$

where t_k^* is the traction vector $\sigma_{kj}^* n_j$ formed from the starred stress tensor and the outward unit normal vector $\mathbf{n}$, and $\hat{\mathbf{v}}$ are the given boundary conditions on S_v. Any solution (σ_{ij}, p, v_i) that satisfies the equations of motion, the mass conservation equation and the boundary condition $v_i = \hat{v}_i$ on S_v will make (8.25) vanish, and hence it is then satisfied for arbitrary $(\sigma_{ij}^*, p^*, v_i^*)$.

Now consider the following expression denoted by $I(^*, 0)$ where

$$I(^*, 0) = \int_\Omega \frac{\partial \sigma_{ij}^*}{\partial x_j} v_i \, dv - \int_S v_k t_k^* \, dS - \int_\Omega p^* \frac{\partial v_i}{\partial x_i} \, dv. \qquad (8.26)$$

Here S is the whole body surface. By using Green's theorem we can show that for a Newtonian fluid there is a reciprocal theorem where starred and unstarred fields are interchanged

$$I(0, ^*) = I(^*, 0).$$

This reciprocal theorem can be used to replace some of the terms in (8.25) finding, when both v_i^* and v_i are incompressible fields,

$$
\int_\Omega \frac{\partial \sigma_{ij}^*}{\partial x_j} v_i \, \mathrm{d}\nu - \int_S v_i t_i^* \, \mathrm{d}S + \int_S v_i^* t_i \, \mathrm{d}S
$$
$$
+ \int_{S_v} (v_i - \hat{v}_i) t_i^* \, \mathrm{d}S + \int_\Omega \rho(f_i - a_i) v_i^* \, \mathrm{d}\nu = 0. \tag{8.27}
$$

We can combine the second and fourth terms to form $-\int_S v_i t_i^* \, \mathrm{d}S$, understanding that $v_i = \hat{v}_i$ on S_v. We shall also suppose here that f_i and a_i are known. For the examples given here, both f_i and a_i are assumed zero, so we have creeping flow under no body forces. We will also restrict the discussion to plane and axisymmetric flows.

We now assume that the (*)-fields are produced by a concentrated unit force in the l-direction. The plane-flow solutions for the $t_k^{*(l)}$ and $v_k^{*(l)}$ in Cartesian coordinates x_i are well-known (Brebbia 1980):

$$
v_k^{*(l)} = \frac{1}{4\pi\eta} \left[(-\ln r)\delta_{lk} + \frac{\partial r}{\partial x_l} \frac{\partial r}{\partial x_k} \right], \tag{8.28}
$$

$$
t_k^{*(l)} = -\frac{1}{\pi r} \left[\frac{\partial r}{\partial n} \frac{\partial r}{\partial x_k} \frac{\partial r}{\partial x_l} \right], \tag{8.29}
$$

where r is the distance from the point of application of the force $(\mathbf{r})$ and the position at which $\mathbf{v}$ is evaluated $(\mathbf{x})$. δ_{lk} is the unit tensor, equal to unity if l equals k, zero otherwise. Hence

$$
r^2 = (x_k - r_k)(x_k - r_k), \tag{8.30}
$$

and by differentiating (8.30) we find

$$
\frac{\partial r}{\partial x_i} = \frac{x_i - r_i}{r}. \tag{8.31}
$$

In eqn (8.29) $\mathbf{n}$ refers to the direction normal to the surface across which the traction is being computed. The corresponding formulas for axisymmetric flows contain Legendre functions and are given by Bush and Tanner (1983).

Consider the body of fluid in Fig. 8.3. Suppose the boundary is discretized into linear 'elements', ab, bc, cd, etc., and at the centre of ab we place node 1. (In this simplest of boundary element schemes no nodes occur where elements join.) We assume a (*)-field which consists of the response to a unit force in the x_1-direction applied at node 1. Then we have (Brebbia 1980)

$$
\frac{\partial \sigma_{ij}^*}{\partial x_j} = -\delta_{1i}, \tag{8.32}
$$

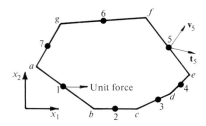

Fig. 8.3 Boundary element discretization. The traction and velocity components are uniform on each segment (element).

where δ_{1i} is a delta-function. In the case of creeping flow with no body forces, substituting (8.32) in eqn (8.27) we find

$$0 = -\frac{1}{2}v_1 - \int_s v_k t_k^* \, dS + \int_s v_k^* t_k \, dS, \tag{8.33}$$

where v_k^* and t_k^* are known exactly from (8.28) and (8.29). If we now assume that the field is uniform over each segment, so that we may speak of $v_1^{(m)}$, $v_2^{(m)}$, $t_1^{(m)}$ and $t_2^{(m)}$ as the uniform components on the segment containing the mth node, then (8.27) becomes a set of linear equations

$$0 = -\frac{1}{2}v_1 + \sum_{m=1}^{N} \left[\int_{S_m} t_1^* \, dS \int_{S_m} t_2^* \, dS \right] \begin{bmatrix} v_1^{(m)} \\ v_2^{(m)} \end{bmatrix} + \sum_{m=1}^{N} \left[\int_{S_m} v_1^* \, dS \int_{S_m} v_2^* \, dS \right] \begin{bmatrix} t_1^{(m)} \\ t_2^{(m)} \end{bmatrix} \tag{8.34}$$

where N is the number of nodes. The $\frac{1}{2}$-factor in (8.34) arises because the point force is applied on the boundary and not in the fluid interior.

The integrals in (8.34) may be computed, since $\mathbf{v}^*$ and $\mathbf{t}^*$ are known, and (8.34) is a linear equation connecting $2N$ components of velocity $2N$ components of traction. Similarly, one can apply a unit force in the x_2 direction at node 1, generating a second equation, and so on for all nodes, finally generating $2N$ equations. There are $2N$ velocity components and $2N$ traction components, but only a total of $2N$ unknowns, since at each node two out of the four unknowns are given as boundary conditions. After eliminating these known quantities, one can solve a set of linear equations for all $2N$ boundary unknowns. To obtain values in the interior, one places a point force where needed as the $(^*)$-field, and uses (8.27) again, thus producing the needed value of $\mathbf{v}$; a different starred field will produce the tractions and stresses. Thus results at all points of the body can be found, but only if needed.

The boundary element method (b.e.m.) uses relatively few unknowns. However, the b.e.m. generates a full matrix. By considering the number of operations needed to solve the equations, we found the b.e.m. is quick for small linear problems. As usual, remarks about relative speed need to be treated with caution and judged for the case in hand.

With this method non-linear problems have to be solved as a series of perturbations about the viscous solutions, and convergence may not always be obtained. Bush and Tanner (1983) found that Reynolds numbers not much greater than 15 can be solved for Newtonian flow in the Hamel (or converging-wall) problem; for complex linear problems (Ramia 1991) the scheme works very well.

8.3.4   Finite element methods

The bulk of the work with non-Newtonian flows has been performed with the finite element method, and the quantity and variety of schemes available for viscoelastic problems is remarkable (Baaijens 1998). Here we will describe the main ideas of the discretization.

The finite element method (f.e.m.) covers the fluid-filled region with an irregular mesh, usually made up from quadrilaterals or triangles. Curvilinear quadrilaterals may also be used (the isoparametric elements, see Zienkiewicz and Taylor (1991)), so that a close fit to irregular boundary shapes may be arranged (Fig. 8.4).

Another advantage of the finite element system is that it permits a closer 'mesh' near singular points. Of course, no finite mesh can exactly capture the singular behaviour there and it would be possible, if one knew the order of the singularity, to insert a special semi-analytical element at the singular point to give a good

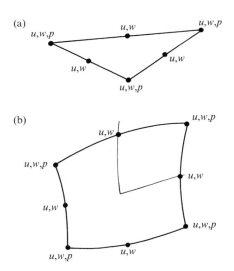

Fig. 8.4   Some successful elements for finite element fluid mechanics. (a) Biquadratic velocity and bilinear pressure area–coordinate triangular element; (b) Combined biquadratic velocity and bilinear pressure isoparametric quadrilateral element.

representation near these points. This device has been used in fracture mechanics and in extrusion studies (Georgiou *et al.* 1989). The finite element method's ability to model the singularity more or less well without special elements is helpful but probably not decisive in favouring finite element methods over finite difference methods; more useful is the ability to have a mesh fit the extrudate shape closely.

Even within the finite element field, a wide choice of computational methods is available. Restricting ourselves to plane or axisymmetric incompressible problems, we have the option of working with a stream function ψ or with the velocity components, u, w and the pressure p as primitive variables; or, we may retain the velocities, pressure, and stresses and use the mixed method. The free boundary conditions are best handled by not using the stream function method (Nickell *et al.* 1974). Briefly, if we consider a computation using a stream function in a Newtonian fluid, then the pressure is eliminated and is recovered afterwards by solving a separate Poisson-type equation. For problems in which velocity boundary conditions are given the scheme works satisfactorily, since boundary conditions are given in terms of ψ and its normal derivative $\partial\psi/\partial n$. For boundaries partially made up of free surfaces, without traction, we require on a part of the boundary (S_t) that the normal (σ_{nn}) and tangential (σ_{ns}) stress components vanish there. For the Newtonian fluid this means that

$$\frac{\partial v_n}{\partial s} + \frac{\partial v_s}{\partial n} = 0; \quad -p + 2\eta\frac{\partial v_n}{\partial n} = 0, \tag{8.35}$$

where v_n and v_s are the components of velocity normal and tangential to the boundary, respectively, and η is the (constant) viscosity. Of course, on the free surface we must have the normal velocity component (v_n) zero, so that $\partial v_n/\partial s$ vanishes. To express eqn (8.35) in terms of the stream function is difficult; it is the presence of p in the normal stress term which is awkward, since p is not known. These problems are by-passed by not using ψ in the computation and electing to work instead with the velocity components u, w, the pressure p, and possibly the stresses, as computational variables. Few completely successful examples of free-surface computations using the stream function are known; for the primitive variable method using u, w and p we proceed as follows.

In a Cartesian tensor representation we have the conservation laws

$$\frac{\partial\sigma_{ij}}{\partial x_j} + \rho(f_i - a_i) = 0, \tag{8.36}$$

and

$$\frac{\partial v_i}{\partial x_i} = 0, \tag{8.37}$$

where v_i is the velocity component in the direction of x_i, f_i are the (known) body force components, and σ_{ij} are stress components; a_i signifies, for steady flow, the acceleration component $v_j\partial v_i/\partial x_j$.

Let the boundary S be composed of two parts; $S = S_v + S_t$. Velocity boundary conditions are given on S_v, while traction (stress) boundary conditions are given on S_t. On S_t we shall assume

$$\sigma_{ij}n_j = t_i, \tag{8.38}$$

where the traction vector components t_i are given and n_j is the outward-pointing normal unit vector on the surface.

We suppose the choice of constitutive equation has been made; most of the difficulties we shall encounter are already present in the convected Maxwell model (4.25). The finite element method now reduces the above system of partial differential equations to a set of (usually non-linear) algebraic equations by one of several approaches. In the displacement method (strictly velocity in this case) all stress variables are expressed in terms of the velocities and the pressures and the stresses are then back-calculated after finding the velocities u, w, and the pressure p. This is the approach followed by Nickell *et al.* (1974). The following steps are made to effect this process:

(i) We assume that the nodal variables (a node usually occurs where two element bounding curves cross but it may also be elsewhere on the element boundary or inside the element) are the unknowns;

(ii) We assume an interpolation function for the variables inside each element. Often linear or quadratic interpolation is used but elements with high interpolation orders which can be changed at will have been very effective (Talwar and Khomami 1995). Thus the representation of the radial velocity component u inside the mth element will take the form

$$u^{(m)} = \sum_{i=1}^{r} N_i^{(m)}(x, y) u_i^{(m)}, \tag{8.39}$$

where the superscript denotes that the field is for the mth element, and r is the number of nodes associated with this element; the shape or weighting functions $N_i^{(m)}$ are chosen in advance and the nodal point variables $u_i^{(m)}$ are the unknowns.

(iii) We substitute the global expression for the variables into the mass-conservation equation and the equation of motion, multiply by the relevant shape functions in turn for each element and integrate over the body. The process then sets the resulting expression equal to zero. This is the Galerkin method used above (see also Finlayson (1972) and Crochet *et al.* 1984). It is usual to eliminate the stresses at this point and to make an integration by parts on the equation of motion (8.36); this reduces by one the number of times one has to differentiate the shape functions. Care must be taken in this step as the shape functions are of limited differentiability.

Taking the mass-conservation equation as an example of the process, we form

$$\int_{\text{body}} \frac{\partial v_j}{\partial x_j} N_i^{(m)} \, dV = 0 \quad (m = 1, N), \tag{8.40}$$

where $N_i^{(m)}$ are the known shape functions for the pressure. Clearly, eqn (8.40), after integration, gives a linear combination of the unknown nodal point velocities and forms part of a system of equations; the remainder comes from the equations of motion.

Thus the result of these integrations is a system of simultaneous equations, usually non-linear, for the nodal point velocities and pressures.

(iv) When integrating the weighted equation of motion by parts we obtain a surface integral which can be evaluated on the parts of the body (S_t) where traction boundary conditions are given; this takes care of these boundary conditions which contribute to the known right-hand side of the system of equations to be solved. The ease with which these traction boundary conditions can be satisfied constitutes a great advantage of the finite element (and boundary element) methods. The remaining (velocity) boundary conditions are now set in; we recognize that some nodal velocities are known and these velocities are removed from the list of unknowns.

(v) Finally, we find a set of non-linear simultaneous equations for the remaining unknown nodal point velocity and pressure components arranged as a vector $\mathbf{V}$

$$(\mathbf{K} + \mathbf{C})\mathbf{V} = \mathbf{F} \qquad (8.41)$$

The matrix $\mathbf{C}$ depends on $\mathbf{V}$ and is often unsymmetric. $\mathbf{K}$ is symmetric and constant and arises from the (linear) creeping flow part of the problem.

There are some constraints on the shape functions. It has been found essential for stability that the pressure field be interpolated with a polynomial one order lower than the velocity terms. [This is the so-called LBB condition; see Huilgol and Phan-Thien (1997)]. Two simple (u, w, p) elements are shown in Fig. 8.4; these have linear pressure and quadratic velocity fields over the element. Further discussions of finite element methods in fluid mechanics are given by Gallagher $et\ al.$ (1975), Kawai (1982), and Johnson (1990).

Crochet and Keunings (1982a) have used the mixed method where no attempt is made to eliminate the stresses; these, plus the pressure and the velocities are used as primitive variables and are interpolated on the elements. Thus one has larger matrices and longer solution vectors for a given number of elements but there are some compensating advantages; for example, the construction of a Newtonian–Raphson equation-solving scheme is facilitated. For inelastic fluids the Picard method gives smooth stress fields. Stability problems at high Weissenberg numbers also occur.

In some cases (Baaijens 1998) the use of pressures and other variables which are not continuous between elements has been advantageous in maintaining stability.

In closure, it appears that the so-called DEVSS-based methods, introduced by Guénette and Fortin (1995) are among the most robust finite element formulations. Some methods of using finite element methods for transient flows have been described by Baaijens (1998).

A great number of books dealing with the finite element method is available; we cite only Finlayson (1972), Crochet *et al.* (1984), Johnson (1990), Reddy and Gartling (1994), Zienkiewicz and Taylor (1991), and Huilgol and Phan-Thien (1997).

8.3.5 *Spectral methods*

In these methods the solution vector does not contain nodal values of the velocity and other variables. Gottlieb and Orszag (1977) describe the method using as an example the solution of the heat diffusion equation by a truncated Fourier series. Consider the equation [for $u(x, t)$]

$$\frac{\partial u}{\partial t} = Lu + f(x, t), \tag{8.42}$$

where $L(x, t)$ is a linear spatial differential operator and f is a known driving function. Appropriate boundary and initial conditions are also required. An approximate solution is sought in the truncated series form

$$u_N(x, t) = \sum_{n=1}^{N} a_n(t)\phi_n(x). \tag{8.43}$$

The ϕ_n are linearly independent functions (Fourier components, polynomials, Chebyshev series, for example) and the Galerkin process applied to (8.42) produces the results ($n = 1, N$)

$$\frac{\mathrm{d}}{\mathrm{d}t}(\phi_n, u_N) = (\phi_n, Lu_N) + (\phi_n, f), \tag{8.44}$$

where $(\phi, u) = \int \phi u \, \mathrm{d}x$ over the space of interest.

Equations (8.44) can then be used to solve for the coefficient functions a_n. Non-linear problems can also be 'discretized' by this method; in a steady flow situation this yields a set of (non-linear) simultaneous equations for the time-independent a_n which needs to be solved. An early example of this method is the solution of a power-law fluid in a square duct by Schechter (1961), who assumed a series solution of the form, for the axial velocity $w(x, y)$,

$$w = \sum_{i=1, j=1}^{N,N} a_{ij} \sin \alpha_i x \sin \beta_j y$$

and used a variational principle to find the a_{ij}. In these early days only a few terms could be used in the expansion and accuracy of order 1 per cent was claimed. Pilitsis and Beris (1989) have used spectral methods for non-Newtonian flows and Beris *et al.* (1987) also used a mixture of spectral and finite element methods. We shall give examples of their work later.

The main advantage of spectral methods lies in their accuracy (for a given number of unknowns) but they are geometrically relatively inflexible and often do not work satisfactorily on problems with sharp boundary wall corners pointing into the flow (salient points); see Section 8.7 below.

8.4 Boundary conditions for viscoelastic flows

Once the discretization has been performed, most methods need to have the simultaneous equations modified to accommodate boundary conditions. In the case of inelastic flows, either of the form (8.1) or of more complex forms, such as the second-order flow model (4.30c), the relevant boundary conditions are that on the boundary of the body of fluid (S in Fig. 8.5) two velocities, two tractions, or a mixture of the two, need to be specified everywhere. If velocity boundary conditions are applied to an incompressible fluid over the entire surface, then one must make sure that the net influx across the boundary is zero; usually it will be necessary to fix the pressure at some arbitrary value at a single point to fully determine the pressure field in these cases.

For viscoelastic fluids, either of the differential [for example, eqn (4.25)] or the integral types [for example, eqn (4.54)] further boundary conditions on the inlet section AB in Fig. 8.5 are needed. Consider the UCM fluid [eqn (4.25)] as an example. For steady flows the equations can be written as

$$v_k \frac{\partial \tau_{ij}}{\partial x_k} = f_{ij}(\mathbf{L}, \boldsymbol{\tau}). \tag{8.45}$$

Given a known velocity field, this is a set of six first-order hyperbolic partial differential equations for the extra stresses τ_{ij}. The characteristics of (8.45) are the streamlines, defined in Section (2.3.1). Such equations (Courant 1972) need initial conditions on the inflow boundary. Hence one expects that all six of the stress components should be specified on AB (Fig. 8.5), or, for plane and axisymmetric flow, three and four components respectively.

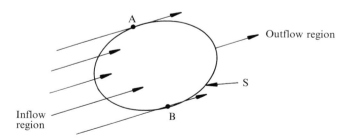

Fig. 8.5 Inflow and outflow regions on the boundary S of a body.

If one elects to solve (8.45) together with the momentum and mass conservation equations, then Renardy (1990) has shown that prescription of all stress components at the entry surface for the UCM model may lead to an over-determined problem. No simple resolution of the boundary data has been proposed, and some ingenuity is needed to generate a correct, compatible set of stresses at the entry surface. In the cases to be reported below, this problem is absent, either because the exact solution is known on the entry plane, or the problem has a periodic character. Another approach is to add time-dependent terms to eqn (8.45) and solve as an initial-value problem, with all extra stresses at zero values in the rest state. Another possibility is to add on an extra entry domain in which the required stresses can be developed; the boundary conditions on the entry domain can then be chosen at first in a fairly arbitrary manner and then they can be iterated as required until suitable stable inlet boundary conditions to the region of interest are obtained.

For other models of a form similar to eqn (8.45), for example the PTT and Giesekus models, similar problems arise. However, addition of a solvent viscosity, as in the Oldroyd-B model, removes the Renardy dilemma.

The above discussion has been in terms of differential viscoelastic models, but clearly the same problems occur with integral constitutive models; a complete (or compatible) set of stresses on the inlet region is needed.

A further discussion of the boundary data problem is given by Huilgol and Phan-Thien (1997), together with a clear discussion of the complex elliptic-hyperbolic nature of the mathematical problems being solved.

With regard to free surfaces, we may look at the finite element case. Here the use of the divergence theorem on the integral $\int_\Omega (\partial \sigma_{ij}/\partial x_i) N_k \, dV$, where N_k is the shape function, yields $\int_\Omega \sigma_{ij} n_j N_k \, dV = \int_s t_i N_k \, dV$. On free surfaces t_i is often zero, so that on areas of the boundary where no boundary conditions are prescribed, a stress-free boundary is automatically generated — this is a *natural* boundary condition for the method. The boundary element method also has this advantage: free surfaces in other methods usually require imposed (forced) boundary conditions on stresses.

In the case of slip at the solid surfaces (Section 3.9) another form of boundary condition is generated, needing special attention in the setting-up phase (Phan-Thien 1988). In other problems it may be necessary to prescribe thermal and other boundary conditions, but these are usually classical in nature and present no unknown problems.

8.5   Solution procedures and stabilization

Once the discretization is complete and boundary conditions are applied, the solution of the simultaneous non-linear equations for the variables is needed. Linearization and iteration are necessary for the nonlinear system. Generally, either a Picard (successive substitution), the Newton–Raphson or a quasi-Newton method can be used to solve the system, and either an iterative method or Gaussian elimination is used on the solution of the linearized equations.

Each discretization method has its adherents. For example, finite difference and finite volume methods usually use successive substitution solution methods and an iterative procedure (or at most, a tridiagonal algorithm) to solve the equations.

Due to the iterative procedures used, updates of the variables are obtained at each iteration. In finite element procedures using differential models, the tendency has been to solve simultaneously for all variables. Alternatively (Sun *et al.* 1996) one can find a velocity field, then compute the stresses, and iterate back and forth between the sets of equations. While this leads to smaller matrices at each step, convergence can be a problem. For integral models, one is forced to use this iterative (Picard) procedure.

Finite element methods (Baaijens 1998) have often used Newton–Raphson methods for problems with differential models, together with Gaussian elimination. Spectral methods have also followed this pattern. On the other hand, with integral models the Picard method is often used, plus Gaussian elimination. Boundary elements with mild non-linearities have generally used a Picard procedure plus Gaussian elimination.

Example: Picard procedures
Consider the single non-linear equation $x^3 - 7x - 6 = R(x) = 0$. The equation has three real solutions $x = 3$, $x = 1$, -2. Several successive substitution algorithms are possible. We can set

$$x = (x^{*3} - 6)/7,$$

where x^* is the current best estimate of x, or $x = 7/x^* + 6/x^{*2}$, or $x = (7x^* + 6)^{1/3}$ obtaining different results. One can show that by starting at different places one may or may not reach a converged solution. Beginning with $x^* = -1.1$ the first algorithm slowly converges to $x = -1$, while the other two reach $x = -2$. If, however, we take $x^* = -2.01$ as the beginning point, then about 12 iterations yield a value of $x < -10^{10}$, so the process diverges. The other two algorithms continue to converge to $x = -2$. Hence the initial guess is very important with non-linear systems.

The extension of the method to many variables is immediate.

Example: Newton iteration
In this process, for N equations in N unknowns, $(x_i, i = 1, N)$ we write the equations as

$$F_k(x_i) = 0 \quad (k = 1, N). \tag{8.46}$$

If an estimate x_i^* is available, then the Newton–Raphson process produces an improved value x_i by expanding (8.46) about x_i^*:

$$F_k(x_i^*) + \frac{\partial F_k}{\partial x_i}\bigg|_{x_i = x_i^*} (x_i - x_i^*) = 0. \tag{8.47}$$

The matrix $\partial F_k/\partial x_i$ is the Jacobian, J_{ki}. Usually one solves the (linear) system (8.47) for x_i (or $x_i - x_i^*$) and obtains a better solution, achieving quadratic convergence in the error $x_i - x_i^*$, unlike the linear convergence of the Picard method.

Applying to the case $N = 1$, with $F = x^3 - 7x - 6$, one finds that two iterations gives $x = -0.99993$, showing rapid convergence. The Newton–Raphson process does not always converge, however. For highly shear-thinning power-law models ($n \leq 0.5$) it is usually safer to employ Picard iteration.

For a discussion on quasi-Newton methods see Dennis and Schnabel (1979).

In all cases a key problem is to get a good initial estimate, so that convergence more readily follows. Of course, one has to respect any inherent limitations on stability due to a large mesh size relative to the time step, as discussed by Richtmyer and Morton (1967). Some schemes (see, for example, Section 8.3.1) actually are more unstable as meshes are refined; these are clearly untrustworthy at any but insignificant Weissenberg numbers.

8.5.1   The high Weissenberg number problem

By late 1970, many workers were becoming frustrated by what quickly became known as the 'High Weissenberg Number problem' (HWNP) — there was an upper limit on (Wi), above which the numerical algorithms failed. In 1984, Crochet et al. referred to 'the outstanding problem in the numerical simulation of non-Newtonian flow' and made the following observations:

 (i) A limit on (Wi) is found in all published work.

 (ii) Minor changes in the constitutive equation and/or the algorithms employed could lead to higher limiting values of (Wi). However, such improvements did no more than delay the breakdown process.

 (iii) Near the critical (Wi), it was often (but not always) observed that spurious oscillations appeared in the field variables.

The presence of boundary corners sticking into the flow (salient corners; commonly referred to as re-entrant corners in many publications) was thought to destabilize the flow.

Solutions to the HWNP have gradually evolved since 1977 (Tanner and Walters 1998). An early idea of importance was that of Crochet and Keunings (1980), who used a system of coupled equations. Here, the unknown pressure, velocity and stress components were solved for simultaneously, using Gaussian elimination. This led to a considerable advance in the critical Weissenberg number attainable. (Crochet and Keunings 1982). The solution method was coupled to a Newton–Raphson scheme, which enabled accurate solutions to be made, once divergence had been avoided. A clear recognition of the source of the convergence problems began to appear. The *hyperbolic* nature of constitutive equations of the differential type was pinpointed and it was recognized as an important factor in the instability. Baaijens (1998) has reviewed progress in this field over the years 1987–97 and gives a long list of references. Generally, several ideas have assisted. For example, Perera and Walters (1977) changed the variables, and instead of solving for the τ_{ik} directly, they split off a Newtonian component, and solved for a new set of variables S_{ik}:

$$S_{ik} = \tau_{ik} - 2\eta_r d_{ik}, \tag{8.48}$$

where η_r is an (arbitrary) reference viscosity. Sun *et al.* (1996) generalized this idea to a variable, adaptive reference viscosity. The use of upwinding, discontinous Galerkin methods, and artificial diffusion have all combined to stabilize computations so that useful (Wi) ranges can now be attained with finite elements (Baaijens 1998), with finite volumes (Xue *et al.* 1998) and with spectral methods (Pilitsis and Beris 1989). The knowledge that very thin stress boundary layers appear with some models, especially the UCM, has also enabled mesh refinement to be made more rational.

8.5.2   *Accuracy of computations*

For a long time it was difficult to achieve *stability* in computations with highly viscoelastic flows, as detailed in the previous section. Now that the HWNP has been at least partially solved, the accuracy of computations needs to be considered.

Generally, in the limit of an infinitesimal mesh size, the algebraic equations produced numerically should reduce to the exact partial differential equations. The accuracy of a solution can usually only be judged by reducing the mesh size, adding more modes in the spectral method, or by using more complex interpolations in the elements (Talwar and Khomami 1995). It is preferable to use at least three meshes, which is sometimes difficult, to establish convergence. Usually, one plots some measure of the change in variables against mesh size (Fig. 8.6). The convergence rate can then be seen, and accurately established. Richardson extrapolation (Roache 1998) can also be useful.

One should not confuse accuracy with the iterative convergence obtained by iteration on a fixed grid, which is necessary to solve the non-linear equations. Often iterative convergence can be driven down to machine accuracy (10^{-15}) and is not related to the overall accuracy of the computation.

Roache (1998) has explained these matters expansively and he notes that it is now editorial policy in some journals to require mesh refinement (or the equivalent) to establish some idea of the accuracy of the computations. He also

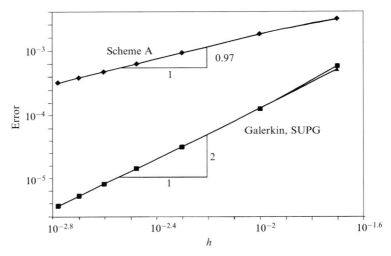

Fig. 8.6 Error analysis for a finite element viscoelastic one-dimensional problem. Ordinate shows $\sum_n |\tau_i - \tau_i^{RK}|/n$ where τ_i is the numerical result at node i and τ_i^{RK} is the Runge–Kutta result at node i; n is the number of elements on $x = 0-1$. The approximate slopes are marked on the curves. The abscissa shows h on a logarithmic scale. The two curves, $O(h)$ and $O(h^2)$, show the effect of changing numerical schemes. The Runge–Kutta solution is essentially exact.

lists a taxonomy of errors:

- Discretization errors
- Errors that are numerical but not errors of discretization (for example, too short a domain)
- Modelling errors
- Errors in a physical parameter
- Programming errors
- Computer round-off errors

The last category is often easily demonstrated to be unimportant.

With these warnings, we now exhibit some uses of computation in rheology.

8.6 Some test problems

While the computation of inelastic response is now routine, that for viscoelasticity is still a delicate matter. One of the hardest equations to compute with is the UCM model [Eqn (4.25)], and we will frequently present results for this model. Addition of a 'solvent' viscosity gives the Oldroyd-B model [Eqn (4.28)] which has also been often used. A very useful feature has been the repeated solving of several test problems of varying severity. Figure 8.7 shows the five most common test problems. Perhaps the simplest, we have discussed the eccentric cylinder

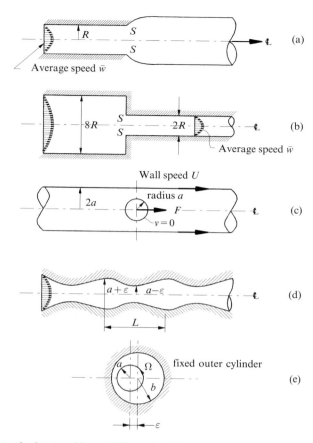

Fig. 8.7 Five standard test problems. (a) Extrusion (plane or axisymmetric). Weissenberg Number $Wi = \lambda\bar{w}/R$ where λ is a relaxation time. Note points of singular stresses S, which are difficult to capture numerically. (b) 4:1 Contraction flow (plane or axisymmetric): the points S cause problems. $(Wi) = \lambda\bar{w}/R$. (c) 2:1 Tube/sphere problem: $(Wi) = \lambda U/a$. No singular points. Drag force F leads to definition of drag coefficient $K = F/6\pi\eta_0 Ua$ where η_0 is the zero-shear rate viscosity. (d) 'Wiggly' tube problem: diameter varies sinusoidally with distance along tube. Amplitude (ε) and wavelength (L) are parameters in problem, besides $\lambda\bar{w}/a$. (e) Eccentric rotating cylinder ('bearing') problem. Two dimensionless criteria are $(Wi) = \lambda\Omega a/(b-a)$ and $(De) = \lambda\Omega$; the eccentricity ratio $\varepsilon/(b-a)$ is also important. (Ω is the angular speed of the inner cylinder).

(Problem (e), Fig. 8.7) in Chapter 6. In this case Beris *et al.* (1983) solved the problem for the UCM model using a combination of finite elements and spectral methods. Huang *et al.* (1996) used a finite volume method. However, the finite volume method did not resolve the thin $O(Wi)^{-1}$ stress boundary layers occurring, despite the fact that the load magnitude and direction agreed with the pseudo-spectral method (Fig. 6.5). No inlet or exit regions are present; only continuity of stresses is needed.

Similarly, Problem (d), the wiggly tube, has periodicity in space.

The solution therefore needs only to be found in a single period of the corrugation. Pilitsis and Beris (1989) used the pseudo-spectral/finite difference method, already used for the eccentric cylinder problem, and again it was successful and relatively stable. A flow resistance f can be defined by

$$f(Re) = 2\pi \Delta p a^4 / \eta L Q, \tag{8.49}$$

where $(Re) = 2\rho Q/\pi \eta a$, ρ is the density, Q is the flow rate, Δp is the pressure drop over a length L, a is the mean radius of the tube, and η is the (constant) shear viscosity. The amplitude of the sinusoidal wall shape is $\varepsilon = \alpha a$; f is a function of α, (Re), (Wi), and the ratio of pitch to radius, L/a.

Pilitsis and Beris (1989) were able to solve the creeping flow of a UCM model quite accurately $[< 0.1$ per cent for $f(Re)]$ and some of their results are shown in Fig. 8.8 for $\alpha = 0.1$, $L/a = 0.16$. Agreement with a perturbation solution is also clear; one sees that $f(Re)$ does not increase with (Wi) and the domain perturbation solution appears to asymptote to a constant $f(Re)$ as $Wi \to \infty$. They emphasized the need for adequate discretization and showed, for $\alpha = 0.01$ (slightly wiggly tube), that for $(Wi) > 40$ inadequate meshing gave an apparent sharp increase in $f(Re)$ which disappeared with mesh refinement. For the Oldroyd-B model, the value of $f(Re)$ initially fell, and then rose as (Wi) increased.

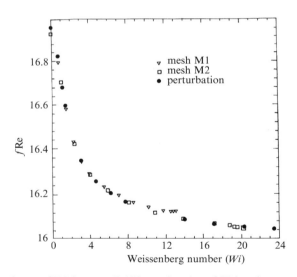

Fig. 8.8 Flow resistance $f(Re)$ [see eqn (8.49)] as a function of Weissenberg number, $(\lambda \bar{w}/a)$, for flow in a tube with sinusoidally varying diameter. Results for the Maxwell (UCM) model. Here the sinusoidal amplitude ε is $0.1a$, where a is the mean tube radius. The pitch/radius ratio (L/a) of the sinusoid is 0.16.

In this work 8–32 Fourier modes were used and up to 200 finite difference modes across the tube; the largest number of unknowns was 38 400. Pilitsis *et al.* (1991) extended this work to various constitutive models, including a PTT model, the power law model, and others, including inertia. They found that their computations for $f(Re)$ did not agree well with experiments and they wondered if there was an instability in the experimental flow which was not captured in the calculations.

While these two problems (d and e) show good results from the spectral method, its geometrical inflexibility is a major problem with complex boundaries.

8.6.1   *Flow around a sphere in a tube—UCM model*

We begin with the UCM model (eqn 4.25).

The test problem 8.7(c) has been widely attacked, and it has proved to be more difficult than one might imagine at first sight. Consider the creeping flow of an upper-convected Maxwell (UCM) model around a fixed sphere of radius a in a long tube of radius R. If the tube speed is U, and the drag on the sphere is F, then

$$K = F/6\pi\eta Ua \equiv f^*[a/R, (Wi)] \tag{8.50}$$

where the force F is compared with the Stokes drag on a sphere in an infinite sea of Newtonian fluid, and the Weissenberg number is $\lambda U/a$. Many computations have been performed with $a/R = 0.5$ [Fig. 8.7(c)] and we will focus on this case first.

Although the problem is geometrically smooth, failures at low $(Wi)(< 0.5)$ occurred up to about 1990 (Tanner and Walters 1998) and no trustworthy results beyond $(Wi) \approx 1.0$ were then available.

Using an EEME (Explicitly Elliptic Momentum Equation) method, Jin *et al.* (1991) produced consistent results up to a (Wi) of 1.5; Lunsmann *et al.* (1993) used EEME and EVSS methods and obtained similar results. Later, Fan and Crochet (1995) used an EVSS method and were able to avoid divergence up to a (Wi) of about 2.1. Baaijens *et al.* (1996) were able to reach a (Wi) of 2.5; Sun *et al.* (1996) reached 2.6 and Luo (1996) published results up to 2.8. The simulations for $(Wi) > 1.6$ are again showing some divergence, but reliable results are now available at medium (Wi) values.

Some recent results are shown in Fig. 8.9 and Table 8.1 shows a comparison of the results of Fan *et al.* (1999) using a finite element code with the ability to resolve thin stress boundary layers, and an average of a group of papers by Sun *et al.* (1996), Lunsmann *et al.* (1993), Luo (1996), Fan (1997), Warichet and Legat (1997), and Luo (1998). At the given (Wi) the highest and lowest K values were discarded and the arithmetic mean of the others is tabulated; the error estimate is formed by studying the remaining maximum differences from the mean. It is clear that as (Wi) increases, the available accuracy goes down by more than one order of magnitude. Most of the computations used in excess of 30 000

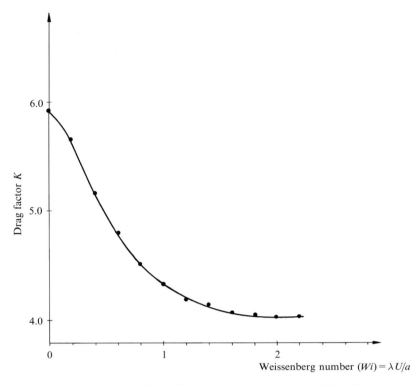

Fig. 8.9   Sphere drag factor $K(=F/6\pi\eta aU)$ versus Weissenberg number $Wi(=\lambda U/a)$ for $a/R=0.5$. Data of Table 8.1.

Table 8.1   Sphere drag coefficient $F/6\pi\eta Ua$

(Wi)	Fan et al. (1999)	Average	Error estimate %
0	5.9476	5.9474	< 0.01
0.2	5.660		
0.4	5.187		
0.6	4.802		
0.8	4.528		
1.0	4.338	4.341	< 0.1
1.2	4.215		
1.4	4.134		
1.6	4.083	4.077	< 0.25
1.8	4.056		
2.0	4.045	4.029	< 0.5
2.2	4.047		

unknowns. The problem with this geometry lies in the thin stress boundary layers, hard to resolve, and in the wake region, where Fan *et al.* (1999) have shown that accuracy and convergence are hard to achieve. Some results up to $(Wi) \approx 4$ have been reported, but they have not been well replicated by others. One can ask whether the drag curve will ultimately rise with increase of (Wi). Walters and Tanner (1992) suggest that experiments with Boger (constant viscosity) fluids show an initial decrease with (Wi) and then an upturn. So far no obvious increase has been noted in the UCM model calculations with $a/R = 0.5$, (see Fig. 8.9) although convergence problems often show up as an increase in the drag coefficient K. There is therefore an argument suggesting that K may stay, for the UCM model and $a/R = 0.5$, at a constant value as (Wi) becomes large.

Most workers have computed the drag by integrating around the sphere, but one may also integrate around the contour ABCD. (Fig. 8.10). Since the extra stress τ is zero at inlet (AB) there is no problem with inlet boundary conditions; a long exit is needed to ensure that there are no residual stresses at exit; Fan *et al.* (1999) used an exit length (Le) of $25a$. As the fluid is carried away at a speed U, it will take a time of about Le/U to reach the exit. In that time, assuming an exponential stress decay, an attenuation of stress of about $\exp(-Le/\lambda U)$ will occur, or $\exp(-Le/a(Wi))$.

Thus for $(Wi) = 3$, $Le/a = 25$, an attenuation of about 2.4×10^{-4} will occur, which is only just sufficient for the claimed accuracy of the computations. At higher (Wi), a very long exit region is clearly needed. Calculating the drag force F by a simple force balance on ABCD, and assuming a zero traction exit, where $p = 0$, one finds

$$\pi R^2 \Delta p + 2\pi R \int_A^B \tau_{\rm w} \, dz = F. \tag{8.51}$$

In the case $R = 2a$ we find, in dimensionless terms,

$$K = \frac{2}{3} \Delta p + \frac{2}{3} \int_A^B \tau_{\rm w} \, dz. \tag{8.51a}$$

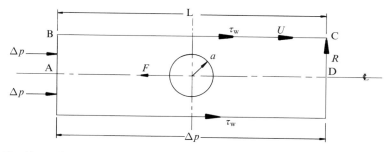

Fig. 8.10 Alternative computation of sphere drag. [see eqn (8.51)]. The pressure drop Δp is made dimensionless with $\eta U/a$.

One can find Δp as a function of (Wi), Fig. 8.11 and Table 8.2 shows this quantity. τ_w is negative, and the stress pattern shifts downstream as (Wi) increases, without much change in the value of the shear integral; K seems to be tending towards a constant. This behaviour would be consistent with the $f(Re)$ results for the UCM sinusoidal tube problem reported above — they reach a constant $f(Re)$ value as (Wi) becomes large.

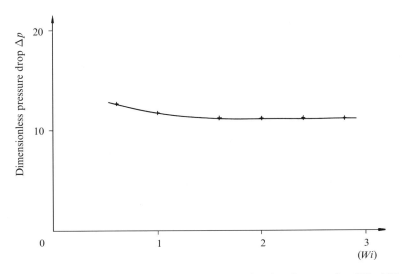

Fig. 8.11   Dimensionless pressure drop Δp as a function of Weissenberg number $Wi(\equiv \lambda U/a)$ for $a/R = 0.5$, UCM model.

Table 8.2   Sphere in tube drag

(Wi)	Δp	K (num)	$\int_{-\infty}^{\infty} \tau\,dz$	$-\int \tau\,dz/\Delta p$	K (calc.) eqn (8.55)	% error
0	14.99	5.947	−6.069	0.405	5.947	
0.6	12.69	4.798	−5.493	0.432	4.63	−4
1.0	11.79	4.330	−5.295	0.449	4.34	−
1.6	11.39	4.077	−5.275	0.463	4.14	+1
2.0	11.33	4.030	−5.285	0.466	4.06	−
2.4	11.35	4.030	−5.305	+0.467	4.01	0.5
2.8	11.31	4.058	−5.223	+0.462	3.97	−2
3.2	11.32	4.04**	−	−0.465(*)	3.94	−1
∞	−	−	−	−	3.71 ± 0.1	

** inferred; (*) assumed.
Drag factor $K = F/6\pi\eta Ua$ in creeping flow for $a/R = 0.5$; UCM model [eqn (4.25)]. τ, Δp made dimensionless with $\eta U/a$; $(Wi) = \lambda U/a$ [Fig. 8.7(c)].

An approximate analysis can be made, as follows.
We assume (see Luo 1996, for example)

(i) The flow field changes little from the Newtonian case

(ii) That, with the UCM model, the pressure drop Δp continues to fall, as (Wi) increases. The plausibility of this argument is strengthened by the asymptotic theory and the sinusoidal tube calculations of Pilitsis and Beris (1989), which show this trend.

(iii) That the two components of drag due to pressure drop Δp and wall shear stress respectively [eqn (8.51a)], are linearly related, so that, for $(Wi) > 1$, one has

$$\int_{-\infty}^{\infty} \tau \, dz = -0.465 \Delta p. \tag{8.52}$$

The actual computed values are shown in Table 8.2.

From Luo's (1996) contour plots, one sees, as (Wi) increases, that the pressure drop is more and more concentrated in the downstream direction. This follows because in the inlet region to the narrowest constriction ($0 > z > -1$) the UCM model cannot respond viscously to a suddenly imposed strain rate and so the pressure needed to drag fluid through the gap is diminished. An approximate computation assumes that the region $z = -1$ to 0 is affected by this lack of response. If a constant shear rate of order U/a, say cU/a, where c is a constant, is imposed at $z = -1$, then if the fluid shear stress response is linear, the dimensionless shear stress τ obeys

$$\frac{(Wi)}{\beta} \frac{d\tau}{dz} + \tau = c, \tag{8.53}$$

where β is a factor which sets the (average) convection speed at U/β instead of U. Solution of (8.53) and insertion of a boundary condition $\tau = \alpha c$ at $z = -1$ gives the result:

$$\tau = c[1 + (\alpha - 1) \exp(-\beta(1 + z)/(Wi))]. \tag{8.53a}$$

The factor αc accounts for initial stresses developed before the entry region is reached. Computation of $\int_{-1}^{0} \tau \, dz$ gives the result

$$-\int_{-1}^{0} \tau \, dz = c\left[1 - (1 - \alpha)\frac{(Wi)}{\beta}(1 - \exp(-\beta/(Wi)))\right]. \tag{8.54}$$

With this result and assumption (iii) above, one can compute the drag factor K when c, α, and β are known, since $K = -0.776 \int_{-\infty}^{\infty} \tau \, dz$, from (8.51a). (Contributions to the integral outside $-1 < Z < 0$ are neglected.)

To fit the data at $(Wi) = 0$, we set $0.776c = 5.947$ or $c = 7.75$. As $(Wi) \to \infty$, $K \to 5.947\alpha$. The limiting K, from Table 8.2, is not known. To find α and β the curve is fitted at $(Wi) = 1$ and 2 giving $1 - \alpha = 0.377$, $\beta = 0.702$, and so

$$K = 5.947[1 - 0.536(Wi)(1 - \exp(-0.702/(Wi)))]. \tag{8.55}$$

Table 8.2 shows the values of K calculated from this formula. The limiting $K(\infty)$ is 3.71 from (8.55), but this value is certainly subject to possible errors of at least ± 2–3% (± 0.1). The drag is reasonably well fitted by (8.55). Since the argument depends on the constancy of Δp and the shear integral, caution is needed for other a/R ratios and other models. The a/R ratios of $1/4$ and $1/8$ have been considered by Lunsmann et al. (1993). The change of K with (Wi) for $a/R = 0.125$ shows an initial decrease of K, then an increase for $(Wi) > 0.8$. For $a/R = 0$, with the UCM model, the drag factor is practically constant up to $(Wi) = 1$ (Gu and Tanner 1985).

Experiments on configurations where $a/R \geq 0.4$ are prone to show the inability of the sphere to stay on the centreline, and so few detailed studies of experimental correlations are available. Also, (Bot et al. 1998) there is evidence that two spheres following one another, in falling ball experiments, assume different speeds. This effect is probably due to destruction of the structure in the fluid and a slow recovery (Walters and Tanner 1992).

8.6.2 Other sphere problems

The flow of inelastic fluids around spheres is of interest. Gu and Tanner (1985) studied the power-law case and showed that wall effects were minimal for $n \leq 0.5$. They compared their computations with upper and lower analytical bounds and also with experiments. (Fig. 8.12).

Butcher and Irvine (1990) studied a modified power-law model and compared their computations with a solution of carboxymethyl cellulose in water. Beaulne and Mitsoulis (1997) studied the flow of Herschel–Bulkley fluids. Beris et al. (1985) studied a true Bingham flow.

Jin et al. (1993) studied the flow of a PTT model using the EEME method and the increased stability of the computation was clear. The reduction of drag due to shear-thinning was also clear; this led to a recirculation behind the sphere.

Lunsmann et al. (1993) studied the Oldroyd-B model and the Chilcott–Rallison (Section 5.5.1) model. Results for the drag factor K, at $a/R = 0.125$, first showed a slight decrease, then an increase for $(Wi) \gtrsim 0.8$. Changes from the Newtonian results were small, of the order of a 2 per cent increase at $(Wi) = 2$. The Chilcott–Rallison model gave a small increase. Thus the $a/R = 0.5$ and $a/R \lesssim 0.125$ cases behave differently; in the former the wall effect dominates the drag, whereas for the latter the extended wake is important. Generally, shear thinning is always important.

8.6.3 Cylinder in a channel

The flow past a cylinder (radius a) in a channel with fixed walls distant 2H apart has been studied and sometimes compared with experiment. The computations

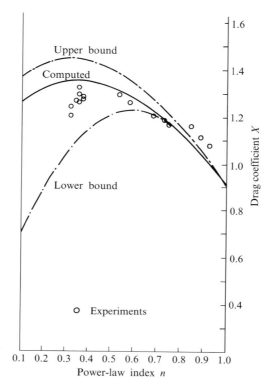

Fig. 8.12   Drag F for spheres in unbounded power-law fluids, where $\tau = k\dot{\gamma}^n$. The drag coefficient X is here defined by $X = 2^n F / 12\pi k U^n a^{2-n}$.

for the UCM model are more difficult than in the sphere case (Fan *et al.* 1999), especially with regard to the stresses behind the sphere. The drag coefficient $K_p = F/\eta\bar{u}$ (here F is the drag force per unit length and $\bar{u}$ is the mean channel velocity) is a function of a/H and $(Wi) = \lambda\bar{u}/a$, and shows the same continuous reduction as in the sphere case for $a/H = 0.5$. The Oldroyd-B model was, as expected, easier to compute with than the UCM model; it is not clear from these computations if there is a minimum value of K_p. Sun *et al.* (1999) investigated the Oldroyd-B result for two a/H values (0.5 and 0.125). For the $a/H = 0.125$ case, there appears to be a shallow minimum value of K_p at about $(Wi) \sim 0.7$, and then an increase in drag. The increase for $a/H = 0.5$ was not seen by Fan *et al.* (1999).

The Giesekus model was also used by Sun *et al.* (1999) and showed a continuous decline in K_p with (Wi), at least up to $(Wi) = 14$. Comparison with experiment has been made by Baaijens *et al.* (1997) who used two 4-mode PTT models and a Giesekus 4-mode model to fit low-density polyethylene data, with auxiliary calculations using one- and eight-mode models. Stress birefringence patterns were compared with computations up to a Weissenberg number of about 5 and reasonably good agreement was noted. For a polyisobutylene solution, see also Baaijens *et al.* (1995).

8.7 Flow near corners and separation points

So far the problems considered have had smooth geometrical outlines. The remaining two [Fig. 8.7(a) and (b)] have sharp corners or separation points. The flow near these points presents additional difficulties because even simple constitutive models generate singular stress fields. We will begin with creeping Newtonian (Stokes) and generalized Newtonian flows.

In Fig. 8.13(a) we show the flow near a sharp corner.

Since there is no length scale, we assume the stream function ψ for the problem can be expressed in the form

$$\psi = r^m f_m(\theta), \tag{8.56}$$

where the exponent m has to be found. In a creeping Newtonian flow the satisfaction of the Stokes equations can be shown to be equivalent to requiring that

$$\nabla^4 \psi = 0, \tag{8.57}$$

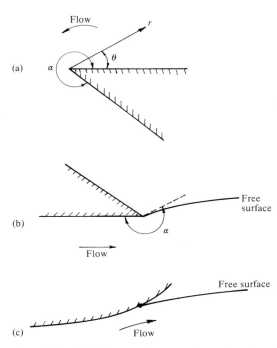

Fig. 8.13 Singular points of flow. (a) Flow round a corner that projects into the fluid. (b) Flow at a sharp, well-defined separation point. (c) General separation point.

where

$$\nabla^2 = \frac{\partial^2}{\partial r^2} + \frac{1}{r}\frac{\partial}{\partial r} + \frac{1}{r^2}\frac{\partial^2}{\partial\theta^2}.$$

Substituting (8.56) in (8.57) one finds an equation for f_m which can be solved to give the general solution

$$f_m = A\cos m\theta + B\sin m\theta + C\cos(m-2)\theta + D\sin(m-2)\theta, \tag{8.58}$$

where A, B, C, and D are arbitrary constants. Case (a) (solid walls) has been extensively treated by Dean and Montagnon (1949). Here we require $\mathbf{v}=\mathbf{0}$ on the solid walls at $\theta = 0, \alpha$. [Fig. 8.13(a)]. This gives four requirements for the determination of the constants; these lead to a characteristic equation for m:

$$\sin\alpha(m-1) = \pm(m-1)\sin\alpha. \tag{8.59}$$

The minus sign in (8.59) gives a series of antisymmetric flow patterns (about a symmetry line $\theta = \alpha/2$) and the positive sign gives symmetric flow patterns.

Dean and Montagnon (1949) have solved (8.59) for the exponent m as a function of α. The results are shown in Table 8.3; these roots correspond to the antisymmetric flow patterns. The important idea here is that corners that point into the flow ($\alpha > 180°$) give singular stresses at the corner.

We prefer to call these solid corners that stick into the flow salient corners; commonly in the literature they are referred to as re-entrant corners, but this can only refer to the fluid, not to the boundaries themselves.

Since the stresses are proportional to r^{m-2}, the stresses behave like $r^{-1/2}$ when $\alpha = 360°$, and the singularity is less powerful when α is smaller. For $\alpha < 180°$, the corner stresses are zero; when $\alpha < 146.3°$ eddies begin to appear in the corner as the exponent m is then complex.

In the case of the separate problem [Fig. 8.13(b)] the angle α of separation is also an unknown quantity. Michael (1958) showed that the separation occurred

Table 8.3 Corner flow exponents

$\alpha°$	m	$\alpha°$	m
360	1.500	220	1.697
340	1.500	200	1.818
320	1.503	180	2.00
300	1.512	170	2.13
280	1.530	160	2.29
270	1.545	155	2.42
260	1.563	150	2.53
240	1.616	146.3	2.76

For angles less than 146.3° the exponent m is complex.

at an angle $\alpha = \pi$. In that case the exponent of the singularity is real and has a minimum value of $m = 3/2$. Thus the stresses (and the pressure) behave like $r^{-1/2}$ near the separation point. Experimentally (and also in computations) separation does not always occur as suggested by Michael. It seems probable that either separation occurs as at Fig. 8.13(c) or that the Newtonian fluid physics is inadequate near the singular point, but further research on this point is needed. (See Jean and Pritchard 1980; Tanner et $al.$ 1985.) Hutchinson (1968) has extended this result to power-law fluids; if

$$\tau = k\dot{\gamma}^n \tag{8.60}$$

in simple shearing, then he found that the stress singularity in such problems behaves like $r^{-n/(n+1)}$, approximately. Tanner and Huang (1993), using the J-integral method of fracture mechanics, confirmed that this is an exact result. Thus as the fluid parameter n decreases from 1 to zero (pseudo-plastic case), the power of the singularity decreases from $-1/2$ to zero. Tanner and Huang (1993) also investigated several other inelastic models (Carreau, biviscosity (Herschel–Bulkley) models), and they computed the intensity of the singular stresses in several cases. See also Henriksen and Hassager (1989) for flow of a power-law fluid round corners.

For the second-order model, the velocity field is the same as the Newtonian case (Section 4.11.1), and one finds that the normal stresses near the singular point behave like r^{-1}, which is non-integrable. One thus reconfirms the unsuitability of this model in rapidly-varying stress fields.

Tanner and Huang (1993) also considered the flow parallel to sharp edge (edge flows); that is, a parallel flow $w(r, \theta)\mathbf{k}$ in Fig. 8.13(a). We shall refer to some of these results in connection with computational results given later.

8.7.1 Viscoelastic models

The methods employed in the previous section do not help with the UCM and Oldroyd-B analyses. Renardy (1993) considered the problem of the UCM model when $\alpha = 270°$. He assumed that the flow field was Newtonian. The stresses were then calculated and were found to behave like $r^{-0.74}$. At the walls, using a theorem of Caswell (1967), the flow is viscometric, and the normal stresses behave as the square of the shear rate, like $r^{-0.91}$. Thus boundary layers appeared and matching was difficult. Hinch (1993) considered the same problem for the Oldroyd-B model, assuming no lip vortex was formed, so that the streamlines continue around the corner. The flow upstream near the wall is viscometric. Rewriting the Oldroyd-B equation as

$$\lambda\frac{\Delta\tau}{\Delta t} + \tau = 2\eta\left[\mathbf{d} + \lambda_2\frac{\Delta\mathbf{d}}{\Delta t}\right], \tag{4.13a}$$

where λ_2 is the retardation time, there has been a tendency to argue that the two $\Delta/\Delta t$ terms dominate, and are of equal importance. This argument then shows that the model behaves, near the singular point, like a Newtonian fluid, with viscosity $\lambda_2\eta/\lambda$. This does not seem to be what occurs (Hinch 1993; Tanner and Huang 1993). In any case, for the UCM model, where $\lambda_2 = 0$, this gives a zero stress. Hinch considered an alternative limit $\Delta\tau/\Delta t = 0$ in a core region, which follows the weak, viscometric flow upstream. Writing this result in Cartesian form, in the core one has

$$v_k \frac{\partial \tau_{ij}}{\partial x_k} - \frac{\partial v_i}{\partial x_k} \tau_{kj} - \tau_{ik} \frac{\partial v_j}{\partial x_k} = 0. \tag{8.61}$$

Letting $\tau_{ij} = h(\psi)v_i v_j$, where h is a function of the stream function ψ, and substituting in (8.61), one finds the left-hand side of this equation reduces to $v_i v_j v_k \,(\partial h/\partial x_k)$, which is zero if h is a function of the stream function, as assumed. Substitution of this solution in the momentum equation, using the incompressibility condition, enables one to eliminate the pressure, and leads to a non-linear eigenvalue problem. The final result gives all the stresses behaving like

$$\sigma \sim r^{-2(1-\pi/\alpha)}, \tag{8.62}$$

where one has chosen the stream function ($\sim \sin^n m\theta$) to be zero at $\theta = 0, \alpha$.

When $\alpha = \pi$, the solution (8.62) is regular. For $\alpha = 2\pi, n = 2\frac{1}{2}$, and the stresses are singular like r^{-1}. When $\alpha = 3\pi/2$, $m = 2/3$, $n = 7/3$, and the stresses are singular like $r^{-2/3}$. Hinch also showed that the solvent stresses were less singular than the elastic stresses.

Renardy (1995) matched upstream and corner flows for the 270° angle case. He found $\sigma \sim r^{-2/3}$, $\dot{\gamma} \propto r^{-4/9}$ in the core region, in agreement with Hinch, and upstream $\sigma \sim r^{-2/3}$, $\dot{\gamma} \propto r^{-1/3}$. During this work with the UCM and Oldroyd-B models Hagen and Renardy (1997) showed that the thin stress boundary layers are of order (Wi^{-1}), but are only of order $(Wi)^{-1/3}$ for the PTT model. These results go far to explaining the greater ease of computing with the PTT model.

For the stick-slip flow, the PTT model gives Newtonian behaviour near the singularity (Tanner and Huang 1993; Renardy 1997).

8.7.2 Numerical results near singular points

It is extremely difficult to achieve a satisfactory numerical solution near singular points because of the steep gradients. We shall see that establishing the exponent of the stresses and the level needs very great mesh refinement.

Newtonian stick-slip problem
Xue *et al.* (1999) and Salamon *et al.* (1995) have studied this problem [Fig. 8.14(a)]. The former group used a finite volume method and the latter a finite element program.

Fluid enters far upstream with a Poiseuille flow in a channel of depth $2H$ and very large width. For $x > 0$ the (thin) walls vanish, and slip occurs, and the velocity eventually settles to a uniform speed $\bar{U}$ far downstream. The point S, at

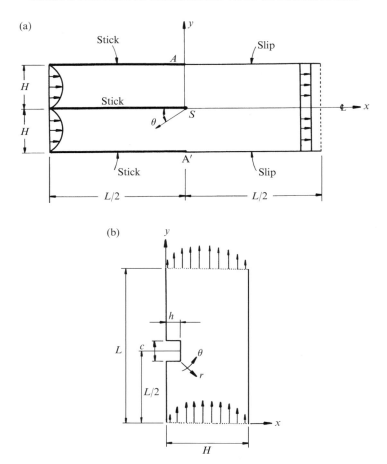

Fig. 8.14 Schematic diagram of: (a) plane stick-slip flow (if a constant pressure gradient is specified along the x-direction or a fully developed Poiseuille flow is imposed at the entrance); and edge flow (if top and bottom plates are moving at a constant speed in z-direction). The singularity is located at the edge $S(0,0,z)$ in the Cartesian coordinate system (x,y,z). A local polar coordinate system (r,θ) with θ corresponding to a ray leading to $x<0$ along the centreline ($y=0$) is used for the analysis of the singularity, and (b) plane flow past a junction in a channel. A local polar coordinate system (r,θ) is used for the analysis of the singularity, and the Cartesian coordinate system (x,y) is for calculation.

the end of the no-slip solid wall, is the singular point. For Newtonian creeping flow the form and strength of the singularity are known exactly (Tanner and Huang 1993).

Upstream, on the solid wall, the shear stress is given by

$$\tau = 2\eta\bar{U}(3/2\pi)^{1/2}(Hr/2)^{-1/2} + \mathrm{O}(r^{1/2}),$$

where η is the viscosity (one should note that the pioneering analysis of Richardson (1970) gives results about 16 per cent too low). The factor $(3/2\pi)^{1/2} \approx 0.691 (\equiv \alpha_{1/2})$ and so in dimensionless terms the factor $\alpha_{1/2}$ represents the intensity of the stress field at S. Note that the length scale used here is $H/2$. The stresses in the filament $y = 0$ downstream near S can be found and the normal stress difference $\tau_{rr} - \tau_{\theta\theta}(= N_1)$ can be found to be

$$N_1 \sim 4\eta\bar{U}\alpha_{1/2}(Hr/2)^{-1/2} \tag{8.63}$$

Three fine meshes were used in the computation, shown in Table 8.4.

Comparisons of the N_1 formula (8.63) and the velocity field are given in Table 8.5, for downstream elements along the centreline. The computed values of N_1 shown in this table are in error by ~ 2 per cent at the (different) radial positions. Figure 8.15 shows the smoothness of the solutions. Salamon $et\ al.$ (1995) used ultrafine meshes to get the result shown in Table 8.5; their minimum

Table 8.4   Meshes used

Mesh	Number of nodes $i\max \times j\max$	Minimum (dimensionless mesh size)
M1	22×106	0.025
M2	44×152	0.0025
M3	70×204	0.00025

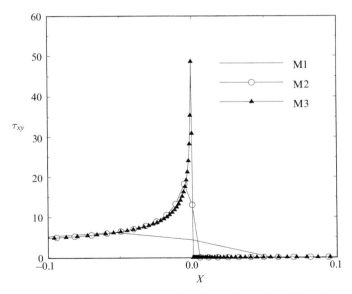

Fig. 8.15   Distribution of τ_{xy} along the centreline for a Newtonian fluid calculated with the different three meshes M1, M2, and M3 in stick-slip flow; $X = x/H$.

Table 8.5 Newtonian stick-slip results

Mesh	M1	M2	M3
Radial position r	0.05618	0.05386	0.05206
u_r (analytic)	0.328	0.321	0.315
u_r (computed)	0.326	0.314	0.309
N_1 (analytic)	11.66	11.91	12.11
N_1 (computed)	11.33	11.59	11.66
N_1 (Salamon *et al.* 1995)	11.68	–	–

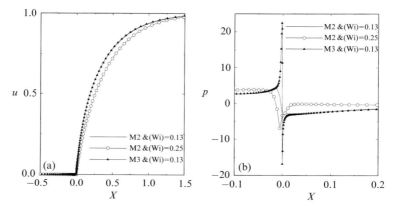

Fig. 8.16 Velocity and pressure fields for UCM model at two Weissenberg numbers ($Wi = 2\lambda \bar{U}/H$) as functions of dimensionless distance $X = x/H$.

mesh element was about 4×10^{-7} units in size. It is clearly very difficult to get accurate results near singular points.

UCM and PTT stick-slip corner and edge problems.
Much more difficult is the UCM stick-slip problem, see Fortin (1992). Here the Weissenberg number is defined as $2\lambda \bar{U}/H$. Two Weissenberg numbers (0.13 and 0.25) have been used and Figs 8.16 and 8.17 show these results. Note the extremely fine meshes needed to get smooth results. The slope appears to be close to the viscous result, -0.5. Unfortunately Hinch's (1993) and Renardy's (1995) results do not give a result for the stick-slip case, and so the problem in Fig. 8.14(b) with a sharp 270° corner was computed. A reasonable agreement with the $-2/3$ prediction for stress, and the $+2/3$ power for the velocity (Renardy 1995) was found (Fig. 8.18). (All of these results are taken along the x-axis). For the PTT model ($\xi = 0, \varepsilon = 0.1$), there seems to be a very small, roughly Newtonian region, as predicted (Renardy 1997.)

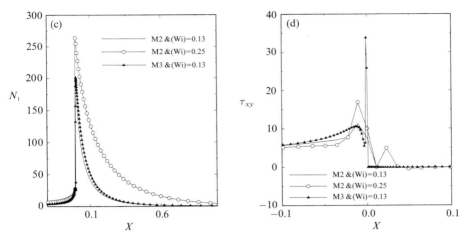

Fig. 8.17 $N_1 (\equiv \tau_{xx} - \tau_{yy})$ and τ_{xy} fields for above.

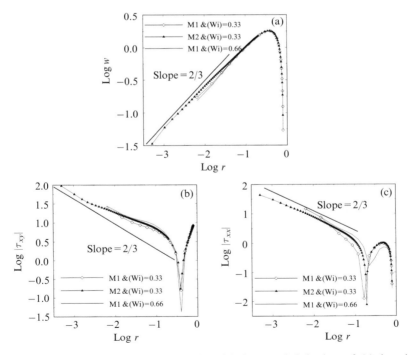

Fig. 8.18 $270°$ corner problem for UCM fluid model. Asymptotic behaviour of: (a) the velocity magnitude w; (b) the shear stress τ_{xy}; and (c) the normal stress τ_{xx} at angle $\theta = \pi/2$ around the salient corner for an UCM fluid at $(Wi) = 0.33$ and $(Wi) = 0.66$ on two different meshes.

For the edge flow, where the flow is *along* the sharp edge, parallel to OZ, exact predictions of the singularity form are available (Tanner and Huang 1993). The results of Xue *et al.* (1999) show the small region of $r^{-1/4}$ behaviour for the PTT model stresses, which was only detected with the finest mesh.

One sees that stable, smooth results can be obtained if fine enough meshes are used in these problems. The question of the validity of the physical description under extreme stress conditions actually must now be faced. One of the few investigations which uses molecular dynamics simulation to study the physics of singularity flows (see Section 8.10) is the work Koplik and Banavar (1997). They were able to find reasonable agreement with the -0.5 Newtonian slope prediction using short molecules, and with chains of around 30 units they found a larger exponent, up to ~ -0.75. Further consideration of the physics near the corner is clearly needed, in view of the opportunities for slip, cavitation and molecular scission there.

In view of the difficulties involved near corners, the two remaining test problems clearly pose great problems, especially for the UCM model. Fortunately, the PTT and similar models are somewhat easier to compute with.

8.8 Entry flow

We now consider the entry flow [Fig. 8.7(b)]. The flow pattern is complex and viscoelasticity makes great changes to the corner vortices (Fig. 8.19). Figure 8.20 shows a sketch of the problem. Often the contraction ratio $\beta(\equiv D_u/D_d)$ has been set at 4, and Boger (1982) and his colleagues have performed many experiments with this geometry. A plane analogue of this problem is also possible, but end effects, due to the finite channel width, then complicate matters.

Aside from the vortex size and strength, the losses at entry are of interest. These need to be defined carefully with non-Newtonian fluids. Usually, in viscoelastic computations, at the upstream entry plane, a fully developed velocity profile is prescribed, plus associated extra stress (τ_{ij}) components. Downstream only the velocity needs to be prescribed, plus possibly a reference pressure at a single point. Alternatively, the exact stress distribution can be imposed across the outlet plane; the point to observe is the variation of stress across the section [see eqn (3.57) above] and, for example, with the UCM model the pressure is constant across the tube section, but the axial stress is not.

The question of how to define the losses in a contraction is probably best resolved by defining

$$\Delta p_e = \Delta p - \Delta p_u - \Delta p_d \qquad (8.64)$$

where Δp_e is the entry loss, Δp is the overall loss and Δp_u and Δp_d are the losses which would ensue if appropriate fully-developed Poiseuille flows persisted up and downstream from the SS' plane. Δp is the value of the pressure difference on the centreline in Fig. 8.20. In addition, there will also be a momentum change which causes an extra loss but this vanishes in creeping flows and in other cases

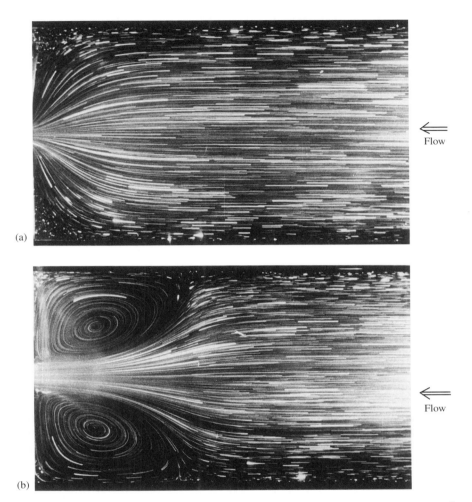

Fig. 8.19 Entry flows; flow is from right to left. (a) Newtonian fluid, $(Re) \sim 1.5 \times 10^{-3}$, $\dot{\gamma}_w = 20\,\text{s}^{-1}$, $(Wi) = 0$. (b) Viscoelastic fluid, $(Re) \sim 1.8 \times 10^{-3}$, $\dot{\gamma}_w = 24\,\text{s}^{-1}$, $(Wi) = 0.108$. Both fluids have comparable viscosity (20 Pa-s). Note small vortex in Newtonian case and large, strong vortex with 'wine-glass' flow in viscoelastic case. Here $(Wi) = \lambda \bar{w}/D$, where D is the downstream diameter and $\bar{w}$ is the average speed downstream. The upstream tube diameter is $7.68D$. (Photo by courtesy of Professor D.V. Boger, University of Melbourne.)

can be readily computed (Problem 8.9). For the UCM model, the overall momentum change is the same for both Newtonian and non-Newtonian flow, in any case.

Other ways of defining the losses are possible: one might prefer to define Δp from measurements at the tube walls; for fluids in which the second normal stress difference N_2 is zero this reduces to the same values as the centreline definition

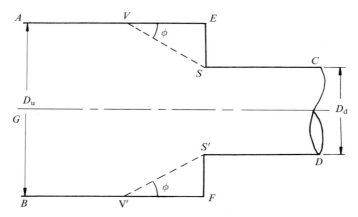

Fig. 8.20   Inlet flow problem. Inside the areas VSE and $V'S'F$ are vortices. ϕ is the vortex opening angle. $S\text{–}S$ are singularity points. Contraction ratio $\beta = D_u/D_d$.

[see eqn (3.57)]. Or, the work necessary to generate the elastic field in the fluid could be considered. Since we will mostly deal with cases where $N_2 = 0$, the simplest definitions will be used.

It is clear that there is no 'flat' or uniform entry profile in any case. Boger (1982) has discussed the theoretical and experimental facts of entry flows; despite the number of entry-length calculations using a flat (uniform) velocity entry profile, there is no evidence that this flow ever occurs. Hence we shall not survey this area of work, especially as it is mainly concerned with higher Reynolds numbers and Newtonian flows.

With the entry flows, the contraction ration β is an important parameter which affects the flow field. Boger (1982) has surveyed the entry lengths for abrupt entries for various β values. He concludes that for axisymmetric contraction ratios greater than 2 one can write

$$\frac{L_e}{R} = 0.49 + 0.11(Re), \qquad (8.65)$$

where the full development is judged to have occurred when the centreline velocity reaches 99 per cent of the fully-developed value. The result is valid up to $(Re) = 2100$, and is a conservative estimate. The creeping flow portion of eqn (8.65) has been subjected to experimental verification with fair results; it is a difficult experiment. Similarly, the entry loss pressure drop Δp_{en} can be written (for $\beta \geq 4$)

$$\frac{\Delta p_{en}}{2\tau_w} = 0.589 + 0.0709(Re). \qquad (8.66)$$

Table 8.6 gives some values for $\Delta p_{en}/2\tau_w$; it is probable that the computer estimates are more accurate than experiments in this case. Exit losses Δp_{ex} are also

Table 8.6   Maxwell fluid losses

Weissenberg number $(Wi)(\lambda\dot{\gamma}_{\mathrm{w}})$	Entry loss $\Delta p_{\mathrm{en}}/2\tau_{\mathrm{w}}$	Exit loss $\Delta p_{\mathrm{ex}}/2\tau_{\mathrm{w}}$	Total loss entry + exit
0.00	0.566	0.234	0.800
0.25	0.470	0.319	0.789
0.50	0.375	0.426	0.801
0.75	0.279	0.525	0.805
1.00	0.185	0.655	0.839

The entry is a sharp-edged axisymmetric 4:1 contraction: for comparison with Fig. 8.19 note $\lambda\bar{w}/R = 0.25\lambda\dot{\gamma}_{\mathrm{w}}$.

Table 8.7   Exit and entry losses for long tubes

Fluid model	Exit loss $\Delta p_{\mathrm{ex}}/2\tau_{\mathrm{w}}$	Entry loss $\Delta p_{\mathrm{en}}/2\tau_{\mathrm{w}}$	Sum $e = (\Delta p_{\mathrm{ex}} + \Delta p_{\mathrm{en}})/2\tau_{\mathrm{w}}$
Newtonian $(\beta \gg 1)$ (no surface tension)	$0.25 - 0.03\ (Re)$	$0.59 + 0.071\ (Re)$	$0.84 + 0.041\ (Re)$
Creeping power-law $(\beta \gg 1)$			
$n = 1.0$	0.25	0.59	0.85
0.9	0.25	0.65	0.90
0.8	0.25	0.73	0.98
0.7	0.25	0.84	1.09
0.6	0.26	0.98	1.24
0.5	0.26	1.17	1.43
0.4	0.27	1.46	1.73
0.3	0.28	1.94	2.22
0.167	0.59	3.54	4.13

given. For shear-thinning fluids less extensive investigations are available, but at higher Reynolds numbers greater entry lengths are found (Boger 1982) for power-law fluids. Table 8.7 gives the computed entry losses for power-law fluids in creeping flow.

Abdali *et al.* (1992) have studied entry and exit flows of an approximate Bingham model. Kim-E *et al.* (1983), with a 4:1 contraction, found $L_{\mathrm{e}}/R \sim 1.15$ for $n = 0.4$ and 1.59 for $n = 0.2$. Thus the figures in Table 8.7 are expected to be upper bounds. It is doubtful if the inelastic fluid model is a proper one for describing shear-thinning fluids in these complex flows (cf. Fig. 8.22), and in experiments it is often difficult to separate elastic and shear-thinning effects. Thus we now consider viscoelastic effects in entry flows.

Computations on Maxwell fluids and second-order fluids at low (Wi) values yield entry pressure losses *less* than the purely viscous case (Caswell and Viriyayuthakorn 1983), see Table 8.6. The total loss coefficient is nearly constant, which is also unrealistic (Boger 1982), see Fig. 8.21; for the second-order fluid the

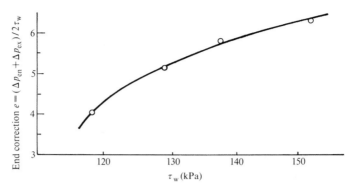

Fig. 8.21   The Bagley and correction (e) versus shear stress for high-density polyethylene at 180 °C (from Han 1976). For comparison, the Newtonian value of *e* is constant and is about 0.8.

plane entry loss can readily be computed (Problem 8.7). Keunings and Crochet (1984) have numerically solved the entry problem for a modified PTT fluid in a 4:1 contraction and the results are shown in Fig. 8.22 (here the Weissenberg number is defined as $\lambda\bar{w}/R$, where R is the radius of the downstream tube). Thus the observed higher losses now become evident with this viscoelastic, shear-thinning model; the elongation of the corner vortices appears (cf. Fig. 8.19), and generally the simulation is more realistic. In Fig. 8.22 the decreasing losses due to the Oldroyd-B model (essentially a Maxwell fluid plus a constant viscosity) are to be noted.

The results with the Maxwell model are reminiscent of the sinusoidal tube losses discussed in Section 8.6. The survey of Baaijens (1998) discusses the then current state of computing with the UCM and Oldroyd-B models; finite element methods are the principal concern of the paper.

The need for the use of realistic constitutive models again emerges clearly in this example. It is also found that more realistic constitutive models are much more stable in computations than the convected Maxwell, Oldroyd-B and second-order models, enabling much higher Weissenberg numbers to be reached.

Aziez *et al.* (1996) were able to show results for the planar 4:1 contraction using the PTT, Giesekus and FENE-P models (see Chapter 5). They used single-mode models and correlated their results with experiments. Their numerical results were obtained on three meshes and they compared reasonably well near the salient corner for (Wi) $(= \lambda\bar{U}/H)$ equal to 2.9; here $\bar{U}$ and H refer to the downstream section. Considerable differences in pressure loss between the three models were noted, and agreement between computed and experimentally measured stresses was only moderate, possibly due to the one-mode models used.

Byars *et al.* (1997) compared computation and experiment for the axisymmetric 4:1 contraction flow. They studied 4-mode PTT and Giesekus models up to $(Wi) = 5$. (Here $(Wi) = 2\lambda_0\bar{U}/D_d$; λ_0 is the relaxation time $[\Psi_1/2\eta]_{\dot{\gamma}\to 0}$; see Section 4.11).

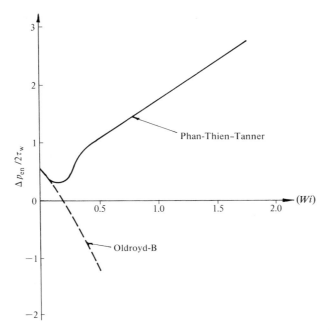

Fig. 8.22 Entry loss as a function of the Weissenberg number $\lambda \bar{w}/R$ computed for the Phan-Thien–Tanner fluid ($\varepsilon = 0.015$, $\xi = 0.1$) and the Oldroyd-B fluid.

The agreement with the PTT model was quite good, as judged by the vortex sizes appearing.

Some work with KBKZ-type integral models has also been done. An early paper is that of Dupont and Crochet (1988). They used the integral model for low-density polyethylene of Luo and Tanner (1988) and were able to show reasonable agreement with some experiments.

Luo (1996) used a finite volume method with the same material models as Dupont and Crochet, with a contraction ratio of 5.75:1 to match with experiments. Eight relaxation times were used and results are plotted versus the stresses ratio (so-called recoverable shear) S_R, which is defined to be

$$S_R = (N_1/2\tau)$$

evaluated at the far downstream wall.

S_R is a measure of the elastic effects. A further quantity is the nominal shear rate, Γ equal to $8\bar{U}/D_d$. Of interest is the vortex opening angle ϕ (Fig. 8.20) and the entrance pressure correction normalized with twice the downstream wall shear stress. Three meshes were used and the results were compared with experiments and other computations. The agreement was good below $10\,s^{-1}$ nominal

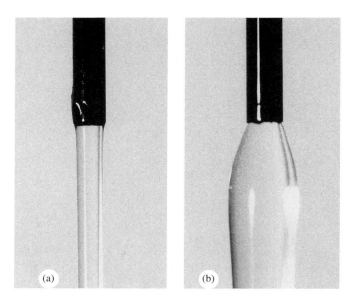

Fig. 8.23 Extrudate swell. At (a) is a Newtonian stream exiting from a capillary and showing little swell; at (b) a viscoelastic liquid swells to three times the capillary diameter.

shear rate. The comparison between the nominal shear rate at the wall, the actual shear rate, and S_R was also given.

In conclusion, one sees that good progress has been made with the more realistic models for this problem, but the UCM model remains elusive because of numerical problems at the salient corner.

8.9   Extrusion

The test problem of Fig. 8.7(a) relates to the simplest extrusion problem. Aside from the exit stress singularity, one also has to find the free surface shape here, as in other extrusion problems.

We have mentioned (Chapter 1) that a stream of fluid emerging from a plain circular tube (or die) does not usually have the same diameter as the tube. With polymeric fluids the extrudate often has a diameter several times that of the tube (Fig. 8.23). If gravitational forces are significant compared with the viscous forces acting on the fluid, then the extrudate shape will partly be governed by gravity, and the details of the orientation of the tube are relevant. To confine the discussion, we suppose here that gravity is negligible or effectively absent; this is often the relevant case for polymer melts. Additionally the flows to be considered will be steady. There is of course an intrinsic interest in finding the mechanism of swelling and, more practically, the production of precision extrusions also needs an accurate prediction of swelling.

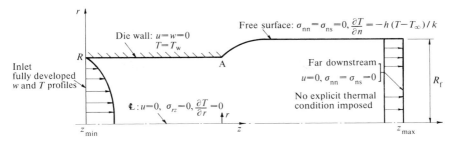

Fig. 8.24 Boundary conditions in extrusion. Typical thermal boundary conditions are included; surface tension and gravity are ignored.

The free surface of the jet makes the problem more complex than the previous non-Newtonian test flows and involves mixed boundary conditions on the fluid. By mixed boundary conditions we mean that part of the fluid S_v has velocity (usually no-slip) boundary conditions and a part S_t has traction (or stress) boundary conditions. (In addition, there may be thermal boundary conditions.) Figure 8.24 shows a typical case; it may be that an external medium or surface tension (to be mentioned later) will cause normal and tangential stresses on the free boundary, or that slip at the wall may occur in the die, but in the main the conditions shown in Fig. 8.24 suffice to describe extrusion. One also has to find where the free surface begins (at A in Fig. 8.24).

The boundary condition problem is basic as free boundaries present funda-mental difficulties even for simple material properties. The creeping (inertia-less) solution was obtained for this problem (Nickell *et al.* 1974) by using a finite element method. The solution shows that at the point of exit where there is a sudden change in boundary condition there is a stress singularity; (see Section 8.7).

8.9.1 Locating the free boundary

On the free surface we have to satisfy three conditions simultaneously

(i)   zero normal velocity;

(ii) zero (or prescribed) shear stress;

(iii) zero (or prescribed) normal stress.

In elementary cases without surface tension, so that the shear and normal stresses are zero, one approach is to ignore the condition (i), set up conditions (ii) and (iii) on an assumed contour, and then calculate the normal velocities on the assumed contour. From these (Fig. 8.25) a new streamline can be constructed, which can serve as a new assumed contour, and the process can be continued until

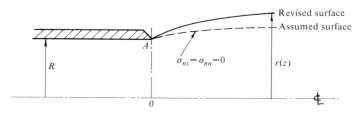

Fig. 8.25 Locating unknown boundary by iteration. Dashed line, current approximation to free streamline. Full line, new streamline. Separation is at A.

a satisfactory level of convergence is obtained, that is, the normal velocity is small enough that the final contour can be regarded as satisfying all conditions. To construct the boundary streamline at each iteration we know that the streamline begins at point A (Fig. 8.25), and we have an estimate of the velocity vector on the assumed boundary (dashed line in Fig. 8.25). If we assume that the velocity components u and w on the true boundary are the same as those on the dashed curve, then the streamline coordinates $r(z)$ are given by

$$r(z) = R + \int_{z=0}^{z} \mathrm{d}z \frac{u}{w}. \tag{8.67}$$

The integral can be found numerically. This defines another bounding curve and a new iteration can start with this curve. This simple approach is adequate for practical extrudate swell calculations where surface tension is not dominant. Caswell and Viryiyayuthakorn (1983) have discussed an improved scheme of this type. If the swelling were very large, it would be useful to employ a more complex scheme to find the free surface, perhaps employing an interpolation scheme to find the values of (u/w) in eqn (8.67); the simple scheme described uses the (u/w) values on the previous surface. It should be noted that the streamline springs from a sharp corner in all these problems; few computations (Tanner *et al.* 1985) have been published with a realistic rounded-corner exit [Fig. 8.13(c)]; since the point of separation is now an unknown, the problem is more difficult.

   For cases when surface tension is very large, the above scheme is unstable. This means that when the dimensionless number $\sigma/\eta\bar{w}$, which gives the ratio of surface tension to viscous forces, becomes very large, another scheme for finding the boundary shape is needed. Orr and Scriven (1978) have in this case set the normal velocity and the shear stress equal to zero and then made the normal stress equal to the surface tension force by adjusting the boundary position. This group has also included special equations treating the free boundary coordinates as unknowns, so that several options are open in difficult cases. Finally, one can include the surface positions as unknowns, along with τ, $\mathbf{v}$ and p in a global solution procedure.

In the next section we report some results and compare the various numerical approaches.

8.9.2 The extrusion process

The object of our studies is to understand the pressure losses at exit, the shape of the extrudate and ultimately the stability of the flow in terms of basic parameters. Although we are discussing extrusion here, the methods and some of the results have obvious relevance to other flow problems such as melt spinning, film-blowing and wire coating.

Let us first concentrate on the the simplest problem of a long (plane slit or axisymmetric circular) die. This will ensure that the flow is a fully developed viscometric flow far upstream of the exit plane. Let us now consider the swelling ratio χ. This ratio is defined (see Fig. 8.24) as

$$\chi = \frac{R_f}{R}. \tag{8.68}$$

The swelling ratio is a function of several parameters. These are:

1. Geometry—R is the only length parameter.

2. Flow kinematics—only $\bar{w}$, the average entry speed, is needed if the die is long enough.

3. Gravity (g). For many polymer melts, ignoring this factor is a good assumption which will be adopted here.

4. Fluid properties. Clearly we need ρ, η for an isothermal incompressible Newtonian fluid (we only consider incompressible fluids). For other fluids, especially viscoelastic fluids, other parameters are needed, and these will be introduced as required. Surface tension (σ) is also a relevant parameter.

5. If heat transfer is important, then the wall temperature (T_w), the surroundings temperature (T_∞), a surface heat transfer coefficient (h), the product of density and specific heat (ρc) and the conductivity (k) need to be specified.

6. Other factors such as slip at the wall can also be considered.

We can study the influence of the above factors by dimensional methods; once the relevant dimensionless quantities have been selected, then computer solutions can be made to explore the range of parameters required.

Newtonian fluids
In this case we have (ignoring gravity)

$$\chi = \chi(R, \bar{w}, \rho, \eta, \sigma). \tag{8.69}$$

Here R is the tube radius and the other parameters are defined above.
Formation of dimensionless groups shows that

$$\chi = \chi(2\rho\bar{w} R/\eta, \sigma/\eta\bar{w}). \tag{8.70}$$

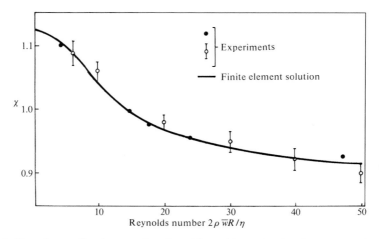

Fig. 8.26   Extrudate swell ratio χ as a function of Reynolds number (Re) for a Newtonian jet with surface tension in axisymmetric flow. Here $\sigma/\bar{w}\eta = 0.3$.

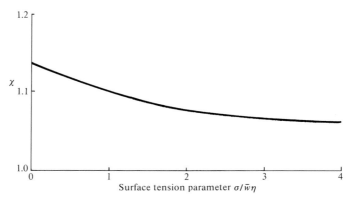

Fig. 8.27   Extrudate swell ratio χ as a function of surface tension group $\sigma/\bar{w}\eta$ for a Newtonian jet in axisymmetric flow; $(Re) = 0$.

The influence of the Reynolds number $[(Re) = 2\rho\bar{w}R/\eta]$ and the surface tension number on χ have been investigated; both produce some change in jet diameter. The results are shown in Figs 8.26 and 8.27, the final diameters agree well with available experiments. The effect of large surface tension is to destabilize the calculation; small 'kinks' in the surface tend to grow, and no convergence was obtained for $\sigma/\eta\bar{w} > 4.0$ (Reddy and Tanner 1978a). As $\sigma/\eta\bar{w} \rightarrow \infty$ the swelling should vanish; Orr and Scriven's (1978) method is useful here.

For the Reynolds number problem the effect of inertia is to slow down the rate of convergence; iteration for the non-linear effects and the free surface correction are done in each iteration cycle. Without surface tension calculations up to

Reynolds numbers of 1000 have been converged without difficulty but they needed a large number of iterations (~ 50 at the highest Reynolds numbers). In Fig. 8.28 the computations were done to match experiments (Gear *et al.* 1983) and higher (Re) calculations were not attempted with $\sigma/\bar{w}\eta = 0.3$.

At high Reynolds numbers one must permit a long jet region to exist; analytical results of the boundary-layer kind for the rates of decay in a jet are useful for setting the field length.

From the polymer processing point of view, the Reynolds number is usually negligible, and in the following we shall ignore it; we shall also ignore surface

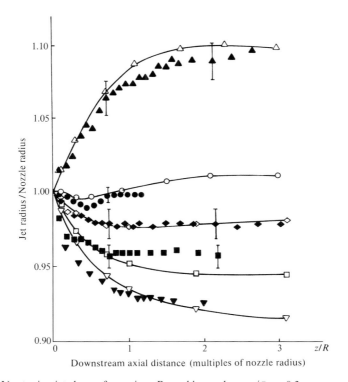

Fig. 8.28 Newtonian jet shapes for various Reynolds numbers; $\sigma/\bar{w}\eta = 0.3$.

Reynolds numbers	Symbol
4.09	Triangle
12.5	Circle
17.2	Diamond
27.3	Square
47.4	Inverted triangle.

Filled symbols are experimental data (note error bars). Open symbols are finite element solutions. (From Gear *et al.* 1983).

tension effects which are also usually small. In brief, inertia tends to make jets smaller, and surface tension tends to pull the jet shape back towards a cylinder of radius R. When we ignore ρ and σ in our list of parameters in (8.69), then no dimensionless group can be formed from R, $\bar{w}$ and η, and we have simply that χ is a constant, χ_0.

A selection of χ_0 values from the literature for the basic Newtonian creeping flow case is shown in Table 8.8. It is noticeable that the f.e.m. results with a greater number of degrees of freedom generally show less swelling. Several programs were involved in the results of Table 8.8. By plotting χ_0 against the reciprocal of the number of degrees of freedom for each family of results the extrapolated values $\chi_0 = 1.127 \pm 0.003$ (axisymmetric) and $\chi_0 = 1.190 \pm 0.002$ (plane) have been estimated for an infinite number of degrees of freedom. The error ranges represent differences in calculation method and grid choice. It is noticeable that the FD (finite difference) method (Ryan and Dutta 1981), the series matching technique of Trogdon and Joseph (1981), and the collocation method of Chang et al. (1979) yield results that differ from the finite element

Table 8.8 Newtonian swelling ratios χ_0

Investigator	Method	Degrees of freedom	χ_0
(a) Axisymmetric			
Batchelor and Horsfall (1971)	Experiment	–	1.135 ± 0.01
Nickell et al. (1974)	f.e.m.	1000	1.128
Allan (1977)	f.e.m.	988	1.132
Tanner (1976)	f.e.m.	254	1.136
Trogdon and Joseph (1981)	Series matching	50	1.111
Ryan and Dutta (1981)	FD	1100	1.120
Crochet and Keunings (1980)	f.e.m.	1178	1.126
Chang et al. (1979)	f.e.m.	306	1.139
		378	1.126
Extrapolated value		∞	1.127 ± 0.003
(b) Plane			
Crochet and Keunings (1982a)	f.e.m.	174	1.227
		338	1.207
		562	1.200
		1178	1.196
Crochet and Keunings (1980)	f.e.m.	1178	1.188
Chang et al. (1979)	f.e.m.	306	1.206
	Collocation	252	1.155
Reddy and Tanner (1978b)	f.e.m	254	1.199
Milthorpe (private communication, 1981)	f.e.m.	2928	1.189
Extrapolated value		∞	1.190 ± 0.002

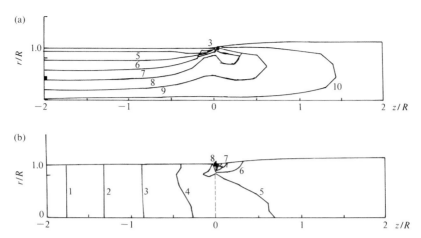

Fig. 8.29 Stresses in Newtonian creeping jet. (a) Contours of constant dimensionless shear stress $\sigma_{rz}R/\eta\bar{w}$. Numbers are marked to identify contours; the values of the contours corresponding to these numbers are: (1) -6.72, (2) -5.98, (3) -5.25, (4) -4.51, (5) -3.77, (6) -3.04, (7) -2.30, (8) -1.56, (9) -0.83, (10) -0.09. (b) Contours of constant dimensionless axial stress $\sigma_{zz}R/\eta\bar{w}$. The values corresponding to the marked numbers are in this case: (1) -16.0, (2) -12.4, (3) -8.77, (4) -5.14, (5) -1.51, (6) 2.12, (7) 5.75, (8) 9.38, (9) 13.0, (10) 16.6. Note the singularity at the exit, $z = 0$.

results. In the remainder of this chapter we shall assume $\chi_0 = 1.13$ and 1.19 for the base cases in axisymmetric and plane flow respectively.

The stress fields σ_{zz} and σ_{rz} are plotted in Fig. 8.29. The singular stress behaviour near the exit lip is prominent. Note also (Fig. 8.30) the region of compressive (negative) axial stress near the axis in the exit plane, and the region of tensile stress near the free surface. These are general features present in all extrusions from a long tube. Pressure losses can be computed for the exiting fluid. Suppose the pressure at a point L units upstream from the exit is known; let it be p_i. The pressure loss per unit length in the fully-developed flow is just $2\tau_w/R$, where τ_w is the fully-developed wall shear stress and R is the tube radius. Then the exit pressure loss Δp can be defined as

$$\Delta p = p_i - 2\tau_w L/R \qquad (8.71)$$

Clearly, Δp is the excess pressure loss (above the Poiseuille fully-developed loss) caused by the non-viscometric flow at the tube exit. It is convenient to compute the effective exit length $\Delta p/2\tau_w$, and this is shown in Table 8.7 for various flow conditions.

Inelastic non-Newtonian fluids
The next level of complexity is to permit the viscosity to vary with shear rate. A realistic viscosity-shear rate curve will usually involve several parameters to

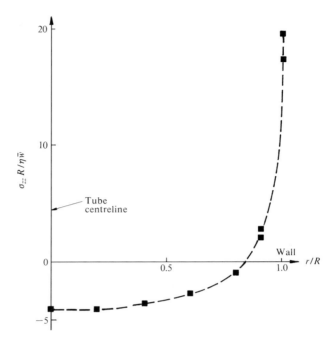

Fig. 8.30 Dimensionless axial stress $\sigma_{zz} R/\eta \bar{w}$ at tube exit plane. Note intense tension at outside $(r = R)$ and compression on axis $(r = 0)$.

describe the Newtonian behaviour at low shear rates $(\dot{\gamma})$ and the power-law behaviour at higher shear rates. The most important features can be investigated by using the simple power-law rule depending on the index n [eqn (4.4)].

When $n = 1$, we have the Newtonian case, when $n < 1$, the flow is 'pseudo-plastic' (shear-thinning) and when $n > 1$, it is shear-thickening. Stability considerations forbid a negative value of n, so that $n = 0$ represents the extreme lower limit for n. In that case, we have slug (or plug) flow in the tube and the extrudate expansion is zero so that $\chi = 1.0$; the 'fluid' just slips at the wall. Dimensional theory shows us that (for zero gravity and surface tension)

$$\chi = \chi(n), \tag{8.72}$$

in the creeping-flow limit. Figure 8.31 shows computed results for this case, and Table 8.7 shows the pressure losses. It is clear that no great change of expansion takes place due to change in viscosity with shear rate. In fact, the usual variation of viscosity with shear rate demands $1 > n$ (shear-thinning) and this yields somewhat less expansion than the Newtonian case.

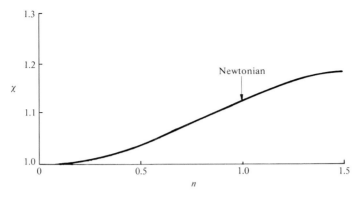

Fig. 8.31 Swell ratio χ as a function of power-law index n for zero inertia and surface tension in axisymmetric flow.

Viscoelastic effects
The Maxwell fluid [eqn (5.57)] has a relaxation time (λ), in addition to the viscosity and density, as a fluid parameter. Under creeping flow conditions, dimensional theory shows that

$$\chi = \chi(\lambda \bar{w}/R). \tag{8.73}$$

The group $\lambda \bar{w}/R$ is a Weissenberg number (Wi). As mentioned above, in the limit of slow flow, a viscoelastic fluid behaves as a second-order fluid. In this case, dimensional analysis gives

$$\chi = \chi\left(\frac{\Psi_2}{\Psi_1}, \frac{\Psi_1 \bar{w}}{\eta R}\right). \tag{8.74}$$

Available experimental evidence (Table 3.9) shows that Ψ_2/Ψ_1 is roughly a constant (about -0.1). If we note that Ψ_1/η has the dimensions of time, then the equivalence of (8.73) and (8.74) is apparent. Hence the stress difference N_1 is proportional to $\Psi_1(\bar{w}/R)^2$, and the group $\Psi_1 \bar{w}/\eta R$ is proportional to $(N_1/\tau)_{\mathrm{w}}$ evaluated at the tube wall; thus the equivalence of the present formulation to those based on 'recoverable shear' $(N_1/2\tau)_{\mathrm{w}}$ is demonstrated. In fact, for equivalence, we set $\lambda = \Psi_1/2\eta$. Reddy and Tanner (1978b) have produced some results for the second order fluid in plane flows but only for low Wi on coarse meshes. There are considerable convergence problems with many computational schemes at high Weissenberg numbers especially with the second-order and Maxwell models, as we have seen in Section 8.5.1. The choice of constitutive equation affects the swelling for a given (Wi) and also the exit pressure losses; see Coleman (1981), Crochet and Keunings (1980, 1982a,b), and Bush *et al.* (1984, 1985). Figure 8.32 shows the effect of choice of constitutive model on swelling for the plane case. Note the difficulty of achieving results for large (Wi) for the Maxwell and Oldroyd-B models (curves 1 and 2).

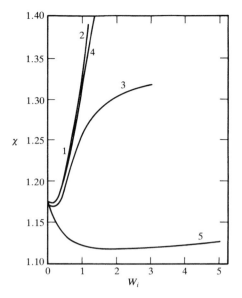

Fig. 8.32 The computed swelling ratio χ for plane flow of several fluid models with a single relaxation time λ as a function of $Wi(=\lambda\dot{\gamma})$

(1) Maxwell ($\eta_s = 0, \eta_m = 1, \lambda = 1$);
(2) Oldroyd-B ($\eta_s = 0.112, \eta_m = 0.888, \lambda = 1$);
(3) PTT ($\varepsilon = 0.01, \xi = 0.1, \eta_s = 0.112, \eta_m = 0.888, \lambda = 1$);
(4) MPTT ($n = 1, \varepsilon = 0.01, \xi = 0, \eta_0 = 1, \eta_s = 0, \lambda = 1$);
(5) Leonov ($\mu = 0.5, \eta_s = 0, \lambda = 1, \beta = 0$).

The High Weissenberg Number Problem has been partially alleviated but is still there with these models. With the Phan-Thien–Tanner (PTT) and modified PTT (Bush *et al.* 1985) models (curves 3 and 4 in Fig. 8.32) it is somewhat easier to achieve convergence. The lack of swelling for the Leonov model [eqn (5.189)] in curve 5 is noted; this is probably due to the lack of elongational flow stiffening; see Fig. 4.16 and eqn (8.95). For the axisymmetric case (Bush *et al.* 1984) the results in Fig. 8.33 were produced, showing over 80 per cent swelling. The curve $\varepsilon = 0.0$ is in fact the Maxwell result; the other curves are for PTT model [eqn (5.161)]. The curve marked 'elastic formula' refers to the semi-analytic formula

$$\chi = 0.13 + \left(1 + \frac{1}{8}\left(\frac{N_1}{\tau}\right)^2\right)^{1/6}, \tag{8.75}$$

which is eqn (8.92) with the Newtonian swelling of 0.13 added. In both Figs 8.32 and 8.33 one sees slight initial reduction of swelling from the Newtonian case,

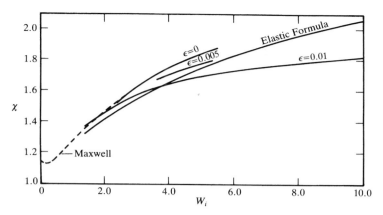

Fig. 8.33 Computed swelling ratio χ for axisymmetric swelling of a PTT model ($\xi = 0, \eta_s = 0,$ $\eta_m = 1.0, \varepsilon$ has various values). $(Wi) = \lambda\dot{\gamma}_w$.

then an expansion as $\lambda\dot{\gamma}_w[(Wi)]$ increases; this is an expected result for plane flows (Tanner 1980a).

Equation (8.75) seems to increase too slowly in the (low-Wi) region, (Fig. 8.33), although it often seems to fit experimental results well at higher Weissenberg numbers.

Expansion begins to increase rapidly for $\lambda\dot{\gamma}_w > 1$. Thus, as expected, we see viscoelasticity as a cause of enhanced swelling.

None of the above computations attempts to model experimental behaviour closely. A very good fit to the IUPAC low-density polyethylene behaviour described in Chapter 3 (Section 3.8) in both shear and elongational flows has been made by Luo (Luo and Tanner 1988) using a KBKZ type integral model. The computations used a finite element method where the streamlines define element boundaries. Results used eight relaxation times as in Table 2.1 and the results are shown in Fig. 8.34. The slight effect of changing the normal stress ratio N_2/N_1 on swelling is also shown. The black dots at decade intervals of nominal shear rate $(4\bar{w}/R)$ show that at low and medium rates the simulation is excellent; at the highest rate $(4\bar{w}/R = 10\,\text{s}^{-1})$ the computation overestimates the swelling. This may be due to wall slip at the die exit (see Section 3.9) or to inadequacies in the constitutive equation.

Phan-Thien (1988) has shown the dramatic reduction in swelling which occurs when partial slip at the wall occurs.

Goublomme *et al.* (1992), and Goublomme and Crochet (1993) simulated a high-density polyethylene (HDPE) extrusion and concluded that the constitutive equation needed to incorporate the irreversible mechanism of Wagner (see Chapter 4) in order to show a reasonable agreement with experiment at high extrusion rates.

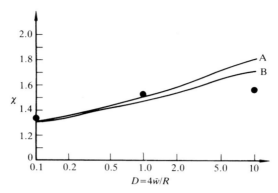

Fig. 8.34 Calculated long die extrusion swelling ratio curves in comparison with experimental data for LPDE.

$$\text{Curve A}: \frac{N_2}{N_1} = -0.1; \quad \text{Curve B}: \frac{N_2}{N_1} = -0.2.$$

• Experimental data. $D = \dfrac{4\bar{w}}{R}$ is the nominal shear rate.

Hence, one needs to consider both mechanisms in order to do realistic simulations. Convergence with mesh size of the swelling ratio has been demonstrated for the KBKZ-type models (see Huilgol and Phan-Thien 1997).

The exit loss Δp_{ex} (Table 8.6) can be defined simply. The radial normal stress at the wall, $\sigma_{rr}(R, 0)$, is the negative of the pressure as measured by a flush-mounted pressure transducer, $p_{\text{w}}(0)$; the reference pressure is established by taking atmospheric pressure to be zero. The pressure varies linearly with distance in the fully-developed flow region, and we define Δp_{ex} as

$$\Delta p_{\text{ex}} = p_{\text{w}}(0) - \frac{2\tau_{\text{w}}L}{R}. \tag{8.76}$$

The notation $p_{\text{w}}(0)$ is used here because the 'pressure' is no longer constant across the section of the tube; the w suffix refers to the wall here and the zero to axial location. τ_{w} is the wall shear stress in fully-developed flow and is taken by convention to be positive. Note that Δp_{ex} is an extrapolation and is not itself measured directly in any experiment.

In cases where inertia is significant this may need to be taken into account by considering the rearrangement of the velocity profiles before and after the exit (Problem 8.9). See also Boger and Denn (1980).

8.9.3   Die geometry and other effects in extrusion

So far only the results for extrusion from long dies have been noted. Tanner (1976), Allan (1977), Crochet and Keunings (1981), Luo and Tanner (1988), Hatzikiriakos and Mitsoulis (1996), and Barakos and Mitsoulis (1995) have

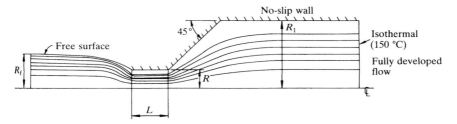

Fig. 8.35 Extrusion geometry for short dies. Flow from right to left.

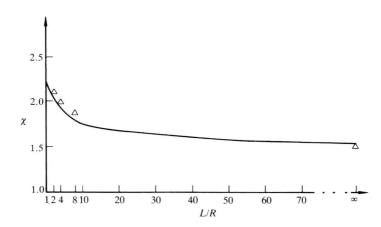

Fig. 8.36 Swelling ratio results of short die calculation at $4\bar{w}/R = 1$ in comparison with the experimental points ($\triangle$) for low density polyethylene at 150 °C.

looked at other geometries including the flow from a reservoir through a short die and flow from a converging or diverging tube die of variable length.

Allan (1977) reported some convergence problems for very short dies (length/ diameter < 0.05). Luo and Tanner (1988) have investigated the short-die geometry shown in Fig. 8.35. For two values of $8\bar{w}/D$ (0.1 and 1.0 s^{-1}) the swelling was computed for the IUPAC LDPE; at $8\bar{w}/D = 10.0$ s^{-1} we have seen over-prediction as in Fig. 8.34 and the results were not accurately computed; it is possible that wall slip at the die exit (Ramamurthy 1986) is important to the result when high wall stresses occur. Fig. 8.36 shows the computed swelling results for LDPE at 150 °C and $8\bar{w}/D = 1.0$ s^{-1} compared with experiment in dies of various lengths.

Hatzikiriakos and Mitsoulis (1996) considered long and short dies and computed the end-corrections for several cases, including zero-length orifice dies. In the latter case, inlet and exit effects clearly interact, and no distinction between these effects can be made. Their results indicated some differences from experiments. Mitsoulis *et al.* (1993) studied an approximation to the Herschel–Bulkley model in various configurations; this is now an easily achievable sort of computation. Swelling was small, as expected. Kiriakidis and Mitsoulis (1993) simulated the flow of HDPE into plane slit and capillary dies.

In the case where the flow is not isothermal, several more dimensionless groups appear (see Chapter 9).

Silliman and Scriven (1980) and Phan-Thien (1988) have also shown that permitting slip between fluid at the wall near the tube lip (where the fluid is under severe shear stress) reduces swelling in Newtonian fluids in the plane case. This is not an unexpected result, since complete slip would produce no swelling at all.

We have shown how to use computational methods to investigate the extrusion process. It is confirmed that viscoelasticity and wall slip are important as factors in extrudate swell, while non-Newtonian behaviour (alone) is not. Surprisingly, we find that variations of temperature, through the agency of temperature-sensitive viscosity, can produce large swelling effects. (See Chapter 9.)

8.9.4   Mechanisms of swelling

The above results are of interest in their own right as computer solutions to complicated boundary value problems, but they can also be regarded as a set of controlled experiments in the search for a physical explanation of swelling. This is in line with the philosophy that one does computing for the sake of enlightenment, not just to obtain numbers. This philosophy is likely to be very valuable in polymer processing operations where actual experiments are hard to perform because of size or instrument problems, for example. One sees three contributions to swelling:

 (i) The small 'Newtonian' swelling due to rearrangement of the velocity field;

 (ii) The 'elastic recovery' swell;

(iii) 'Inelastic' swelling due to, for example, thermal effects (Chapter 9).

The first mechanism has been described above. The elastic recovery idea is based on the idea of unconstrained recovery. To explain the basic idea we may consider the diagram (Fig. 8.37) showing the trajectory of a 'particle' of fluid as it emerges from the die, passing from a stressed viscometric state to an unstressed state. Lodge (1964) has shown how such a sheared piece of material exhibits elastic recovery and sideways swelling when released from stress suddenly. We assume that the exit from the tube is sudden and an instantaneous elastic strain takes place which unloads the fluid. (It is more accurate to regard the present calculation as the short-term response to the sudden abolition of the walls in a shearing flow.) We shall calculate the instantaneous recovery after shearing to illustrate the mechanism.

As a constitutive equation we assume a KBKZ form

$$\sigma = -p\mathbf{I} + \int_{-\infty}^{t} U'\mathbf{C}^{-1}\,\mathrm{d}t',\tag{8.77}$$

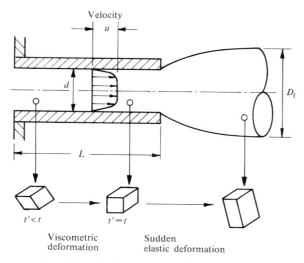

Fig. 8.37 Elastic recovery swelling mechanism.

where $U' = \partial U / \partial I_{c^{-1}}$ is a function of $t - t'$ and tr $\mathbf{C}^{-1}$. This ignores the second normal stress difference, but it shows elastic effects and also realistic viscosity and first normal stress-difference functions.

Suppose that the material is undergoing a simple shearing up until $t = 0$, and then is released from constraint. Thus, for $t < 0$

$$\mathbf{v} = \dot{\gamma} y \mathbf{i},$$

and we have, in a viscometric flow

$$\tau_{xx} - \tau_{yy} = N_1 = \dot{\gamma}^2 \int_0^\infty U' s^2 \, ds \tag{8.78}$$

$$\tau_{xy} = \dot{\gamma} \int_0^\infty U' s \, ds,$$

where $s = t - t'$.

In the present problem the kinematic history consists of a jump strain from the present time $t = 0^+$ to a time $t = 0^-$, then a viscometric history. Let the (unknown) jump strain be described by a deformation gradient $\mathbf{F}_0$. Then the deformation gradient can be written

$$\mathbf{F} = \mathbf{F}^{(v)} \mathbf{F}_0, \tag{8.79}$$

where $\mathbf{F}^{(v)}$ is a viscometric history. Suppose there is a deformed state $\mathbf{x}^*$ just before the jump strain. Then we can write

$$\mathbf{F}^{(v)} = \frac{\partial \mathbf{r}}{\partial \mathbf{x}^*}, \mathbf{F}_0 = \frac{\partial \mathbf{x}^*}{\partial \mathbf{x}}, \tag{8.80}$$

where $\mathbf{r}$ is the particle position at time t'; $\mathbf{x}^*$ is the particle position before the jump $(t = 0^-)$ and $\mathbf{x}$ is the current particle position $(t > 0)$. Then if $s = -t'$ we have

$$\mathbf{F}^{(v)} = \begin{bmatrix} 1 & -\dot{\gamma}s & 0 \\ 0 & 1 & 0 \\ 0 & 0 & 1 \end{bmatrix} \tag{8.81}$$

For $t' > 0$, we have $\mathbf{F} = \mathbf{I}$, and for $t' < 0^-$ we have $\mathbf{F} = \mathbf{F}^{(v)}\mathbf{F}_0$. Now consider the material for $t > 0$, when the deviatoric stress tensor is zero, but the pressure is non-zero. Then

$$p\mathbf{I} = \int_{-\infty}^{t} U'(\mathbf{F}_0^T \mathbf{F}^{(v)T}\mathbf{F}^{(v)}\mathbf{F}_0)^{-1}\, \mathrm{d}t', \tag{8.82}$$

or, since $\mathbf{F}_0$ is a constant matrix,

$$p\mathbf{F}_0\mathbf{F}_0^T = \int_0^{\infty} U'\mathbf{C}^{(v)-1}\, \mathrm{d}s, \tag{8.83}$$

where U' is now a function of $\mathrm{tr}(\mathbf{F}_0^{-1}\mathbf{C}^{(v)-1}\mathbf{F}_0^{-T})$ and s; $\mathbf{C}^{(v)} = \mathbf{F}^{(v)T}\mathbf{F}^{(v)}$.

To begin with, suppose U' is independent of $\mathrm{tr}\,\mathbf{C}^{-1}$, as in the Lodge fluid. Then

$$p\mathbf{F}_0\mathbf{F}_0^T = \begin{bmatrix} G + N_1 & \tau & 0 \\ \tau & G & 0 \\ 0 & 0 & G \end{bmatrix} \tag{8.84}$$

where

$$\dot{\gamma}^n \int_0^{\infty} s^n U'\, \mathrm{d}s = G, \tau, N_1, \tag{8.85}$$

for $n = 0, 1, 2$ respectively; G is a modulus, τ is the shear stress in the shear flow, and N_1 is the first normal stress-difference. Now the strain matrix $\mathbf{F}_0\mathbf{F}_0^T$ must be volume-preserving, so its determinant must be 1. Taking the determinant of (8.84), we get

$$p^3 = G(G^2 + N_1 G - \tau^2) \tag{8.86}$$

as a condition on the residual pressure p. The jump strain $\mathbf{F}_0$ is constructed from an elastic displacement field

$$x^* = x/\lambda^2 + \gamma y \quad y^* = \lambda y \quad z^* = \lambda z, \tag{8.87}$$

where λ is a sideways expansion and γ is a shear so that

$$\mathbf{F}_0 = \begin{bmatrix} 1/\lambda^2 & \gamma & 0 \\ 0 & \lambda & 0 \\ 0 & 0 & \lambda \end{bmatrix}. \tag{8.88}$$

Substituting in (8.84) gives three relations

$$p(\lambda^{-4} + \gamma^2) = G + N_1 \quad p\gamma\lambda = \tau \quad p\lambda^2 = G. \tag{8.89}$$

Eliminating γ and p from these gives the result for $1/\lambda$, the sideways swelling:

$$\frac{1}{\lambda} = \left(1 + \frac{N_1}{G} - \frac{\tau^2}{G^2}\right)^{1/6}. \tag{8.90}$$

For a single relaxation time,

$$\frac{1}{\lambda} = \left\{1 + \left(\frac{N_1}{2\tau}\right)^2\right\}^{1/6} \tag{8.91}$$

which evaluates the sideways swelling in terms of quantities measurable in viscometric flow. Lodge *et al.* (1965) have investigated the response for multiple relaxation times.

One can apply similar methods to the flow through a tube. If the tube is suddenly removed, then elastic recoil takes place as above. One can calculate the resulting swelling as (Tanner 1970)

$$\chi = \frac{D_f}{d} = \left[1 + \frac{1}{8}\left(\frac{N_1}{\tau}\right)^2\right]_w^{1/6}, \tag{8.92}$$

or in the plane flow case

$$\chi = \frac{D_f}{d} = \left[1 + \frac{1}{12}\left(\frac{N_1}{\tau}\right)^2\right]_w^{1/4}, \tag{8.93}$$

where N_1/τ is evaluated at the wall.

The coincidence between eqn (8.92) and the computed swelling results is shown in Fig. 8.33. The computations are for a Maxwell fluid, eqn (5.57). The formulae above also agree with some experiments on polymer melts (Utracki *et al.* 1975); Huang and White (1978) have considered the case where N_1 is proportional to τ^a, where a is a constant, and have replaced the coefficients $1/8$ and $1/12$ in the above formulae by $1/4a$ and $1/4(2a-1)$ respectively, so as to improve the fit of their experimental data.

This mechanism of swelling, though plausible, inadequately describes the kinematics of die-swell. For example, where N_1/τ is largest, one expects, from (8.91), the largest swelling. In fact, as inspection of Fig. 8.30 shows, this must in reality occur on the centreline, where N_1/τ is zero. Pearson and Trottnow (1978) have (not quite convincingly) attempted to rectify this problem; their result is not very different in kind from (8.92). Despite these reservations, Allain *et al.* (1997) have shown that these formulas are useful in correlating their experiments.

Finally, there are swellings which cannot be explained by either of the above mechanisms. In a Newtonian fluid whose viscosity depends on temperature, large swellings can take place when the viscosity near the centreline is lower than that at the outside. It has been proposed that a two-fluid model be used to explain such swellings. Figure 8.38 shows a sketch of the proposed system. By making a force balance we obtain (Tanner 1980b)

$$\chi^2 = \frac{\eta_0}{\eta_i}\left[1 - \left(\frac{R_i}{R_o}\right)^2\left(1 - \frac{\eta_i}{\eta_o}\right)\right] \tag{8.94}$$

in the axisymmetric case, and

$$\chi = 1 + \left(\frac{\eta_o}{\eta_i} - 1\right)\left(\frac{h_o - h_i}{h_o}\right) \tag{8.95}$$

in the plane case.

Here the subscripts i and o refer to the inner and outer layers respectively. As an example, if the outer layer has double the viscosity of the inner layer, and $R_i/R_o = 0.8$, then the axisymmetric and plane formulas give respectively $\chi = 1.166$ and 1.2, relative to the Newtonian case. This idea seems to give a reasonable correlation with numerical calculations.

The mechanism of delayed die-swell, whereby a fluid exiting from a tube takes some distance before swelling occurs, has been discussed by Joseph (1990), in

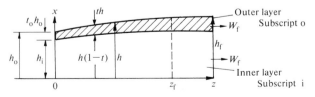

Fig. 8.38 Plane two-layer extrusion-definition sketch.

terms of a change of type of the governing equations. One is reminded that a change of type requires the presence of significant inertia in the flow.

8.9.5 Three-dimensional flows

Relatively speaking, few investigations of 3-D flows have been made. Schoonen *et al.* (1998) report a flow in a cross-slot apparatus and compared results using 4-mode PTT and Giesekus models; however, the velocity computation used a generalized viscous model. Beverly and Tanner (1992) studied a bi-viscosity Bingham-type model and looked at 3-D extrusions.

The boundary-element method using an iterative decoupled technique has been used, breaking the problem into the boundary element solution of the elliptic field equations with the non-linear terms treated as pseudo-body forces, and the integration of the constitutive model using the fixed velocity profile obtained in the previous iteration. The inclusion of free surfaces is also treated iteratively, by up-dating the geometry using the previous fixed velocity field. Separate solution of the momentum, constitutive, energy (if needed), and free surface update allows different techniques to be used in each stage of the iterative procedure, which permits some optimization in the solution process. The main disadvantage of the method is that the iterations, usually of the Picard-type, converge at best linearly, and the radius of convergence is often limited. Despite this, the flexibility of the decoupled boundary element method makes it a good vehicle for mildly nonlinear behaviour. Huilgol and Phan-Thien (1997) give a large number of references, including some three-dimensional extrusion problems. Tran-Cong and Phan-Thien (1988a,b) give the example of boundary-element computation displayed in Fig. 8.39, showing the extrudate shape from a square die [$(Wi) = 0.9$], and from a triangular die [$(Wi) = 1.6$] for the Maxwell model. The amount of swelling is of the order 30 per cent, at $(Wi) = O(1)$, to be compared with 13 per cent for a Newtonian circular extrudate. The Weissenberg number shown is based on the wall shear rate at the centre of the die far upstream.

In the die-design problem, one wishes to find the shape of the die for a given extrudate profile. In this inverse problem, one updates the free surface from the given cross-sectional profile of the extrudate, and follows a particle path until one gets to the die exit station. The die profile is then updated and the cycle of iterations continues. In practice, extrusion dies usually have a gradual transition from a simple cross-section, usually circular, to the final profile forming section. This is also the adopted method in the numerical simulation; it makes the task of specifying the inflow boundary conditions much simpler, and the algorithm much more robust, since the inflow boundary conditions are fixed. To date, only simple die designs have been attempted; for example square and triangular extrudates. The die profile to produce a square extrudate compares well with existing dies in use, but there is no published information on a die profile that produces a triangular extrudate. At low (Wi), one finds that the die profiles for the PTT model are similar in shape to those designed for the Newtonian fluid.

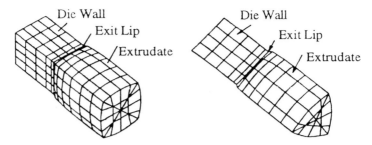

Fig. 8.39   Extrudate shape from a square die ($Wi = 0.9$), and from a triangular die ($Wi = 1.6$) for the Maxwell model. The Weissenberg number is based on the wall shear rate far upstream. (Courtesy Professor Phan-Thien)

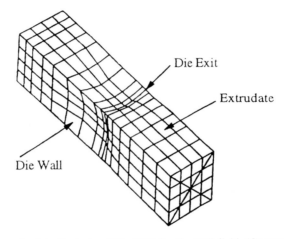

Fig. 8.40   Square extrudate die design ($Wi = 0.34$) for the PTT fluid. (Courtesy N. Phan-Thien)

Thus, the Newtonian die design could form an excellent starting point for the corresponding viscoelastic die design. Figure 8.40 shows a die design for a square extrudate, for the PTT model at (Wi) = 0.34, based on the shear rate at the wall far upstream. The amount of shrinkage, relative to the extrudate, is about 24 per cent.

Finally, we mention the work Xue *et al.* (1998) on the flows through 3-D contractions. These are very large finite volume computations ($\sim 700\,000$ unknowns) and a typical grid used is shown in Fig. 8.41. Convergence with mesh refinement was shown, with UCM, PTT, and Oldroyd-B models up to Weissenberg numbers of 2.9; after that the time of computation was excessive, but solutions were still smooth. A variety of vortices were found near the sharp inlet corner, flowing from the wide channel to the narrow one.

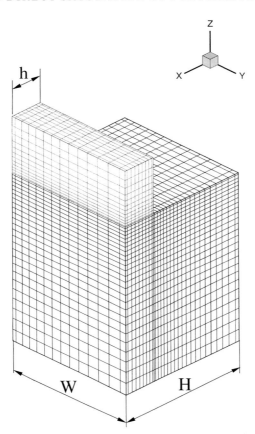

Fig. 8.41 3D planar contraction geometry and graded meshes (one quarter of the domain). (Courtesy Dr S.-C. Xue)

8.10 Direct simulation of polymer flow

The previous sections have dealt with the use of computers to solve mathematical problems for polymeric liquids using the continuum approach. Another possibility is to use computing for more direct simulation taking 'microstructure' into account explicitly. Whilst this idea was tried in the 1970s, the lack of speed and memory frustrated early attempts. However, in 1978, with the growth of computer power, simulations were employed by various workers to study the relaxation and flow of polymer chains; see Tanner and Walters (1998). Since then, there has been an explosion of activity, not only for polymeric liquids, but also in colloidal suspensions and in flows with reactions.

Simulations for the Green–Tobolsky–Lodge–Yamamoto network model were made by Petruccione and Biller (1988a,b) and Biller and Petruccione (1996).

Termonia (1991) has developed an interesting polymer network model that behaves quite realistically in multiaxial deformations, allowing him to find strain energy functions for these rubbery materials.

A volume edited by Elizabeth Colbourn (1994) describes some of the wide field now being attacked in this way: computer-aided molecular design; molecular dynamics; modelling of amorphous polymers; Monte Carlo studies of collective phenomena in dense polymer systems; crystalline polymers; failure mechanisms of networks and bio-polymers. Öttinger (1996) has championed the calculation of viscoelastic flow using molecular models and Kremer and Grest (1990) have discussed the interaction of polymers with walls. In the mid 1980s, Brownian Dynamics was developed for suspensions. The particles are damped [Langevin equations here in place of Newton's equations used for molecular dynamics (MD)] and many papers on dispersions now use these or similar methods (for example, Heyes and Melrose 1993; Grassia *et al.* 1995). Keunings (1997) has carried out a study of Brownian dynamics in non-Hookean dumbbells and has shown the strengths and weaknesses of earlier approaches based on simplified linearizations of the Peterlin type (cf. Chapter 5).

Heyes (1992) gives a short history of molecular simulation applied to rheology, emphasizing that the declining cost of computation makes it ever more sensible to use direct simulation methods. By direct simulation, we imagine a group of model particles, which could be small molecules or even macromolecules, that are set up in the computer, usually in a box with porous walls, so that periodic arrays of molecules can be dealt with in repeated images. The molecules are allowed to move around under Newton's laws, and stresses and rates of deformation can be computed for the models. This procedure leads to Molecular Dynamics (MD) simulations. Brownian Dynamics (BD) deals with the Langevin equations and is more suited to dealing with molecules in solution.

According to Heyes (1992), the first appearance of molecular simulation in rheology occurred in the 1960s, when the Green–Kubo formula was introduced to compute the zero-shear rate viscosity, without needing to shear the sample! In 1975, Ashurst and Hoover actually did shear the sample between walls, but later Evans and Morriss (1990) used the Green–Kubo method to deal with non-Newtonian flow, including normal-stress effects. Shear thinning in *n*-butane has been investigated; the main problems with the scheme are the limitation to small molecules (hexadecane, for example) and the very large shear rates ($\dot{\gamma} \sim 10^{12}\,\text{s}^{-1}$) at which one sees shear thinning. At such rates, 'thermostatting' is necessary and discussion on best methods is still progressing.

The effects of walls have been considered by Jabbarzadeh *et al.* (1999); as long ago as 1986, Heyes reported shear thinning in a Lennard-Jones potential liquid. We have already mentioned the salient corner simulation of Koplik and Banaver (1997).

While it is clear that, at present, these methods are not generally competitive with continuum-based methods, they may become so soon, at least for research purposes. The growth of parallel computing is clearly important here and will assist in the 'computer-aided physics' phenomenon.

8.11 Conclusion

This chapter has attempted to show the state of computation versus experiment for complex flows; see also Pearson and Richardson (1983), Crochet *et al.* (1984), Huilgol and Phan-Thien (1997), Tanner and Walters (1988), and Baaijens (1998) for further references. There is clearly need for much improvement especially in free surface and corner problems and in 3-D and transient flows. There are also unsteady flows of the kind occuring in mould-filling which have begun to be attacked computationally. The opportunities for using direct simulations are also to be noted. While the discussion gives some hope for optimism in the practical resolution of flow behaviour, we must now turn to a most important aspect of fluid behaviour — the effect of temperature on rheology and flow patterns.

References

Abdali, S. S., Mitsoulis, E., and Markatos, N. C. (1992). *J. Rheol.*, **36**, 389.

Allian, C., Cloitre, M., and Perrot, P. (1997). *J. Non-Newtonian Fluid Mech.*, **73**, 51.

Allan, W. (1977). *Int. J. numer. Meth. Engng*, **11**, 1621.

Ames, W. F. (1977). *Numerical methods for partial differential equations*, (2nd edn). Nelson, London.

Ashurst, W. T. and Hoover, W. G. (1975). *Phys. Rev.*, *A* **11**, 658.

Azaiez, J., Guénette, R., and Ait-Kadi, A. (1996). *J. Non-Newtonian Fluid Mech.*, **62**, 253.

Baaijens, F. P. T. (1998). *J. Non-Newtonian Fluid Mech.*, **79**, 361.

Baaijens, F. P. T., Schoonen, J. Peters, G. W. M., and Meijer, H. E. H. (1996). *Proc. 12th Intl. Cong. Rheol.* (Quebec; ed. A. Ait-Kadi *et al.*), 419–20.

Baaijens, F. P. T., Selen, S. H. A., Baaijens, H. P. W., Peters, G. W. M., and Meijer, H. E. H. (1997). *J. Non-Newtonian Fluid Mech.*, **68**, 173.

Baaijens, H. P. W., Peters, G. W. M., Baaijens, F. P. T., and Meijer, H. E. H. (1995). *J. Rheol.*, **39**, 1243.

Barakos, G. and Mitsoulis, E. (1995). *J. Rheol.*, **39**, 193.

Batchelor, J. and Horsfall, F. (1971). Rubber and Plastics Research Assoc. of Great Britan, *Report No. 189, Die swell in elastic and viscous fluids.*

Beaulne, M. and Mitsoulis, E. (1997). *J. Non-Newtonian Fluid Mech.*, **72**, 55.

Beris, A. N., Armstrong, R. C., and Brown, R. A. (1983). *J. Non-Newtonian Fluid Mech.*, **13**, 109.

Beris, A. N., Armstrong, R. C., and Brown, R. A. (1987). *J. Non-Newtonian Fluid Mech.*, **22**, 129.

Beris, A. N., Tsamopoulos, J. A., Armstrong, R. C., and Brown, R. A. (1985). *J. Fluid Mech.*, **158**, 219.

Beverley, C. R. and Tanner, R. I. (1992). *J. Non-Newtonian Fluid Mech.*, **42**, 85.

Biller, P. and Petruccione, F. (1996). *J. Chem. Phys.*, **92**, 6322.

Boger, D. V. (1982). *Adv. Transport Processes*, **2**, 43.

Boger, D. V. and Denn, M. M. (1980). *J Non-Newtonian Fluid Mech.*, **6**, 183.

Bot, E. T. G., Hulsen, M. A., and van den Brule, B. H. A. A. (1988). *J. Non-Newtonian Fluid Mech.*, **79**, 191.

Brebbia, C. A. (1980). *The boundary element method for engineers*, 2nd edn. Pentech Press, London.

Bush, M. B., Milthorpe, J. F., and Tanner, R. I. (1984). *J. Non-Newtonian Fluid Mech.*, **16**, 37.

Bush, M. B. and Tanner, R. I. (1983). *Int. J. numer. Meth. Fluids*, **3**, 71.

Bush, M. B., Tanner, R. I., and Phan-Thien, N. (1985). *J. Non-Newtonian Fluid Mech.*, **18**, 143.

Butcher, T. A. and Irvine, T. F. (1990). *J. Non-Newtonian Fluid Mech.*, **36**, 151.

Byars, J. A., Binnington, R. J., and Boger, D. V. (1997). *J. Non-Newtonian Fluid Mech.*, **72**, 219.

Caswell, B. (1967). *Arch. Rat. Mech. Analy.*, **26**, 385.

Caswell, B. and Viriyayuthakorn, M. (1983). *J. Non-Newtonian Fluid Mech.*, **12**, 13.

Chang, P. W., Patten, T. W., and Finlayson, B. A. (1979). *Computers & Fluids*, **7**, 285.

Chorin, A. J. (1967). *J. Comput. Phys.*, **2**, 12.

Colbourn, E. (ed.) (1994). *Computer simulation of polymers*. Longmans, Harlow.

Coleman, C. J. (1981). *Non-Newtonian Fluid Mech.*, **8**, 261.

Courant, R. (1962). *Methods of mathematical physics*, Vol. 2. Interscience, New York.

Crochet, M. J., Davies, A. R., and Walters, K. (1984). *Numerical simulation of non-Newtonian flow*. Elsevier, Amsterdam.

Crochet, M. J. and Keunings, R. (1980). *J. Non-Newtonian Fluid Mech.*, **7**, 199.

Crochet, M. J. and Keunings, R. (1981). *Proc. 2nd World Congr. Chem. Engng*, **6**, 285.

Crochet, M. J. and Keunings, R. (1982*a*). *J. Non-Newtonian Fluid Mech.*, **10**, 85.

Crochet, M. J. and Keunings, R. (1982*b*). *J. Non-Newtonian Fluid Mech.*, **10**, 339.

Cruse, T. A. and Rizzo, F. J. (1968). *J. math. Analysis Applic.*, **22**, 244.

Dean, W. R. and Montagnon, P. E. (1949). *Proc. Camb. Phil Soc.*, **45**, 389.

Dennis, J. E. and Schnabel, R. B. (1979). *SIAM Review* **21**, 443.

Dupont, S. and Crochet, M. J. (1988), *Non-Newtonian Fluid Mech.*, **29**, 81.

Evans, D. J. and Morriss, G. P. (1990). *Statistical mechanics of non-equilibrium liquids*. Academic Press, London.

Fan, Y. (1997). *Comp. Meth. Appl. Mech. Eng.*, **141**, 47.

Fan, Y. and Crochet, M. J. (1995). *J. Non-Newtonian Fluid Mech.*, **57**, 177.

Fan, Y., Phan-Thien, N., and Tanner, R.I. (1999). *J. Non-Newtonian Fluid Mech.*, **84**, 233.

Finlayson, B. A. (1972). *The method of weighted residuals and variational principles*. Academic Press, New York.

Fortin, A. (1992). *J. Non-Newtonian Fluid Mech.*, **45**, 209.

Gallagher, R. H., Oden, J. T., Taylor, C., and Zienkiewicz, O. C. (eds) (1975). *Finite elements in fluids*, Vols 1–4. Wiley, London.

Gear, R. L., Keentok, M., Milthorpe, J. F., and Tanner, R. I. (1983). *Physics of Fluids*, **26**, 7.

Georgiou, G. C., Olson, L. G., Schultz, W. W., and Sagan, S. (1989). *Int. J. Num. Meth. Fluids*, **9**, 1353.

Gottlieb, D. and Orszag, S. A. (1977). *Numerical analysis of spectral methods: theory and applications*. SIAM, Philadelphia.

Goublomme, A. and Crochet, M. J. (1993). *J. Non-Newtonian Fluid Mech.*, **47**, 281.

Goublomme, A., Draily, B., and Crochet, M. J. (1992). *J. Non-Newtonian Fluid Mech.*, **44**, 171.

Grassia, P. S., Hinch, E. J., and Nitsche, P. (1995). *J. Fluid Mech.*, **282**, 373.

Gu, D. and Tanner, R. I. (1985). *J. Non-Newtonian Fluid Mech.*, **17**, 1.

Guénette, R. and Fortin, M. (1995). *J. Non-Newtonian Fluid Mech.*, **60**, 27.

Hagen, T. and Renardy, M. (1997). *J. Non-Newtonian Fluid Mech.*, **73**, 181.

Hatzikiriakos, S. G. and Mitsoulis, E. (1996). *Proc. 1st Hellenic Conf. on Rheol.*, Cyprus, p. 61.

Henriksen, P. and Hassager, O. (1989). *J. Rheol.*, **33**, 865.

Heyes, D. M. (1992). *Bull. Br. Soc. Rheol.*, **35**, 41.

Heyes, D. M. and Melrose, J. R. (1993). *J. Non-Newtonian Fluid Mech.*, **46**, 1.

Hill, G. A., Shook, C. A., and Esmail, M. N. (1981). *Can. J. chem. Eng.*, **59**, 100.

Hinch, E. J. (1993). *J. Non-Newtonian Fluid Mech.*, **50**, 161.

Huang, D. C. and White, J. L. (1978). *University of Tennessee Polymer Sci. & Engng Rept. No. 113.*

Huang, X., Phan-Thien, N., and Tanner, R. I. (1996). *J. Non-Newtonian Fluid Mech.*, **64**, 71.

Huilgol, R. R. and Phan-Thien, N. (1997). *Fluid mechanics of viscoelasticity.* Elsevier, Amsterdam.

Hunt, B. W. (1968). *J. Fluid Mech.*, **31**, 361.

Hutchinson, J. W. (1968). *J. Mech. Phys. Solids*, **16**, 13.

Jabbarzadeh, A., Atkinson, J. D., and Tanner, R. I. (1999). *J. Chem Phys.*, **110**, 2612.

Jean, M. and Pritchard, W. G. (1980). *Proc. R. Soc.*, **A370**, 61.

Jin, H., Phan-Thien, N., and Tanner, R. I. (1991). *Comp. Mech.*, **8**, 409.

Jin, H., Phan-Thien, N., and Tanner, R. I. (1994). *Comp. Mech.*, **13**, 443.

Johnson, C. (1990). *Finite element methods.* CUP, Cambridge.

Joseph, D. D. (1990). *Fluid dynamics of viscoelastic liquids.* Springer-Verlag, New York.

Kawai, T. (ed.) (1982). *Finite element flow analysis.* University of Tokyo Press.

Keunings, R. (1997). *J. Non-Newtonian Fluid Mech.*, **68**, 85.

Keunings, R. and Crochet, M. J. (1984). *J. Non-Newtonian Fluid Mech.*, **14**, 279.

Kim-E., M. E., Brown, R. A., and Armstrong, R. C. (1983). *J. Non-Newtonian Fluid Mech.*, **13**, 241.

Kiriakidis, D. G. and Mitsoulis, E. (1993). *Adv. In Polymer Technology*, **12**, 107.

Koplik, J. and Banaver, J. R. (1997). *J. Rheol.*, **41**, 787.

Kremer, K. and Grest, G. R. (1990). *J. Chem. Phys.*, **92**, 5057.

Leal, L. G. (1992). *Laminar flow and convective transport processes*, Butterworth-Heinemann, Boston.

Leonard, B. P. (1979). *Comp. Meth. Appl. Mech. Eng.*, **19**, 59.

Levich, V. G. (1962). *Physicochemical hydrodynamics.* Prentice-Hall, Englewood Cliffs, NJ.

Lodge, A. S. (1964). *Elastic liquids.* Academic Press, London.

Lodge, A. S., Evans, D. J., and Scully, D. B. (1965). *Rheol. Acta*, **4**, 140.

Luo, X.-L. (1996). *J. Non-Newtonian Fluid Mech.*, **64**, 173.

Luo, X.-L. (1998). *J. Non-Newtonian Fluid Mech.*, **79**, 57.

Luo, X.-L. and Tanner, R. I. (1986). *Int. J. Num. Meth. Eng.*, **25**, 9.

Lunsmann, W. J., Genieser, L., Armstrong, R. C., and Brown, R. A. (1993). *J. Non-Newtonian Fluid Mech.*, **48**, 63.

Michael, D. H. (1958). *Mathematika*, **5**, 82.

Mitsoulis, E., Abdali, S. S., and Markatos, N. C. (1998). *Can. J. Chem. Eng.*, **71**, 147.

Nickell, R. E., Tanner, R. I., and Caswell, B. (1974). *J. Fluid Mech.*, **65**, 189.

Orr, F. M. and Scriven, L. E. (1978), *J. Fluid Mech.*, **84**, 145.

Öttinger, H. C. (1996). *Stochastic processes in polymeric fluids.* Springer-Verlag, Berlin.

Patankar, S. V. (1980). *Numerical heat transfer and fluid flow.* Hemisphere, New York.

Pearson, J. R. A. and Richardson, S. M. (eds) (1983). *Computational analysis of polymer processing.* Applied Science Publications, London.

Pearson, J. R. A. and Trottnow, R. (1978). *J. Non-Newtonian Fluid Mech.*, **4**, 195.

Perera, M. G. N. and Walters, K. (1977). *J. Non-Newtonian Fluid Mech.*, **2**, 49; 191.

Petruccione, F. and Biller, P. (1998*a*). *J. Chem. Phys.*, **89**, 577.
Petruccione, F. and Biller, P. (1998*b*). *Rheol. Acta*, **27**, 557.
Phan-Thien, N. (1988). *J. Non-Newtonian Fluid Mech.*, **26**, 327.
Pilitsis, S. and Beris, A. N. (1989). *J. Non-Newtonian Fluid Mech.*, **31**, 231.
Pilitsis, S., Souvaliotis, A., and Beris, A. N. (1991). *J. Rheol.* **35**, 605.
Ramamurthy, A. V. (1986). *J. Rheol.*, **30**, 337.
Ramia, M. (1991). *Biophys. J.*, **60**, 1057.
Reddy, J. N. and Gartling, D. K. (1994). *The finite element method in heat transfer and fluid dynamics*. CRC Press, Florida.
Reddy, K. R. and Tanner, R. I. (1978*a*). *Computers & Fluids*, **6**, 83.
Reddy, K. R. and Tanner, R. I. (1978*b*). *J. Rheol.*, **22**, 661.
Renardy, M. (1990). *J. Non-Newtonian Fluid Mech.*, **36**, 419.
Renardy, M. (1993). *J. Non-Newtonian Fluid Mech.*, **50**, 127.
Renardy, M. (1995). *J. Non-Newtonian Fluid Mech.*, **58**, 83.
Renardy, M. (1997). *J. Non-Newtonian Fluid Mech.*, **69**, 99.
Richardson, L. F. (1910). *Trans. Roy. Soc. Lond.*, **A 210**, 307.
Richardson, S. (1970). *Proc. Camb. Phil. Soc.*, **67**, 477.
Richtmyer, M. D. and Morton, K. W. (1967). *Difference methods for initial value problems*. Wiley, New York.
Roache, P. J. (1976). *Computational fluid dynamics*. Hermosa Publishers, Albuquerque, New Mexico.
Roache, P. J. (1998). *Verification and validation in computational science and engineering*. Hermosa Publishers, Albuquerque, New Mexico.
Runge, C. (1908). *Z. Math. Phys.*, **56**, 225.
Ryan, M. E. and Dutta, A. (1981). *Proc. 2nd World Congr. Chem. Engng, Montreal*, **6**, 277.
Salamon, T., Bornside, D. E., Armstrong, R. C., and Brown, R. A. (1995). *Phys. Fluids*, **7**, 2328.
Schechter, R. S. (1961). *AIChEJ*, **7**, 445.
Schoonen, J., Swartijes, F. H. M., Peters, G. W. M., Baaijens, F. P. T., and Meijer, H. E. H. (1998). *J. Non-Newtonian Fluid Mech.*, **79**, 529.
Silliman, J. J. and Scriven, L. E. (1980). *J. comput. Phys.*, **34**, 287.
Southwell, R. V. (1946). *Relaxation methods in theoretical physics*, Oxford University Press.
Stoker, J. J. (1957). *Water waves*. Interscience, New York.
Sun, J., Phan-Thien, N., and Tanner, R. I. (1996). *J. Non-Newtonian Fluid Mech.*, **65**, 75.
Sun, J., Smith, R. D., Armstrong, R. C., and Brown, R. A. (1999). *J. Non-Newtonian Fluid Mech.*, **86**, 281.
Talwar, K. K. and Khomami, B. (1995) *J. Non-Newtonian Fluid Mech.*, **59**, 49.
Tanner R. I. (1963). *J. Fluid Mech.*, **17**, 161.
Tanner R. I. (1970). *J. Polym. Sci. Mech.*, (A.2), **8**, 2067.
Tanner R. I. (1976). In *Proc. 7th int. Congr. Rheol. (Göthenburg)* (eds C. Klason and J. Kubat). p. 140. Swedish Society of Rheology, Göthenburg.
Tanner R. I. (1980*a*). *J. Non-Newtonian Fluid Mech.*, **7**, 265.
Tanner R. I. (1980*b*). *J. Non-Newtonian Fluid Mech.*, **6**, 289.
Tanner R. I. (1982). *J. Non-Newtonian Fluid Mech.*, **10**, 169.
Tanner R. I. and Huang, X. (1993). *J. Non-Newtonian Fluid Mech.*, **50**, 135.
Tanner R. I., Lam, H. and Bush, M. B. (1985). *Physics Fluids*, **28**, 23.
Tanner R. I. and Walters, K. (1998). *Rheology: an historical perspective*. Elsevier, Amsterdam.

Termonia, Y. (1991). *Macromolecules*, **24**, 1128.

Tran-Cong, T. and Phan-Thien, N. (1988*a*). *J. Non-Newtonian Fluid Mech.*, **30**, 37.

Tran-Cong, T. and Phan-Thien, N. (1988*b*). *Rheol. Acta*, **27**, 639.

Trogdon, S. A. and Joseph, D. D. (1981). *Rheol. Acta*, **20**, 1.

Utracki, L. A., Bakerdjian, Z., and Kamal, M. R. (1975). *J. appl. Polymer Sci.*, **19**, 481.

Walters, K. and Tanner R. I. (1992). In Chabbra, R. P., and De Kee, D. (eds), *Transport processes in bubbles, drops and particles*, p. 73. Hemisphere Publ. Co., New York.

Warichet, V. and Legat, V. (1997). *J. Non-Newtonian Fluid Mech.*, **73**, 95.

Xue, S.-C., Phan-Thien, N., and Tanner R. I. (1998). *J. Non-Newtonian Fluid Mech.*, **74**, 195.

Xue, S. -C., Tanner R. I., and Phan-Thien, N., (1999). *Comp. Meth. Appl. Math. Eng.*. (To appear.)

Young, D. M. and Wheeler, M. F. (1964). *In non-linear problems in engineering* (ed. W.F. Ames) p. 220. Academic Press, New York.

Zienkiewicz, O. C. and Taylor, R. L. (1991). *The finite element method* 4th edn Wiley, New York.

Problems

1. Derive a formula based on the stream function in plane creeping, Newtonian flow which is equivalent to the boundary condition $\sigma_{nn} = 0$ on a free surface.

2. Show that the solutions (8.28), (8.29), when $l = 1$, are the flow fields produced by a point force in the x_1 direction.

3. Using the data of Table 8.8 form your own estimate of the basic Newtonian swelling ratios by making a least-squares fit of χ as a function of $1/($Number of degrees of freedom).

4. Using the shear-thinning data of Table 8.7 and the Maxwell data of Table 8.6 can you estimate the entry losses in a viscoelastic system? Compare your result with Fig. 8.22. Is it qualitatively correct?

5. Suppose an elastic swelling mechanism of the Lodge type is observed in which the transverse strains are held in the ratio 2:1. Find what the formula corresponding to (8.91) would be in this case.

6. If a multiple-layer extrusion is taking place in an inelastic material (Fig. 8.42) estimate the swelling χ when the outer layer has an effective viscosity 10 times that of the inner layer.

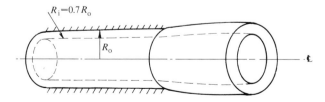

Fig. 8.42 Problem 6—Multilayer extrusion.

7. Find the entry losses due to elastic effects for the plane creeping entry of a second-order fluid.

8. Consider the problem of an abrupt expansion in a pipe flow. Can you analyse this problem with the data given here? Discuss the results you obtain.

9. Form an estimate of inertia effects in entry (contraction) flows and also in exit flow from a long, straight, circular tube.

9
TEMPERATURE AND PRESSURE EFFECTS

9.1 Introduction

So far we have looked at the effects of viscoelasticity on flows without considering the great importance of temperature changes; we note that Newtonian hydrocarbon oils often change their viscosity by about 10 per cent per degree Celsius near room temperature. Aside from the quantitative effects on flow, it is often also necessary to be sure that the maximum temperature in the material is held below the temperature at which degradation sets in. In Newtonian liquids the viscosity decreases with temperature rise as thermal expansion of the fluid permits the molecules increased mobility. By an extension of such an argument one can expect that an increase of pressure will decrease the molecular mobility and lead to an increase in viscosity. This is observed in both Newtonian and non-Newtonian systems and is a significant factor in high-pressure polymer processing and in some lubrication problems. Therefore this effect must also be considered. To begin we take the 'pressure' to mean $-\frac{1}{3}\sigma_{ii}$ when discussing these effects; it is probably that in a highly-anisotropic stress field some other measure of 'pressure' would be important, see Section 9.2.1. There are numerous other potentially important effects (effect of temperature and flow on crystallization) which should be borne in mind when considering applications: these aspects are considered only briefly here.

9.1.1 Heat transfer

The basic mechanisms of heat transfer by conduction, convection, and radiation are generally thought to be independent of whether or not the material is Newtonian and this will be assumed in what follows. In particular, we shall usually suppose that Fourier's law of heat conduction holds, so that, for thermally isotropic materials

$$q_i = -k\frac{\partial T}{\partial x_i}, \tag{9.1}$$

where $\mathbf{q}$ is the heat flux vector and k is the thermal conductivity, which may be a function of temperature.

In convection, the heat-transfer coefficient (h) will be used

$$q_w = h(T_w - T_f), \tag{9.2}$$

where the flux of heat normal to a bounding surface (q_w) which is at a temperature T_w is given by (9.2); q_w is the flux from surface to fluid (bulk temperature T_f). Equation (9.2) amounts to a definition of h. The heat transfer coefficient depends on the fluid properties and flow field and may be locally or globally defined.

Radiant transfer can be important in some processes, for example in glass-fibremaking (Stehle and Bruckner 1979), because of the high temperatures (~ 1500 K) involved, but for polymers (~ 500 K) it is less important. Radiation can often be allowed for approximately by taking an appropriate heat-transfer coefficient (see, for example, Huynh and Tanner 1983) or an augmented conductivity and we shall not consider radiative transfer effects in detail.

In summary, the heat transfer to be considered here is classical. There exist, at least theoretically, the possibilities of thermomechanical effects. For example, one might find that the heat flux within a material depended on the strain or the flow field.

This problem has been considered by van den Brule (1995), who discusses the effect of flow and deformation on heat conduction. It is difficult to obtain values of thermal conductivity to better than about ± 10 per cent, so any effect on the conductivity has to be larger than this to be really significant.

Van den Brule argues that molecular orientation affects conductivity because conduction along a molecule is easier than conduction betwen molecules. He considers a network theory. Instead of a scalar heat conduction, a tensor $\mathbf{k}$ appears, such that

$$q_i = -k_{ij}\partial T/\partial x_j \quad \text{where}$$
$$\mathbf{k} = k_1\mathbf{I} + n_0\beta\langle\mathbf{RR}\rangle$$

Here n_0 is the number of chains per unit volume, β is a coefficient, and $\langle\mathbf{RR}\rangle$ is the structure tensor $\mathbf{A}$, Section 5.5.1. The term $k_1\mathbf{I}$ describes the usual isotropic conduction. Data for stretched rubber samples quoted by van den Brule show a considerable variation in the stretch and compression directions. The conductivity tensor $\mathbf{k}$ was described by the form

$$\frac{\mathbf{k}}{k_0} = (1 - C)\mathbf{I} + C\mathbf{B},$$

where C is another constant and $\mathbf{B} = \mathbf{FF}^T$, the Finger strain tensor (Chapter 2): here $\mathbf{F}$ is the deformation tensor referred to the undeformed (or original) state as a reference.

For a uniaxial elongation along the z-direction,

$$\mathbf{B} = \begin{bmatrix} \varepsilon^{-1} & 0 & 0 \\ 0 & \varepsilon^{-1} & 0 \\ 0 & 0 & \varepsilon^2 \end{bmatrix}$$

where $\varepsilon = L/L_0$, the extension ratio. However, tension and compression needed very different C values (0.25 and 0.75 respectively) which is difficult to explain.

He also discusses bead-spring models in dilute solutions and deduces the change of the conductivity tensor due to the flow. For Hookean dumbbells one finds

$$\lambda_H \frac{\Delta \mathbf{k}}{\Delta t} + \mathbf{k} = 2\lambda_H k_0 \mathbf{d}.$$

This expression will clearly predict infinite conductivity in an elongational flow when $\lambda_H \dot{\varepsilon} > 1/2$ (see Section 4.3.1) where λ_H is a thermal relaxation time. Also, in dilute solutions these effects must be small; van den Brule reports some experiments where conductivity changes of 5–10 per cent were noted. In addition, Wapperom and Hulsen (1995) have found that additions of the flow-induced anisotropy made little difference to their computed results, especially at high Peclet numbers.

Further changes of conductivity with the crystalline state have also been seen, see Kennedy (1995) and van Krevelen and Hoftyzer (1976). Thermal conductivity can also vary with temperature. In view of the uncertainties in experiments and theory for fluids, we shall often assume that an appropriate isotropic tensor, as in eqn (9.1), can be chosen.

A general exposition is given by Truesdell and Noll (1965) in which the question of reversibility of thermomechanical action is discussed, and the question of the proper application of the second law of thermodynamics is addressed. A point of view closer to the one adopted here is expressed by Woods (1975) whose book is, however, mainly directed towards non-polymeric fluids and gases.

9.1.2   The energy equation

The energy equation (2.62) is valid for both incompressible and compressible materials. For the former, we write

$$\sigma_{ij} = -p\delta_{ij} + \tau_{ij}, \tag{2.66}$$

and note that the pressure does no work, since $\nabla \cdot \mathbf{v} = 0$. Also, we shall write, for the internal energy e

$$e = C_v T. \tag{9.3}$$

In this case, although we have written C_v as the symbol for the specific heat the subscript v is not necessary, because the specific heat at constant volume and constant pressure are clearly equal for truly incompressible media.

For compressible media we may split the stress as in eqn (2.66) and then p may be defined as the thermodynamic pressure, so that $p = p(\rho, T)$. While this is adequate for media at rest, there seem to be no crucial tests of this matter; in the absence of further information we shall assume that p is the thermodynamic pressure, and that the constitutive relations used are compatible with this assumption. Then the energy equation becomes, after some calculation

$$\rho \frac{\mathrm{D}}{\mathrm{D}t}(e) + \rho p \frac{\mathrm{D}}{\mathrm{D}t}\left(\frac{1}{\rho}\right) = \tau_{ij}d_{ij} - \frac{\partial q_i}{\partial x_i} + \rho s. \tag{2.62a}$$

If we introduce the enthalpy h, so that

$$h = e + p/\rho, \tag{9.4}$$

and define the specific heat at constant pressure C_p then we find the energy equation in the form (Goldstein 1960):

$$\rho C_p \frac{\mathrm{D}T}{\mathrm{D}t} + \frac{T}{\rho}\left(\frac{\partial \rho}{\partial T}\right)_p \frac{\mathrm{D}p}{\mathrm{D}t} = \tau_{ij}d_{ij} - \frac{\partial q_i}{\partial x_i} + \rho s, \tag{9.5}$$

Generally, the second term on the left-hand side of (9.5) can be neglected. For example, for polyethylene, from Tables 9.6 and 9.7 we have $\rho \sim 860\,\mathrm{kg/m^3}$, $C_p \sim 2260\,\mathrm{J/kg\,K}$, $(\partial\rho/\partial T)_p \sim -0.6\,\mathrm{kg/m^3\,K}$. If $T = 400\,\mathrm{K}$, and characteristic temperature and pressure changes in the flow are ΔT and Δp respectively, then the first term in (9.5) is of order $2 \times 10^6\,\overline{\Delta T}$ and the second is of order $0.3\,\overline{\Delta p}$. Hence very large pressure change rates ($\sim 100\,\mathrm{atm/s}$) only produce about the same effect as a temperature rate of change of about $1\,^\circ\mathrm{C/s}$. These figures are typical, and hence we shall omit the second term in (9.5) here.

One can now regard the incompressible limit as the case when $(\partial\rho/\partial T)_p$ is negligible; this justifies our use of C_p as the relevant specific heat in the incompressible case, as in (9.5). We shall always omit the 'radiant' transfer term ρs here and where it is justified the dissipation term $\tau_{ij}d_{ij}$ will also be ignored.

9.1.3   Other forms of the energy function

While the above discussion has assumed $e = C_v T$, a function of temperature only, there are many more options for e, which essentially describes the sum of the heat and work inputs to a particle. We have already seen in eqn (9.5) how the energy equation is modified from (2.62) by using the enthalpy h to account for the elastic work of compression in a compressible material. In many materials one can have stored shear-elastic energy, changes of phase, work-hardening, chemical reactions or other physical changes to make the energy balance more complex. To illustrate the problem, we consider elastic-plastic flow as an example. In this case the deviatoric stresses are of order of the yield stress (τ_y) and if the elastic modulus is G, then the rate of plastic energy dissipation is of order $\tau_y\dot\gamma$, while the rate at elastic energy change is of order $\tau_y\dot\gamma_e$, where γ_e is the elastic strain, of order τ_y/G. In this case $\gamma \gg \gamma_e$, usually, and the elastic component can be neglected. Experiments by Taylor and Quinney (1931) nevertheless showed that not all of the external work on plastically deforming specimens appeared as heat (only about 90 per cent did in their experiments on metals). The rest of the losses is generally ascribed to energy associated with dislocations, and defects in the crystal structure, which are not recoverable in the short term. There appear to be similar opportunities for external work producing structural changes in soft materials (Zdilar and Tanner 1994), but the physics is evidently complex.

In the case of ideal rubberlike behaviour, one supposes that the internal energy is entirely entropic, caused by network rearrangement in a reversible manner. In this case it is useful to use the free energy, defined as $e - Ts$, where s is now the specific entropy.

Wapperom and Hulsen (1995) have included in their computations the elastic-stored energy theory of Peters (1995), who has considered these questions from a microstructural network viewpoint. They again found, especially at high Peclet numbers, only a few per cent of change in their results. Hence we shall generally omit these refinements. In view of the irreversibility associated with polymeric deformation (see Section 4.9), effects may be smaller than Peters's estimate.

9.2 Pressure and temperature-induced variations of viscosity

One can consider the effects of viscosity variation for liquids by using the free-volume concept or by reaction-rate concepts. The latter give a temperature dependence of η (for Newtonian fluids) in the form

$$\eta = B\exp(E/RT), \tag{9.6}$$

where E is an activation energy, R is the gas constant $(8.314\,\mathrm{JK^{-1}mol^{-1}})$ and T is the absolute temperature. This type of equation is often adequate to describe the variation of viscosity with temperature except over very large changes of temperature. Alternatively, eqn (9.6) can be put in a form containing the density (ρ):

$$\nu = \frac{\eta}{\rho} = B'\exp(E/RT), \tag{9.6a}$$

where ν is the kinematic viscosity; this equation, because of the small variation in ρ with temperature, behaves similarly to (9.6). Over small ranges of temperature we can also write $\eta = \eta_0\exp-\alpha(T - T_0)$. This expression is sometimes useful in analytical work. A better description is given by the Walther equation

$$\log_{10}\log_{10}(\nu + a) = m\log_{10}T + b, \tag{9.6b}$$

where a, m, and b are constants. This equation has been widely used for lubricating oils; in this case the value of a can be taken as about 0.6 to $0.8 \times 10^{-6}\,\mathrm{m^2/s}$ when $\nu > 1.5 \times 10^{-6}\,\mathrm{m^2/s}$, thus leaving only two disposable constants. Then the equation can be fitted using temperatures at two points only.

Another empirical equation that seems more satisfactory than the Walther equation in that it is more accurate over wider ranges of temperature is the

Vogel equation

$$\eta = A_1 \exp B_1/(T - T_0), \tag{9.7}$$

where A_1, B_1, and T_0 are constants.

Although these equations can be made sufficiently accurate for Newtonian fluids, they may need to be improved for the non-Newtonian case. Consider the power-law model $\eta = k\dot{\gamma}^{n-1} \exp -\alpha(T - T_0)$. Taking natural logarithms one finds

$$\ln \eta = A_0 + (n - 1)\ln \dot{\gamma} - \alpha T.$$

In this equation all the $(\eta, \dot{\gamma})$ lines of constant temperature are shifted by a constant amount proportional to the temperature differences. However, many tests show a convergence of constant shear-rate lines as temperature increases, and a convergence of constant temperature lines with increasing shear rate. Also, as we have already seen, the exponential temperature dependence is not very accurate, and there also seems to be a decrease in the value of n as $\dot{\gamma}$ increases. To remedy these problems Kennedy (1995) uses a second-order model containing squares and products of $\dot{\gamma}$ and T:

$$\ln \eta = A_0 + A_1 \ln \dot{\gamma} + A_2 T + A_3 (\ln \dot{\gamma})^2 + A_4 T \ln \dot{\gamma} + A_5 T^2 \tag{9.7a}$$

where A_0 to A_5 are constants. This model fits some observed data well, but care must be taken to ensure realistic behaviour of the model. Kennedy (1995) does not consider that this model can describe liquid crystal polymers, and so in this and other cases one simply tabulates data for $(\eta, \dot{\gamma}, T)$, and uses numerical interpolation for computations.

While these rules can deal with fluid behaviour above the melting point, there are significant complications near this point. Figure 9.1 (Kennedy 1995) shows a sketch of the situation; the extrapolated viscosities predicted by the first and second order models are too low. While it is possible to use an 'infinite' viscosity

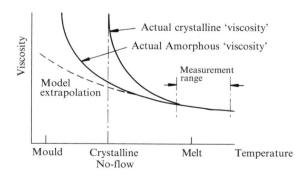

Fig. 9.1   Variation and extrapolation of viscosity data at low temperatures (after Kennedy 1995).

below a no-flow temperature (Kennedy 1995), one might also use a Vogel-type equation (9.7) to produce a sudden rise in viscosity.

9.2.1 Pressure effects and cavitation

Batchinsky (see Vinogradov and Malkin 1977) introduced the free-volume (V_f) concept, where

$$V_f = V - V_m.$$

Here V is the specific volume of the fluid (equal to $1/\rho$) and V_m is the specific volume of the molecules. The experiments of Doolittle on n-alkanes showed that

$$\eta = A' \exp(CV_m/V_f), \tag{9.8}$$

where A' and C are constants. Often $V_m \sim V$ and these quantities are interchangeable in (9.8)

For isothermal conditions we can write, for a change in density due to pressure (p),

$$V = V_0(1 - p/K), \tag{9.9}$$

where V_0 is the specific volume at zero pressure and K is the bulk modulus. Use of (9.9) in (9.8), noting the definition of V_f above, yields the result, when $p \ll K$,

$$\eta = \eta(0) \exp(p/\beta), \tag{9.10}$$

where $\eta(0)$ is the viscosity at zero pressure. This often appears to describe the variation of viscosity with pressure adequately. Values of the constant β are given in Table 9.1. Since β is very large, we may measure p from atmospheric pressure as datum.

Many results describing the effect of pressure on lubricating oil viscosity have been given. A sample of these results is shown in Fig. 9.2 which is for a silicone fluid.

These curves may be accurately fitted by semi-empirical equations if required; see Pywell (1973).

Table 9.1 Pressure–viscosity exponents

Substance	Temperature (K)	β (MPa)
Mineral oils	310–370	30–50
Polystyrene	455	28
Polyethylene	475	76
Polydimethysiloxane	375	44
	495	83
Polyisobutylene	375	49
	495	83
Polybutene	400	23

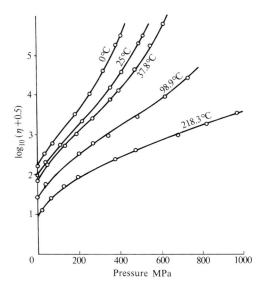

Fig. 9.2 Viscosity–pressure curves at various temperatures for ASME silicone oil 55H. Circles, experimental points; lines are fitted curves.

Binding *et al.* (1998) have done experiments to determine the pressure dependence of shear and elongational properties of a number of polymers (HDPE, LDPE, polypropylene, polymethyl methacrylate, and polystyrene). They used three temperatures for each material and a range of shear rates, with mean pressures from atmospheric to 70 MPa. They used capillary data for measuring shear viscosities and an orifice pressure drop method to infer elongational properties. See also Jacobson (1991) for a discussion of lubrication rheology at high pressures.

Cavitation

While the above considerations apply for increased pressure above atmospheric, when the pressure drops below atmospheric, cavitation may occur either due to the expansion of entrained or dissolved air, or due in extreme cases to vaporization. Brennen (1995) discusses these phenomena mainly for water and similar low-viscosity liquids, where the pressure is shown to be the factor governing cavitation. See also Frenkel (1955) for the effect of ultrasound on cavitation. Carvalho and Scriven (1996) have discussed the formation of cavities at the exit of roller nips, including the effect of surface tension; see also Chapter 6. In these problems film-splitting occurs at exit from the nip, and viscoelasticity affects the number and width of the 'fingers' produced in the exit stream.

However, both of the above effects are more in the realm of fluid mechanics than rheology. Bair and Winer (1992) have done an interesting experiment where a sheared fluid with various ambient pressures gave the results shown in Fig. 9.3. Generally, one would not expect any effect of such low pressures on

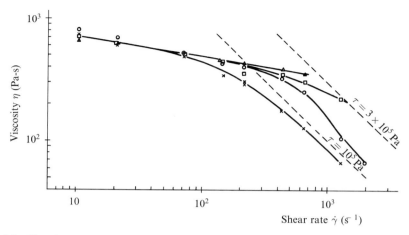

Fig. 9.3 Viscosity versus shear rate for a polybutene at 22 °C for various ambient pressures P.
$P = 0.01$ MPa ($\times$); 0.1 MPa ($\bigcirc$); 0.3 MPa ($\square$); 1.0 MPa ($\triangle$).

the viscosity, but this figure clearly shows that when the shear stress is of order 1 atmosphere (0.1 MPa), and the ambient pressure is of similar order, there is a marked effect of pressure on viscosity. Bair and Winer (1992) suggested that in a shear field the maximum tensile stress in the fluid causes cavitation and consequent loss of shear stress. The maximum tensile stress with an ambient pressure p_a and a shear stress τ is given by $\tau - p_a$, from Mohr's circle (Section 2.2), hence the argument is plausible. Archer *et al.* (1997) have also found that the shear stress can collapse at around 0.1–0.3 MPa. In one case the fluid lost its bonding to the driving surface, and in polystyrene bubbles were observed to open up. Both fluids were made of molecules which were too small to entangle, but did show normal stress differences. Streator *et al.* (1986) also noted 'lubricant rupture'.

Joseph (1998) has also considered the maximum tensile stress as the criterion for 'failure' of the liquid. If the largest principal stress is σ_1 then he postulates that when $\sigma_1 > \sigma_c$ (σ_c is a critical tensile stress), cavitation occurs. He also considers elongational flows.

Hence when stresses approach O(0.1 MPa) one needs to consider the possibility of such cavitation. See also Chapter 10, Section 10.8.

9.2.2 Amorphous and semi-crystalline polymers

We need to recognize the states that polymers find themselves in at low temperatures. Some polymers are almost completely amorphous at low temperatures and some are semi-crystalline. Amorphous polymers are composed of randomly-packed chains and have a characteristic temperature where they change from brittle, glassy, solids to rubberlike materials. This temperature is called the glass transition temperature (T_g). At this temperature there is a change in the

Table 9.2 Critical temperatures for polymers

Polymer	T_g (°C)	T_m (°C)	T_u (°C)
Linear polyethylene (high density polyethylene, HDPE)	−110	134	404[†]
Branched polyethylene (low density polyethylene, LDPE)	−110	115	404[†]
Polystyrene (PS)	90–100	240	
Polypropylene (PP)	−10	165	387[†]
Nylon 6–6	50	240	
Polyethylene terephtalate (PET)	70	260	
Polymethyl methacrylate (PMMA)	90–100	–	
Natural rubber	−73	28	
Polytetrafluroethylene (PTFE)	−150	327	∼404[‡]
Polyisobutylene	−70	–	
Polyethylene oxide (PEO)	−75	–	345[†]
Polyvinylchloride (PVC)	87	–	130
Polycarbonate (PPO)	150	–	

[†]Temperature at which polymer loses half of its mass in 30 minutes.
[‡]Degrades to monomer above this temperature.

slope of the specific volume/temperature curve, and the glass transition temperature may be described as a second-order transition; above this temperature the molecules are flexible chains; below it, they are rigid. Semi-crystalline polymers, below their melting temperature (T_m), have patches of both crystalline and amorphous structure. The amorphous regions behave as we have described above, and the notion of glass transition is relevant there. Thus there is observed a brittle region below T_g, then a region of tough, flexible behaviour up to the melting point of the crystalline phase (T_m), then the melt region itself. There is a jump of specific volume on melting (a first-order transition) with these materials. We have not defined so far a melt temperature for amorphous polymers. As the temperature rises above T_g we find the viscosity diminishes rapidly; often the temperature $T_g + 100$ °C is taken as being the onset of the truly fluid region. As the temperature rises still higher, decomposition accelerates and there is an upper temperature (T_u) beyond which the material should not be heated. Table 9.2 gives these temperatures for some common polymers. It should be noted that in practice T_g will have a range of uncertainty of the order of 5–10 °C, and the other temperatures have at least this amount of latitude.

We shall be concerned with temperatures above T_g (for amorphous materials) or above T_m (semi-crystalline materials).

9.2.3 The Williams–Landel–Ferry shift factor

We shall pursue the free-volume idea for temperature changes and assume the free volume V_f to obey

$$V_f = V_{fg}[1 + \alpha_f(T - T_g)], \tag{9.11}$$

where α_f is the thermal expansion coefficient appropriate to the free volume V_{fg} at the glass temperature. We also have

$$V = V_g[1 + \alpha_1(T - T_g)], \tag{9.12}$$

where α_1 is the coefficient of cubical expansion of the liquid. Equation (9.8) then yields

$$\log_{10}\left(\frac{\eta}{\eta_g}\right) = \log a_T = \frac{-C_1^g(T - T_g)}{C_2^g + T - T_g}, \tag{9.13}$$

where C_1^g and C_2^g are constants. This is the Williams, Landel, and Ferry (see Ferry 1981) shift formula which turns out to be widely applicable to amorphous polymers in the range T_g to about $T_g + 100\,°C$.

At first it was thought that the constants were universal for all polymers. It is now recognized that a better fit is obtained when the constants are specially chosen for each polymer (Ferry 1981). One may also alter the datum temperature in (9.13) to some other temperature, T_0 say, and write

$$\log\left(\frac{\eta}{\eta_0}\right) = \log a_T = \frac{-C_1^0(T - T_0)}{C_2^0 + (T - T_0)}. \tag{9.14}$$

The subscripts imply that the C_1 and C_2 factors depend on the reference temperature; we find that

$$\left.\begin{array}{l} C_2^0 = C_2^g + T_0 - T_g \\ \text{and} \\ C_1^0 = C_1^g C_2^g / C_2^0 \end{array}\right\}. \tag{9.15}$$

A selection of recent values of these constants is given in Table 9.3 together with the 'universal' constants. The data are those of Ferry (1981). These values are

Table 9.3   Shift factors for polymers

Polymer	Reference temperature T_0 (°C)	Glass temperature T_g (°C)	C_1^0 (°C^{-1})	C_2^0 (°C)	$C_1^0 C_2^0$ —
Polystyrene	100	97	12.7	49.8	632
Polymethyl methacrylate	115	115	32.2	80	2576
Polyvinylchloride	74	66	11.2	35	388
Polyurethane rubber	−42	−52	16.7	68	1136
Natural rubber	−25	−73	8.86	101.6	900
Polydimethyl siloxane	30	−123	1.90	222	422
'Universal values'	T_g	T_g	17.44	51.6	900

See eqn (9.14) for definition of a_T.

usable up to $T_g + 100\,^\circ\mathrm{C}$. The 'universal values' should only be used in the absence of relevant experimental data. When some data are available one can take $C_1^0 = 8.86\,\mathrm{K}^{-1}$, $C_2^0 = 101.6\,\mathrm{K}$ and then choose T_0 to give a best fit.

For values of T larger than $T_g + 100\,^\circ\mathrm{C}$, or where T_g is irrelevant we can use the Andrade form (9.6); we have already shown in Chapter 3 that this is a useful form for low-density polyethylene. By taking logarithms of (9.6) we may again define a_T for viscosity change in the form

$$\log\left(\frac{\eta}{\eta_0}\right) = \log a_T = \frac{0.434E}{R}\left\{\frac{1}{T} - \frac{1}{T_0}\right\}. \tag{9.16}$$

Typical values of E/R are given in Table 9.4. The values depend somewhat on the molecular parameters of the polymer. For LDPE, a useful formula is

$$\frac{E}{R} = a_0 + a_1\rho_{20}, \tag{9.17}$$

where $a_0 = 82770\,\mathrm{K}$ and $a_1 = -83.3\,\mathrm{K\,m^3/kg}$; here ρ_{20} is the density $(\mathrm{kg/m^3})$ at $20\,^\circ\mathrm{C}$.

Similarly, we may use a shift factor to describe the behaviour of solutions of polymers (Ferry 1981) and hence the concept is generally useful.

9.2.4   Time–temperature shifting

So far we have only looked at the quantity a_T as a device for correlating the zero-shear-rate viscosities of polymers. However, it has a much greater significance in that it also gives a shift of material time-scale with temperature.

Let us consider a material undergoing stress relaxation. The rate of relaxation is determined by an internal time-scale (or clock) within the material. As temperature rises, so does the amount of molecular motion occurring in one unit of an observer's time; the material's time-scale shortens so that relaxation proceeds faster. Let us suppose that the material time-scale changes so that one unit of material time is now equivalent to $a(T)$ units of observer time; $a(T)$ is a decreasing function of temperature T. Let $G(t, T)$ be the stress relaxation modulus at constant temperature T, and let $G(t) = G(t, T_0)$ be the modulus at a

Table 9.4   Flow activation energy ratio E/R

Fluid	η_0 (Pa-s)	T_0 (K)	E/R (K)
High-impact polystyrene	1.45×10^5	463	3707
Polystyrene	9.2×10^3	483	4954
Polypropylene	3.2×10^3	463	$5.1\text{--}5.6 \times 10^3$
High-density polyethylene	1520	473	$2.8\text{--}3.3 \times 10^3$
Low-density polyethylene	3200	453	6840
Polymethyl methacrylate	6000	513	9855

reference temperature T_0. We let $a(T_0) = 1$, and then assume

$$G(t, T) = \left(\frac{\rho T}{\rho_0 T_0}\right) G(t/a(T)), \qquad (9.18)$$

where ρ is the density at T, and ρ_0 is the density at T_0. This represents a scaling such that $G(t, T)$ has the same shape, but not the same scale, at all temperatures. Now we have a connection between the zero-shear-rate viscosity $\eta_0(T)$ and G [eqn (2.91)]:

$$\eta_0(T) = \int_0^\infty G(t, T)\, dt = a(T)\left(\frac{\rho T}{\rho_0 T_0}\right) \int_0^\infty G(t')\, dt'$$

$$= a(T)\left(\frac{\rho T}{\rho_0 T_0}\right) \eta_0(T_0). \qquad (9.19)$$

Thus

$$\log\left(\frac{\eta_0(T)}{\eta_0(T_0)}\right) = \log a(T) + \log\left(\frac{\rho T}{\rho_0 T_0}\right). \qquad (9.20)$$

In (9.20) $\log a(T)$ is always much greater than $\log(\rho T/\rho_0 T_0)$, and the latter factor can usually be ignored. Except for the small $\rho T/\rho_0 T_0$ factor (9.20) agrees with the previous definition of a_T. To see that the $a(T)$ is a time-shift factor we may calculate the mean relaxation time $\bar{\lambda}(T)$

$$\bar{\lambda}(T)\eta_0(T) = \int_0^\infty t G(t, T)\, dt$$

$$= \eta_0(T_0)\bar{\lambda}(T_0)\left(\frac{\rho T}{\rho_0 T_0}\right) a^2(T). \qquad (9.21)$$

Thus

$$\bar{\lambda}(T) = \bar{\lambda}(T_0) a(T). \qquad (9.22)$$

More generally, molecular theories (Bird *et al.* 1977) lead us to expect that for non-Newtonian fluids a plot of $\log|\eta(\dot{\gamma}, T)\eta_0(T_0)/\eta_0(T)|$ against $\log(\dot{\gamma} a_T)$ will collapse all the separate $\eta(\dot{\gamma}, T)$ curves on to a single master curve. Figure 9.4 shows a set of data and the master curve for polyisobutylene. (In the absence of η_0 data, one can often still construct such a master curve by careful fitting of the curves or by estimating η_0). Note that use of eqn (9.18) gives a direct method of finding $a(T)$.

The shift factor is of the form (Bird *et al.* 1977), for dilute polymer solutions

$$a(T) = a_T = \frac{[\eta_0 - \eta_s]_T T_0 \rho(T_0)}{[\eta_0 - \eta_s]_{T_0} T \rho(T)} \qquad (9.23)$$

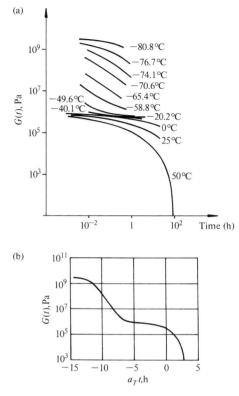

Fig. 9.4 Time–temperature superposition. (a) Stress-relaxation of polyisobutylene at various temperatures. (b) Stress-relaxation master curve for polyisobutylene found by shifting data in (a).

where η_s is the solvent viscosity. This form may readily be derived from dilute solution theory of the type discussed in Chapter 5. From eqns (5.44) and (5.57) we have an expression for the viscosity of the form

$$\eta - \eta_s = \lambda n_0 k T, \qquad (9.23\text{a})$$

where λ is the time constant, n_0 is the number of molecules per unit volume, k is the Boltzmann constant and T is the absolute temperature. Equation (9.23) then follows immediately, since n_0 is proportional to the polymer density, if we form the ratio of the time constant at two temperatures T and T_0. For concentrated solutions and melts we ignore η_s. The $T_0\rho_0/T\rho$ factor is also nearly unity; the variations in T and ρ tend to cancel one another. When this factor is ignored we return to the definition of (9.14) which can be represented by the WLF or Andrade forms given above in (9.13) and (9.16) respectively.

The previous section was based on the idea that the relaxation modulus $G(t)$ varied mainly through the time shift [eqn (9.18)]. It follows from the linear viscoelastic theory that the characteristic quantities G' and G'' (and their deriva-tives η' and η'') can be regarded as a function of the reduced frequency ωa_T when the temperature varies; it is usual to write

$$G'(\omega, T) = \frac{\rho T}{\rho_0 T_0} G'(\omega a_T),$$ (9.24)

and so on. Hence the problem of predicting linear viscoelastic response given data at a reference temperature is solved.

Some idea of the effect of temperature changes can be gathered if one takes the universal WLF curve, and considers a change in temperature from $T_g + 30\,^\circ\text{C}$ to $T_g + 10\,^\circ\text{C}$. We find a_T changes by a factor of about 3800, so that processes that took one second at the higher temperature take more than one hour at the lower temperature; a further lowering of temperature by $20\,^\circ\text{C}$ will increase the time scale by a factor greater than 10^7 so that the material is then well and truly 'frozen' as would be expected since it is below the glass temperature.

9.2.5   The Morland–Lee hypothesis

Although time–temperature shifting gives one a method of determining fluid properties at a temperature T given a master curve at T_0 one still has to consider that a particle in a flow will encounter a series of temperature states. Morland and Lee (1960) showed how to incorporate time–temperature shifting into linear viscoelastic boundary-value problems. In this case a pseudo-time ξ can be introduced where ξ is the time measured by the particle's own internal 'clock'. The amount of time that elapses during an interval $d\xi$ of pseudo-time is given by $dt/a(T)$ where $a(T)$ is the time-shift factor. Then we define

$$\xi = \int_0^t a^{-1}[T(t')]\,dt',$$ (9.25)

and reformulate the problem in terms of ξ. (Note that Morland and Lee used a where we use a^{-1}.) Thus if we have a simple shear-strain situation, we have, in the isothermal case [eqn (1.24)] that the contribution to the stress at time t is $G(t - t')\,d\gamma$ due to the strain $d\gamma(t')$ applied at time t' and in the interval dt'. In the non-isothermal case the stress-contribution, from the above, using eqn (9.18) and the Morland–Lee hypothesis is

$$d\tau = \left(\frac{\rho T}{\rho_0 T_0}\right)_t G[\xi(t) - \xi(t')]\,d\gamma(t').$$

Hence, the total shear stress at time t is

$$\tau = \left(\frac{\rho T}{\rho_0 T_0}\right)_t \int_{-\infty}^t G[\xi(t) - \xi(t')]\,d\gamma(t').$$ (9.26)

Note that the ρT-factor represents the change in the material modulus with temperature (it is a small factor which could be permitted to be more general than we have assumed). The ρT factor should not be inside the integral as it is the current temperature that determines the current modulus of the contribution of the strain $\mathrm{d}\gamma$ which was applied at time t'. (In their original paper Morland and Lee (1960) assumed $\rho T/\rho_0 T_0$ was always unity; it is not a very important factor.)

The inverse of (9.26) is

$$\gamma(t) = \int_{-\infty}^{t} J(\xi(t) - \xi(t'))\, \mathrm{d}[\rho_0 T_0 \tau/\rho T]_{t'}. \tag{9.27}$$

Thus, the stress and temperature are now both involved in (9.27), but Pipkin (1972) points out that experimental data would also be consistent with

$$\gamma(t) = (\rho_0 T_0/\rho T)_t \int_{-\infty}^{t} J\, \mathrm{d}\tau, \tag{9.28}$$

since the front factor is not very important. We may therefore take either (9.27) or (9.28) as the relevant form.

9.2.6   Non-linear materials

The concept of a time–temperature shift factor can clearly be applied to the non-linear single integral models discussed in Chapter 4 by replacing the time constants by shifted time constants and making the necessary adjustments to the moduli. For example, the special form of the KBKZ model [eqn (4.63)],

$$\tau_{ij}(t) = \int_{-\infty}^{t} \mu(t - t') H_{ij}(t')\, \mathrm{d}t', \tag{9.29}$$

can be reorganized to read, for the non-isothermal case:

$$\tau_{ij}(t) = F(T/T_0) \int_{-\infty}^{t} \mu(\xi - \xi') H_{ij}(\xi')\, \mathrm{d}\xi', \tag{9.30}$$

where the front factor $F(T/T_0)$ is not much different from unity, and $F(1) = 1$. Phan-Thien (1979) has arranged a differential PTT model of the type (5.163) to cope with temperature variations. He shows, from the dumbbell theory of Chapter 5 [especially eqns (5.41) and (5.51)] assuming the bead friction forces are proportional to the solvent viscosity η_s, that we should regard the reduced stress $\tau^{(m)}(\rho_0 T_0/\rho T)$ as a function of the reduced time $t(T\eta_s(T_0)/T_0\eta_s(T))$. When $\rho = \rho_0$ and $\eta_s = \eta_0$ this agrees with eqn (9.20). Where important we shall, for consistency, use (9.20) to connect η, ρ and a_T. Since the model involves instantaneous quantities only, the instantaneous temperature shift for each factor is involved; (and the instantaneous front factor $F(T/T_0)$ if required) and thus the non-linear Maxwell model

$$\lambda \frac{\Delta \boldsymbol{\tau}}{\Delta t} + \boldsymbol{\tau} = 2\eta \mathbf{d}, \tag{4.25}$$

becomes, since $F^{-1}\boldsymbol{\tau}$, from (9.30), is always the same function of the shifted time

$$\lambda a_T \frac{\Delta}{\Delta t} (F^{-1}\boldsymbol{\tau}) + F^{-1}\boldsymbol{\tau} = 2\eta a_T \mathbf{d}. \tag{9.31}$$

This model predicts that in a steady isothermal shearing flow the normal stress difference N_1 is proportional to τ^2;

$$\frac{N_1}{\tau^2} = \frac{2\lambda}{\eta}. \tag{9.32}$$

Applying the shift theory, one finds

$$\frac{N_1}{\tau^2} = \frac{2\lambda}{\eta} F^{-1}. \tag{9.33}$$

Hence, this relation between N_1 and τ is predicted to be a very weak function of temperature. This is borne out in experiments by Vinogradov *et al.* (1970) on polyisobutylene (Fig. 9.5); a similar relation between N_2 and τ has been confirmed by M. Keentok and the author. By applying such a consistent set of rules we may convert any isothermal constitutive equation to the variable temperature form as shown in Table 9.5. For integral equations it is necessary to introduce the pseudo-time variable ξ as indicated. As examples, if we have a power-law

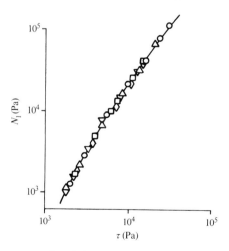

Fig. 9.5 Relation between normal and shear stresses in shearing flow at five temperatures for polyisobutylene: ($\circ$) 22 °C, ($\triangle$) 40 °C, ($\square$) 60 °C, (∇) 80 °C, ($\diamond$) 100 °C.

Table 9.5   Shift factors

Quantity	Shifted quantity
Time t	t/a_T
Frequency ω	ωa_T
Stress τ	$F^{-1}\tau \; (\approx \tau \rho_0 T_0/\rho T)$
Strain $\mathbf{C}$	$\mathbf{C}$
Strain rate $\mathbf{d}, \dot{\gamma}, \mathbf{A}$	$a_T \mathbf{d}, a_T \dot{\gamma}, a_T \mathbf{A}$
Constants, $\eta_0, \lambda_0, n''',$	$\eta_0, \lambda_0, n'''.$

$\tau = K\dot{\gamma}^n$, then the shifted form must be $F^{-1}\tau = K(a_T\dot{\gamma})^n$ where K and n are evaluated at T_0.

Similarly, for a Carreau form (Table 1.6) we have the standard form

$$\tau = \eta_0 \dot{\gamma} |1 + (\lambda_0 \dot{\gamma})^2|^{(n-1)/2}, \tag{9.34a}$$

and the shifted form

$$F^{-1}\tau = \eta_0 a_T \dot{\gamma} |1 + (\lambda_0 a_T \dot{\gamma})^2|^{(n-1)/2}. \tag{9.34b}$$

Some cautions have been sounded by workers over the too-enthusiastic use of the shifting ideas. In order of increasing difficulty:

(i) Collins and Metzger (1970) report the following viscosity data for a relatively low molecular weight PVC resin:

Temperature (°C)	Viscosity (Pa-s)	
	At 'zero' shear rate	At $10\,\mathrm{s}^{-1}$
190	4.2×10^3	3.0×10^3
160	1.0×10^5	3.5×10^4

One can calculate the ratios of the tabulated viscosities at the two temperatures compared with the prediction of the WLF equation; since unplasticized PVC has $T_g = 87\,°C$, using eqn (9.14)

$$a_T(160\,°C) = \log\left(\frac{\eta_{160}}{\eta_{87}}\right) = -10.2$$

$$a_T(190\,°C) = \log\left(\frac{\eta_{190}}{\eta_{87}}\right) = -11.6.$$

Subtracting the second from the first expression above, we obtain the WLF prediction

$$\frac{\eta_{160}}{\eta_{190}} = 25.3.$$

The ratio of viscosities from the reported data is at zero shear rate 23.8, whereas at a shear rate of $10\,\text{s}^{-1}$ the ratio is 11.7.

However, looking at (9.34b), we see that it is quite wrong to use $\dot{\gamma} = 10\,\text{s}^{-1}$ in each case; for a proper comparison one must take corresponding shear rates $a_T\dot{\gamma}$.

(ii) Christensen (1971), following Crochet and Naghdi (1968) develops a theory for small strains which considers small temperature changes. This theory does not seem to obey time–temperature superposition for constant temperature states.

(iii) Matsumoto and Bogue (1977) have shown that when temperatures change rapidly, a cooling rate effect is noticeable, and the shifting theory appears to fail. Gupta and Metzner (1982) have also considered this matter.

Further exploration of this idea is needed; it is a difficult field of experimentation.

(iv) The equation (9.7a) lies outside the framework.

We now discuss some theoretical work of Wiest (1995).

9.2.7   Non-linear effects in the Giesekus model

Wiest (1995) has considered the bead-spring theory and has shown that for the Giesekus model [eqn (5.111)] one finds a temperature-shifted equation of the form

$$a_T\lambda\frac{\Delta\tau}{\Delta t} + \tau + \frac{1}{nkT}\boldsymbol{\tau}\cdot\boldsymbol{\tau} + \tau\lambda\frac{\text{D}\ln T}{\text{D}t} = 2\eta_0 kTa_T\lambda\mathbf{d}. \tag{9.35}$$

Comparing (9.35) with (9.31) (setting the front factor $F = 1$) shows that the extra term $\lambda\text{D}\ln T/\text{D}t$ has appeared. If we compare this term with the $\text{D}/\text{D}t$ component of the first term, we see that for the new term to be significant, one needs

$$\tau\text{D}\ln T/\text{D}t \approx a_T\text{D}\tau/\text{D}t$$

or

$$\text{D}\ln T/\text{D}t \approx a_T\text{D}\ln|\tau|/\text{D}t$$

Now the possible variation of $\ln|\tau|$ is much greater than that of $\ln T$. If the maximum and minimum temperatures are 500 K and 300 K respectively, then $\ln T_{\max}/\ln T_{\min} \approx 1.09$, whereas the shear variation is essentially unlimited. Hence it is improbable that the extra term is often important. These conclusions are reinforced by Wiest's own calculations, which found a few per cent change with the new terms included.

It therefore seems safe to use the classical time-shift in many cases, and we shall not consider more complex ideas here.

9.3   Other thermophysical properties

To solve the complete set of equations we need to know (in addition to the constitutive equation) the density (ρ), specific heat (C_p) and thermal conductivity (k) of the polymer. A useful source of data is the book by van Krevelen and Hoftyzer (1976). Some data on common polymers are given in Table 9.6. Surface tension and radiative properties may also be important in some cases; generally they are not important in the flow of molten polymers. The surface tension of polymers varies between 19 and 46 mN/m. (See van Krevelen and Hoftyzer 1976.)

9.3.1   Density

The density does not vary very much with temperature and pressure; nevertheless, it is sometimes important to take account of compressibility. Table 9.6 gives values of the rubbery-phase density at 25 °C and also the rate of change of density with temperature at constant pressure $(\partial \rho / \partial T)_p$; this quantity is almost independent of temperature (Bondi 1968). For the variation of density with pressure and temperature we write (van Krevelen and Hoftyzer 1976):

$$1 - \frac{\rho(0)}{\rho(p)} = C \ln \left| 1 + \frac{p}{B'} \right|,$$

where

$$B' = b_1 \exp(-b_2 T'). \tag{9.36}$$

Here T' is the temperature in degrees Celsius (°C) and the values of C, b_1 and b_2 are given in Table 9.6. One can see that pressures of the order of 0.1 GPa (10^3 bar) are needed for pressure to have an appreciable effect on density; a pressure of 0.1 GPa (100 bar) will increase the density of low-density polyethylene by about 8 per cent at 150 °C. Similarly, a change of temperature of 100 °C will decrease the density by about 7 per cent.

Table 9.6   Density data for polymers in the amorphous state

Polymer	Density ρ (25 °C) kg/m^3	$-(\partial\rho/\partial T)_p$ (kg/m^{3}K)	Values for eqn (9.36)		
			100 °C	b_1 (GPa)	$b_2 10^{-3}$K^{-1}
Polyethylene (LDPE)	855	0.548–0.701	9.70	0.199	5.10
Polypropylene	850	0.397–0.679	–	–	–
Polyisobutylene	840	0.395–0.496	8.71	0.191	4.15
Polyethylene oxide	1130	0.792–0.843	–	–	–
Polystyrene	1050	0.474–0.750	8.94	0.244	4.14
Polyvinylchloride	1385	0.806–0.997	8.94	0.352	5.65
Polymethyl methacrylate	1170	0.712–0.753	8.94	0.385	6.72
Nylon 6	1084	0.658	–	–	–
Nylon 6–6	1070	–	–	–	–

(Van Krevelen and Hoftyzer 1976).

9.3.2 Specific heat

The specific heat can be found either at constant pressure (C_p) or at constant volume (C_v). The former is the relevant one for the energy equation, but since liquids are not very compressible, one may set $C_p \approx C_v$.

The difference between the two specific heats is exactly $K\alpha^2 T/\rho$, where α is the thermal expansion coefficient, T is the absolute temperature, and K is the bulk modulus. For molten low-density polyethylene, $C_p - C_v$ is about 300 J/kg K, while $C_p \sim 2000$ J/kg K.

The general shape of $C_p - T$ curves is shown in Fig. 9.6 and values are given in Table 9.7. Aqueous solutions behave similarly to the water base (Cho and Hartnett 1982); see Table 9.7.

9.3.3 Thermal conductivity

Polymer melts are poor conductors. Sample values of the thermal conductivity k are given in Table 9.7. The key feature to note in Table 9.7 is that the values

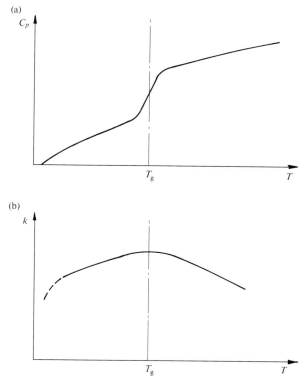

Fig. 9.6 Sketch of polymer thermal property variation. (a) Specific heat as a function of temperature for amorphous polymers. (b) Thermal conductivity as a function of temperature for amorphous polymers.

Table 9.7 Specific heat and thermal conductivity of amorphous polymers and other fluids

	C_p (J/kg K) (296 K)	$(1/C_p) (dC_p/dT)(K^{-1})$ (298 K)	k (W/m K)[b] (298 K) (except where indicated)
Polyethylene[a]	2260	1.0×10^{-3}	~0.25 (423 K)
Polyisobutylene	1970	2.2×10^{-3}	0.130
Polystyrene	1720	1.2×10^{-3}	0.176 (373 K)
Polymethyl methacrylate	~1800	1.5×10^{-3}	0.195 (423 K)
Polyethylene oxide	2050	0.5×10^{-3}	0.205
Nylon 6	2140/2470	$\sim 1.0 \times 10^{-3}$	0.21 (473 K)
Glycerol	2400	–	0.283
Water	4200	–	0.590

[a] For partly crystalline polymers the thermal conductivity is increased. Eiermann (1965) gives k crystalline/k amorphous $\sim 1 + 5.8(\rho_c/\rho_a - 1)$ where ρ_c/ρ_a is the ratio of crystalline to amorphous densities. Lohe (1965) gives $k \sim 0.25$ W/m K at 100 °C for LDPE and Greig and Sahoto (1978) give $k \sim 0.41$ W/m K at 100 °C for a partly crystalline isotropic sample; they also discuss thermal anisotropy of extruded polyethylene.
[b] Roughly we may take $(T_g/k_g)(dk/dT) \sim -0.2$ for amorphous polymers.

of k are similar in magnitude to those of glycerol, and about 1/3 those of water. As a result of this, heat transfer to polymers is difficult.

Molten polymers at rest appear to be isotropic with respect to heat conduction so that the values in Table 9.7 apply to conduction in all directions. As we have discussed above, deformation can change this; drawn or extruded polymer samples show higher conductivities in the direction of strain than transverse to the strain. These differences are quite noticeable in crystallizable materials where order of magnitude differences are possible in the two directions (Greig and Sahota 1978). However, in glassy polymers, the effect of orientation on k is not nearly as dramatic and this is expected to be more indicative of conduction in amorphous polymer melts. Where the Weissenberg number $(\lambda\dot{\gamma})$ is much greater than one the conductivity will depend somewhat on direction (van den Brule (1995)). However, few data are available to verify this, and in any event the ratios of conductivities in different directions in amorphous materials are expected to be only of order two or three.

It is important to note that k is insensitive to the chemical nature of the polymer, molecular weight, temperature, and pressure. Thus for many typical polymers k is between 0.12 and 0.27 W/m K. Conductivity data are frequently given as the thermal diffusivity k

$$\kappa = k/\rho C_p, \tag{9.37}$$

where ρ is the density and C_p is the heat capacity per unit mass. Shoulberg (1963) has shown that κ for polymers is nearly independent of T, p, and chemical characteristics. It should be kept in mind that there are fewer data available for k and κ than for viscosity, so there may well be exceptions to trends indicated here.

Because of the insensitivity of k or κ to temperature, accurate representations, which are useful for design calculations, can be given for k, ρ, and C_p by low order polynomials (van Krevelen and Hoftyzer 1976); often a linear variation with temperature is adequate. Further work on the thermal conductivity and diffusivity of polyethylene is to be found in Kamal *et al.* (1983) who show that the conductivity reaches a minimum at the melt temperature and thereafter increases slowly.

9.4 Effect of pressure on flows

The effect of pressure on a fluid shows in two ways: the variation of density affects the flow via the mass-conservation equation, and the pressure also affects the viscosity via the momentum conservation law. We can illustrate both of these points by considering creeping flow through a circular tube. We shall suppose that the flows are nearly-viscometric.

In the first case the pressure-drop problem can be treated by writing a Poiseuille law at each axial location and then applying the appropriate average mass conservation law $\dot{m} = \pi\rho\bar{w}R^2 = $ constant; here R is the local tube radius, which may be a slowly-varying function of z; and $\bar{w}$ and ρ are also functions of z only. Thus the problem resembles lubrication theory; and one can show that compressibility reduces the pressure drop over that of a similar incompressible fluid. An analysis of calendaring in the compressible case has been made by Chung (1982).

In the second case (viscosity a function of pressure) one can treat the case of a lubrication-type flow readily by introducing a modified pressure function. In these cases, when the pressure is a function of z only, and the viscosity is given by $\eta = \eta_0 \exp(p/\beta)$ then use of the function $\beta\{1 - \exp(-p/\beta)\}$ in place of p reduces the problem to the isothermal form (Problem 9.3). The pressure drop through a tube is increased over the case $\beta = \infty$, other factors being equal (Problem 9.4). Dowson and Higginson (1966) have applied this method to ball and roller bearing lubrication problems.

In both of these situations one can estimate the effects as being (roughly) of order p/K and p/β, respectively, where K is the bulk modulus. They will be ignored in the following pages. In many cases the effects of temperature are more important, and these are considered next.

9.5 Effect of temperature on flows

The effect of temperature on material properties, especially viscosity, is large, and can often mask non-Newtonian effects in flow. As an example, Denn (1983) has shown that for spinning calculations both non-Newtonian isothermal and non-isothermal Newtonian cases can be made to fit the data almost equally well with small adjustments of the surface heat transfer coefficient (h) (Fig. 9.7). Therefore it is often most important to consider temperature variations in flow.

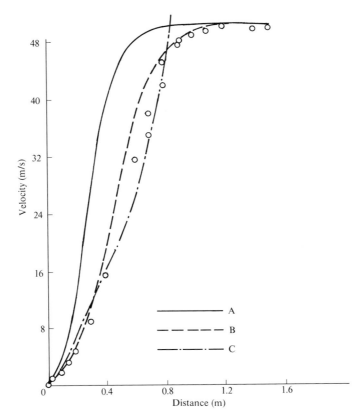

Fig. 9.7 Spinning data (circles) showing fit of isothermal viscoelastic calculation (curve C, PTT model) and non-isothermal Newtonian calculations (A and B) using (A) full value of heat transfer coefficient h and (B) 75 per cent of h used in A.

Temperature variations can be caused by temperatures external to the fluid, for example a hot tube wall, or they can be caused by self-heating due to viscous dissipation. It is often useful to make a rough estimate of the latter. In a shear flow between two conducting walls, spaced h apart and at a constant temperature T_0, provided the viscosity η_0 is constant, the rate of energy dissipation is $\eta_0 U^2/h^2$, where U is the speed difference. The rate of heat loss from the interior is of order $k\Delta T/h^2$, where k is the thermal conductivity of the fluid and ΔT is the maximum temperature rise in the centre of the fluid mass. Equating these fluxes, one finds

$$\Delta T = C\eta_0 U^2/k. \tag{9.38}$$

The constant of proportionality C is found to be $1/8$ (Problem 9.5). Equation (9.38) is an estimate of self-heating in experiments.

In many cases the effect of temperature rise has a great effect on rheology via the viscosity function, and we now consider some examples. One can also consider many other heat transfer problems where the flow field is not disturbed by the temperature rise (temperature-independent viscosity); see Bird *et al.* (1977), p. 236, for a collection of these heat transfer coefficients.

Aside from the self-heating analysis (9.38) which leads to the definition of the Brinkman number $\eta_0 U^2/k\Delta T$, there are problems in which ΔT is imposed by an external heat source. One may still define a Brinkman number in this case which will give an indication of the importance of self versus external heating. For the case where the viscosity varies with $\exp - \alpha T$, then the Nahme–Griffith number $\eta_0 U^2 \alpha/k$ is a relevant dimensionless group. A large Nahme–Griffith number indicates strong coupling between the momentum and energy equations. In case convection dominates conduction, one has the Peclet number $(Pe) \gg 1$; here $(Pe) = \rho C_p UL/k$.

These numbers should be formed from the energy and momentum equations taking into account relevant length scales; for example, in lubrication and boundary-layer flows there are two length scales that are relevant (h and L). In this way one finds the Graetz number $[(Pe)h/L]$ which is relevant to some processing operations which are nearly viscometric. We now consider some illustrative examples.

9.5.1 Circular tube flow

Solutions of the laminar heat transfer problem with both Newtonian and non-Newtonian fluids can be divided into three categories as follows:

(a) Asymptotic solutions which apply far from the duct entrance where the velocity and temperature profiles are both fully developed.

(b) The thermal entry length solutions which assume that the velocity profile is fully developed at the inlet to the heated section of the duct but consider the developing temperature profile.

(c) The combined hydrodynamic and thermal entry length solutions which take into account the development of both the velocity and temperature profiles.

The Prandtl number $C_p\eta/k$, which is the ratio of momentum diffusivity to thermal diffusivity for a fluid, is greater than 50 for most of the viscous fluids we are concerned with. Hence, the hydrodynamic entry length is small and there is therefore little difference between type (b) and (c) solutions. However, type (a) solutions which give asymptotic Nusselt numbers can lead to underestimates of average heat flux, as the mean Nusselt number can be substantially greater than the limiting value.

In this subsection and the next we will consider treatments based on the assumptions that the physical properties of the fluid are independent of temperature and that viscous shear heating effects are negligible.

The solutions are thus somewhat unrealistic and hence we shall not develop results at great length.

For the asymptotic zone, where velocity and the temperature profiles are fully-developed, neglecting energy dissipation and axial conduction ($k \, \partial^2 T/\partial z^2$ term) then the energy equation becomes, in cylindrical geometry,

$$\frac{\partial^2 T}{\partial r^2} + \frac{1}{r}\frac{\partial T}{\partial r} = \frac{w(r)}{\kappa}\frac{\partial T}{\partial z}, \tag{9.38a}$$

where κ, the thermal diffusivity, is constant. In the fully-developed region we assume $\partial T/\partial z$ is constant everywhere, and hence

$$T = Az + g(r). \tag{9.39}$$

Then we find, for a tube of radius a,

$$g(r) = -\frac{2\bar{w}Aa^2}{\kappa}\left[\frac{3}{16} - \frac{1}{4}\left(\frac{r}{a}\right)^2 + \frac{1}{16}\left(\frac{r}{a}\right)^4\right] \tag{9.40}$$

which ensures $g(r) = 0$ at $r = a$ and $g'(r) = 0$ at $r = 0$; the flow is a Poiseuille flow with a mean speed $\bar{w}$. This gives

$$T_{\max} = Az - \frac{3}{8}\frac{\bar{w}Aa^2}{\kappa}, \tag{9.41}$$

and if we compute the averaged temperature T_{m} where

$$T_{\mathrm{m}} = 2/\bar{w}a^2 \int_0^a Twr\,\mathrm{d}r = Az - \frac{11}{48}\frac{\bar{w}Aa^2}{\kappa}, \tag{9.42}$$

then the local heat transfer coefficient can be computed. The rate of heat transfer q_{w} through the wall is $-k\,\partial T/\partial r|_{r=a}$ out of the fluid, and is $\bar{w}kAa/2\kappa$ here. Thus

$$\bar{w}kAa/2\kappa = h(T_{\mathrm{w}} - T_{\mathrm{m}}) = \frac{11}{48}h\frac{\bar{w}Aa^2}{\kappa}, \tag{9.43}$$

or

$$(Nu) = \frac{hd}{k} = \frac{48}{11} = 4.364, \tag{9.44}$$

where $d = 2a$.

Other cases can be studied, see Kays (1966). The limiting Nusselt number for constant wall temperature is 3.658.

For several types of non-Newtonian fluid the corresponding limiting Nusselt numbers have been derived by Beek and Eggink (1962). These are illustrated in Fig. 9.8. For example, for power-law fluids with constant wall flux we have

$$(Nu)_\infty = \frac{8(3n+1)(5n+1)}{31n^2 + 12n + 1} \tag{9.45}$$

and this reduces to $(Nu)_\infty = 4.364$ for a Newtonian fluid when $n = 1$ and when $n = 0$, the condition of limiting pseudo-plasticity or plug flow, we get $(Nu)_\infty = 8$ and for highly dilatant fluids $(Nu)_\infty = 3.87$. For constant wall temperature the limiting cases are $(Nu)_\infty = 6$, when $n = 0$ and $(Nu)_\infty = 3.30$ when $n = \infty$.

For the thermal entry region, making the assumption that the flow is fully-developed, properties are constant, and axial conduction and dissipation are ignored, eqn (9.38a) still holds.

Solutions can be obtained for the appropriate boundary conditions to give the Nusselt number, hd/k, as a function of the Graetz number $(Gz) = \dot{m}C_p/kz$, in the form ($\dot{m}$ is the mass flux)

$$(Nu) = f(Gz), \tag{9.46}$$

where the Nusselt number can be the mean value $(Nu)_m$ from the start of the heated section to any axial position z, which is useful for the case of constant wall temperature, or alternatively the local value, $(Nu)_z$, at a particular value of z which is useful for constant wall heat flux. Solutions for the mean Nusselt number with constant wall temperature are given in Fig. 9.9 with the power law index n as a parameter.

An extension of the Lévêque approximation to non-Newtonian fluids has been given by Pigford (1955). This approximation takes the form:

$$(Nu)_m = 1.75\delta^{1/3}(Gz)^{1/3} \tag{9.47}$$

where δ is the ratio of the wall shear rate for a Newtonian fluid to that for a non-Newtonian fluid at the same flow rate. For example, for power-law fluids, it can

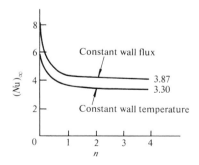

Fig. 9.8 Asymptotic Nusselt numbers for power-law fluids in tube flow.

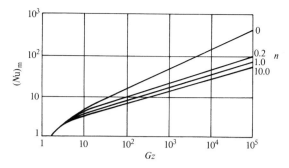

Fig. 9.9 Nusselt number $(Nu)_m$ as a function of Graetz number (Gz) for power-law fluids with n as parameter ($n = 1$ is Newtonian case). Constant physical properties and constant wall temperatures are assumed.

be shown that

$$\delta = (3n + 1)/4n. \tag{9.48}$$

The limiting cases become:

$$\left.\begin{array}{ll} n = 0; & (Nu)_m = \dfrac{8}{\pi} + \dfrac{4}{\pi} Gz^{1/3} \\[2mm] n = 1; & (Nu)_m = 1.75 Gz^{1/3} \\[2mm] n = \infty; & (Nu)_m = 1.59 Gz^{1/3} \end{array}\right\} \cdot \tag{9.49}$$

These are given in Fig. 9.9. Further discussion of tube heat transfer is given by Cho and Hartnett (1982) and Gottifredi *et al.* (1983).

9.5.2 Flat plate flow

The simplest form of the energy equation for heat transfer in laminar flow between flat plates is:

$$u(y)\frac{\partial T}{\partial z} = \kappa \frac{\partial^2 T}{\partial y^2}, \tag{9.50}$$

where z is the axial distance and y the distance from the centreline; viscous dissipation, heat generation and expansion terms are neglected.

Tien (1962) has solved this problem for power-law fluids with constant material properties, constant plate temperature, fully-developed inlet velocity profile and constant inlet temperature. The results can be used to give the average fluid temperature at axial distance z, $\bar{T}_z$, in terms of the wall temperature T_w and the fluid inlet temperature T_i as follows:

$$\frac{T_w - \bar{T}_z}{T_w - T_i} = \sum_{j=0}^{\infty} A_j \exp[-\tfrac{8}{3} a_j^2 z^+].$$ (9.51)

where A_j and a_j are tabulated functions of the power-law index, n and z^+ is a reduced axial distance, equivalent to a reciprocal Graetz number given by

$$z^+ = \frac{\kappa z}{\bar{u} H^2},$$ (9.52)

H being the plate separation.

Tien (1962) has also considered the heat transfer to power-law fluids between parallel plates when one plate is stationary and the other is moving at a constant velocity.

Yan and Tien (1963) and Agur and Vlachopoulous (1981) have considered the simultaneous development of velocity and temperature profiles for laminar flow of a non-Newtonian power law fluid in the entrance region of flat ducts, the former authors assuming constant material properties and constant wall temperature.

The Lévêque approximation has also been applied to the flat plate problem at high values of (Gz) by Metzner (1965) and with this approach, by taking the hydraulic mean diameter, D_e, as $2H$ one obtains:

$$\frac{h_m D_e}{k} = 1.86 \delta^{1/3} \left[\frac{D_e^2 \rho \bar{u} C_p}{kL} \right]^{1/3},$$ (9.53)

where δ is the ratio of the shear rates at the plate surfaces for the non-Newtonian fluid and a Newtonian fluid at the same mean velocity; for a power-law fluid $\delta = (2n + 1)/3n$.

The viscosity being very sensitive to temperature the assumption of constant physical properties is not realistic. Empirical correction factors of the Sieder–Tate form can be applied to eqn (9.53) but these are not very satisfactory.

Theoretical solutions are complicated by the fact that the equations of motion and energy become coupled since temperature is involved in both but solutions have been obtained by Christiansen et al. (1966) for non-Newtonian fluids of the power-law type with a temperature dependent constitutive equation of the form

$$\tau = K\{\exp(E/RT)\dot{\gamma}\}^n.$$ (9.54)

By neglecting viscous shear heating and assuming thermal diffusivity constant solutions for constant wall temperature for pipe flow were obtained in the form

$$(Nu)_m = f(Gz, n, \psi(E)),$$ (9.55)

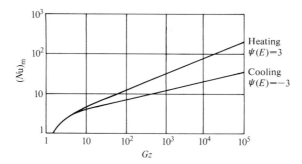

Fig. 9.10   Nusselt number $(Nu)_m$ as a function of Graetz Number (Gz) for temperature-dependent power-law fluid with $n=0.3$. For definition of ψ see eqn (9.56).

where

$$\psi(E) = \frac{E}{R}\left(\frac{1}{T_i} - \frac{1}{T_w}\right), \tag{9.56}$$

T_i being the (constant) inlet fluid temperature and T_w the (constant) wall temperature. Some results are shown in Fig. 9.10 for both heating (ψ positive) and cooling (ψ negative) for a shear thinning fluid with $n=0.3$.

It should be noted that the effects of the temperature-dependence of the rheological properties can be much more important than the degree of non-Newtonian behaviour and the effect of the latter decreases as $\psi(E)$ increases. This can be seen by comparing $(Nu)_m$ for a Newtonian fluid and the highly pseudo-plastic fluid with $n=0.3$ at $(Gz)=1000$, as follows:

n	$\psi(E)$	$(Nu)_m$ at $(Gz)=1000$
1	0	17
1	3	26
0.3	0	19
0.3	3	28

Clearly, most of these flows are best handled by computation at the present time.

9.5.3   Simple shearing flows with heating

If we consider a shearing flow between parallel planes at constant temperature, (Fig. 9.11) then one can see some of the effects of temperature dependent viscosity readily.

To begin with, suppose the flow field is described by $\mathbf{v} = \mathbf{i}u(z)$, and that pressure gradients are absent. Then mass conservation is satisfied, and the momentum equations reduce to the statement $\sigma_{xz} = \tau = $ constant. The energy

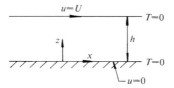

Fig. 9.11   Self-heating shearing flow.

equation reduces to $\{T = T(z) \text{only}\}$

$$k\frac{\mathrm{d}^2 T}{\mathrm{d}z^2} + \tau\frac{\partial u}{\partial z} = 0. \tag{9.57}$$

We now assume that the thermal conductivity k is constant, and that τ is related to the shear rate and temperature by

$$\tau = \mu_0 \exp(-\alpha T)\left(\frac{\partial u}{\partial z}\right)^n, \tag{9.58}$$

where μ_0 is the value of the consistency at the zero reference temperature. We can eliminate $\partial u/\partial z$ using (9.58) to obtain

$$\frac{\mathrm{d}^2 T}{\mathrm{d}z^2} + \left(\frac{\tau}{k}\right)\left(\frac{\tau}{\mu_0}\right)^{1/n}\exp\left(\frac{\alpha T}{n}\right) = 0. \tag{9.59}$$

Integration yields

$$\exp\frac{\alpha T}{n} = \frac{\alpha C_1^2}{2n}\left(\frac{k}{\tau}\right)\left(\frac{\mu_0}{\tau}\right)^{1/n}\mathrm{sech}^2\left\{\frac{\alpha C_1}{2n}(z + C_2)\right\}, \tag{9.60}$$

where C_1 and C_2 are constants which may be evaluated from the boundary conditions on temperature. Suppose that $T=0$ at $z=0$. The surface $z=h$ (Fig. 9.11) could be at any temperature; or it could be insulated; suppose it is also at $T=0$. Then it follows that $C_2 = -h/2$, and we find that C_1 must satisfy

$$1 = \frac{\alpha C_1^2}{2n}\left(\frac{k}{\tau}\right)\left(\frac{\mu_0}{\tau}\right)^{1/n}\mathrm{sech}^2\left(\frac{\alpha h C_1}{4n}\right). \tag{9.61}$$

In this relation τ is still not known, but we can see that there are definite limits on the shear rate which can be applied. Letting $x = \alpha h C_1/4n$, we see that (9.61) can be rewritten as

$$\frac{\alpha h^2}{8n}\left(\frac{\tau}{k}\right)\left(\frac{\tau}{\mu_0}\right)^{1/n} = x^2 \mathrm{sech}^2 x \tag{9.62}$$

and that the right-hand side never can exceed about 0.43923, which limits the possible shear stress that can be applied; this occurs when $x \approx 1.1997$. This thermal run-away at a critical shear stress occurs because the $\exp - \alpha T$ variation of viscosity vanishes faster than an Arrhenius law variation ($\exp A/T$) when T is large; it is not necessarily a physical limitation. Accepting this limitation, one can now find the velocity field from (9.57). We find

$$u = C_3 + C_1 \frac{k}{\tau} \tanh\left\{ \frac{\alpha C_1}{2n} (z - h/2) \right\} \tag{9.63}$$

where C_3 and C_1 are to be evaluated from the boundary conditions on u. Let (Fig. 9.11) $u = U$ on $z = h$, $u = 0$ on $z = 0$. Then we find $C_3 = U/2$ and also an equation connecting U, τ, and C_1:

$$\frac{U}{2} = C_1 \left(\frac{k}{\tau} \right) \tanh\left(\frac{\alpha C_1 h}{4n} \right). \tag{9.64}$$

In terms of $x = \alpha h C_1 / 4n$, (9.64) yields

$$\frac{\alpha h \tau U}{8nk} = x \tanh x, \tag{9.65}$$

or, using (9.62), $\left(\dfrac{h}{U} \right) \left(\dfrac{\tau}{\mu_0} \right)^{1/n} = x / \sinh x \cosh x.$ \hfill (9.66)

Thus one can choose x, compute the corresponding shear stress from (9.62) and then find U from (9.64). Alternatively, one can choose x, find τ from (9.62) knowing U/h and μ_0, and then find the corresponding value of α from (9.65). A further procedure is to eliminate τ and find an equation for x which depends only on the group $(\alpha h^2 \mu_0 / 8nk)(U/h)^{n+1}$; when $n = 1$, this group is independent of h. In summary, this problem shows how non-linearity of the velocity profile can develop due to thermal effects. The complexity of the solution procedure is also to be noted, and also the limiting shear stress. The special power-law and exponential viscosity form has been used here for analytical convenience but it would be possible to compute numerically a solution for any separable form; normal stresses do not enter the problem. Kearsley (1962) and Martin (1967) have investigated other viscometric flows using similar methods. Agur and Vlachopoulos (1981) have shown how to use finite difference techniques in problems of heat transfer to molten polymers in tubes.

9.5.4   Thermal runaway in a viscoelastic slab

Let us suppose that a slab of (linear viscoelastic) material is being sheared sinusoidally. If the rate of work, $\tau \dot{\gamma}$, is integrated over a complete cycle, we find that the work per cycle (per unit volume) is $\pi G'' \gamma_0^2$ or $\pi J'' \tau_0^2$, where τ_0 and γ_0 are the stress amplitude and the shear amplitude, and G'' and J'' are the loss modulus and loss compliance respectively. Multiplied by the number of cycles per second, $\omega/2\pi$, this gives the average rate of dissipation as

$$D = \tfrac{1}{2}\omega G''\gamma_0^2 = \tfrac{1}{2}\omega J''\tau_0^2. \tag{9.67}$$

If the slab is insulated so that no heat can escape, this energy input will cause a rise in temperature. So far as the storage modulus is concerned, this temperature rise will 'soften' the material, unless the frequency is so small that the material is effectively in thermodynamic equilibrium at all times. Let us suppose that the loss modulus also decreases as the temperature increases.

If the shearing is carried out at a constant shear amplitude γ_0, then the stress amplitude gradually decreases as the material becomes hotter. Consequently, the rate of dissipation also decreases. The temperature rise will become slower and slower, and, if heat is allowed to flow out of the slab, the temperature will approach some steady-state value.

On the other hand, suppose that the stress amplitude τ_0 is held constant. Then as the material heats, the compliance increases, the shear amplitude increases, and so the rate of dissipation also increases. There is a feed-back effect. The hotter the material is, the faster the temperature goes up. The situation is unstable, and there will be some sort of a catastrophe; even if heat is allowed to flow out of the slab temperatures may rise greatly.

Now consider the energy equation,

$$\rho \frac{De}{Dt} = -\frac{\partial q_i}{\partial x_i} + \sigma_{ij}\frac{\partial v_i}{\partial x_j}. \tag{9.68}$$

We suppose that the internal energy density e is determined by the history of deformation and temperature. In a quasi-steady-state cyclic oscillation, the histories of deformation and temperature are characterized by their cycle-averaged values and the amplitudes and frequency of their fluctuations. If we consider only the average rate of increase of internal energy over a complete cycle, then the term $\rho\dot{e}$ can be replaced by $\rho C_p(\partial T/\partial t)$ where $\partial T/\partial t$ is the rate of increase of average temperature; C_p will be considered constant.

For the heat flux we use Fourier's law, $q_i = -k(\partial T/\partial x_i)$. We treat the conductivity k as constant here. In the present problem, if we let the direction normal to the slab be the y-direction, then the average temperature will depend only on y and t, so $\partial q_i/\partial x_i$ will take the form $-k(\partial^2 T/\partial y^2)$.

For the average rate of work $\sigma_{ij}(\partial v_i/\partial x_j)$ we use the rate of dissipation D discussed previously. We consider the case in which the stress amplitude is constant, so the expression for D in terms of J'' is convenient. We finally obtain

$$\rho C_p\frac{\partial T}{\partial t} = k\frac{\partial^2 T}{\partial y^2} + \tfrac{1}{2}\omega\tau_0^2(T_0/T)J''[a(T)\omega]. \tag{9.69}$$

The methods to be used can be carried out just as well if J'' and $a(T)$ are empirically-determined functions, given graphically or numerically. However, to be explicit, let us suppose that J'' has a power-law form, $J''(\omega) = J\omega^{-p}$, and that $a(T) = A \exp(-BT)$. To simplify matters, let us omit the small factor T_0/T.

Let the boundaries of the slab be at $y = \pm h$. We scale y with respect to h, so that the boundaries are $y = \pm 1$ in the new variable. We also scale t and T in such a way as to combine as many parameters as possible, and translate the temperature origin so that the new dimensionless temperature ϕ is initially zero. The equation can thus be brought into the form

$$\frac{\partial \phi}{\partial t} = \frac{\partial^2 \phi}{\partial y^2} + \tfrac{1}{2}g^2 \exp(\phi), \tag{9.70}$$

where g^2 is $J_1 B\omega^{1-p}\tau_0^2/h^2 kA$.

Let us consider the case of insulated boundaries. In that case the temperature remains uniform at all times if it is uniform initially, so the equation yields

$$t = (2/g^2) \int_0^\phi \exp(-\phi)\, \mathrm{d}\phi. \tag{9.71}$$

The integration can be carried out explicitly; but suppose the integrand is given in the form of data. The main thing to notice about the integral is that it converges as $\phi \to \infty$, so the temperature diverges to infinity within a finite time t (equal to $2/g^2$ in the present case); the material has 'run-away'.

If heat escapes from the boundary then there may or may not be run-away. Pipkin (1972) discusses this matter in more detail: see also Problem 9.8.

9.5.5   Flow round a sphere

The previous flows considered were viscometric or nearly-viscometric flows. Morris (1982) has shown the changes in flow patterns around a hot sphere. The solution was obtained for both Newtonian and shear-thinning (power-law) fluids. If the viscosity at the sphere temperature T_0 is η_0 then if the temperature field is determined by conduction the sphere drag is proportional to $\eta_0 U$, where U is the speed of the sphere through the medium; this is a small Peclet-number limit. The Peclet number is here Ua/κ, where κ is the thermal diffusivity. If forced convection is important then one finds that the drag is proportional to $\eta_0 U^4$. For larger Peclet numbers one finds that the low-viscosity lubricated layer cannot build up to a reasonable thickness and the motion is governed by the viscosity (η_∞) at large distances from the sphere where the temperature is T_∞; the drag is proportional to $\eta_\infty U$. The analysis assumed that $|\alpha(T_0 - T_\infty)| \gg 1$; the asymptotic method of analysis used has also been applied to channel flow of polymeric fluids; here α is defined by $\eta = \eta_0 \exp -\alpha(T - T_0)$.

9.5.6   *Tension in a cooling rod*

The above illustrations have not, in the main, dealt with viscoelastic systems where a particle has a complex thermal history. The present example (Pipkin 1972) shows how complex the matter of thermal behaviour is in viscoelastic materials.

In this example, the temperature varies both in time and space. Consider that an extending force F is applied to a rod of radius R at time zero. Suppose that at time zero, the temperature of the rod is everywhere T_1, but its surface $r = R$ is lowered to a reference temperature T_0 at time zero and held there. The ends of the rod are insulated. The temperature is then a function of r and t only, at least away from the ends. It satisfies the initial and boundary conditions

$$T(r,0) = T_1, \quad T(R,t) = T_0. \tag{9.72}$$

We neglect dissipation, and thus suppose that the energy equation can be written as

$$\rho C_p \frac{\partial T}{\partial t} = \frac{k}{r}\frac{\partial}{\partial r}\left(r\frac{\partial T}{\partial r}\right). \tag{9.73}$$

The temperature can now be computed explicitly (if we had not ignored dissipation this could not be done).

We suppose that far enough away from the ends of the rod, the extension is uniform, a function $\varepsilon(t)$ of time only. We omit consideration of thermal expansion and suppose that the material is incompressible. The tensile stress $\sigma\,(\equiv\sigma_{zz})$ is then (ignoring the small front factor)

$$\sigma(r,t) = \int_0^t E[\xi(r,t) - \xi(r,t')]\,\mathrm{d}\varepsilon(t'), \tag{9.74}$$

where $E(t)$ is the elongational (Young's) relaxation modulus. Here the material time at the radius r is, from (9.25)

$$\xi(r,t) = \int_0^t \phi[T(r,t')]\,\mathrm{d}t'. \tag{9.75}$$

where $\phi(T) = a^{-1}(T)$. The condition of equilibrium under the specified load F is then

$$\int_0^R \sigma(r,t)2\pi r\,\mathrm{d}r = FH(t). \tag{9.76}$$

By combining the stress–strain relation with the equilibrium equation, we obtain

$$\int_0^t E^*(T,t')\,\mathrm{d}\varepsilon(t') = \sigma_0 H(t), \tag{9.77}$$

where $\sigma_0 = F/\pi R^2$ and where the effective modulus E^* is defined by

$$E^*(t, t') = (2/R^2) \int_0^R E[\xi(r, t) - \xi(r, t')] r \, \mathrm{d}r. \tag{9.78}$$

Note that the equation that $\varepsilon(t)$ satisfies is not a simple convolution product.

It is possible to solve the problem numerically, with given functions $\phi(T)$ and $E(\xi)$ but it is helpful to obtain an approximate solution, as follows.

Just after time zero, the temperature is T_1 everywhere except at the surface, so the rod extends uniformly according to the tensile compliance $D(\phi_1 t)$ where $\phi_1 = \phi(T_1)$. The stress is σ_0 almost everywhere. Right at the surface $r = R$, the modulus is $E(t)$ rather than $E(\phi_1 t)$, because the surface temperature is T_0. With the quasi-elastic approximations $\varepsilon(t) \simeq D(\phi_1 t)\sigma_0 \cong \sigma_0/E(\phi_1 t)$ and $\sigma(R, t) \cong E(t)\varepsilon(t)$, we find that the stress at the surface is

$$\sigma(R, t) \cong \sigma_0 E(t)/E(\phi_1 t). \tag{9.79}$$

Assuming that ϕ_1 is large, $E(\phi_1 t)$ relaxes much faster than $E(t)$ does, so the surface stress quickly becomes a good deal larger than σ_0. If $E(\phi_1 t)$ and $E(t)$ can both be approximated by the same power law $Ct^{-p}/(-p)!$ (after a little time has elapsed, so that both are somewhat below the glass modulus), then we obtain $\sigma(R, t) \cong \sigma_0 \phi_1^p$. If $T_1 - T_0 = 100°C$, then ϕ_1 might be about 10^4, but p is likely to be small, say $p = 0.1$, so ϕ_1^p is only about 2.5.

The surface stress seems to go up to about 2.5 times the nominal stress (with these figures) very quickly. It will then come back down, more slowly. As the lower temperature T_0 penetrates into the material, the annulus of stiff material on the outside grows thicker, and it will rapidly start supporting most of the load, both because it is stiffer and because a layer of given thickness has a much bigger area near the outside than a layer of the same thickness has near the middle due to the factor r in the integral defining E^*. Let us suppose that there is a circle $r(t)$ such that the temperature is roughly T_0 for $r > r(t)$ and T_1 for $r < r(t)$. As a further approximation suppose that

$$E^*(t, 0) = \{1 - [r(t)/R]^2\}E(t) = [r(t)/R]^2 E(\phi_1 t). \tag{9.80}$$

Then, with $\sigma(R, t) = \sigma_0 E(t)/E^*(t, 0)$, we can visualize how the surface stress begins to decrease as the load-carrying (cool) region grows larger.

9.6 Thermally-induced extrudate swell

Using a finite element method, Phuoc and Tanner (1980) investigated the swelling of Newtonian extrudates where the viscosity varies with $\exp -\alpha \Delta T$. Dissipation of energy created a hot fluid near the centreline (Fig. 8.24) while the die wall (a long tube of radius R) was maintained at T_w. Ambient temperature $T_\infty < T_w$ caused a heat loss from the extrudate. If the maximum temperature on the centreline is T_{max}, then the swelling ratio χ for a given material can be plotted in terms of $\alpha(T_{max} - T_w)$ (Fig. 9.12). In this figure the thermal properties of polyethylene were used, but no non-Newtonian properties are included. The swelling is remarkably high (~70 per cent). Repetition of this calculation with power-law shear-thinning (Huynh 1983) reduces the thermal swelling as would be expected. Correlation with the results of the inelastic swelling mechanism (Section 8.9.4) is good and it is clear that the inelastic swelling mechanism is important in the present case; it does not depend on energy dissipation but only on the temperature distribution across the die exit plane; if the outside layers are cool and the inner layers are hot, one expects increased swelling.

9.7 Injection moulding

We now briefly discuss the process of injection moulding as an example of a flow field which depends on (transient) rheology and heat transfer and is, consequently, very complex (Kennedy 1995). Injection moulding involves two processes: production of the flow of molten polymer, and shaping the product in the die. We shall only consider the latter aspect here; see Tadmor and Gogos (1979) for a discussion of the first part.

The stream of molten polymer is forced into the mould under pressure from the ram action of the moulder. A typical injection mould (Fig. 9.13) is made in two parts, one of which can be opened at the end of the time needed to finish one item. The mould temperature is below the melt (T_m) or glass transition temperature (T_g). The flow is through the sprue, runner, and gate system into the mould cavity; heat transfer to the mould then solidifies the fluid. Initially flow rates are high, but freezing of a 'skin' next to the mould surface slows the flow towards the end of the mould-filling cycle. (If cooling is too rapid, the mould may not

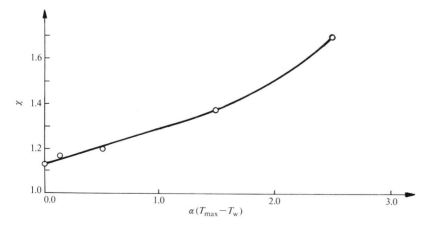

Fig. 9.12 Extrudate swelling ratio χ for a temperature-thinning Newtonian fluid as a function of temperature difference. T_{max} is the temperature on the centreline of the upstream tube and T_w is the tube wall temperature; α is the temperature–viscosity exponent, $\eta/\eta_0 = \exp - \alpha(T - T_0)$.

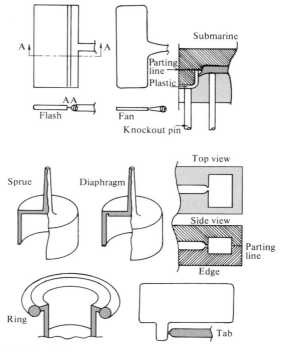

Fig. 9.13   Typical gate designs and locations for injection moulding.

be properly filled.) After the cavity has been filled, the pressure is kept on the melt to pack it securely into the cavity and to compensate for contraction during the cooling phase. Back-flow may take place upon removal of the pressure unless prevented or solidification has taken place. The component is then ejected, completing the cycle.

In the mould-filling part of the cycle, the flow is often nearly-viscometric for simple moulds [Fig. 9.14(a)]. This figure shows the position of the polymer front at different times during the filling. The 'fountain-effect' flow at the front is also shown. This flow is largely determined by kinematical considerations and the non-slip condition at the walls.

There has been a number of mathematical simulations of this flow. Figure 9.14(b) shows the frozen 'skin' which constricts the flow. Near the entry there is enough heat transfer to keep the skin thin; further into the mould it first gets thicker, and is then thinner near the polymer front. In view of the complexity, numerical simulation is needed. We shall mention the work of Isayev and Hieber (1980) who have discussed the problem of solidification and the subsequent prediction of residual stresses in the solid product. They used a Leonov model of the melt, thereby including viscoelastic effects; Dietz and White (1978) used an inelastic power-law fluid in their study with some success. Isayev and Hieber

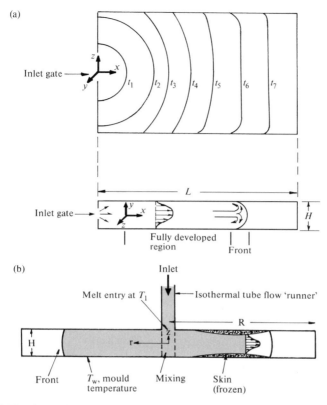

Fig. 9.14 (a) Flow into a simple mould showing front position at different times, fully developed region of flow and 'fountain effect' in front region. (b) Cross-sectional view of a centre-fed disc-shaped mould cavity. Indicated schematically are the frozen skin layer that may form during filling as well as the nipple-shaped velocity profile.

(1980) considered the flow in a Poiseuille (channel-flow) geometry, which is a two-dimensional problem. Let the gap width be $2h$ and the initial temperature be T_0 everywhere. At time $t = 0$ the wall temperatures are lowered to T_w while the flow is maintained at an average speed $\bar{w}$. Later, at a time (t_f) the flow is stopped. A time-marching numerical (finite-difference) procedure was used to find the results for the flow field, temperature and stresses at the end of the flow cycle (time t_f). Then the calculation was continued until all the material had dropped below T_g and was frozen.

The material constants and geometry were chosen to match the experiments of Wales (1976) on polystyrene. Results are shown in Fig. 9.15. In order to maintain constant inflow rate, the speed in the core increases with time, and the point of maximum shear rate moves inwards from the wall.

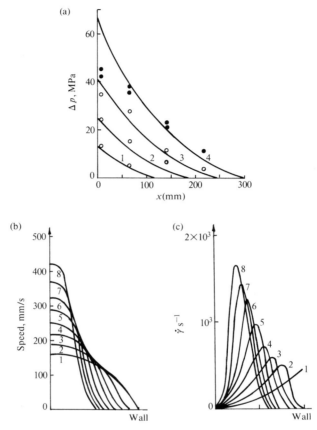

Fig. 9.15 Idealized moulding problem and results. (a) Pressure distribution along cavity for various flow times (s) (1) 1.0 (2) 1.5 (3) 2.0 (4) 2.5. Curves are theoretical predictions. Experimental results for short (○) and flash (●) conditions are shown. (b) Velocity profiles, and (c) shear rate profiles. Number on curves refer to times in seconds: (1) 0 (start), (2) 0.05, (3) 0.2, (4) 0.5, (5) 1.0, (6) 1.5, (7) 2.0, (8) 2.5. $T_{inlet} = 503$ K, $T_w = 323$ K, $\bar{w} = 120$ mm/s; thickness of moulding 10 mm.

This idealized problem can be applied to injection moulding. The thickness of the frozen zone is over-estimated in the present calculations and the flow, of course, does not model the fountain flow in the advancing nose. Despite the simplicity of the model, comparisons with experimental work of Dietz and White (1978) for polystyrene are promising. The overall pressure drop is well predicted (slightly over-predicted). Predictions of the residual stresses are somewhat erratic, but generally plausible in a qualitative sense.

Since the flow is nearly-viscometric, one might query what the effect of elasticity is. In the present case the difference between the value of $\partial p/\partial x$ in the elastic case as compared to an inelastic model is only about 2 per cent as a maximum. With other mould shapes this may not be an accurate conclusion.

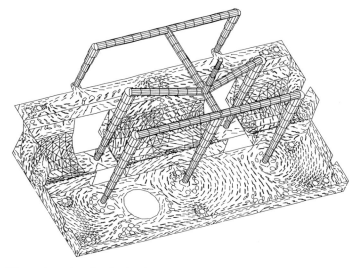

Fig. 9.16 Fibre orientation distribution in a multi-gated injection moulded part. Here, the planar projections of three-dimensional orientation ellipsoids are used to show the predicted fibre orientation state. Circles represent a random orientation whereas lines indicate a perfect alignment.

Some comments on injection-moulding modelling have been made by Richardson (Pearson and Richardson 1983). Kennedy (1995) has also devoted his book to injection moulding.

A much more sophisticated approach to the simulation of thermally and pressure induced stresses in injection moulding is the work of Zheng *et al.* (1999). This involves the numerical (finite element) solution of the equations of momentum, mass, and energy conservation for short-fibre-reinforced thermoplastics. A flow-orientation picture is shown in Fig. 9.16. Predictions of shrinkage and warpage are made in this computation, with reasonable agreement with experiment.

9.8 Summary

This chapter has tried to emphasize the great importance of fluid property changes with temperature. It is often the most important part of the process simulation. The combination of variable temperature with viscoelasticity practically dictates that numerical methods be used to study these flows and compare with experiments. The effects of pressure can also be important. Much remains to be done, especially with regard to freezing processes in materials. A survey and a beginning in this area is that of Kulkarni and Beris (1998); see also Larson (1999).

References

Agur, E. E. and Vlachopoulos, J. (1981). *J. Polym. Sci.*, **26**, 765.
Archer, L. A., Ternet, D. and Larson, R. G. (1997). *Rheol. Acta*, **36**, 579.
Bair, S. and Winer, W. O. (1992). *J. Tribology*, **114**, 1.
Beek, N. J. and Eggink, R. (1962). *De Ingenieur*, **74**, 81.
Binding, D. M., Couch, M. A. and Walters, K. (1998). *J. Non-Newtonian Fluid Mech.*, **79**, 137.
Bird, R. B., Armstrong, R. C., and Hassager, O. (1977). *Dynamics of polymeric liquids*, Vol. 1. Wiley, New York.
Bondi, A. (1968). *Physical properties of molecular crystals, liquids and glasses*. Wiley, New York.
Brennen, C. E. (1995). *Cavitation and bubble dynamics*. Oxford University Press.
Carvalho, M. S. and Scriven, L. E. (1996). *J. Tribology*, **118**, 872.
Cho, Y. I. and Hartnett, J. P. (1982). *Adv. Heat Transfer*, **15**, 60.
Christensen, R. M. (1971). *Theory of viscoelasticity*. Academic Press, New York.
Christiansen, E. B., Jensen, G. E., and Tao, F. S. (1966). *Am. Inst. chem. Engrs. J.*, **12**, 1196.
Chung, T. S. (1982). *J. appl. Polym. Sci.*, **28**, 2119.
Collins, E. A. and Metzger, A. (1970). *Polymer Eng. Sci.*, **10**, 57.
Crochet, M. J. and Naghdi, P. M. (1968). *Proc. IUTAM Symp. on Viscoelasticity* (*Scotland*). Springer-Verlag, Berlin.
Denn, M. M. (1983). In J. R. A. Pearson, and S. M. Richardson (ed.). *Computational analysis of polymer processing*, p. 179. Applied Science Publications, London.
Dietz, W. and White, J. L. (1978). *Rheol. Acta*, **17**, 676.
Dowson, D. and Higginson, G. R. (1966). *Elastohydrodynamic lubrication*, Pergamon, Oxford.
Eiermann, K. (1965). *Kolloid-Z.*, **201**, 3.
Ferry, J. D. (1981). *Viscoelastic properties of polymers* (3rd edn). Wiley, New York.
Frenkel, J. (1955). *The kinetic theory of liquids*. Dover, New York.
Goldstein, S. (1960). *Lectures on fluid dynamics*. Interscience, London.
Gottifredi, J. C., Quiroga, O. P., and Flores, A. F. (1983). *Int. J. Heat Mass Transfer*, **26**, 1215.
Greig, P. and Sahoto, M. (1978). *Polymer*, **19**, 503.
Gupta, R. K. and Metzner, A. B. (1982). *J. Rheol.*, **26**, 181.
Huynh, B. P. (1983). *J. Non-Newtonian Fluid Mech.*, **13**, 1.
Huynh, B. P. and Tanner, R. I. (1983). *Rheol. Acta*, **23**, 1.
Isayev, A. I. and Hieber, C. A. (1980). *Rheol. Acta*, **19**, 168.
Jacobson, B. O. (1991). *Rheology and elastohydrodynamic lubrication*. Elsevier, Amsterdam.
Joseph, D. D. (1998). *J. Fluid Mech.*, **366**, 367.
Kamal, M. R., Tan, V., and Kashani, F. (1983). *Adv. in Polymer Tech.*, **3**, 89.
Kays, W. M. (1966). *Convective heat and mass transfer*. McGraw-Hill, New York.
Kearsley, E. A. (1962). *Trans. Soc. Rheol.*, **6**, 253.
Kennedy, P. (1995). *Flow analysis of injection molds*. Hanser, Munich.
Kulkarni, J. A. and Beris, A. M. (1998). *J. Rheol.*, **42**, 971.
Larson, R. G. (1999). *The structure and rheology of complex fluids*. Oxford University Press, New York.
Lohe, P. (1965). *Koll.-Z. u. Ze.f. Polym.*, **203**, 115.
Martin, B. (1967). *Int. J. nonlinear Mech.*, **2**, 285.
Matsumoto, T. and Bogue, D. C. (1977). *Trans. Soc. Rheol.*, **21**, 453.
Metzner, A. B. (1965). *Adv. Heat Transfer*, **2**, 357.

Morland, L. W. and Lee, E. H. (1960). *Trans. Soc. Rheol.*, **4**, 233.

Morris, S. (1982). *J. Fluid Mech.*, **124**, 1.

Pearson, J. R. A. and Richardson, S. M. (ed.) (1983). *Computational analysis of polymer processing*, p. 139. Applied Science Publications, London.

Peters, G. W. M. (1995). *In Numerical simulation of non-isothermal flow of viscoelastic liquids*. (ed. J. F. Dijksman and G. D. C. Kuiken), p. 21. Kluwer, Amsterdam.

Phan-Thien, N. (1979). *J. Rheol.* **23**, 451.

Phuoc, H. B. and Tanner, R. I. (1980). *J. Fluid Mech.*, **98**, 253.

Pigford, R. L. (1955). *Chem. Eng. Progr. Symp.*, **51**, 79.

Pipkin, A. C. (1972). *Lectures on viscoelasticity theory*. Springer-Verlag, New York.

Pywell, R. F. (1973). In T. C. Davenport (ed.) *The rheology of lubricants*, p. 118. Applied Science Publications, London.

Shoulberg, R. H. (1963). *J. appl. Polymer Sci.*, **7**, 1597.

Stehle, M. and Bruckner, R. (1979). *Glastech. Ber.*, **52**, 82, 105.

Streator, J. L., Gerhardstein, J. P., and McCollum, C. B. (1986). *J. Tribology*, **116**, 119.

Tadmor, Z. and Gogos, C. G. (1979). *Principles of polymer processing*. Wiley, New York.

Tayor, G. I. and Quinney, H. (1931). *Phil. Trans. Roy. Soc. London*, **A 230**, 323.

Tien, C. (1962). *Can. J. Chem. Eng.*, **40**, 130.

Truesdell, C. and Noll. W. (1965). *The nonlinear field theories of mechanics*. Springer-Verlag, Berlin.

van den Brule, B. H. A. A. (1995). In *Numerical simulation of non-isothermal flow of viscoelastic liquids* (ed. J. F. Dijksman, and G. D. C. Kuiken), p. 11. Kluwer, Amsterdam.

Van Krevelen, D. W. and Hoftyzer, P. J. (1976). *Properties of polymers*. Elsevier, Amsterdam.

Vinogradov, G. V. and Malkin, Y. (1977). *Rheology of polymers*. Mir Publ., Moscow.

Vinogradov, G. V. and Shumsky, V. F. (1970). *Rheol. Acta*, **9**, 155.

Wales, J. L. S. (1976). *The application of flow birefringence studies to rheological studies of polymer melts*. Delft University Press.

Wapperom, P. and Hulsen, M. A. (1995). In *Numerical simulation of non-isothermal flow of viscoelastic liquids* (ed. J. F. Dijksman and G. D. C. Kuiken), p. 37. Kluwer, Amsterdam.

Wiest, J. (1995). In *Numerical simulation of non-isothermal flow of viscoelastic liquids* (ed. J. F. Dijksman and G. D. C. Kuiken), p. 1. Kluwer, Amsterdam.

Woods, L. C. (1975). *Thermodynamics of fluid systems*. Oxford University Press.

Yan, J. and Tien, C. (1963). *Can. J. Chem. Eng.*, **41**, 139.

Zdilar, A. M. and Tanner, R. I. (1994). *J. Rheol.*, **38**, 909.

Zheng, R., Kennedy, P., Phan-Thien, N., and Fan, X-J (1999). *J. Non-Newtonian Fluid Mech.* **84**, 159.

Problems

1. Using the data of Fig. 9.4 find an appropriate time–temperature shift curve. Compare this curve with the 'universal' shift curve.

2. Suppose polyisobutylene is cooled at a uniform rate from $50\,°C$ to $-100\,°C$ over a period of 30 min. Plot a graph of pseudo-time versus time for this interval.

3. Show that by using the function $\beta\{1 - \exp -p/\beta\}$ instead of p in Newtonian isothermal incompressible lubrication problems, one can reduce variable viscosity problems to the standard constant-viscosity lubrication problem. Assume $\eta = \eta_0 \exp(p/\beta)$.

4. Assuming a fixed pressure drop Δp in a tube of length L and diameter D, and an incompressible fluid whose viscosity varies exponentially with pressure according to the law $\eta = \eta_0 \exp(p/\beta)$, compute the mass flux.

5. Find the temperature field in a Newtonian shear flow between two plates when one plate is insulated and stationary and the other is at a constant temperature T_0 and moves at a speed U. What change occurs if the stationary plate is held at T_0?

6. Do a dimensional analysis of the extrusion problem including thermal effects. How many dimensionless groups affect the swelling ratio?

7. Use a computer program to solve the shear-heating problem for a general (non-Newtonian) variation of viscosity function and thermal properties with temperature. Check your results using (a) the constant property case and (b) the results of Section 9.5.3.

8. Integrate the slab-heating problem (Section 9.5.4) for the case when the edges of the slab are kept at a temperature T_0. Is there a steady-state solution?

10
STABILITY OF FLOW AND TURBULENCE

10.1 Flow instabilities

In Newtonian fluid flowing in a channel, we know that at a certain Reynolds number the simple laminar flow pattern breaks down and turbulent flow results. The prediction of the critical Reynolds number and the flow mode just beyond this point are the domain of hydrodynamic stability theory (Drazin and Reid 1981). Soon after the critical point the flow becomes chaotic and turbulence sets in; generally the study of turbulent flow proceeds by a close link between experiment and theory because it is very difficult to do completely a priori calculations of turbulence in complex flows, beginning from the Navier–Stokes equations. In some other simple flows, the pattern of transition to turbulence is more complex. At a critical Reynolds number in a Couette flow (for definiteness, with a rotating inner cylinder and a stationary outer cylinder) a steady secondary flow sets in having the form of toroidal vortices (Fig. 10.1). As the speed of the inner cylinder increases (and consequently the Reynolds number increases) the flow becomes more and more chaotic until it ultimately begins to look more like channel turbulence. Thus the transition to chaos proceeds in at least two steps in this case. In this chapter we seek to look at the effect of non-Newtonian fluid properties on instabilities, and also at the effect of non-Newtonian fluid properties on turbulent flows. Readers most interested in turbulence may go directly to Section 10.9.

The instabilities in Newtonian flows are characterized by a critical Reynolds number. If we regard this number as the ratio of inertia to viscous forces, then we see instability when the inertia forces become large relative to the viscous forces. One naturally expects to see, in a viscoelastic fluid characterized by, for example, a relaxation time λ, that the flow transition would occur when

$$f[(Re), (Wi)] = 0, \tag{10.1}$$

where (Wi) is the Weissenberg number, $\lambda V/d$, say; the Reynolds number (Re) is $\rho V d/\eta$. Similarly, the characteristics of a turbulent flow would be expected to depend on both (Re) and (Wi).

Generally, in polymer processing, one does not get inertia-driven instabilities because (Re) is very small ($\ll 1$). However, there is the possibility of instabilities arising when (Wi) comes to a critical magnitude; if we can find these, then we have a special non-Newtonian effect, not present in purely viscous flows; this is an elastic instability where elastic forces dominate viscous forces. Conversely, at least one purely viscous [$(Re) = 0$] instability in fibre-spinning can be damped out

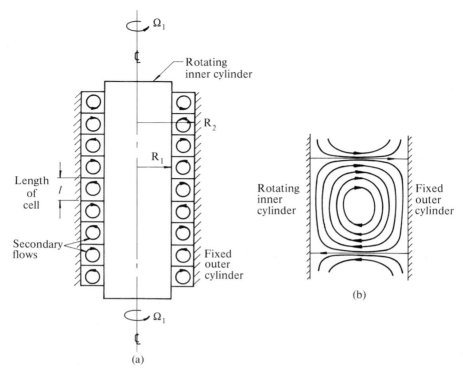

Fig. 10.1 Sketch of Taylor–Couette vortex flow field showing post-critical secondary flow. (a) Flow between two cylinders showing periodic secondary flow cells along axial direction-superimposed on basic Couette motion. (b) Details of cell.

by elastic effects and consequently care is needed in making predictions about the general effects of viscoelasticity on stability.

10.1.1   *Instability built into constitutive models*

We have noted already an inbuilt instability in the second-order fluid equation (Section 4.5.1) and there is the possibility of a work-producing cycle with the single integral model [eqn (4.63)] unless an elastic potential exists for the h_1 and h_{-1} functions in this equation; when such a potential exists then one returns to a form of the KBKZ model, eqn (4.54). Tanner and Simmons (1967) reported an instability when working with a special form of the Oldroyd equation (4.13) in a steady shearing flow disturbed by an orthogonal sinusoidal shearing flow. None of these instabilities is expected to reflect real fluid behaviour and one must be careful to distinguish such spurious modes of instability, due to poor or inappropriate modelling, from real instabilities.

As a simple example of instability we can consider the shearing motion of a purely viscous fluid. Let

$$\tau = \eta(\dot{\gamma})\dot{\gamma}. \tag{10.2}$$

We suppose the (unidirectional) velocity field is

$$u(y, t) = \dot{\gamma}_0 y + \varepsilon V(y, t), \tag{10.3}$$

and that the motion is confined between the planes $y = 0$ and $y = h$ with velocities of zero and $\dot{\gamma}_0 h$ respectively. The equations of motion reduce to

$$\rho \frac{\partial u}{\partial t} = \frac{\partial \tau}{\partial y}. \tag{10.4}$$

Using (10.3) and (10.2), assuming $\varepsilon \ll 1$, we find

$$\rho \frac{\partial V}{\partial t} = \frac{\partial^2 V}{\partial y^2} \left\{ \eta(\dot{\gamma}_0) + \dot{\gamma}_0 \frac{\partial \eta}{\partial \dot{\gamma}_0} \bigg|_{\dot{\gamma}_0} \right\}. \tag{10.5}$$

Now assume

$$V = \sum_{m=1}^{\infty} e^{\alpha_m t} \sin \frac{m\pi y}{h}, \tag{10.6}$$

which satisfies the boundary conditions at $y = 0, h$. Using (10.6) in (10.5) we find

$$\alpha_m = -\frac{m^2 \pi^2}{\rho h^2} \left[\eta + \dot{\gamma} \frac{\partial \eta}{\partial \dot{\gamma}} \right]_{\dot{\gamma}_0}. \tag{10.7}$$

Thus the motion is unstable if the bracketted term is negative; this gives as a criterion for stability

$$\frac{d\tau}{d\dot{\gamma}} > 0. \tag{10.8}$$

If a real material exhibits a flow curve in which $d\tau / d\dot{\gamma} < 0$, it will rapidly become unstable; one must avoid violating (10.8) if one wishes to have a stable material model.

When applying the numerical methods described in Chapter 8 one frequently finds that convergence is not attained beyond a certain Weissenberg number $(Wi)_c$. If the flow itself is unstable with the chosen constitutive model, then the computation should reflect this behaviour; a program that can only compute steady-state solutions may fail to reflect the complexity of the real flow. On the other hand, lack of convergence is most often just a numerical artifact. We now consider the stability of some basic flows.

10.2   Perturbations about a state of rest

The method of stability analysis usually adopted consists in applying an infinitesimal perturbation to the basic flow as demonstrated above. The more complex is the basic flow, the more difficult is the stability analysis. The simplest possible basic 'flow' is a state of rest, and the next most complex is a state of rigid body motion. We shall consider the former; for rigid body motions of viscous fluids one can prove that the motion is stable (Huilgol and Phan-Thien 1997).

An incompressible, isothermal fluid in a state of rest must be stable, and small disturbances must die away for all realistic constitutive equations. In order to feed a potential instability with energy, a driving force is needed. A simple physically interesting case is the Bénard problem (Fig. 10.2) where a fluid is contained between two large planes spaced a distance d apart. If the lower plate is hotter than the upper plate, thermal expansion lowers the density of the fluid there and the hot fluid tries to rise. At a certain critical temperature-difference motion sets in and this motion intensifies and develops as the temperature difference increases. The motion is thus buoyancy-driven. An account of the Newtonian problem is given by Busse (in Swinney and Gollub 1981). The viscoelastic case may be treated in a fairly general manner (Sokolov and Tanner 1972) and illustrates many aspects of classical stability theory.

The analysis follows that of Drazin and Reid (1981) for the Newtonian case. In particular, we assume all fluid properties, except the density, are constant. The usual Boussinesq approximation is also used; this means that *except in the buoyancy terms* we assume the density is constant; any resulting flow is thus assumed to be incompressible. Hence the equations of motion are

$$\rho_0 \frac{D\mathbf{v}}{Dt} = -\boldsymbol{\nabla}p + \nabla \cdot \boldsymbol{\tau} - (\rho_0 + \delta\rho)g\mathbf{k}, \tag{10.9}$$

where ρ_0 is the undisturbed fluid density, g is the gravitational acceleration, $\mathbf{k}$ is a unit vector pointing along the z axis, $\mathbf{v}$ is the velocity vector, p is the pressure,

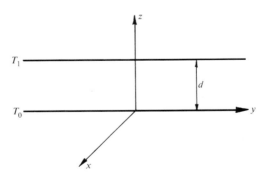

Fig. 10.2 Coordinate system for Bénard problem. Lower plate temperature T_0 exceeds the upper plate temperature T_1.

and $\boldsymbol{\tau}$ is the extra stress tensor, which is assumed symmetric. The mass-conservation equation is

$$\nabla \cdot \mathbf{v} = 0, \tag{10.10}$$

and the energy equation, ignoring viscous dissipation and other small terms, reduces to

$$\frac{\mathrm{D}T}{\mathrm{D}t} = \kappa \nabla^2 T, \tag{10.11}$$

where κ is the thermal diffusivity and T is the temperature; in the above $\mathrm{D}/\mathrm{D}t \equiv \partial/\partial t + (\mathbf{v} \cdot \nabla)$ as usual. In eqn (10.9) the variation of density $\delta\rho$ is assumed to be given by

$$\delta\rho = -\rho_0 \alpha (T - T_0), \tag{10.12}$$

where α is the coefficient of volumetric expansion and T_0 is a properly chosen mean temperature at which all fluid properties ρ_0, κ, and α are evaluated. For the constitutive equation we assume the fluid is an incompressible simple fluid [eqn (4.29)], so that the stress at a particle is a functional only of the history of the strain seen by that particle; the strain is conveniently measured relative to the present configuration as reference. In order to make an analysis of a given viscoelastic flow problem it is usually necessary to have, in advance, a good idea of the flow field to be expected before choosing a relevant approximation to the general simple fluid constitutive equation. In the present problem we are concerned with small velocity perturbations about a state of rest and it is this fact which makes the problem tractable. In stability problems it is often found that one may assume the small test disturbance to be of the separated form

$$\mathbf{v} = e^{\sigma t}\mathbf{f}(x, y, z), \quad T = e^{\sigma t}g(x, y, z). \tag{10.13}$$

The analysis then seeks to find any eigenvalue σ which permits a non-zero solution of the disturbance equations, and hence a bifurcation from the zero-velocity pure conduction basic solution. Several possibilities exist for σ, which is in general a complex number:

(a) The real part of σ, $\mathrm{re}(\sigma)$, is greater, equal to or less than, zero.
(b) Similar choices exist for the imaginary part of σ, $\mathrm{im}(\sigma)$.

For stable disturbances, which die away, $\mathrm{re}(\sigma) < 0$; for $\mathrm{re}(\sigma) > 0$ the disturbances will grow until arrested by non-linearities; thus the case $\mathrm{re}(\sigma) = 0$ is the interesting one as it signals the onset of instability. When $\mathrm{im}(\sigma)$ is zero, one has a steady disturbance flow as in the classical Taylor vortices (Fig. 10.1), otherwise the eigensolution is oscillatory. In the former case it is usual to refer to an 'exchange of stability' and the latter is termed 'over-stability'.

In the overstable case each particle undergoes a small-amplitude vibration about a mean position and strains in the fluid are small. In the exchange of stability case the strains are not necessarily small, but the strain rates are, and in these circumstances the simple fluid with fading memory behaves like a Newtonian fluid; the important idea here is that no large strain is undergone in a time comparable with the characteristic memory time of the fluid. These arguments may be formalized as follows. The general simple fluid constitutive equation for incompressible fluids, may be represented in the form (4.29)

$$\boldsymbol{\tau}(\mathbf{x}, t) = \underset{t=-\infty}{\overset{t}{\mathcal{F}}} [\mathbf{C}(t')].$$

Here, $\boldsymbol{\tau}$ is the extra stress tensor at the Eulerian place $\mathbf{x}$ at the current time t which is a functional of the whole past history of the strain tensor $\mathbf{C}$. The particle which arrives at $\mathbf{x}$ at time t was at the coordinate station $\mathbf{r}(\mathbf{x}, t')$ at time t' ($t > t'$). Then, the relative deformation gradient matrix $\mathbf{F}$ is defined as

$$\mathbf{F} = \frac{\partial \mathbf{r}}{\partial \mathbf{x}}, \tag{2.28}$$

whence $\mathbf{C}$ is defined as $\mathbf{F}^T \mathbf{F}$; note that $\mathbf{F}^T$ is the transpose of $\mathbf{F}$. Let $\mathbf{r} - \mathbf{x} = \mathbf{U}(\tau, x)$, so that $\mathbf{U}$ is the relative displacement. We now suppose that $\partial \mathbf{U}/\partial \mathbf{x} \ll 1$, so that squares and higher powers of $\partial \mathbf{U}/\partial \mathbf{x}$ may be neglected. Then, the small-strain tensor becomes

$$2\mathbf{E} = \frac{\partial \mathbf{U}}{\partial \mathbf{x}} + \left(\frac{\partial \mathbf{U}}{\partial \mathbf{x}}\right)^T + \mathrm{O}\left|\frac{\partial \mathbf{U}}{\partial \mathbf{x}}\right|^2. \tag{2.77}$$

Now, if we assume that $\boldsymbol{\tau}$ depends linearly on $\mathbf{E}$ at small enough strains we get the linear form (2.81)

$$\boldsymbol{\tau}(t) = \int_{-\infty}^{t} G(t - t')\mathbf{A}(t') \, \mathrm{d}t', \tag{2.81}$$

where G is the memory function describing the fluid and the components of the Rivlin–Ericksen tensor $\mathbf{A}$ ($\equiv 2\mathrm{d}\mathbf{E}/\mathrm{d}t'$) in this case are given in terms of the velocity $\mathbf{v}$ by

$$A_{ij} = \frac{\partial v_i}{\partial x_j} + \frac{\partial v_j}{\partial x_i}. \tag{2.35a}$$

Here again, t is the present time and t' is a time in the past; note that (2.81) is an isotropic linear functional of the hereditary type, so that the argument of G is $t - t'$, and not t, t' separately. For steady motions, where $\mathbf{A}$ is not a function of t', eqn (2.81) reduces to the Newtonian case. Provided $|\mathbf{E}|$ is small (2.81) holds irrespective of the size of $|\mathbf{A}|$, so that we are not confined to slow flows. The memory function G is chosen to coincide with the value of this function at the mean temperature T_0; it is not otherwise assumed to be a function of temperature. Although this assumption will affect the details of the analysis it will not make qualitative changes, and it simplifies the analysis greatly; however, it is not very realistic, see Chapter 9 for discussion. Consistent with the assumption of small motions about a state of rest we may drop the convective term $(\mathbf{v} \cdot \nabla)\mathbf{v}$ in the momentum equations.

The steady-state (primary) solution is now written for the situation in Fig. 10.2 where we have a slab of fluid, infinite in the x, y directions and of thickness d in the z direction; there is a constant mean-temperature gradient (hot below, cool above) of magnitude β, the stress distribution is hydrostatic, and the velocity is zero; all $\partial/\partial x$, $\partial/\partial y$ are zero for the static field. We now consider small velocity and temperature perturbations on this state of rest.

Let the disturbed quantities be denoted by a prime, the steady (primary) values by an overbar. Then, we have

$$\mathbf{v} = \mathbf{v}', \quad T = \bar{T} + T', \quad p = \bar{p} + p', \quad \rho = \bar{\rho} + \rho', \quad \boldsymbol{\tau} = \boldsymbol{\tau}'. \tag{10.14}$$

Thus we find, from (10.9)–(10.14)

$$\frac{\partial \mathbf{v}'}{\partial t} = g\alpha T'\mathbf{k} - \rho_0^{-1}\nabla p' + \rho_0^{-1}\nabla \cdot \boldsymbol{\tau}', \tag{10.15}$$

$$\tau'_{ij} = \int_{-\infty}^{t} G(t - t')\left(\frac{\partial v'_i(t')}{\partial x_j} + \frac{\partial v'_j(t')}{\partial x_i}\right) dt', \tag{10.16}$$

which we can write as a convolution product $\tau'_{ij} = G * A'_{ij}$ for convenience;

$$\nabla \cdot \mathbf{v}' = 0 \tag{10.17}$$

$$\left(\frac{\partial}{\partial t} - \kappa\nabla^2\right)T' = \beta(\mathbf{v}' \cdot \mathbf{k}), \tag{10.18}$$

where $\beta = -\partial\bar{T}/\partial z$. By twice taking the curl of (10.15) we eliminate p' and two of the velocity components to obtain the following equations for the z component of velocity (v'_z) and the perturbation T':

$$\left(\frac{\partial}{\partial t} - \rho_0^{-1}G * \nabla^2\right)\nabla^2 v'_z = g\alpha\left(\frac{\partial^2 T'}{\partial x^2} + \frac{\partial^2 T'}{\partial y^2}\right), \tag{10.19}$$

$$\left(\frac{\partial}{\partial t} - \kappa\nabla^2\right)T' = \beta v'_z. \tag{10.20}$$

We now look for non-trivial solutions of (10.19–10.20) with appropriate boundary conditions. Assume solutions of the form

$$v'_z = \frac{\kappa}{d}f(x, y)w(z)e^{\sigma t}, \tag{10.21}$$

$$T' = \beta\,df(x, y)\theta(z)e^{\sigma t}, \tag{10.22}$$

where σ is a complex parameter. According to linear stability theory the sign of the real part of σ governs whether the motion is stable or not. Evaluation of the term involving the convolution product in (10.19) shows that, setting $t - t' = s$,

$$- \rho_0^{-1} G * \nabla^4 v_z' = -\rho_0^{-1} \frac{\kappa}{d} \nabla^4 (fw)e^{\sigma t} \int_0^\infty G(s)e^{-\sigma s} \, ds. \tag{10.23}$$

It is advantageous to define the dimensionless complex quantity $q(\sigma)$, where

$$\eta(\sigma) \equiv \eta_0 q(\sigma) \equiv \int_0^\infty G(s)e^{-\sigma s} \, ds, \tag{10.24}$$

η_0 is the zero-shear rate reference viscosity and $q(\sigma)$ is dimensionless. In the case when σ is pure imaginary, $\sigma = i\omega$ say, then

$$\eta(i\omega) = \eta_0 q(i\omega) \equiv \eta'(\omega) - i\eta''(\omega). \tag{10.25}$$

Thus, $\eta(i\omega)$ is the complex viscosity as usually defined in linear viscoelasticity (see Chapter 2). When $\sigma = 0$, the Newtonian values are recovered and $\eta'' = 0$, $\eta'(0) = \eta(0) = \eta_0$. After substitution of (10.21) and (10.22) in (10.20) and separating variables, we find

$$-f^{-1}\left(\frac{\partial^2 f}{\partial x^2} + \frac{\partial^2 f}{\partial y^2}\right) = \theta^{-1}\frac{d^2\theta}{dz^2} + \frac{w}{d^2\theta} - \frac{\sigma}{\kappa} = a^2, \tag{10.26}$$

where a^2 must be a (real) constant with the dimensions of (length)$^{-2}$. Thus, we may eliminate f; note that the equation for f, governing the eigenfunction (cell shape), is the same as in the Newtonian case. The problem of finding cell shapes is discussed extensively by Chandrasekhar (1961) and will not be repeated here. We make various quantities non-dimensional in the following way:

$$\zeta = \frac{z}{d}, \quad D = d\frac{\partial}{\partial z} = \frac{\partial}{\partial \zeta}, \quad \sigma^* = \sigma d^2/k, \quad (Pr) = \eta_0/\rho_0\kappa, \quad a^* = ad, \tag{10.27}$$

and find from (10.19) and (10.20) after using (10.26)

$$[\sigma^* - (D^2 - a^{*2})]\theta = w, \tag{10.28}$$

$$[\sigma^* - q(\sigma)(Pr)(D^2 - a^{*2})][D^2 - a^{*2}]w = -(Pr)Ra^{*2}\theta, \tag{10.29}$$

where the Rayleigh number $R \equiv g\alpha\beta\rho_0 d^4/\kappa\eta_0$. Note that (Pr) (Prandtl number) is based on the reference viscosity; conveniently this may be taken as the viscosity for motions in which $\sigma = 0$. We now need to consider boundary conditions on w and θ.

The boundary conditions on θ, the temperature perturbation, are that $\theta = 0$ on $\zeta = 0, 1$. The no-slip boundary conditions on $\zeta = 0, 1$ are $w = \partial w/\partial \zeta = 0$ on these planes. Generally, one needs numerical methods to solve the resulting eigenproblem; see Vest and Arpaci (1969); Sokolov and Tanner (1972). To illustrate the process of solution when this is not necessary we shall assume a no-shear stress boundary condition at $\zeta = 0, 1$ so that we set $w = \partial^2 w/\partial \zeta^2 = 0$ on these planes (instead of the no-slip conditions). Equations (10.28) and (10.29) can now be combined by operating with $[\sigma^* - (D^2 - a^{*2})]$ on both sides of (10.28) and substituting in (10.29). The resulting equation is

$$[D^2 - a^{*2}][\sigma^* - (D^2 - a^{*2})][\sigma^* - q(\sigma)(Pr)(D^2 - a^{*2})]w$$
$$= -(Pr)Ra^{*2}w. \tag{10.30}$$

For the no-shear boundary conditions we can assume a solution of the form

$$w = \sum_{n=1}^{\infty} w_n \sin n\pi\zeta \quad n = 1, 2, 3 \ldots . \tag{10.31}$$

Substitution of (10.31) into (10.30) gives the characteristic equation for R:

$$(n^2\pi^2 + a^{*2})(n^2\pi^2 + a^{*2} + \sigma^*)[\sigma^* + q(\sigma)(Pr)(n^2\pi^2 + a^{*2})]$$
$$= R(Pr)a^{*2}. \tag{10.32}$$

For the free boundary condition, if $\sigma^* = 0$ (exchange of stabilities), then we recover the Newtonian problem and we can immediately show that for free boundaries

$$R \approx 657.5, \tag{10.33}$$

which is the minimum R occurring when $n = 1$, $a^* = \pi/\sqrt{2}$. However, in the viscoelastic case we should not assume $\sigma^* = 0$; it is of most interest here to investigate overstability. In the Newtonian case it can be shown (Chandrasekhar 1961) that $\sigma^* = 0$. In the viscoelastic case $q(\sigma) = q(i\omega)$ and this can be expressed, from (10.25), in terms of $\eta'(\omega)$ and $\eta''(\omega)$, the components of the complex viscosity, which are the commonly measured quantities in viscoelastic fluids. In terms of these quantities, on equating the imaginary part of (10.32) to zero and letting

$$a^{*2} + n^2\pi^2 = b_n^2, \tag{10.34}$$

$$b_n^2 = \omega^*[\eta_0 + (Pr)\eta'(\omega)]/(Pr)\eta''(\omega), \tag{10.35}$$

from the real part of (10.32) we find, in terms of ω^*,

$$R(Pr)a^{*2} = b_n^2\{[(Pr)b_n^2/\eta_0](b_n^2\eta' + \omega^*\eta'')\} - \omega^{*2}. \tag{10.36}$$

We could now find $R = R(a^{*2})$ by eliminating ω^* and then minimizing with respect to a^{*2}; however, it is easier to obtain $R = R(\omega^*)$ and then find the frequency of oscillation which minimizes R; a^{*2} can then be found from (10.35).

We find that $n = 1$ gives the lowest value of R:

$$R = \frac{m\omega^{*3}\eta'(m^2 + (Pr^2)\eta''^2)}{(Pr^2)\eta_0\eta''^2[\omega^* m - \pi^2(Pr)\eta'']}, \tag{10.37}$$

where $m = \eta_0 + (Pr)\eta'$ and

$$a_{cr}^{*2} = [\omega^* m/(Pr)\eta''] - \pi^2. \tag{10.38}$$

For rigid boundaries no simple exact solution corresponding to (10.31) exists but we may use the Galerkin method of solution to find an approximate value of R. This process gives eigenvalues different from the free-boundary case; see Vest and Arpaci (1969). The numerical results of Sokolov and Tanner (1972) for the rigid wall Maxwell case are in error.

In most cases we need to assume a specific form for η', η'', but in case the Prandtl number of the viscoelastic fluid is high $[O(10^3)]$ a look at the limit $(Pr) \to \infty$ is interesting. Then, for the free-boundary case we get

$$a^{*2} = \omega^*\eta'/\eta'' - \pi^2, \tag{10.39}$$

$$R = \omega^{*3}|\eta'|^2(\eta'^2 + \eta''^2)/\eta_0|\eta''|^2(\omega^*\eta' - \pi^2\eta''). \tag{10.40}$$

Letting $\tan \delta = \eta'/\eta''$ (δ is the loss angle) we see that (10.39) implies $\tan \delta > \pi^2\omega^*$ for oscillation to be possible, or, in terms of dimensional quantities,

$$\frac{\tan \delta}{\pi^2} > \frac{\text{oscillation time}}{\text{thermal relaxation time}} = \frac{1/\omega}{d^2/\kappa}. \tag{10.41}$$

For $(Pr) \to 0$, the heat conductivity of the fluid must be very large and the fluid cannot support large temperature gradients. As a result there will not be any strong buoyancy forces and the field will always tend to be stable. This fact can be observed mathematically by letting $(Pr) \to 0$ in (10.37), the result of which will be $R \to \infty$ for free (and also rigid) boundaries.

A further refinement of this problem permits the free surface to deform so that the upper surface is truly free. The critical Rayleigh number then drops considerably from the value found from (10.37).

There are a great many papers on the Bénard problem (Larson 1992), but most consider the constant property case, as we have done. More realistic variations, including variations of η_0 with temperature, usually need computation for solution.

The analysis of the Bénard problem is typical of classical stability theory. We shall therefore omit details in later problems where possible. Generally, it is not possible to complete solutions and find the relevant critical condition as easily as in this model problem.

Finally, one asks whether viscoelasticity is a stabilizing or destabilizing influence. In the Bénard problem, when $\sigma = 0$ (exchange of stability) there is no change in the critical Rayleigh number, and hence there is no influence of viscoelasticity on stability. In those cases where over-stability occurs (oscillatory transition) then viscoelasticity is destabilizing, in that the critical Rayleigh number is less than the $\sigma = 0$ case, which is the Newtonian result.

10.3 Couette flow stability

The classical stability problem of flow between two long cylinders has much in common with the Bénard problem; the flow is destabilized by centrifugal forces acting on a fluid particle in much the same way that the Bénard problem was destabilized by buoyancy forces. A survey of progress for Newtonian fluids was given in Swinney and Gollub (1981). Here we will confine attention to the simplest case where the cylinders are supposed to be of infinite length and where the difference in the cylinder radii ($R_2 - R_1$) is very small compared with the outer radius (narrow-gap case). We will also consider in detail only the case where the inner cylinder rotates; when the outer cylinder rotates in the opposite direction to the inner cylinder the transition mode is more like the channel flow transition (Section 10.4). Figure (10.1) shows the type of steady secondary flow (exchange of stabilities) characteristic of the flow just beyond the critical rotation rate of the inner cylinder. We will assess the effect of viscoelasticity in two case—first, as an addition to the classical inertia-driven problem, and second, in the case where the effects of inertia are negligible and one has purely elastic instabilities.

10.3.1 Couette flows with inertia and viscoelasticity

Some of the first papers to consider the stability of these flows date back to about 1962 (see Tanner and Walters (1998) for a short history). The so-called Dean problem, the Taylor–Couette problem, and the analogue of the Orr–Sommerfeld equation for elastic liquids of the Oldroyd type were investigated.

Early work generally used the assumption of exchange of stabilities (Miller and Goddard 1967; Goddard 1979) and we shall initially make this assumption. Beard *et al.* (1966) showed that overstability was a possibility with the upper convected Maxwell fluid but one can justify the choice of $\sigma = 0$ by noting that observations of transition in Couette flow of viscoelastic fluids often show the exchange of stabilities pattern. Miller and Goddard (1967) used an unusual co-rotational integral constitutive model expansion which was general enough for the flow under consideration to give the same results as if one had begun from the simple fluid case, eqn (4.29). There are a number of unidentifiable coefficients in their analysis, however, and hence we shall proceed here by considering an analysis based on the KBKZ theory, eqn (4.54). Since the flow under consideration is a small deviation from a viscometric flow, it is expected to give accurate results. It should be mentioned that Datta (1964) and Rao (1964) used a second-order fluid theory to study the stability of the flow, and they showed that the sign of N_2 is important for stability. The present analysis follows that of Karlsson *et al.* (1971).

We suppose that the KBKZ constitutive relation for the stress tensor can be written as

$$\boldsymbol{\sigma} = -p\mathbf{I} + 2\eta_s \mathbf{d} + \int_{-\infty}^{t} \left[\frac{\partial U}{\partial I_B} \mathbf{B}(t') - \frac{\partial U}{\partial I_C} \mathbf{C}(t') \right] \mathrm{d}t'. \tag{10.42}$$

Here, $\mathbf{B} = \mathbf{C}^{-1}$ and $2\eta_s \mathbf{d}$ is the solvent contribution; U is a function of I_B and I_C, the traces of $\mathbf{B}$ and $\mathbf{C}$, respectively, and of $t - t'$. In Chapter 5 we used a special form of the KBKZ theory in which a memory function $m(t - t')$ was factored out of the kernel U. This form arose naturally in some concentrated solution theories and it enabled us to find $(\partial U/\partial I_B)$ and $(\partial U/\partial I_C)$ from viscometric flow measurements and small-strain experiments. We shall not need to make this assumption here. We will denote the integral part of eqn (10.47) by $\boldsymbol{\tau}$. For small homogeneous strains $\boldsymbol{\tau}$ becomes a linear functional; $\partial U/\partial I_B$ and $\partial U/\partial I_C$ become functions only of $t - t'$, hence

$$\boldsymbol{\tau} = \int_{-\infty}^{t} m(t - t')\mathbf{E}(t')\,\mathrm{d}t', \tag{10.43}$$

where $\mathbf{E}$ is a multiple of the infinitesimal strain tensor used in classical elasticity theory (see Chapter 2); usually $m(t)$ is chosen to be of the form

$$m(t) = \sum_n \frac{a_n}{\lambda_n^2} \exp(-t/\lambda_n), \tag{10.44}$$

where the a_n are constants, so that a series of discrete relaxation times λ_n appears; we shall not need to specify m here.

For steady homogeneous flows $\boldsymbol{\tau}$ reduces to

$$\boldsymbol{\tau} = \int_0^{\infty} \left[\frac{\partial U}{\partial I_B} \mathbf{B} - \frac{\partial U}{\partial I_C} \mathbf{C} \right] \mathrm{d}s = \boldsymbol{\tau}(\mathbf{L}), \tag{10.45}$$

where the $\mathbf{L}^T$ is the constant velocity gradient matrix $\partial \mathbf{v}/\partial \mathbf{x}$ and $s = t - t'$. If we are interested in a basic shear flow, $\mathbf{v} = (\dot{\gamma}y, 0, 0)$, then using known results for $\mathbf{B}$ and $\mathbf{C}$, we easily find

$$\tau_{xy} = \dot{\gamma} \int_0^{\infty} s \left\{ \frac{\partial U}{\partial I_B} + \frac{\partial U}{\partial I_C} \right\} \mathrm{d}s, \tag{10.46}$$

$$N_1 \equiv \tau_{xx} - \tau_{zz} = \dot{\gamma}^2 \int_0^{\infty} s^2 \left\{ \frac{\partial U}{\partial I_B} + \frac{\partial U}{\partial I_C} \right\} \mathrm{d}s, \tag{10.47}$$

$$N_2 \equiv \tau_{yy} - \tau_{zz} = -\dot{\gamma}^2 \int_0^{\infty} s^2 \frac{\partial U}{\partial I_C} \mathrm{d}s, \tag{10.48}$$

where $\partial U/\partial I_{B,C}$ are now functions of $\dot{\gamma}^2 s^2$ and s in general.

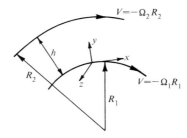

Fig. 10.3 Notation for Couette stability problem.

In the classical Couette stability problem the base flow (Fig. 10.3) is represented, in the narrow-gap case treated here ($h/R_1 \ll 1$), by

$$V(y) = -\Omega_1 R_1 (1 + by/h), \tag{10.49}$$

where v is the azimuthal (θ) velocity component, and

$$b \equiv -1 + \Omega_2 R_2 / \Omega_1 R_1. \tag{10.50}$$

It is a viscometric flow. We now assume a small perturbation of the flow field, so that

$$\mathbf{v} = \left\{ V + u(y, z), \frac{\partial \psi}{\partial z}, -\frac{\partial \psi}{\partial y} \right\}, \tag{10.51}$$

where $\psi = \psi(y, z)$. This velocity field satisfies the mass conservation equation in an incompressible fluid; it is an axisymmetrical perturbation and later we shall assume that it is periodic in the z direction.

To calculate the $\mathbf{B}$, $\mathbf{C}$ matrices in eqn (10.42) we note that

$$\mathbf{C} = \mathbf{F}^T\mathbf{F}, \quad \mathbf{B} = \mathbf{C}^{-1}, \tag{10.52}$$

where [see Chapter 2, eqn (2.28)]

$$\mathbf{F} = \partial\mathbf{r}/\partial\mathbf{x}. \tag{10.53}$$

It may be noted that $\mathbf{C}$ and $\mathbf{B}$ are measures of particle and material plane separation, respectively, in the incompressible fluid; thus they represent related measures of the strain undergone by line elements and area elements in the fluid. The position vector $\mathbf{r}$ is connected to the velocity vector:

$$\frac{d\mathbf{r}}{dt'} = \mathbf{v}(\mathbf{r}, t'). \tag{10.54}$$

In general, the solution of eqn (10.54) is difficult. For a small steady perturbation from a viscometric flow, we may set

$$\mathbf{r} = \mathbf{r}^{(0)} + \mathbf{r}^{(1)}, \tag{10.55}$$

where $\mathbf{r}^{(0)}$ is the basic viscometric flow and $\mathbf{r}^{(1)}$ is the perturbation. We will ignore squares of $\mathbf{r}^{(1)}$ and all higher powers. Then, using eqns (10.54) and (10.55), we find, using the boundary condition $\mathbf{x} = \mathbf{r}$ at $t' = t$

$$\mathbf{r} = x\mathbf{i} - (t - t')V(y)\mathbf{i} + \mathbf{r}^{(1)}, \tag{10.56}$$

where

$$\mathbf{r}^{(1)} = \left\{ \frac{dV}{dy} \frac{\partial \psi}{\partial z} \frac{(t-t')^2}{2} - (t-t')u(y,z), \quad -(t-t')\frac{\partial \psi}{\partial z}, \quad (t-t')\frac{\partial \psi}{\partial y} \right\}. \tag{10.57}$$

Now we suppose u and ψ are periodic in the z direction, so that

$$u = \hat{v}(y)\cos \lambda z, \quad \psi = \hat{u}(y)\sin \lambda z/\lambda. \tag{10.58}$$

This is exactly the 1923 G. I. Taylor form. Calculation of the **B** and **C** components is now straightforward, giving the following formulas, where $s = t - t'$; we also define the basic shear rate

$$\dot{\gamma} \equiv \frac{dV}{dy} = -\Omega_1 R_1 b/h \quad \text{(a constant)} \tag{10.59}$$

$$
\left.
\begin{aligned}
C_{11} &= 1 \\
C_{12} &= C_{21} = -s\dot{\gamma} + s\cos \lambda z(\dot{\gamma}s\hat{u}'/2 - \hat{v}') \\
C_{13} &= C_{31} = \lambda s(\hat{v} - \lambda\hat{u}s/2)\sin \lambda z \\
C_{22} &= 1 + s^2\dot{\gamma}^2 + s\cos \lambda z(2\dot{\gamma}\hat{v}'s - 2\hat{u}' - \dot{\gamma}^2\hat{u}'s^2) \\
C_{23} &= C_{32} = \lambda s(\hat{u}''/\lambda^2 + \hat{u} - \dot{\gamma}\hat{v}s + \dot{\gamma}^2s^2\hat{u}/2)\sin \lambda z \\
C_{33} &= 1 + 2\hat{u}'s\cos \lambda z \\
B_{11} &= 1 + \dot{\gamma}^2s^2 + \dot{\gamma}s^2(2\hat{v}' + \dot{\gamma}\hat{u}'s)\cos \lambda z \\
B_{12} &= B_{21} = \dot{\gamma}s + s(\hat{v}' + \tfrac{3}{2}\hat{u}'\dot{\gamma}s)\cos \lambda z \\
B_{13} &= B_{31} = -\lambda s(\hat{v} + \hat{u}\dot{\gamma}s/2 + \hat{u}''\dot{\gamma}s/\lambda^2)\sin \lambda z \\
B_{22} &= 1 + 2\hat{u}'s\cos \lambda z \\
B_{23} &= B_{32} = -\lambda s(\hat{u}''/\lambda^2 + \hat{u})\sin \lambda z \\
B_{33} &= 1 - 2\hat{u}'s\cos \lambda z
\end{aligned}
\right\} \tag{10.60}
$$

where $\hat{u}' = \mathrm{d}\hat{u}/\mathrm{d}y$, $\hat{v}' = \mathrm{d}\hat{v}/\mathrm{d}y$. It is now straightforward to calculate the stresses from eqn (10.42). In general case we note that

$$I_{B,C} = I_{B,C}^{(0)} + \Delta I_{B,C}, \qquad (10.61)$$

where $I_{B,C}^{(0)}$ refers to the viscometric base flow and $\Delta I_{B,C}$ are small deviations. Thus

$$\frac{\partial U}{\partial I_B} = \frac{\partial U}{\partial I_B}\bigg|_0 + \frac{\partial^2 U}{\partial I_B^2}\bigg|_0 \Delta I_B + \frac{\partial^2 U}{\partial I_B \partial I_C}\bigg|_0 \Delta I_C + \mathrm{O}(\Delta I^2) \qquad (10.62)$$

where the subscript 0 refers to the viscometric base flow. In the present case we easily find that

$$
\begin{aligned}
I_B^{(0)} &= I_C^{(0)} = 3 + \dot{\gamma}^2 s^2 \\
\Delta I_B &= 2\dot{\gamma}s^2(\hat{v}' + \dot{\gamma}s\hat{u}'/2)\cos \lambda z \\
\Delta I_C &= 2\dot{\gamma}s^2(\hat{v}' - \dot{\gamma}s\hat{u}'/2)\cos \lambda z.
\end{aligned}
\qquad (10.63)
$$

Thus, the complete expression for the stresses involves five partial derivatives of U in the form of moments with respect to s. Knowledge of these five partial derivatives will enable us to write down a complete expression for the stress perturbation in terms of $\hat{u}$, $\hat{v}$, and their derivatives. Using the equations of motion, we may now eliminate the stress perturbation and the pressure, finally obtaining two equations for $\hat{u}$, $\hat{v}$.
    One finds

$$
\begin{aligned}
\boldsymbol{\sigma}^{(1)} = {} & 2\eta_s \mathbf{d}^{(1)} + \int_0^\infty \frac{\partial^2 U}{\partial I_C \partial I_B}(-\Delta I_B \mathbf{C}^{(0)} + \Delta I_C \mathbf{B}^{(0)})\,\mathrm{d}s \\
& + \int_0^\infty \left(\frac{\partial^2 U}{\partial I_B^2}\Delta I_B \mathbf{B}^{(0)} - \frac{\partial^2 U}{\partial I_C^2}\Delta I_C \mathbf{C}^{(0)}\right)\mathrm{d}s \\
& + \int_0^\infty \left(\frac{\partial U}{\partial I_B}\mathbf{B}^{(1)} - \frac{\partial U}{\partial I_C}\mathbf{C}^{(1)}\right)\mathrm{d}s
\end{aligned}
\qquad (10.64)
$$

$$\boldsymbol{\sigma}^{(0)} = -p\mathbf{I} + 2\eta_s\,\mathbf{d}^{(0)} + \int_0^\infty \left(\frac{\partial U}{\partial I_B}\mathbf{B}^{(0)} - \frac{\partial U}{\partial I_C}\mathbf{C}^{(0)}\right)\mathrm{d}s, \qquad (10.65)$$

where all the pressure terms are lumped into eqns (10.65). $\boldsymbol{\sigma}^{(1)}$ may be put in the form

$$
\begin{aligned}
\boldsymbol{\sigma}^{(1)} = {} & \mathbf{A}^{(1)}\hat{u}'' \sin \lambda z / \dot{\gamma}\lambda + \mathbf{A}^{(2)}\hat{u}' \cos \lambda z / \dot{\gamma} \\
& + \mathbf{A}^{(3)}\hat{u}\lambda \sin \lambda z / \dot{\gamma} + \mathbf{A}^{(4)}\hat{v}' \cos \lambda z / \dot{\gamma} + \mathbf{A}^{(5)}\hat{v}\lambda \sin \lambda z / \dot{\gamma},
\end{aligned}
\qquad (10.66)
$$

Table 10.1　Stress–perturbation matrices

The values of the shear rate-dependent ($\dot{\gamma}$-dependent) symmetric matrices in eqn (10.66) are as follows in terms of the moments

$$U_B^i \equiv \int_0^\infty \frac{\partial U}{\partial I_B}(\dot{\gamma}s)^i\,\mathrm{d}s, \quad U_{BC}^i \equiv \int_0^\infty \frac{\partial^2 U}{\partial I_B I_C}(\dot{\gamma}s)^i\,\mathrm{d}s,$$

etc. [eqn (10.67)]

$$\mathbf{A}^{(1)} = \left\{\begin{matrix} 0 & 0 & -U_B^2 \\ & 0 & -(U_B^1 + U_C^1 + \eta_s\dot{\gamma}) \\ & & 0 \end{matrix}\right\}$$

$$\mathbf{A}^{(2)} = \left\{\begin{matrix} U_B^3 + U_{BB}^5 - U_{BC}^5 & U_{BB}^4 - U_{CC}^4 + \frac{3}{2}U_B^2 - \frac{1}{2}U_C^2 & 0 \\ & 2U_C^1 + U_C^3 + 2U_B^1 + 2\eta_s\dot{\gamma} - U_{BC}^5 + U_{CC}^3 & 0 \\ & & -2(U_B^1 + U_C^1 + \eta_s\dot{\gamma}) \end{matrix}\right\}$$

$$\mathbf{A}^{(3)} = \left\{\begin{matrix} 0 & 0 & \frac{1}{2}(U_C^2 - U_B^2) \\ & 0 & -(U_C^1 + \frac{1}{2}U_C^3 + U_B^1 + \eta_s\dot{\gamma}) \\ & & 0 \end{matrix}\right\}$$

$$\mathbf{A}^{(4)} = \left\{\begin{matrix} 2(U_B^2 + U_{BB}^4 + U_{BC}^4) & \eta_s\dot{\gamma} + U_C^1 + U_B^1 + 2U_{CC}^3 + 2U_{BB}^3 + 4U_{BC}^3 & 0 \\ & -2(U_C^2 + U_{CC}^4 + U_{BC}^4) & 0 \\ & & 0 \end{matrix}\right\}$$

$$\mathbf{A}^{(5)} = \left\{\begin{matrix} 0 & 0 & -(U_B^1 + U_C^1 + \eta_s\dot{\gamma}) \\ & 0 & U_C^2 \\ & & 0 \end{matrix}\right\}$$

where the components of the matrices $\mathbf{A}^{(n)}$ ($n = 1-5$) are given in Table 10.1; note that these matrices are not the Rivlin–Ericksen kinematic matrices denoted previously by $\mathbf{A}^{(n)}$. All elements of the $\mathbf{A}^{(n)}$ matrices are functions of $\dot{\gamma}$ through the two integral forms, each having the dimensions of stress

$$U^i_{\alpha\beta} \equiv \int_0^\infty (\dot{\gamma}s)^i \frac{\partial^2 U}{\partial I_\alpha \partial I_\beta}\, \mathrm{d}s; \quad U^m_\alpha \equiv \int_0^\infty (\dot{\gamma}s)^m \frac{\partial U}{\partial I_\alpha}\, \mathrm{d}s \tag{10.67}$$
$$(\alpha, \beta = B \text{ or } C, \quad i = 0, 1, 2, 3\ldots),$$

where all the derivatives of U are evaluated at the viscometric history. The equations of motion, after eliminating p, give [with the notation $\sigma_{(11)} = \sigma_{xx}$, etc., and $\partial(\ )/\partial y = (\ )_{,y}$, etc.]

$$\sigma_{yz,yy} + \sigma_{zz,zy} = -2\rho V \lambda \hat{v} \sin \lambda z / R + \sigma_{yy,zy} + \sigma_{yz,zz} - \frac{1}{R}\sigma_{xx,z} \tag{10.68}$$

$$\sigma_{xy,y} + \sigma_{xz,z} = \rho \hat{u} \cos \lambda z (\dot{\gamma} + V/R). \tag{10.69}$$

Here, R is the mean radius of the cylinders $\frac{1}{2}(R_1 + R_2)$. Note that the component $-a^{(1)}_{23}$ of $\mathbf{A}^{(1)}$ is equal to $\sigma^{(0)}_{xy} \equiv \dot{\gamma}\eta(\dot{\gamma})$. This gives the value of viscosity η used in defining the Taylor (Ta) and Reynolds numbers (Re) below.

It is now straightforward to write down the equations for the disturbances $\hat{u}$, $\hat{v}$. We normalize y with respect to $h\ (\equiv R_2 - R_1)$ and write $\mathrm{d}/\mathrm{d}y \equiv D$, $l = h\lambda$, finding

$$(\beta_1 D^2 - l^2)\hat{v} = \left(\frac{(Ta)}{2(Re)} + \beta_2 D^2 - \beta_3 l^2\right)\hat{u} \tag{10.70}$$

$$(D^4 - \beta_4 l^2 D^2 + 2\beta_9 l^2 \frac{hD}{R} + \beta_5 l^4)\hat{u}$$
$$= l^2\left(\beta_6 D^2 + \beta_8 \frac{hD}{R} + \beta_7 l^2 - 2(Re)(1 + by)\right)\hat{v} \tag{10.71}$$

where

$$(Ta) = -\frac{2R}{h}(Re)^2\left(\frac{\Omega_2 - \Omega_1}{\Omega_1}\right) \tag{10.72}$$

$$(Re) = \rho\Omega_1 h^2/\eta, \tag{10.73}$$

where $b \sim (\Omega_2 - \Omega_1)/\Omega_1$ from eqn (10.50); note (Ta) and (Re) are not independent.

Here the small-gap approximation has again been used in eqns (10.73) and (10.74) and the term V/R neglected; clearly this will not be valid for Ω_2 very close to Ω_1. Goddard (1979), following Miller and Goddard (1967), finds the complete set of equations for the simple fluid case to be, in our notation

$$(\beta_1 D^2 - l^2)\hat{v} = \left[\frac{(Ta)}{2(Re)} + \beta_2 D^2 - \beta_3 l^2 + \beta_{10}\frac{hD}{R}\right]\hat{u};$$

$$\left\{D^4 - \beta_4 l^2 D^2 + 2\beta_9 l^2 \frac{hD}{R} + \beta_5 l^4 + 4\beta_{11}l^2\left(\frac{h^2}{R^2}\right)\right\}\hat{u}$$
$$= l^2\left\{\beta_6 D^2 + \beta_8 \frac{h}{R}D + \beta_7 l^2 - 2(Re)(1 + by)\right\}\hat{v}. \tag{10.74}$$

The terms with odd derivatives make little contribution to the result and may be ignored providing their rheological coefficients are not too large (Goddard 1979). If we do this and also ignore the β_{11} term as being of order $(h/R)^2$ times a rheological coefficient, then (10.70–71) and (10.74) agree and one has seven coefficients (Lockett and Rivlin (1968) also ignored a term corresponding to the β_{11} term here, as we have done, justifying the omission for the small-gap case). Table 10.2 shows the relation among the KBKZ coefficients, Goddard's (1979) coefficients, the viscometric quantities $\eta\dot\gamma$, N_1, and N_2, and the moments U_B^m, etc. We thus see that the number of coefficients appearing in our special theory equals the number found from the general simple fluid analysis for the small-gap case. The values of the coefficients in the Newtonian case are also given in Table 10.2.

The boundary conditions for eqns (10.74) are

$$\hat u = D\hat u = \hat v = 0 \quad \text{on} \quad y = 0, 1, \tag{10.75}$$

defining an eigenvalue problem for (Ta), (l). The Galerkin method provides a convenient method for estimating (Ta) for a given l (see Chapter 8); then we search for the l which minimizes (Ta), giving the first appearance of the type of secondary flow postulated. Because of the complexity arising from the nine coefficients $\beta_1–\beta_9$, it is convenient to take only one term in a polynomial expansion and use the Galerkin method; Denn (1975, p. 134) shows that for $\Omega_2/\Omega_1 > 0$ one term is adequate for the Newtonian case. Let

$$\hat v = y(1 - y), \quad \hat u = C\hat v^2, \tag{10.76}$$

in the eqns (10.74). These forms satisfy the boundary conditions. Multiplying the first equation by $y(1 - y)$ and the second by $y^2(1 - y)^2$ and integrating over the gap gives two equations to determine C and the eigenvalue (Ta).

Table 10.2 Stability coefficients in differential equations

Coefficient	Value (KBKZ)	Value (Simple fluid)	Value (Newtonian)
β_1	$1 + d\ln\eta/d\ln|\dot\gamma|$	$1 + d\ln\eta/d\ln|\dot\gamma|$	1
β_2	$+\left(\dfrac{N_1}{2} + N_2\right)\left(1 - \dfrac{d\ln(N_1 + 2N_2)}{d\ln|\dot\gamma|}\right)/\eta\dot\gamma$	undetermined	0
β_3	$-\left(\dfrac{N_1}{2} + N_2\right)/\eta\dot\gamma$	undetermined	0
β_4^*	$2 + (U_C^3 - 2U_{BC}^5 + 2U_{CC}^5)/2\eta\dot\gamma$	undetermined	2
β_5^*	$1 + U_C^3/2\eta\dot\gamma$	undetermined	1
β_6	$\dfrac{N_2}{\eta\dot\gamma}\left(-1 + \dfrac{d\ln|N_2|}{d\ln|\dot\gamma|}\right)$	$\dfrac{N_2}{\eta\dot\gamma}\left(-1 + \dfrac{d\ln|N_2|}{d\ln|\dot\gamma|}\right)$	0
β_7	$-\dfrac{N_2}{\eta\dot\gamma}$	$-\dfrac{N_2}{\eta\dot\gamma}$	0

* Note: See eqn (10.67) for the definition of U_{BC}^5, U_C^3, etc.

This yields the following relation between (Ta) and l

$$(Ta) = \frac{28}{27l^2}\left(1 + \frac{b}{2}\right)^{-1}$$

$$\times \frac{(504 + 12\beta_4 l^2 + \beta_5 l^4)(10\beta_1 + l^2)}{\left\{1 - 2(Re)\left(\frac{28}{3}\beta_2 + \beta_3 l^2\right)/(Ta)\right\}\left\{1 + \left(\frac{28}{3}\beta_6 - \beta_7 l^2\right)/2(Re)(1 + b/2)\right\}}$$

$$(10.77)$$

Thus, to a first approximation β_8 and β_9 do not have any effect on the stability of the flow. It may be confirmed numerically, using a second term in the Galerkin representation that very large values of β_8 and β_9 (order 10^2) have a very small effect on the critical Taylor number, mostly because of the multiplying factor h/R; both of these coefficients will be ignored henceforth; Table 10.2 shows that the coefficients β_4 and β_5 cannot be found from viscometric data. We now express the change in the critical Taylor number in the form

$$(Ta) = (Ta)_N\{1 + g_i\Delta\beta_i + O(\Delta\beta_i)^2\},\tag{10.78}$$

where $\Delta\beta_i$ is the change of β_i from the Newtonian value.

The coefficients g_i are given in Table 10.3 where we also give the sign of $\partial l/\partial\Delta\beta_i$. Looking at Tables 10.2 and 10.3, we see that for the case when $\dot{\gamma} > 0$, β_1, β_2, and β_3 are less than the Newtonian values and are therefore destabilizing. The deviations in $\beta_4 - \beta_7$ depend on $\partial U/\partial I_C$ and $\partial^2 U/\partial I_B\partial I_C$, $\partial^2 U/\partial I_C^2$. Since $\partial U/\partial I_C$ is small and positive (corresponding to N_2 negative), we find, ignoring the second derivatives $\partial^2 U/\partial I_C^2$, $\partial^2 U/\partial I_B\partial I_C$, that the changes in β_4, β_5, and β_7 are positive

Table 10.3 Influence of Taylor number (Ta) and wave number of small departures of β_i from the Newtonian values $(Ta)_N \sim 1,760/(1 + b/2)$, $l_N = 3.12$

Parameter	$g_i = [\partial(Ta)/\partial\Delta\beta_i]/(Ta)_N$	$\partial l/\partial\Delta\beta_i$	Expected sign $(\Delta\beta_i)$ $(b<0)$	Remarks
β_1	$+0.505$	$+$	$-$	Destabilizing
β_2	$+0.222\sqrt{\left(\frac{h}{R}\right)}\sqrt{\left(\frac{2+b}{(-b)}\right)}$	0	$-$	Destabilizing
β_3	$+0.233\sqrt{\left(\frac{h}{R}\right)}\sqrt{\left(\frac{2+b}{(-b)}\right)}$	$-$	$-$	Destabilizing
β_4	$+0.140$	$-$	prob. $+$	Stabilizing if $N_2<0$
β_5	$+0.114$	$-$	prob. $+$	Stabilizing if $N_2<0$
β_6	$-0.222\sqrt{\left(\frac{R}{h}\right)}\sqrt{\left(\frac{-b}{(2+b)}\right)}$	0	$-$	Stabilizing if $N_2<0$
β_7	$+0.232\sqrt{\left(\frac{R}{h}\right)}\sqrt{\left(\frac{-b}{(2+b)}\right)}$	$-$	$+$	Stabilizing if $N_2<0$

and those of β_6 are negative relative to the Newtonian case. All of these changes are stabilizing. The effects of β_7 and β_3 can be inferred directly from eqns (10.74) by noting whether or not the additional terms add or subtract from the inertia effect terms. In case U is not a function of I_C, then β_4–β_7 have Newtonian values and it is clear that a KBKZ fluid is less stable than a Newtonian fluid of the same density and viscosity (at the mean shear rate). Thus a dependence of U on I_C is essential for flow stabilization; that is, N_2 must be negative.

In the limit $h/R \rightarrow 0$, $\dot{\gamma}$ fixed, we see that β_6 and β_7 are the only terms of importance for very dilute solutions (concentration $\rightarrow 0$). This is because all $\Delta\beta_i$ are multiplied by the concentration, which is small, but for the β_6 and β_7 terms there is also the large multiplying factor $\sqrt{(R/h)}$. In this limiting case if we drop all non-Newtonian terms except β_6 and β_7 in eqn (10.77), we find

$$(Ta) = \frac{28}{27l^2(1+b/2)} \frac{504 + 24l^2 + l^4}{\left\{1 + \left(\frac{28}{3}\beta_6 - \beta_7 l^2\right)/2(Re)(1+b/2)\right\}}. \tag{10.79}$$

For the case when $b = -1$ (outer cylinder at rest) and setting $d \ln |N_2| / d \ln |\dot{\gamma}| = m$ in β_6 (Table 10.2), we find the results shown in Fig. 10.4. It is clear from eqn (10.79) that a negative $N_2/\eta\dot{\gamma}$ is stabilizing. Figure 10.4 shows the critical Taylor number and wavelength as a function of $(N_2/\eta\dot{\gamma})\sqrt{(R/h)}$ for various m values.

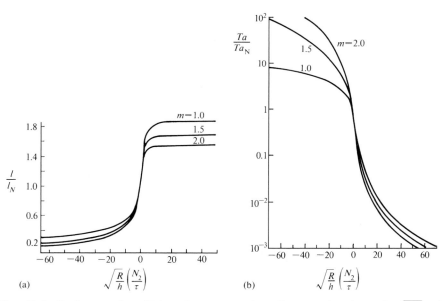

Fig. 10.4 Taylor number (Ta) and wave number (l) as a function of $\sqrt{R/h} N_2/\tau$ for various values of m, where $m = d \ln |N_2| / d \ln |\dot{\gamma}|$. In these results $R/h = 23.1$, $b = -1$ (hence τ is positive), $\beta_1 = 1$, $\beta_4 = 2$, $\beta_5 = 1$.

It is not difficult to see that if $N_2 < 0$, then it is easy to obtain stabilization as $R/h \to \infty$. Unfortunately, in most reported reliable experiments the asymptotic limit $R/h \to \infty$ applies only for much larger R/h values than those actually used. For example, there are few experiments on dilute solutions for which both the Taylor and wave numbers were observed; no drastic change in wave number has been observed experimentally, but it is predicted theoretically by Fig. 10.4. In the work of Denn et al. (1971) and of Bailey (1969) no measurements of wave number are given.

Denn and Roisman (1969) used a second-order theory to explain the stabilizing effect of polymers but since some terms were omitted there, the work of Sun and Denn (1972) is preferred. They permitted viscosity and normal stress difference to vary with shear rate, but still some of the terms in (10.74) are omitted. Using their experimental data and Tables 10.2–3 one finds that the β_7 term in (10.74) dominates and that a value of N_2/N_1 of about -0.05 explains their observation of a 29 per cent increase in Taylor number. However, the 3 per cent increase in wave number remains unexplained; Sun and Denn (1972) advanced the idea that this is a finite amplitude phenomenon. One can note that the infinitesimal amplitude case is a nearly-viscometric flow, and the viscometric constitutive equation and the KBKZ analysis are almost equivalent here as expected.

Jones et al. (1976a) have investigated the stability of three polymers in fairly dilute aqueous solutions for three values of the radius ratio (R_1/R_2), namely 0.9, 0.925, and 0.95. The concentration of polymer was up to 500 p.p.m. in each case. It is not easy to draw any conclusions from these results as the uncertainty in the Taylor number changes (± 4 per cent) is of the same order as the changes in Taylor number. There appears to be a slight stabilization effect of less than 1 per cent on average over these results, but it is hard to pick a trend with the parameters. Similarly, it is hard to pick trends in the wave number changes.

More recent experiments with dilute solutions of xanthan gums in water (Yi and Kim 1997) have shown a decrease in cell size (increase in wave number) and a very slight increase in Taylor number for dilute solutions. The gap/radius ratio was 6–14 per cent in these experiments. There was also evidence of a decrease in (Ta) in some of their cases, especially at higher solute concentrations with polyacrylic acid solutions. They also inferred the values of N_2/N_1 which would be needed to explain their results, and found values of about -0.015, -0.03 for polyacrylamide and xanthan gum respectively; for polyacrylic acids the ratio was about -0.03 for a molecular weight of 450 000, about -0.05 for a molecular weight of 750 000, and approaching -0.1 for a molecular weight of 1.25×10^6.

We see that the analysis of this problem is fairly satisfactory, but that more careful comparisons with experiment, using independently-determined material properties, would be welcome. The question of whether or not viscoelasticity is stabilizing (see Table 10.3) is clearly more complex than in the Bénard case. Also, the problem of the relation between computed and observed wave numbers remains somewhat unsatisfactory; it seems possible that effects due to finite gap, finite amplitude of the disturbance or finite cylinder length are important in setting this wave number (Swinney and Gollub 1981) and preventing large changes from the Newtonian value.

Denn *et al.* (1971) have considered a non-linear analysis of this problem and have shown that when the onset of instability is delayed, so also is the post-onset torque reduced relative to the Newtonian case. This is in agreement with observation.

10.3.2 Purely elastic instabilities

In 1966, Giesekus studied the Taylor–Couette flow experimentally at very small Reynolds numbers, so that the inertia-driven instabilities discussed above did not occur. He concluded that there might be a purely elastic instability, and Larson *et al.* (1990) later predicted such a result using an Oldroyd-B model. Avgousti and Beris (1993) have obtained numerical results for this case and the results are quite complex. Larson (1992) and Shaqfeh (1996) have given surveys.

Unlike the classical case, it has been found that axisymmetric disturbances may not be the most unstable. For the UCM model [eqn (4.25)] Avgousti and Beris (1993) showed that this was indeed the case under certain combinations of inertia and elastic parameters. Also, an overstability (oscillatory instability) was found by Larson *et al.* (1990) using the Oldroyd-B model, and this was confirmed by Muller *et al.* (1993). In common with the classical problem results above, it was found that (positive) N_1 is destabilizing and (negative) N_2 is stabilizing.

10.4 Parallel shear flows

Much of the classical theory of hydrodynamic stability is concerned with parallel shear flows and related nearly viscometric flows (Drazin and Reid 1981). In the Newtonian case, simple shearing and Poiseuille flow in a circular tube are stable against all small disturbances. By contrast, plane channel flow does exhibit a critical Reynolds number $\rho V_{\mathrm{m}} h/\eta$ equal to 5772.2, where V_{m} is the maximum velocity on the centreline and $2h$ is the channel width; using a definition of the critical Reynolds of

$$(Re)_c = 2\rho \bar{V} h/\eta$$

where $\bar{V}$ is the mean velocity in the channel, we find $(Re)_c = 7696$, which is a large number. For finite disturbances, one finds (Denn 1975) a lower value of the critical Reynolds number; experimentally the values found are 1–2000.

In these flows there is no obvious driving mechanism, like gravity or centrifugal force, and intuitive arguments are more difficult than in previous cases. Also, the onset of instability occurs when σ is complex, instead of zero as in the Couette and Bénard flows. Ho and Denn (1978) have reviewed the earlier computations on plane Poiseuille flow for a convected Maxwell fluid and have confirmed that viscoelasticity is destabilizing. Careful numerical work was necessary to avoid spurious eigenvalues; some previous work contains erroneous results. For a Reynolds number $\rho V_{\mathrm{m}} h/\eta$ equal to 2320 (that is, less than half that causing instability in the Newtonian case) Ho and Denn (1978) found that a Weissenberg number ($\lambda V_m/h$) of 2.28 was sufficient to cause instability.

Jones *et al.* (1976*b*) have studied the stability of dilute solutions of poly-acrylamide and have concluded that the addition of polymer did not change the critical Reynolds number for stability of flow when the mean shear rate ($\bar{u}/R$) was less than $650\,\mathrm{s}^{-1}$; for ($\bar{u}/R$) from $890\,\mathrm{s}^{-1}$ to $1500\,\mathrm{s}^{-1}$ they saw stabil-ization, and destabilization above that. It is difficult to correlate these results with theory because the relaxation time of the material is not given. The changes in Reynolds number were of order $+10$ per cent (stabilizing) to -50 per cent (destabilizing).

For lower Reynolds numbers extensive computations by Ho and Denn (1978) using even eigenfunctions failed to find unstable modes; for $\rho V_m h/\eta$ equal to 500 no instabilities were found at a Weissenberg number ($\lambda V_m/h$) of 70; large values of the product Weissenberg number $\times$ dimensionless wave number would be required for instability. It appears that unstable wave numbers of order $1/h$ occur in practice, so that large unstable wave numbers are unlikely to be important. Hence it was concluded that plane channel convected Maxwell flow is likely to be stable to infinitesimal disturbances at this Reynolds number. The work of Rothenberger *et al.* (1973) is now considered erroneous. They found instability at zero Reynolds number and quite low (O(1)) Weissenberg number in both tube and channel flow for Maxwell fluids. Gorodtsov and Leonov (1966) con-sidered the stability of a plane shear flow of a UCM model, and found that both inertia and viscoelasticity was needed to trigger instability, but no numerical data were given.

For the zero-inertial case of (UCM) plane Couette flow, Renardy (1992) proved that the flow was stable at all Weissenberg numbers. If the Oldroyd-B fluid model is used, then linear stability is assured only in the presence of inertia (Guillope and Saut 1990). It is hard to forecast the effects of inertia and large amount of elasticity on the stability of this class of flows for other models. For example, Ganpule and Khomami (1999) studied the effect of transient viscoe-lastic properties on interfacial instabilities in a two-fluid pulsating channel flow, as part of a number of articles studying this kind of instability. They used multiple mode Giesekus and White–Metzner models in this complex problem, and found that the choice of model affected the stability boundaries, although the steady state viscometric responses of the models were essentially identical.

In connection with two-fluid layer plane shearing problems, it is known that for Newtonian fluids a difference of viscosity between layers can promote instability. Y. Renardy (1988) has shown that a jump of N_1 across the interface can produce a purely elastic instability in this geometry.

10.5 Parallel-plate and cone-plate flows

Because of their widespread use in measurements, it is important to discover if inertia-less instabilities in such flows exist. As a step in this direction Phan-Thien

(1983) has analysed the flow between coaxial discs (torsional flow) of infinite extent separated by a distance d along the z-axis.

The Newtonian flow was solved by von Kármán in 1921 and has been widely discussed.

Phan-Thien (1983), using an exact solution in the von Kármán form for the Maxwell fluid, studied the stability of the flow. We follow his discussion here, omitting inertia.

Suppose the plane $z = 0$ rotates at an angular speed Ω_0 and the plane $z = d$ rotates at Ω_d. Both plates are set in motion at $t = 0^+$, and the flow is transient. The constitutive equation used is the Oldroyd-B fluid:

$$\boldsymbol{\tau} = 2\eta_s \mathbf{d} + \mathbf{S}, \tag{10.80}$$

where $\mathbf{S}$ is a Maxwell stress [eqn (4.25)] and η_s is the (Newtonian) solvent viscosity; when $\eta_s = 0$ we return to the Maxwell case. We let the viscosity associated with the Maxwell element be η_p and the associated time constant be λ; let $\eta = \eta_p + \eta_s$. The retardation time is then

$$\lambda_2 = \lambda \eta_s / \eta = (1 - \beta)\lambda, \tag{10.81}$$

and β is the retardation parameter (η_p/η) ranging from 0 (Newtonian) to 1 (Maxwell).

The steady-state solution in general (including inertia) uses a velocity field of the form

$$\mathbf{v} = \left(r\frac{\partial H}{\partial z}, \; rG, \; -2H \right), \tag{10.82}$$

Where G, H are functions of z and t; in the inertia-less case $H = 0$ and G is linear in z (torsional flow). We assume a form (10.82) and a development of the solution in series

$$\left.\begin{aligned}
g &= \xi + g_1 + \text{higher-order terms} \\
h &= h_1 + \cdots \\
\tau_{\theta\theta} &= 2\beta(\text{Wi}) + \widehat{tt} + \cdots \\
\tau_{\theta z} &= \beta + \widehat{tz} + \cdots
\end{aligned}\right\} \tag{10.83}$$

where $\xi = z/d$, $t' = (\Omega_d - \Omega_0)t$, $G = (\Omega_d - \Omega_0)g + \Omega_0$, and $H = d(\Omega_d - \Omega_0)h$; h and g are functions of ξ and t'. The Deborah number (De) is defined here as $\lambda(\Omega_d - \Omega_0)$ and the stresses $\tau_{\theta\theta}$ and $\tau_{\theta z}$ are dimensionless; they are formed by dividing the relevant components by $\eta(\Omega_d - \Omega_0)$. Substituting the forms (10.83) in the equations of motion and the constitutive equations and linearizing about the steady solution one finds a set of eight linear ordinary differential equations. A normal mode Fourier analysis of the form

$$h_1 = \hat{h}[\exp(2\pi i k\xi) - 1], \tag{10.84}$$

shows that the kth mode has a critical Weissenberg number $(Wi)_{c,k}$ of

$$(De)_{c,k} = \frac{\pi k}{\sqrt{\beta(2\beta + 3)}}.$$ (10.85)

The most unstable mode, $(Wi)_c$, is when $k = 1$, hence if

$$(De) > \frac{\pi}{\sqrt{\beta(2\beta + 3)}}$$ (10.86)

the flow is unstable to infinitesimal disturbances of the form assumed; Note that $(De) \to \infty$ as $\beta \to \infty$ (Newtonian case) which is consistent with expectations. When $\beta = 1$ (Maxwell case). $(De)_c = 1.405$.

Note that the (steady flow $\sigma = 0$) instability is a purely elastic one (no inertia) and that $N_2 = 0$ in this model. More details are given by Huilgol and Phan-Thien (1997).

The steady flow is in the form of secondary vortices between the plates, increasing the measured torque.

Phan-Thien (1985) considered the (infinite) cone-plate flow in a similar way and found a critical Weissenberg number in this case also.

Experimentally, Jackson et al. (1984) observed what appeared to be an unexpected 'antithixotropic' shear-thickening in experiments on Boger fluids. Up to that time, the attraction of Boger fluids had been their almost constant shear viscosity. The group observed that, at high shear rates, measurements of the torque and the normal force in a Weissenberg Rheogoniometer increased steadily at a fixed shear rate over a long period of time; the data for torsional flow (reproduced in Fig. 10.5) showed nearly double the torque expected for the final long-term 'equilibrium' values. Jackson et al. noted that halving the rheometer gap, keeping the maximum shear rate constant, removed the 'antithixotropy'.

Research by Magda and Larson (1988) and McKinley et al. (1991) showed that the antithixotropy was a manifestation of an instability, similar to that predicted by Phan-Thien. However, the flow patterns did not resemble those of the analysis and so they concluded that the agreement between the predicted and observed Deborah numbers was fortuitous. Further work by Avagliano and Phan-Thien (1996), who analysed the stability of torsional flow between plates of finite radius, gave more satisfactory agreement with experiment. See Huilgol and Phan-Thien (1997) for further details and comparison with experiments.

The form of the disturbance flow-field found by Avagliano and Phan-Thien is very similar to that observed by McKinley et al. (1991), except that in this work, the disturbances are large near the outer edge and small near the centreline. In the linearized stability region the roll cells would first be seen at the outer edge and then appear to travel inwards, as the amplitude of the inner cells increases exponentially in time. In the work of McKinley et al. cells of large amplitude are observed initially at both the outer edge and at the centreline. A possible reason

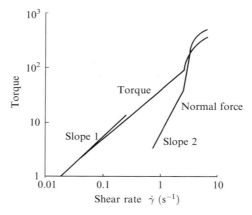

Fig. 10.5 Torsional-flow data for a Boger fluid obtained from a Weissenberg Rheogoniometer. Here, $\dot{\gamma}$ is the shear rate (from Jackson *et al.* 1984); note the increase in torque and normal force at a critical shear rate of about $2\,\mathrm{s}^{-1}$.

for this discrepancy is the presence of inertia, or some shear-thinning, not accounted for in the analysis.

These results show that elastic instabilities can occur in very common flows. Hence, in these cases numerical schemes may find steady solutions where no such stable solutions exist. Physically, it appears that the large $\tau_{\theta\theta}$ is the cause of the instability; there is no such instability in the corresponding (stable) plane shearing flow where $\tau_{\theta\theta}$ is absent.

10.5.1   Edge fracture

Frequently, the fluid in a cone-plate or parallel plate device 'fractures' at a certain shear rate and one observes a lessening of the measured torque, often followed by ejection of the sample (Fig. 10.6). Although it is completely different in kind to the cases so far discussed in this chapter, we shall regard it as an instability; it severely limits the shear rates measurable.

Typically, [Fig. 10.6(b)] an indentation appears in the surface and grows in towards the axis.

An early attempt to explain the phenomenon was made by Hutton (1965) on the basis of elastic energy associated with the first normal stress difference N_1. Tanner and Keentok (1983) put forward an alternative explanation starting from the idea of a semi-circular 'flaw' at the edge; we follow their analysis.

Near the edge of the plates it seems reasonable to ignore the curvature, so in Fig. 10.7 one has $(R/h \gg 1)$ a simple shear flow with a notch on groove of semicircular form in it. The groove radius a is supposed to be much less than h. Using the second-order model eqn (4.30c) we assume the disturbed flow is of the form $w(x, y)$, a parallel flow.

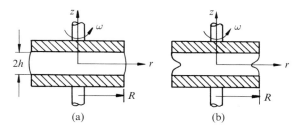

Fig. 10.6 Edge fracture in rheometry.

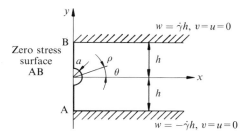

Fig. 10.7 Boundary values for w; a is the 'notch' radius.

Far from the 'crack' ($x \to \infty$) the flow becomes a simple shear flow of magnitude $\dot{\gamma}$.

From our considerations in Chapter 4 we know that

$$p = \tfrac{1}{2} N_2 + \text{constant}, \tag{4.108}$$

where the constant is set by boundary conditions on the straight edge (AB, Fig. 10.7). For a small crack, $h \gg a$, one finds, in polar coordinates (ρ, θ) (Fig. 10.7)

$$w = \dot{\gamma}\rho \sin \theta (1 + a^2/\rho^2).$$

Note that when $\rho \to \infty$, this reduces to a simple shear, and on the surface $\rho = a$ one has $\partial w/\partial n = 0$, or no shear stress. The tensile stress on the circle $\rho = a$ is

$$\sigma_{\rho\rho}|_{\rho=a} = \tfrac{1}{2} \Psi_2 \dot{\gamma}^2 (1 - 4\cos^2 \theta). \tag{10.87}$$

When $\sigma_{\rho\rho}$ is positive, the groove surface will tend to open up against the pull σ/a of the surface tension σ; if $\Psi_2 = 0$ (no N_2) then the 'crack' will disappear because of surface tension. The maximum tensile stress occurs when $\theta = 0$, since Ψ_2 is assumed negative. Hence the crack will deepen if

$$|N_2| > 2\sigma/3a. \tag{10.88}$$

If we assume that $a = kh$ then it follows that there is a critical N_2 when

$$|N_{2c}| = \frac{2\sigma}{3kh} \tag{10.89}$$

Elaborations of this analysis are given by Huilgol and Phan-Thien (1997) and Keentok and Xue (1999), but the essential result (10.89) is unchanged. The latter paper suggests that fracture be used to measure N_2.

Although Tanner and Keentok (1983) provided some experimental evidence for the theory, a more definite proof was given by Lee *et al.* (1992), who found that N_2 did control the phenomenon. See also Keentok and Xue (1999).

This instability might be termed a static instability, because two 'static' forces are compared.

10.6 Non-viscometric flow stability

So far we have concentrated on viscometric flow stability. We now turn first to general homogeneous flows and then to the case of fibre spinning. The first case has been investigated by Lagnado *et al.* (1984) for Newtonian fluids and by the same group (1985) for Maxwell–Oldroyd-B and more general fluids. Let the basic flow be

$$\mathbf{v} = \mathbf{Lx},$$

where $\mathbf{L}$ is independent of $\mathbf{x}$. We will also assume $\mathbf{L}$ is independent of time.

We can select a coordinate system where

$$\mathbf{L} = \frac{L}{2} \begin{bmatrix} 1 + \mu & 1 - \mu & 0 \\ -(1 - \mu) & -(1 + \mu) & 0 \\ 0 & 0 & 0 \end{bmatrix}. \tag{10.90}$$

Here, $L \geq 0$ is the magnitude of the velocity gradient and μ is a parameter varying between $+1$ and -1. The cases $\mu = -1$, $\mu = 0$ and $\mu = 1$ correspond to pure rotational flow, simple shearing flow, and pure extension (pure shearing) respectively. For the inertialess case the flow is stable but there are instabilities for inertial cases (Lagnado *et al.* 1984, 1985). For the Maxwell fluid, the Weissenberg number is λL and there is no steady base solution for $(Wi) > 0.5$. For the other cases $[(Wi) < 0.5]$ the results depend on the ratio λ_2/λ_1, where λ_2/λ_1 is the ratio of retardation to relaxation time for the Oldroyd-B model. If the disturbance wave number is α, where α lies *across* the axis of stretching, then there is a critical value of $\eta\alpha^2/\rho L$ below which the flow is unstable. Hence as $\rho \to 0$, the flow is always stable, since $\alpha \neq 0$. These results show the importance of the flow kinematics on stability; the method of solution is also interesting since the decomposition (10.13) could not be used (Lagnado *et al.* 1984).

10.6.1 *Stability of fibre spinning. Draw resonance*

We now turn to a flow that is non-viscometric and also non-uniform in space.

We shall distinguish between *stability* and *sensitivity*. A system is unstable to infinitesimal disturbances if a steady state cannot be maintained following any arbitrarily small disturbance. The sustained oscillations characterizing draw resonance (Fig. 7.6), which occur reproducibly at a critical draw ratio, represent such a process instability. Sensitivity is a reaction to the propagation of disturbances through a stable system. If the system shows high sensitivity it may not be practical to operate such a system even if it is stable in a formal sense. We shall mainly be concerned with instability here. Spinning instability has been reviewed by Petrie and Denn (1976), Pearson (1976), Denn (1980), and Denn and Pearson (1981).

It is possible to identify the onset of an instability using computer simulation methods. Kase and Denn (1978) have shown that stability can be analysed in terms of the frequency response relation (*transfer function*) between velocity and tension at the solidification point. However, the traditional way of studying stability is through the type of stability analysis discussed above.

A set of spinning conditions is stable to infinitesimal disturbances if all eigenvalues have negative real parts, since the perturbations then decay to zero. The spinning conditions are unstable if any one eigenvalue contains a positive real part. Thus, in principle, all eigenvalues in the infinite set must be examined. In practice, the stability behaviour of this system seems always to be governed by the eigenvalue with the smallest modulus; however, this is not a general result for non-self-adjoint operators, and counter-examples exist (cf. Porteous and Denn 1972).

Several approaches have been used for solving the eigenvalue problem in these non-homogeneous flows. One method is to replace the spatial derivatives by finite differences, leading to a matrix eigenvalue problem that can be solved by a standard library programme, and then to use extrapolation to obtain the limiting value as the number of discretization points tends to infinity. A sequence of such calculations for the eigenvalue of smallest modulus gives the critical value as 20.218 for D_R in the Newtonian spinning case.

The orthogonal collocation approach to the transient equations used by Gupta and Ballman (1982) is quite similar in concept. Here, the discretization is done first by transforming to a set of linear ordinary differential equations at the spatial collocation points. The eigenvalues of the system matrix then determine the stability characteristics. Gupta and Ballman found a transition at a draw ratio between 20 and 21 using 7 approximating functions in the collocation expansion. Chang *et al.* (1981) have directly integrated the equations for non-isothermal flow both with and without viscoelasticity, using a Runge–Kutta method, the equations being formulated as an initial-value problem. Iteration to the eigenvalue satisfying the downstream boundary condition was done interactively, and convergence was rapid. The interaction is aided by the observation that the real part of the first eigenvalue is nearly linear in D_R. A stability 'map' of the critical draw ratio (D_{R_c}) as a function of $\lambda_{max} W(0)/L$ is shown in Fig. 10.8 for low speed isothermal spinning of a PTT model fluid. Here one or two time constants were used ($N = 1$, 2), the modulus for each relaxation time was the same (that is, $G_1 = G_2$) and $\lambda_1/\lambda_2 = 5$ (Chang and Denn 1980). The values of the parameters ε and ξ are shown (see Chapter 5). The stabilization at large values of $\lambda_{max} W(0)/L$, and at large draw ratios is in agreement with experiment. Some questions in the application of linear stability theory to spinning have been discussed by Denn and Pearson (1981).

Gupta *et al.* (1996) showed the profound effect of temperature variations in the spinline for glass fibre drawing. They show that cooling along the spin line is stabilizing, but not completely so; much higher draw ratios are possible than in the isothermal case. They attributed previous results which appeared to give unconditional stability to inaccurate numerical work. Schultz *et al.* (1996) also permitted their fibres to be non-isothermal and viscoelastic, using a UCM model. They compared exact, Galerkin and finite difference methods for finding critical

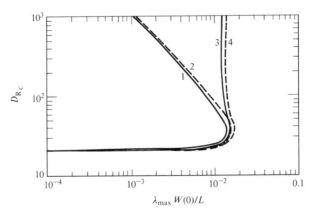

$\lambda_{max} W(0)/L$

Fig. 10.8 Critical draw ratio for onset of draw resonance as a function of Weissenberg number $\lambda W(0)/L$ for isothermal spinning of Phan-Thien–Tanner fluid. Either one or two relaxation times were used ($N = 1$ or 2) and the parameter ε controlling the extensional behaviour was either 0 or 0.015. In all cases the shear parameter ξ was 0.1.

The combinations used are indicated below with a key to the relevant curve.

Curve	N	ε
1	1	0
2	2	0
3	1	0.015
4	2	0.015

Thus ε has a major effect on stability (from Chang and Denn 1980).

draw ratios. They found that the combination of viscoelasticity and cooling was sufficient to permit draw ratios $> 10^3$, which is close to the practical range $(10^4 - 10^5)$. There seems to be some question that glass is viscoelastic (at the low stretch rates involved), but the results are of interest.

The stability to finite disturbances and the approach to a limit cycle with draw resonance has been studied numerically by Ishihara and Kase (1975, 1976) using the dynamic simulation methods discussed previously. Fisher and Denn (see Denn 1980) have used an approximate method based on an expansion in the spatial eigenfunctions of the linear system, using the method of weighted residuals to obtain a set of ordinary differential equations for the time-dependent coefficients in the expansion. A procedure described by Hyun (Denn 1980) for stability to finite disturbances does not reduce to the result of the linear theory as the disturbance amplitude becomes vanishingly small.

In spinning flows viscoelastic effects stabilized the system. Hence it is difficult to generalize about the effect of viscoelasticity on stability; often it is destabilizing, because new degrees of freedom and greater possibilities for instability are opened up, but when a Newtonian-like instability is found, then viscoelasticity can be stabilizing.

10.7   General remarks on stability

In non-Newtonian stability analyses neither Squire's theorem nor energy methods, which are commonly used in Newtonian fluid mechanics, has proved to be useful. (Huilgol and Phan-Thien 1997.) On the other hand, some guidance is needed to avoid very difficult and delicate analyses and experiments wherever possible. The work of McKinley et al. (1996) offers some hope of simplifying matters. For purely elastic instabilities, they proposed a new dimensionless criterion.

Let the critical Deborah number $(De)_c$ be $\lambda\Omega_c$ where Ω_c is the inverse of the critical dwell time of a particle in the system, and let the critical Weissenberg number be $\lambda\dot{\gamma}_c (\equiv \lambda U_{crit}/L)$ where $\dot{\gamma}_c$ is the critical shear rate; λ is a suitable relaxation time, U and L are measures of velocity and length respectively. McKinley et al. conclude, from studying various elastic instabilities, that the combination of streamline curvature and large normal stresses is sufficient to cause elastic instability. The flows considered include the Taylor–Couette flow, cone-plate and torsional flow, eccentric rotating cylinders, cylinder in a channel, the planar contraction, and a lid-driven cavity problem. The latter is a rectangular box in which plane flows with closed streamlines occur; two side walls are of length H, the bottom is of length L, and a sliding lid of length L moving with speed U completes the geometry. The (De) and (Wi) numbers are not always independent—for example in the cone-plate flow (angle α) $(De) = \lambda\Omega$ whereas $(Wi) = \lambda\Omega/\alpha$. The 'aspect ratio' (Λ) of the flow is the ratio $(Wi)/(De) = \alpha^{-1}$ in this case. For other flows an appropriate geometric ratio Λ can be similarly defined. The dimensionless criterion proposed is, simply, that the dimensionless product below exceeds a critical value, so one has

$$\frac{\lambda U}{\rho_s}\frac{\tau_f}{\eta_0\dot{\gamma}} > M_c^2 \tag{10.91}$$

for the onset of instability, where M_c is a critical value. Here ρ_s is the characteristic radius of curvature of the streamlines and τ_f is the extra stress in the flow direction; $\dot{\gamma}$ is a locally relevant shear rate. The criterion uses the concept of local stability, which is sometimes difficult to appreciate: when a portion of the flow goes unstable, it usually spreads rapidly throughout the flow. In eqn (10.91) it is supposed that ρ_s and $\dot{\gamma}$ are both functions of the geometry through the parameter Λ.

One can apply these ideas to problems where the answers are known. For the Taylor–Couette flow, with a narrow gap h and an inner radius R_i (Fig. 10.2), $\Lambda = h/R_i$; the radius of curvature $\rho_s = R_i$, assuming $h/R_i \ll 1$. (For wider gaps there is an additional dependence on h/R_i.) The stability criterion 10.91 then reduces to (assuming $\tau_f = 2\eta_0\lambda\dot{\gamma}^2$)

$$2\lambda^2\Omega_i\dot{\gamma} > M_c^2$$

or

$$[2(De)(Wi)]^{1/2} > M_c,$$

where $(De) = \lambda\Omega_i$ and $(Wi) = \lambda\Omega_i R_i/h$. Hence one finds

$$\Lambda^{1/2}(Wi) > M_c/\sqrt{2}.$$

For the UCM model $M_c \approx 8.37 \pm 0.03$. This is a considerable simplification of the problem, since one has covered a whole range of geometries and Weissenberg numbers in one formula. The extension to the Oldroyd-B model is almost immediate. Let θ be the ratio of solvent viscosity to total viscosity; when $\theta = 0$ one has the UCM model; when $\theta = 1$, the fluid is Newtonian. The result is

$$\sqrt{(De)(Wi)} \geq M_c/[2(1-\theta)]^{1/2} \tag{10.92}$$

which is a crude, but reasonable approximation to the full stability computations, especially for $\theta \to 1$.

McKinley *et al.* (1996) further develop these ideas for the flows described above, and discuss their application to more difficult flow geometries. They draw an analogy with the Görtler number (Go) for Newtonian boundary-layer flows, where $Go = [(Re)\delta/\rho_s]^{1/2}$. Here the Reynolds number is based on the local boundary layer thickness δ, and is equal to $U\delta/\nu$, where ν is the kinematic viscosity and ρ_s is again the radius of curvature of the streamlines.

There are problems as yet unresolved—the role of N_2 as a stabilizing factor, for example, needs further developments.

10.8 Melt fracture mechanisms

Melt fracture instability occurs when a critical throughput rate occurs in a capillary or die. Petrie and Denn (1976) and Denn (1990) have reviewed the extensive literature on this subject; see also Boudreaux and Cuculo (1978), Tordella (1969), De Kee and Wissbrun (1998), and Wang (1998). Inertia is not a plausible cause of instability; instability has been noted when (Re) was $O(10^{-15})$. The term melt fracture was coined by Tordella because of noises associated with gross extrudate distortion; these noises do not always occur, nor do very large distortions always occur. However, Legrand *et al.* (1998) have seen cracks at the die exit. It seems likely that the term melt fracture will continue to be applied to various instabilities (Fig. 1.6 shows some typical phenomena). Sharkskin is a high-frequency roughness on the surace which appears to begin at the die exit; in contrast, melt fracture also may occur within the die (Petrie and Denn 1976). Much experimental work on these and related stability problems appears in the literature and one can trace significant reviews and sources through the articles cited above.

Most observers agree that extrusion flow instability occurs at a critical value of the recoverable shear S_R, where S_R is defined as

$$S_R = \lambda \tau_w / \eta, \tag{10.93}$$

where λ is a time constant and η is a viscosity; the ratio η/λ is a modulus, from whence the term recoverable shear comes. Clearly, S_R is equivalent to a Weissenberg number (equal to $\lambda \dot{\gamma}_w$ for a Maxwell fluid) and we can consider that a critical Weissenberg number exists for stability. The critical value of S_R appears to lie between 1 and 10 usually; 5–8 is the most common range. The existence of a critical Weissenberg number is a result of dimensional analysis. If inertia, gravity, and surface tension are ignored, then it follows that the critical average velocity $\bar{W}_c$ depends only on the die size (d) and relevant fluid properties (λ, η) for the isothermal case. Hence $(Wi)_c = $ constant. We shall not consider thermal effects; they can be used to mitigate instability (for example, by heating die lips) but do not provide a convincing explanation for instability. If 'material failure' occurs, then we have some explanation for the grosser forms of distortion (Fig. 1.6) but it is not an explanation of the onset of instability.

The possibility that melt fracture is initiated by slip at the die wall deserves careful consideration; if slip does occur then a powerful mechanism for inducing instability exists (stick-slip mechanism) but it is clearly not the only source of instability since various workers have noticed irregular flow before the onset of slip. Opinion remains divided on this issue. Other factors possibly leading to instability are shear acceleration waves (Coleman and Gurtin 1968), but there is no evidence that these waves are important.

El Kissi *et al.* (1997) consider that sharkskin can appear when the upstream flow is stable, with polymer adhering to the wall, and that it is an exit effect, with cracking under high tensile stress.

Ghenta *et al.* (1999) extruded a linear low-density polyethylene (LLDPE) through capillary dies of brass and stainless steel. They confirmed the result of Ramamurthy (1986), that sharkskin can be eliminated by using a brass die. However, their results depended crucially on a nitrogen blanket at the hopper (intake to extruder) and on the application of an abrasive to the die before extruding the LLDPE; this probably removed oxide layers. Above a wall shear stress of 0.17 MPa sharkskin appears using the steel die; the brass die flow curve deviates from the steel die curve at this stress. Stick-slip flow in the steel die occurs at $\tau_w = 0.37$ MPa, but the brass die shows slip without oscillations at this stress, and has a much larger throughput. Thus these results do not support Ramamurthy's (1986) conclusion that brass inhibits slip. These and Ramamurthy's results do indicate, however, that surface chemistry and/or condition is important in considering sharkskin. See also Yang *et al.* (1998) for some discussion at the molecular level.

Evidently, the sharkskin phenomenon is not yet fully understood. Beyond this point, where gross extrudate distortion sets in (Fig. 1.6) the instability is clearly out of the range of linear stability theory; the fluid 'spurts' out. However, approximate descriptions can be given. Den Doelder *et al.* (1998) produced a simple quantitative model of 'spurting' in capillary tubes, and compared it with experimental data on HDPE. Compressibility of the melt is assumed. A switch from a no-slip to a slip boundary condition over a range of shear rate is postulated, and melt elasticity is neglected. Good agreement is shown by this complex one-dimensional computer model with experiments.

10.9   Turbulence

Beyond the stability limit turbulence rapidly develops; the flow becomes more and more chaotic. Each particle therefore undergoes a rich kinematic experience. We shall not look into (grossly) different macroscopic flows; we shall concentrate on turbulent pipe and channel flows. These are the most studied cases, but they may be somewhat special kinematically.

Turbulent flows are irregular, diffusive, three-dimensional and dissipative; turbulence is a flow property, not a fluid property. Scales of structures in turbulence range from lengths much greater than molecular size (even with macromolecules in solution) to scales of the order of the flow-channel. No complete analytical theory exists and frequent recourse to experiment as a check on hypotheses is necessary. One of the characteristics of turbulence is the rapidity with which mixing takes place. For example, in a static fluid or parallel laminar flow we can consider the time (T_m) taken to mix (for example) salt and water by molecular diffusion in a channel of characteristic dimension L with a diffusivity α. We find (Tennekes and Lumley 1972, Chapter 1)

$$T_{\mathrm{m}} \sim L^2/\alpha. \tag{10.94}$$

When $L \sim 1$ m and $\alpha \sim 10^{-8}\,\mathrm{m}^2/\mathrm{s}$ we get $T_{\mathrm{m}} \sim 10^8$ s. In a turbulent flow in the same channel, where the mean speed is $\bar{V}$, then the turbulent fluctuations of velocity, which transport and mix the salt, will often be of order $0.1\bar{V}$. Hence the mixing time T_{t} is of order $10L/\bar{V}$ in this case. The ratio of $T_{\mathrm{m}}/T_{\mathrm{t}}$ is thus $0.1\bar{V}L/\alpha$, a Peclet number, which will generally be very large. The flow problem is more complex than diffusion, but is similar; in this case the relevant molecular diffusivity is the kinematic viscosity $\eta/\rho (\equiv \nu)$ and the ratio $T_{\mathrm{m}}/T_{\mathrm{t}} \sim 0.1$ (Re), which is large ($\sim 10^4$). The length-scale at which viscosity is important is usually smaller than the scale of mixing, which is on the order of L. One can estimate the viscous length-scale l by noting that the viscous term in the Navier–Stokes equation is of order $\eta V/l^2$, while the inertia term is of order $\rho V^2/L$. Assuming they are equally important at some point, we get

$$\frac{l^2}{L^2} = \frac{\eta}{\rho V L} = (Re)^{-1}. \tag{10.95}$$

In turbulent channel flow (Re) is large, so the ratio of the length-scales is small.

A major difficulty in turbulent flow analysis is the following. Suppose we split the velocity and pressure fields (v_i, p) into a mean part (V_i, P) and a fluctuating part (v_i', p') so that

$$v_i = V_i + v_i', \quad p = P + p', \tag{10.96}$$

where

$$V_i = \lim_{T \to \infty} \frac{1}{T} \int_{t_0}^{t_0 + T} v_i\, \mathrm{d}t = \bar{v}_i, \tag{10.97}$$

defines the averaging process. We shall only consider mean values which are independent of t_0 (steady turbulent flows). Substituting (10.96) into the Navier–Stokes equations and averaging over time yields the Reynolds equations

$$\rho V_j \frac{\partial V_i}{\partial x_j} = -\frac{\partial P}{\partial x_i} + \eta \frac{\partial}{\partial x_j} \left\{ \frac{\partial V_i}{\partial x_j} + \frac{\partial V_j}{\partial x_i} \right\} - \frac{\partial}{\partial x_j}(\overline{\rho v_i' v_j'}), \tag{10.98}$$

where the overbar on the last term in (10.98) denotes a time average [see eqn (10.97)]. Equation (10.98) shows that the macroscopic quantities (V_i, P) depend on the details of the v_i' field via the averages. The quantities $-\overline{\rho v_i' v_j'}$ are called the Reynolds stresses; in general they cannot be simply expressed in terms of the macroscopic variables.

The simplest approach to this problem is to assume the Reynolds stresses are proportional to the $(\partial V_i/\partial x_j + \partial V_j/\partial x_i)$; one has then a viscous fluid with (possibly non-Newtonian) augmented viscosity, usually termed the eddy viscosity. This can be a useful calculational device especially if the eddy viscosity varies from point to point in the flow, but it is not a fundamental approach to turbulence.

One can illustrate the use of (10.98) in the case of fully developed plane channel flow. Suppose the channel is bounded by the planes $z = 0$ and $z = 2h$ and the mean flow is in the direction x; all mean values except P are functions of z alone. Then (10.98) reduces to ($x_i \to x$, y, z, $V_i \to U$, V, W, etc.)

$$\rho \frac{\partial}{\partial z} (\overline{u'w'}) = -\frac{\partial P}{\partial x} + \eta \frac{\partial^2 U}{\partial z^2},$$

$$0 = -\frac{\partial P}{\partial y},$$

$$0 = -\frac{\partial P}{\partial z} - \rho \frac{\partial \overline{w'^2}}{\partial z}, \tag{10.99}$$

where P and U are the mean values of the pressure and the x-component of velocity respectively. The second and third equations in (10.99) can be integrated to give

$$P + \overline{\rho w'^2} = P_0(x), \tag{10.100}$$

and, since the fluctuations are zero on the rigid walls, P_0 is the measurable wall pressure. The first equation yields, for the shear stress $\tau (\equiv \sigma_{xz})$

$$\tau = \eta \frac{\partial U}{\partial z} - \rho \overline{u'w'} = \tau_w + z \frac{\mathrm{d}P_0}{\mathrm{d}x}, \tag{10.101}$$

where τ_w is the shear stress at the wall. We define the friction velocity u_* as

$$u_* = \sqrt{\tau_w / \rho}.$$

Then, from (10.101) we find, since the shear stress is zero on the centreline, $z = h$, and $\mathrm{d}P_0/\mathrm{d}x$ is constant,

$$-\overline{u'w'} + \frac{\eta}{\rho} \frac{\mathrm{d}U}{\mathrm{d}z} = u_*^2 \left(1 - \frac{z}{h} \right). \tag{10.102}$$

We can divide the region near the wall into three parts, a viscous sublayer, a buffer layer and an inertial sublayer. The no-slip condition near the wall means that $\overline{u'w'}$ is zero when z is zero. In the viscous sublayer $-\overline{u'w'}$ and z are very small so that we have

$$\nu \frac{\mathrm{d}U}{\mathrm{d}z} = u_*^2. \tag{10.102a}$$

Integrating, we find a linear profile

$$U = \frac{u_*^2 z}{\nu},$$ (10.103)

or

$$\frac{U}{u_*} = \frac{u_* z}{\nu},$$ (10.103a)

or, defining new dimensionless quantities,

$$U_+ = z_+.$$ (10.103b)

Experimentally, this relationship only holds up to $z_+ \cong 5$. If we remain near the wall so that the shear stress is constant, but sufficiently far away so that the viscous stress $\eta(dU/dz)$ is small compared with the Reynolds stress, then (10.102) becomes

$$-\overline{u'w'} \cong u_*^2,$$ (10.104)

when $z/h \ll 1$, as hypothesized. Thus the Reynolds stress scales with the friction velocity squared in this region. In this constant-stress layer we argue that if the length-scale is l, and the velocity scale is u_*, then the mean velocity variation

$$\frac{dU}{dz} \propto u_*/l.$$ (10.105)

The wall constrains the turbulent motion so that transport of momentum downward from some level z is restricted to distances smaller than z. Thus one expects the scale l to be proportional to z itself. Eliminating l from (10.105) we get the equation for the inertial sub-layer

$$\frac{dU}{dz} = \frac{u_*}{\kappa z},$$ (10.106)

where κ is the von Kármán constant. This equation has also been deduced from a turbulent energy balance (Townsend 1980). Integrating, we find, for large z_+

$$U_+ = \frac{U}{u_*} = \frac{1}{\kappa} \ln z + \text{constant}.$$ (10.107)

One cannot determine the constant by setting $U=0$ at $z=0$ because (10.106) does not hold in the viscous sub-layer. One can scale z as we did in (10.103) to get the main result of this section, the logarithmic profile law

$$U_+ = A \ln z_+ + B,$$ (10.108)

where A and B are constants and

$$A = \kappa^{-1}$$ (10.109)

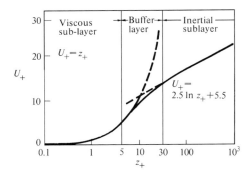

Fig. 10.9   Velocity profile near the wall for Newtonian turbulent pipe flow showing sub-layers.

Although we have discussed a constant-stress layer, similar arguments may be applied to the near-wall regions for smooth tube flow, with a similar result for the velocity (Townsend 1980). In Newtonian fluids $A \sim 2.5$ and $B \sim 5.5$. To join the regions where (10.103b) and (10.108) hold we have a buffer layer. The various regions are shown in Fig. 10.9 the viscous sub-layer or wall layer, the buffer layer, and the inertial sub-layer. Eventually, when $z/h \sim 1$, we must have a dependence on z/h; this is in the core region (not shown on Fig. 10.9). This will be further discussed in the next sub-section.

   We have not considered rough walls where the viscous sub-layer is removed; for this and other aspects of turbulence see Tennekes and Lumley (1972) and Townsend (1980).

10.9.1   Friction factor as a function of Reynolds number

Measurements on pipe and channel-flow yield relations between the mean flow speed $\bar{u}$, the pressure drop, and the tube dimensions and fluid properties. We shall consider tube flow, and use the wall shear stress τ_w as a convenient parameter for analysis. The Reynolds number (Re) is $\rho \bar{u} d / \eta$ and the friction coefficient (f) is defined by

$$f = \tau_w / \tfrac{1}{2} \rho \bar{u}^2. \qquad (10.110)$$

We need to relate $\bar{u}$ to the turbulent velocity field. In the core region (away from the wall) we find the velocity defect law (Tennekes and Lumley 1972)

$$U_0 - U = u_* g(z/R), \qquad (10.111)$$

where R is the tube radius, z is the distance from the wall and U_0 is the centreline velocity. Equation (10.111) asserts that curves of velocity distribution plotted against z/R condense on to a single curve for all Reynolds numbers. This curve is also found to be logarithmic and one finds

$$\frac{U - U_0}{u_*} = 2.5 \ln\left(\frac{z}{R}\right). \tag{10.112}$$

Agreement between theory and experiment is good (Tennekes and Lumley 1972). We can integrate (10.112) over the cross-section to get the mean speed $\bar{u}$; we find, ignoring the thin viscous and buffer layers near the wall, where the logarithmic law does not hold,

$$\bar{u} = U_0 - 3.75 u_*. \tag{10.113}$$

In the intermediate range, (10.108) and (10.111) must overlap, and hence, from (10.108) with $z = R$,

$$\frac{\bar{u}}{u_*} = 2.5 \ln\left(\frac{R u_*}{\nu}\right) + 1.75. \tag{10.114}$$

Now from (10.110) $\frac{1}{2}f = (u_*/\bar{u})^2$, hence (10.114) becomes

$$\frac{1}{\sqrt{f}} = 4.07 \log_{10}[(Re)\sqrt{f}] - 0.60. \tag{10.115}$$

This is close to the experimentally-observed line

$$\frac{1}{\sqrt{f}} = 4.0 \log_{10}[(Re)\sqrt{f}] - 0.4. \tag{10.116}$$

(If we had chosen $A = 2.46$ in (10.112), we would have factors of 4.0 and -0.5 in 10.115.) The analysis can be repeated with the law of the wall in the form

$$\frac{U}{u_*} = A \ln\left(\frac{z u_*}{\nu}\right) + B, \tag{10.117}$$

and the velocity defect law (10.112) written as

$$\frac{U}{u_*} = A \ln\left(\frac{z}{R}\right) + B'. \tag{10.118}$$

We then find

$$\frac{\bar{u}}{u_*} = B' - 1.5A = A \ln\frac{R u_*}{\nu} + B - 1.5A, \tag{10.119}$$

or

$$\frac{1}{\sqrt{f}} = \frac{1}{\sqrt{2}}\{A \ln[(Re)\sqrt{f}] + B - A[1.5 + \ln(2\sqrt{2})]\}. \tag{10.120}$$

Thus, from the logarithmic velocity profile the von Kármán form (10.115) is recovered. Because of the approximations made there are minor differences in the constants but these are often within experimental variability, and we shall accept the classical view of the turbulent scaling laws given above.

10.9.2 Power-law fluids

We now consider how non-Newtonian fluid properties affect turbulent flow. Because a high Reynolds number is usually necessary to get a turbulent flow (hence we have excluded low Reynolds number instabilities) inertia forces are considerably larger than viscous forces over nearly all of the flow. Near the walls the viscous properties are important and one can readily modify the standard Newtonian correlations by using a modified Reynolds number. For the viscous sub-layer we found $u_+ = z_+$ for the Newtonian case. In the power-law case, for example, we would replace (10.102a) by

$$\frac{k}{\rho}\left(\frac{\mathrm{d}U}{\mathrm{d}z}\right)^n = \frac{\tau_{\mathrm{w}}}{\rho} = u_*^2, \tag{10.121}$$

or

$$u_+ = \frac{U}{u_*} = \left\{\frac{\rho z^n}{k}u_*^{2-n}\right\}^{1/n} = z_+^{1/n}, \tag{10.122}$$

which defines z_+ in this case.

For the inertial sub-layer we would again find eqn (10.108)

$$U_+ = A_n \ln z_+ + B_n, \tag{10.123}$$

where the constants A_n and B_n depend only on n; the argument leading to this form was independent of the viscous forces.

Given the expression (10.123) for the mean velocity one can find the total average bulk velocity $\bar{u}$ over the whole tube, and then find (see Section 10.9.1) the Kármán-type relationship derived by Dodge and Metzner (1962)

$$\frac{1}{\sqrt{f}} = \left(\frac{4.0}{n^{0.75}}\right) \log_{10}[(Re)_{\mathrm{c}} f^{1-n/2}] - 0.4/n^{1.2}, \tag{10.124}$$

where f is the friction factor and

$$(Re)_{\mathrm{c}} = \rho D^n \bar{u}^{2-n}/8^{n-1}k' \tag{10.125}$$

with the modified power-law consistency

$$k' = k[(3n+1)/4n]^n \tag{10.126}$$

The modified Reynolds number $(Re)_c$ was introduced by Metzner and Reed (1955) and has the advantage that in the laminar flow region the Newtonian rule $f = 16/(Re)_c$ is recovered; when $n = 1$ (10.124) reduces to the familiar Newtonian correlation. It is not the only possible definition of Reynolds number (Cho and Hartnett 1982); for example, one might use the viscosity at the wall to form a Reynolds number, or even the solvent viscosity, in, for example dilute solutions. The correlation (10.124) seems to be adequate (± 3 per cent) for inelastic solutions in turbulent flow (Wilkinson 1960). For fluids not obeying the power-law relation, one may use values of n and k appropriate to the wall shear stress. Extension of this work to rough pipes is also available (Wilkinson 1960). Onset of turbulence in all cases seems to occur at $(Re)_c \sim 2-4000$, not very different from the Newtonian value (Wilkinson 1960).

10.9.3   Drag-reducing fluids

Addition of quite small amounts (typically 5–100 weight parts per million) of long-chain polymer to turbulent flows in a tube can reduce the pressure-loss dramatically (Virk 1975). This phenomenon was discovered in the 1940s and an account is given by Toms (1977) who published the first paper on the subject in 1949; see Tanner and Walters (1998) for an historical account of drag reduction.

In order to describe the physical changes in the flow associated with polymer drag-reduction, comparisons with ordinary turbulent flow will be made. For flow in pipes the solid lines in Fig. 10.10 give the dependence of f on Reynolds number, (Re), in laminar flow and turbulent flow for smooth walls. For ordinary fluids the transition to turbulence is almost always complete at $(Re) = 4000$, unless special efforts are made to remove disturbances, and the turbulent friction factor for smooth pipes is related to the Reynolds number by the semi-empirical von Kármán equation (10.116).

Drag-reduction is indicated by a reduction of the friction factor such that the pressure drop for a solution at the same flow rate is lower than that for the solvent itself. For polymer solutions, friction factors are shown by the dotted lines in Fig. 10.10 which indicates typical behaviour for reduction of friction factor for various pipe diameters. If negligible viscosity increase occurs when the polymer is dissolved in the fluid, the 'degree of drag-reduction' is given by one minus the ratio of solution to solvent friction factors at the same Reynolds number. However, with a viscosity change, the comparison must be made at the 'solvent Reynolds number', the Reynolds number computed using the solvent viscosity.

The degree of drag-reduction is a function of both flow and fluid variables. Flow variables are flow rate, pipe diameter, and roughness. Fluid variables are viscosity, density, and viscoelastic properties. The viscosity and viscoelastic properties are functions of the polymer type and concentration, and of the interaction between the polymer and the fluid (solvent). Velocity profiles for turbulent flow in pipes are as shown by the solid line in Fig. 10.11 from the buffer region to the core region. Each profile has a slight hump above the straight line

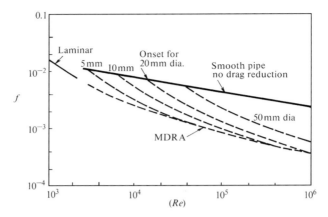

Fig. 10.10 Typical friction factor behaviour of a drag-reducing polymer solution showing effect of pipe diameter on onset and maximum drag reduction asymptote (MDRA).

correlation. The hump always occurs at the pipe centre and is, therefore, at a value of z_+ which depends on the flow variables.

The buffer region is the transition from the viscous sublayer on the wall to the turbulent region, represented by the straight line. The equation for the turbulent region is, in the Newtonian case

$$U_+ = 2.5 \ln z_+ + 5.5. \tag{10.127}$$

Velocity profiles for drag-reduced flow of polymer solutions are shown by the dotted lines in Fig. 10.11. These lines correspond to the dotted lines in Fig. 10.10 of the same pipe diameter and at the indicated Reynolds number. It is evident that the degree of deviation from the ordinary fluid velocity profile increases as the degree of drag-reduction increases. At a low degree of drag-reduction the velocity profile curves seem only to be displaced above the solid curve, but at a high degree of drag-reduction the curves show an extension of the buffer layer until a limit [Virk's (1975) drag-reduction asymptote] seems to be reached. As the level of drag reduction increases, the buffer layer increases in thickness for most observed cases. This is considered to be a result of the viscoelastic alteration of the turbulence near the wall.

The movement of fluid in turbulent flow was previously thought of as a completely chaotic motion, best described by statistical quantities and a stochastic theory. Research has shown, however, that organized events of a describable nature are present in the seemingly chaotic fluid movements. Some of these events are frequently called 'bursts', but are better described as a sequence of processes all of which contribute to a burst. In a burst there is a sudden uplifting of low velocity fluid near the wall into the main stream; see Achia and

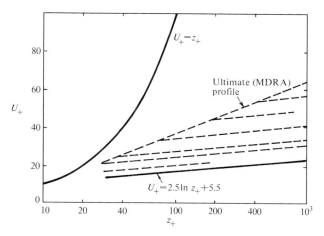

Fig. 10.11 Typical velocity profiles corresponding to results of Fig. 10.10. As one moves up from the Newtonian line one sees increasing drag-reduction without change of slope in the core region. Ultimate maximum drag-reducing profile is also shown.

Thompson (1977) and Tiederman *et al.* (1985) for a discussion of bursts in drag-reducing flow.

The temporal and spatial distribution of the turbulent bursts have been widely studied. It has been found that the non-dimensional frequency of bursts at a given location is about 80 to 100 when expressed as $t_B u_*^2/\nu$ where t_B is the time between bursts. The lateral spacing of bursts is close to $y_+ = 100$, where $y_+ = yu_*/\nu$. For cases of drag-reducing flow of polymer solutions, evidence shows the rate of turbulent bursts decreasing with drag reduction, and lateral burst spacing increasing with drag-reduction (see Achia and Thompson 1977 and Tiederman *et al.* 1985).

The burst rate and spacing findings suggest that the intensity of turbulence defined as r.m.s. (root mean square) velocity fluctuation divided by a reference velocity, is generally reduced for drag-reduction. However, other evidence (Virk 1975) shows that the changes taking place near the wall are more in the nature of a 'decoupling' of axial and radial turbulent flow components rather than a simple reduction in intensity.

One of the most important phenomenological observations for drag-reduction is the maximum drag-reduction asymptote (MDRA), (Virk 1975). The MDRA is given by:

$$\frac{1}{\sqrt{f}} = 19.0 \log_{10}(Re)\sqrt{f} - 32.4. \qquad (10.128)$$

The corresponding velocity profile is

$$U_+ = 11.7 \ln z_+ - 17.0. \qquad (10.129)$$

The MDRA effectively gives the minimum friction factor one expects for a given Reynolds number. Few observations of lower friction factors have been reported, and many of these could be explained by normal experimental error. However, Berman (1977) has queried the idea of a maximum drag-reduction by pointing out that at very high concentrations the buffer layer can continue to expand and it may ultimately reach the centreline. Then the flow will consist of large eddies which do not resemble a normal turbulent flow. Under these conditions one does not expect a natural limit to the drag-reduction. Nevertheless, these cases are regarded as abnormal and for more dilute solutions the MDRA is a useful concept.

Another observation made by Virk and others is that there is an onset wall shear stress, i.e., a lowest shear stress at which perceptible drag-reduction occurs. For a given polymer the onset wall shear stress is essentially the same for pipes of various diameters. Virk found that for polyethylene oxide solutions in water and benzene the following correlation holds:

$$\tau_{\mathrm{w}}^* = 4.4 \times 10^6 / R_G^3 \tag{10.130}$$

where τ_{w}^* is the onset shear stress, R_G is the r.m.s. radius of the polymer molecule (see Chapter 5) and the constant has units of $(\mathrm{nm})^3$ $\mathrm{N/m^2}$. For several other randomly-coiling polymers in solution, the constant in eqn (10.130) was 13×10^6 $(\mathrm{nm})^3$ $\mathrm{N/m^2}$.

The velocity profile in turbulent flow is directly related to the friction factor, both for ordinary and drag-reducing fluids. A simple way to quantify the effect of drag-reduction on velocity profiles for moderate amounts of drag-reduction (not close to the MDRA) is by use of a modified equation. We incorporate a modification of the intercept constant, indicating a line shifted upward as follows:

$$U_+ = 2.5 \ln z_+ + 5.5 + \Delta B. \tag{10.131}$$

The value of ΔB depends on the degree of drag-reduction which occurs. As the MDRA is approached, the velocity profiles plotted as U_+ versus $\ln z_+$ become curved, and have increased slope near the viscous sub-layer, so the above approach is inadequate; Virk (1975) has given the velocity profile for MDRA flow as $U_+ = 11.7 \ln z_+ - 17.0$.

Values of ΔB for the velocity profiles shown in Fig. 10.11 can be found. For any given velocity profile of the logarithmic form $U_+ = A \ln z_+ + B$, the constants of the friction factor equation, $1/\sqrt{f} = C \log_{10}(Re)\sqrt{f} + D$, are given approximately by:

$$\begin{bmatrix} C \\ D \end{bmatrix} = \begin{bmatrix} 1.626 & 0 \\ -1.796 & 0.707 \end{bmatrix} \begin{bmatrix} A \\ B \end{bmatrix}. \tag{10.132}$$

The converse transformation is:

$$\begin{bmatrix} A \\ B \end{bmatrix} = \begin{bmatrix} 0.615 & 0 \\ 1.562 & 1.414 \end{bmatrix} \begin{bmatrix} C \\ D \end{bmatrix}. \tag{10.133}$$

From the above it can be seen that $\Delta B = B - 5.5$.

If the transformation of eqn (10.133) is used, however, the resulting velocity profile equation does not correspond to experimental data. The slope A increases significantly in drag-reducing cases if determined by (10.133), but experimentally it remains about equal to the 2.5 of the Newtonian case in the core-flow region. Values of ΔB may be computed from the following relationship from Virk (1975):

$$\Delta B = (\Delta C / 1.626) \ln [(Re)f^{1/2}/(Re)^* f^{*1/2}]. \tag{10.134}$$

The computation of ΔC, $(Re)^*$ and f^* are explained in the following passages; $(Re)^*$ and f^* are onset-of-drag-reduction parameters.

Virk has correlated what he calls the 'slope increment' for drag-reduction with polymer concentration and molecular weight for a given polymer. The 'slope increment' is the increase in slope of the $1/\sqrt{f}$ ordinate on a plot of $1/\sqrt{f}$ vs. $\log_{10}(Re)\sqrt{f}$ for a drag-reducer over the slope for an ordinary fluid. If the equation $1/\sqrt{f} = C \log_{10}(Re)\sqrt{f} + D$ applies, then the slope increment, ΔC, is $C - C_{\text{Newt}}$. The slope increment correlates as follows:

$$\Delta C = 70 \times 10^{-6} \left(\frac{c}{M}\right)^{1/2} \left(\frac{ML}{M_{\text{mono}}}\right)^{3/2}, \tag{10.135}$$

where c is the polymer concentration in weight parts per million, M is the average molecular weight, L is the number of chain links per monomer unit, and M_{mono} is the monomer unit molecular weight.

As discussed above, the onset of drag-reduction may be correlated for many polymers. Through use of the onset shear stress, τ_{w}^*, which is related to the onset friction factor as follows:

$$f^* = 2\tau_{\text{w}}^*/\rho\bar{u}^2 \tag{10.135a}$$

it is possible to determine the equation for the friction factor as a function of the Reynolds number for drag-reduction by many polymers. The result is:

$$\frac{1}{\sqrt{f}} = \left[4.0 + 70 \times 10^{-6} \left(\frac{c}{M}\right)^{1/2} \left(\frac{ML}{M_{\text{mono}}}\right)^{3/2}\right]$$

$$\times [\log_{10}(Re)\sqrt{f} - (1/\sqrt{f^*} + 0.4)/4.0] + 1/\sqrt{f^*}, \tag{10.136}$$

where f^* is given by eqn (10.135a) and τ_{w}^* is given by eqn (10.130).

The detailed mechanism of drag-reduction remains elusive; some progress on the Newtonian problem was given by Perry and Chong (1982) who consider a vortex-stretching model of the turbulent boundary layer; a less refined, but similar idea was applied to drag-reduction by Black (1969). Drag-reduction has also been considered by Goldshtik *et al.* (1982), who give a theoretical development for the Maxwell fluid. In this paper there is a recognition of two length-scales—the viscous scale ν/u_* and the elastic scale λu_*, where λ is the relaxation time. There are some arbitrary assumptions in the paper (for example, the assumption of maximum stability of the turbulent flow) but it does give the right trends in behaviour.

McComb and Rabie (1982) have considered the drag-reduction resulting from injecting polymer solutions into the wall region; thus a gradient of polymer concentration exists in the flow. They were able to demonstrate that the region $15 \le z_+ \le 100$ is the important region for initiating drag-reduction. Sellin and Ollis (1983) show the difficulty of scaling up drag-reducing turbulent pipe flows from one pipe size to another. Further aspects of turbulent flow behaviour in various geometries are reviewed by Sellin *et al.* (1982), Gyr and Bewersdorff (1995) and Hoyt (1999).

A review of heat transfer for turbulent tube and channel flow is given by Cho and Hartnett (1982). Most analyses have followed a similar path to that of the Newtonian case.

The turbulent heat transfer performance of drag-reducing viscoelastic fluids flowing through circular tubes is characterized by thermal entrance lengths as long as 400–500 pipe diameters (Cho and Hartnett 1982). This is in contrast to Newtonian fluids in turbulent flow, which have thermal entrance lengths of the order of 10–15 pipe diameters. Many of the early investigators were not aware of this critical fact, and accordingly most of the experimental turbulent heat-transfer results for viscoelastic fluids involved relatively short tubes. In such circumstances, the experimental data are in the thermal entrance region. See also Metzner and Friend (1959), and Wilkinson (1960).

10.9.4   *Computer simulation of drag reduction*

Dimitropoulos *et al.* (1998) performed a direct numerical simulation of drag reducing flow using a FENE-P constitutive model [eqn (5.97)] and a Giesekus model (eqn (5.111)). They investigated a plane channel flow, but with three-dimensional time-dependent motion. The friction velocity $u_* = \sqrt{\tau_w/\rho}$ is used as a velocity scale; ν_0/u_* is taken as a length scale and ν_0/u_*^2 as a time scale; ν_0 is the kinematic viscosity (Chapter 9) of the resting fluid. The extra (polymeric) stresses are made dimensionless using $(1 - \beta)\tau_w$, where β is the ratio of the solvent viscosity (η_s) to the total solution viscosity at zero shear rate (η_0). The Reynolds number is defined as $(Re) = hu_*/\nu_0$, where h is the channel half-width and the Weissenberg number $(Wi) = \lambda u_*^2/\nu_0$. The computation was stabilized by adding a stress diffusion term to the constitutive model, which is assumed, from previous work, to have only a small effect. However, this may need further investigation, since the results show a reduction of the streamwise vorticity even for a $\beta\,(=\eta_s/\eta_0)$ value of 0.98874. A critical (Wi) was found for the onset of drag reduction, somewhere in the range 12–25. The mean flow profiles $\{(u/u_*)$ vs. $yu_*/\nu_0 = y^+\}$ show little deviation from the Newtonian results, even though the fluid model with $\beta = 0.9$ is highly non-Newtonian. It was concluded that the extensional viscosity was very important in drag reduction, as we have mentioned in previous sections. The numerical experiments used two β values (0.9 and 0.98874) with different dumbbell maximum lengths, respectively 10 and 30 units, so that the elongational viscosities were the same. The computed drag reductions were about 44 per cent and 10 per cent respectively, and the important role of elongational viscosity was confirmed. Clearly, this complex and difficult study does not contradict, but rather complements, the experimental work on drag reduction.

References

Achia, B. U. and Thompson, D. W. (1977) *J. Fluid Mech.*, **81**, 439.

Avagliano, A. and Phan-Thien, N. (1996). *J. Fluid Mech.*, **271**, 173.

Avgousti, M. and Beris, A.N. (1993). *J. Non-Newtonian Fluid Mech.*, **50**, 225.

Bailey, B. J. (1969). *Nature, Lond.*, **222**, 373.

Beard, D. W., Davies, M. H., and Walters, K. (1966). *J. Fluid Mech.*, **24**, 321.

Berman, N. S. (1977). *Ann. Rev. Fluid Mech.*, **10**, 47.

Black, T. J. (1969). In *Viscous drag reduction* (ed. C. S. Wells), p. 383. Plenum Press, New York.

Boudreaux, E. and Cuculo, J. A. (1978). *J. macromol. Sci., Rev. macromol. Chem., C.*, **16**(1), 39.

Chandrasekhar, S. (1961). *Hydrodynamic and hydromagnetic stability*. Clarendon Press, Oxford.

Chang, J. C. and Denn, M. M. (1980). In *Rheology*, Vol. 3 (ed. G. Astarita, G. Marruci, and L. Nicolais), p. 9. Plenum Press, New York.

Chang, J. C., Denn, M. M., and Geyling, F. T. (1981). *Ind. Eng. Chem. Fund.*, **20**, 147.

Cho, Y. I. and Hartnett, J. P. (1982). *Adv. Heat Transfer*, **19**, 60.

Coleman, B. D. and Gurtin, M. E. (1968). *J. Fluid Mech.*, **33**, 165.

Datta, S. K. (1964). *Phys. Fluids*, **7**, 1915.

De Kee, D. and Wissbrun, K. F. (1988). *Phys. Today*, **24**, June issue.

Den Doelder, C. F. J., Koopmans, R. J., and Molenaar, J. (1998). *J. Non-Newtonian Fluid Mech.*, **79**, 503.

Denn, M. M. (1975). *Stability of reaction and transport processes*. Prentice-Hall, Englewood Cliffs, NJ.

Denn, M. M. (1980). *Ann. Rev. Fluid Mech.*, **12**, 365.

Denn, M. M. (1990). *Ann. Rev. Fluid Mech.*, **22**, 13.

Denn, M. M. and Pearson, J. R. A. (1981). In *Proc. 2nd World Congr. Chem. Engrs*, p. vi–354.

Denn, M. M. and Roisman, J. J. (1969). *Am. Inst. Chem. Engrs J.*, **15**, 454.

Denn, M. M., Sun, Z.-S., and Rushton, B. D. (1971). *Trans. Soc. Rheol.*, **15**, 415.

Dimitropoulos, C. D., Sureshkumar, R., and Beris, A. N. (1998). *J. Non-Newtonian Fluid Mech.*, **79**, 433.

Dodge, D. W. and Metzner, A. B. (1962). *Am. Inst. Chem. Engrs J.*, **18**, 143.

Drazin, P. G. and Reid, W. H. (1981). *Hydrodynamic stability*. Cambridge University Press.

El Kissi, N., Piau, J.-M., and Toussaint, F. (1997). *J. Non-Newtonian Fluid Mech.*, **68**, 271.

Ganpule, H. K. and Khomami, B. (1999). *J. Non-Newtonian Fluid Mech.*, **80**, 217.

Ghanta, V. G., Riise, B. L., and Denn, M. M. (1999). *J. Rheol.*, **43**, 435.

Giesekus, H. (1966). *Rheol. Acta*, **5**, 29.

Goddard, J. D. (1979). *Adv. Appl. Mech.*, **19**, 143.

Goldshtik, M. A., Zametalin, V. V., and Shtern, V. N. (1982). *J. Fluid Mech.*, **119**, 423.

Gorodtsov, V. A. and Leonov, A. I. (1966). *PMM*, **31**, 310.

Guillope, C. and Saut, J. C. (1990). *Nonlin. Anal.*, **15**, 8.

Gupta, G. K., Schultz, W. W., Arruda, E. M., and Lu, X. (1996). *Rheol. Acta*, **35**, 384.

Gupta, R. K. and Ballman, R. L. (1982). *Chem. Eng. Commun.*, **14**, 23.

Gyr, A. and Bewersdorff, H. W. (1995). *Drag Reduction in turbulent flow by additives*. Kluwer, London.

Ho, T.-C. and Denn, M. M. (1978). *J. Non-Newtonian Fluid Mech.*, **3**, 179.

Hoyt, J. W. (1999). In *Advances in the flow and rheology of non-Newtonian fluids* (ed. D. A. Siginer, D. de Kee, and R. P. Chhabra) p. 797. Elsevier, Amsterdam.

Huilgol, R. R., and Phan-Thien, N. (1997). *Fluid mechanics of viscoelasticity*. Elsevier, Amsterdam.

Hutton, J. F. (1965). *Proc. Roy. Soc.*, **A 287**, 222.

Ishihara, H. and Kase, S. (1975). *J. Appl. Polym. Sci.*, **19**, 557.

Ishihara, H. and Kase, S. (1976). *J. Appl. Polym. Sci.*, **20**, 169.

Jackson, K. P., Walters, K., and Williams, R. W. (1984). *J. Non-Newtonian Fluid Mech.*, **14**, 173.

Jones, W. M., Marshall, D. E., and Walker, P. C. (1976*a*). *J. Phys.*, **D9**, 735.

Jones, W. M., Thomas, A. H., and Thomas, M. C. (1976*b*). *J. Phys.*, **D9**, 1967.

Karlsson, S. K. F., Sokolov, M., and Tanner, R. I. (1971). *Drag reduction, Am. Inst. Chem. Engrs. Symposium Series*, **67**, 11.

Kase, S. and Denn, M. M. (1978). In *Proc. 1978 Joint Automatic Control Conf.*, **2**, 71.

Keentok, M. and Xue, S.-C. (1999). *Rheol. Acta*, **38**, 321.

Lagnado, R., Phan-Thien, N., and Leal, G. (1984). *Phys. Fluids*, **27**, 1094.

Lagnado, R., Phan-Thien, N., and Leal, L. G. (1985). *J. Non-Newtonian Fluid Mech.*, **18**, 25.

Larson, R. G. (1992). *Rheol. Acta*, **31**, 213.

Larson, R. G., Shaqfeh, E. S. G., and Muller, S. J. (1990). *J. Fluid Mech.*, **218**, 573.

Lee, C. S., Tripp, B. C., and Magda, J. J. (1992). *Rheol. Acta*, **31**, 306.

Legrand, F., Piau, J. M., and Hervet, H. (1998). *J. Rheol.*, **42**, 1389.

Lockett, F. J. and Rivlin, R. S. (1968). *J. Mech.*, **7**, 475.

Magda, J. J. and Larson, R. G. (1988). *J. Non-Newtonian Fluid Mech.*, **30**, 1.

McComb, W. D. and Rabie, L. H. (1982). *Am. Inst. Chem Engrs, J.*, **28**, 547.

McKinley, G. H., Byars, J. A., Brown, R. A., and Armstrong, R. C. (1991). *J. Non-Newtonian Fluid Mech.*, **40**, 20.

McKinley, G. H., Pakdel, P., and Öztekin, A. (1996). *J. Non-Newtonian Fluid Mech.*, **67**, 19.

Metzner, A. B. and Friend, P. S. (1959). *Ind. Eng. Chem. J.*, **51**, 879.

Metzner, A. B. and Reed, J. C. (1955). *Am. Inst. Chem Engrs, J.*, **1**, 434.

Miller, C. and Goddard, J. D. (1967). *University of Michigan Tech. Report* (NASA Grant, NSG-659).

Muller, S. J., Larson, R. G., and Shaqfeh, E. S. G. (1993). *J. Non-Newtonian Fluid Mech.*, **46**, 315.

Pearson, J. R. A. (1976). *Ann. Rev. Fluid Mech.*, **8**, 163.

Perry, A. E. and Chong, M. S. (1982). *J. Fluid Mech.*, **119**, 173.

Petrie, C. J. S. and Denn, M. M. (1976). *Am. Inst. Chem Engrs J.*, **22**, 209.

Phan-Thien, N. (1983). *J. Non-Newtonian Fluid Mech.*, **13**, 325.

Phan-Thien, N. (1985). *J. Non-Newtonian Fluid Mech.*, **17**, 37.

Porteous, R. and Denn, M. M. (1972). *Trans. Soc. Rheol.*, **16**, 295.

Ramamurthy, A. V. (1986). *J. Rheol.*, **30**, 337.

Rao, P. B. B. (1964). *Appl. Sci. Res., A* **14**, 199.

Renardy, M. (1992). *Euro. J. Mech., B*, **11**, 511.

Renardy, Y. (1988). *J. Non-Newtonian Fluid Mech.*, **28**, 99.

Rothenberger, R., McCoy, D. H., and Denn, M. M. (1973). *Trans. Soc. Rheol.*, **17**, 259.

Schultz, W. W., Gupta, G. K., Arruda, E. M., and Lu, X. (1996). In *Proc. Hell. Soc. Rheol.*, Cyprus, p. 15.

Sellin, R. H. J., Hoyt, J. W., and Scrivener, O. (1982). *J. Hydraulic Res.*, **20**, 29.

Sellin, R. H. J., Ollis, M. (1983). *I. and E. C. Product Res. Developm.*, **22**, 445.

Shaqfeh, E. S. G. (1996). *Ann. Rev. Fluid Mech.*, **28**, 129.

Sokolov, M. and Tanner, R. I. (1972). *Phys. Fluids*, **15**, 534.

Sun, Z.-S. and Denn, M. M. (1972). *Am. Inst. Chem Engrs J.*, **18**, 1010.

Swinney, H. L. and Gollub, J. B. (eds) (1981). *Hydrodynamic instabilities and the transition to turbulence.* Springer-Verlag, New York.

Tanner, R. I. and Keentok, M. (1983). *J. Rheol.*, **27**, 47.

Tanner, R. I. and Simmons, J. M. (1967). *Chem. Eng. Sci.*, **22**, 1079.

Tanner, R. I. and Walters, K. (1998). *Rheology: an historical prespective.* Elsevier, Amsterdam.

Tennekes, H. and Lumley, J. L. (1972). *A first course in turbulence.* MIT Press, Cambridge, Mass.

Tiederman, W. G., Luchik, T. S., and Bogard, D. G. (1985). *J. Fluid Mech.*, **156**, 419.

Toms, B. A. (1977). *Phys. Fluids*, **20**, 53.

Tordella, J. P. (1969). In *Rheology*, Vol. 5 (ed. F. R. Eirich), Ch. 2. Academic Press, New York.

Townsend, A. A. (1980). *The structure of turbulent shear flow*, 2nd edn. Cambridge University Press.

Vest, C. M. and Arpaci, V. S. (1969). *J. Fluid Mech.*, **36**, 316.

Virk, P. S. (1975). *Am. Inst. Chem Engrs J.*, **21**, 625.

Wang, S. Q. (1998). *Adv. Polymer Sci.*, **138**, 227.

Wilkinson, W. L. (1960). *Non-newtonian fluids.* Pregamon Press, Oxford.

Yang, X., Ishida, H., and Wang, S.-Q. (1998). *J. Rheol.*, **42**, 163.

Yi, M. K. and Kim, C. (1997). *J. Non-Newtonian Fluid Mech.*, **72**, 113.

Problems

1. Consider a rigid-body motion of a Newtonian fluid. Show, by considering the kinetic energy of the motion, that a small disturbance will die away.

2. Do an analysis of the classical Taylor problem for the second-order fluid model. Show that the results are a special case of those given in Table 10.2.

3. Show that the inviscid Couette problem produces the equation

$$\frac{\mathrm{d}^2\phi}{\mathrm{d}y^2} + l^2\left[-1 + \frac{2}{\sigma^2}\frac{Mh}{R}(1 - My)\right]\phi = 0$$

where $e^{\sigma t}$ is the time-dependent part of the infinitesimal disturbance, $M = 1 - \Omega_2/\Omega_1$, and the radial velocity u is

$$u = e^{\sigma t}e^{ilz}\phi(y)$$

where $y = (r - R_1)/h$.

Show that $\Omega_2 > \Omega_1$ is the necessary and sufficient condition for stability.

4. Sun and Denn (1972) experimented with a Couette flow and found, for $R/h = 32.52$ and a stationary outer-cylinder that the Taylor number went up from 3230 for a Newtonian fluid to 4060 for a non-Newtonian fluid. The critical shear rate in the latter case was $3800\,\mathrm{s}^{-1}$. Assume $\tau = k\dot{\gamma}^{0.85}$, $N_1 = \alpha\dot{\gamma}^{1.04}$ and $N_2 = \beta N_1$, where k, α, and β are constants. Suppose $\tau = 120\,\mathrm{Pa}$ and $N_1 = 4200\,\mathrm{Pa}$ when $\dot{\gamma} = 4000\,\mathrm{s}^{-1}$. Estimate the value of β needed to explain the stabilizing effect. Comment on the observed wave-number change from 3.10 to 3.22.

5. Find the equations equivalent to (10.99) in a fully-developed turbulent pipe flow.

6. Using the logarithmic form (10.108) for the velocity distribution in the entire pipe, find a relation for the friction coefficient f similar to (10.115) in terms of A and B.

7. (i) Examine the stability of the rest state of the first-order, single-integral fluid

$$\boldsymbol{\tau} = \alpha\mathbf{A}_1 + \int_0^\infty G(s)\mathbf{A}_1(t-s)\,\mathrm{d}s,\text{ where } \mathbf{A}_1 \text{ is the first Rivlin–Ericksen tensor.}$$

(ii) Let the base motion be rigid; that is, the velocity field has the form $\mathbf{v} = \omega \times \boldsymbol{x}$, where ω is the constant angular velocity. Examine the stability of this motion for the fluid model above.

APPENDIX
FORMULAS IN CARTESIAN, CYLINDRICAL, AND SPHERICAL COORDINATES

ALTHOUGH there are many possibilities, the bulk of rheological work uses Cartesian and cylindrical (r, θ, z) coordinate systems, a few papers use spherical (r, θ, ϕ) coordinates, and other systems are rare. Spherical systems will usually be most advantageous when the problem in hand has a symmetric, r-only dependence, and this case only will be set out here; for the full spherical polar equations see Bird *et al.* (1987). Cartesian and cylindrical forms are given below; some are difficult to find in the literature.

To begin we nominate unit vectors **i**, **j**, and **k** in the (Cartesian) x, y, and z directions. In the cylindrical (r, θ, z) system the unit vectors are $\mathbf{i}_r$, $\mathbf{i}_\theta$ and **k**. The rates of change of these unit vectors are zero except in the θ-direction, where one has

$$\frac{\mathrm{d}(\mathbf{i}_r)}{\mathrm{d}\theta} = \mathbf{i}_\theta; \quad \frac{\mathrm{d}(\mathbf{i}_\theta)}{\mathrm{d}\theta} = -\mathbf{i}_r. \tag{A1}$$

A1   The gradient operator

The Cartesian gradient operator is

$$\mathbf{\nabla} = \mathbf{i}\frac{\partial}{\partial x} + \mathbf{j}\frac{\partial}{\partial y} + \mathbf{k}\frac{\partial}{\partial z}. \tag{A2}$$

In cylindrical coordinates (r, θ, z) we have (Jeffreys and Jeffreys 1956)

$$\mathbf{\nabla} = \mathbf{i}_r\frac{\partial}{\partial r} + \frac{\mathbf{i}_\theta}{r}\frac{\partial}{\partial \theta} + \mathbf{k}\frac{\partial}{\partial z}. \tag{A3}$$

Conversion from Cartesian to polar coordinates can be effected by using

$$\begin{aligned}
\frac{\partial}{\partial x} &= \cos\theta\frac{\partial}{\partial r} - \frac{\sin\theta}{r}\frac{\partial}{\partial \theta} \\
\frac{\partial}{\partial y} &= \sin\theta\frac{\partial}{\partial r} + \frac{\cos\theta}{r}\frac{\partial}{\partial \theta}.
\end{aligned} \tag{A4}$$

The definition (A3) can be used to find any operator in cylindrical coordinates; one must respect the relations (A1).

A2   Mass conservation

For the incompressible case $\nabla \cdot \mathbf{v} = 0$ ($\partial v_i / \partial x_i = 0$ in component form) and one has, if $\mathbf{v} = u\mathbf{i} + v\mathbf{j} + w\mathbf{k}$,

$$\frac{\partial u}{\partial x} + \frac{\partial v}{\partial y} + \frac{\partial w}{\partial z} = 0, \tag{A5}$$

and if $\mathbf{v} = v_r \mathbf{i}_r + v_\theta \mathbf{i}_\theta + v_z \mathbf{k}$ in the cylindrical case, applying (A3) gives

$$\frac{\partial v_r}{\partial r} + \frac{v_r}{r} + \frac{1}{r}\frac{\partial v_\theta}{\partial \theta} + \frac{\partial v_z}{\partial z} = 0. \tag{A5a}$$

One can compute $\nabla \cdot \nabla \phi (\equiv \partial^2 \phi / \partial x_i \partial x_i)$ using these results.

A3   Acceleration

The acceleration vector $\mathbf{a}$, written $D\mathbf{v}/Dt$, is defined as

$$a_i = \frac{\partial v_i}{\partial t} + v_k \frac{\partial v_i}{\partial x_k}; \text{ or } \mathbf{a} = \frac{\partial \mathbf{v}}{\partial t} + (\mathbf{v} \cdot \nabla)\mathbf{v}. \tag{A6}$$

In Cartesian form

$$
\begin{aligned}
a_x &= \frac{\partial u}{\partial t} + u\frac{\partial u}{\partial x} + v\frac{\partial u}{\partial y} + w\frac{\partial u}{\partial z} \\[2mm]
a_y &= \frac{\partial v}{\partial t} + u\frac{\partial v}{\partial x} + v\frac{\partial v}{\partial y} + w\frac{\partial v}{\partial z} \\[2mm]
a_z &= \frac{\partial w}{\partial t} + u\frac{\partial w}{\partial x} + v\frac{\partial w}{\partial y} + w\frac{\partial w}{\partial z},
\end{aligned}
\tag{A6a}
$$

and one can compute the results in cylindrical coordinates.
   The term $(\mathbf{v} \cdot \nabla)\mathbf{v}$ becomes

$$= \left(v_r \frac{\partial}{\partial r} + \frac{v_\theta}{r}\frac{\partial}{\partial \theta} + v_z \frac{\partial}{\partial z} \right)(\mathbf{i}_r v_r + \mathbf{i}_\theta v_\theta + \mathbf{k} v_z)$$

$$= \mathbf{i}_r \left\{ v_r \frac{\partial v_r}{\partial r} + \frac{v_\theta}{r}\frac{\partial v_r}{\partial \theta} - \frac{v_\theta^2}{r} + v_z \frac{\partial v_r}{\partial z} \right\} + \mathbf{i}_\theta \left\{ v_r \frac{\partial v_\theta}{\partial r} + \frac{v_\theta v_r}{r} + \frac{v_\theta}{r}\frac{\partial v_\theta}{\partial \theta} + v_z \frac{\partial v_\theta}{\partial z} \right\}$$

$$+ \mathbf{k} \left\{ v_r \frac{\partial v_z}{\partial r} + \frac{v_\theta}{r}\frac{\partial v_z}{\partial \theta} + v_z \frac{\partial v_z}{\partial z} \right\}. \tag{A7}$$

A4   Velocity gradient

In this case $\mathbf{L}^T = \nabla \mathbf{v}$, is a tensor. In Cartesian form $L_{ij} = \partial v_i / \partial x_j$, and writing $\mathbf{L}$ as a matrix, one finds

$$
\mathbf{L} = \begin{bmatrix}
\dfrac{\partial u}{\partial x} & \dfrac{\partial u}{\partial y} & \dfrac{\partial u}{\partial z} \\[2mm]
\dfrac{\partial v}{\partial x} & \dfrac{\partial v}{\partial y} & \dfrac{\partial v}{\partial z} \\[2mm]
\dfrac{\partial w}{\partial x} & \dfrac{\partial w}{\partial y} & \dfrac{\partial w}{\partial z}
\end{bmatrix}. \tag{A8}
$$

$\mathbf{L}$ is generally not symmetric. In cylindrical form, the result is conveniently represented as a matrix:

$$
\mathbf{L} = \begin{bmatrix}
\dfrac{\partial v_r}{\partial r} & \dfrac{1}{r}\dfrac{\partial v_r}{\partial \theta} - \dfrac{v_\theta}{r} & \dfrac{\partial v_r}{\partial z} \\[2mm]
\dfrac{\partial v_\theta}{\partial r} & \dfrac{v_r}{r} + \dfrac{1}{r}\dfrac{\partial v_\theta}{\partial \theta} & \dfrac{\partial v_\theta}{\partial z} \\[2mm]
\dfrac{\partial v_z}{\partial r} & \dfrac{1}{r}\dfrac{\partial v_z}{\partial \theta} & \dfrac{\partial v_z}{\partial z}
\end{bmatrix}. \tag{A8a}
$$

or as a dyadic

$$
\begin{aligned}
\mathbf{L} = &\, L_{rr}\mathbf{i}_r\mathbf{i}_r + L_{r\theta}\mathbf{i}_r\mathbf{i}_\theta + L_{rz}\mathbf{i}_r\mathbf{k} + L_{\theta r}\mathbf{i}_\theta\mathbf{i}_r \\
&+ L_{\theta\theta}\mathbf{i}_\theta\mathbf{i}_\theta + L_{\theta z}\mathbf{i}_\theta\mathbf{k} + L_{zr}\mathbf{k}\mathbf{i}_r + L_{z\theta}\mathbf{k}\mathbf{i}_\theta + L_{zz}\mathbf{k}\mathbf{k}.
\end{aligned} \tag{A8b}
$$

The form (A8b) is often convenient when doing further gradient operations.

By taking the symmetric and antisymmetric parts of (A8) we may find the rate of deformation and vorticity tensors; the Rivlin–Ericksen tensor $\mathbf{A}_1 \equiv 2\mathbf{d} \equiv \mathbf{L} + \mathbf{L}^T$. One finds, using $2\mathbf{d} = \mathbf{L} + \mathbf{L}^T$,

$$
\mathbf{d} = \frac{1}{2}\begin{bmatrix}
2\dfrac{\partial u}{\partial x} & \dfrac{\partial u}{\partial y} + \dfrac{\partial v}{\partial x} & \dfrac{\partial u}{\partial z} + \dfrac{\partial w}{\partial x} \\[2mm]
+ & 2\dfrac{\partial v}{\partial y} & \dfrac{\partial v}{\partial z} + \dfrac{\partial w}{\partial y} \\[2mm]
+ & + & 2\dfrac{\partial w}{\partial z}
\end{bmatrix} \tag{A9}
$$

where $+$ denotes a symmetric entry. In cylindrical coordinates

$$
\mathbf{d} = \frac{1}{2}\begin{bmatrix}
2\dfrac{\partial v_r}{\partial r} & \dfrac{\partial v_\theta}{\partial r} + \dfrac{1}{r}\dfrac{\partial v_r}{\partial \theta} - \dfrac{v_\theta}{r} & \dfrac{\partial v_r}{\partial z} + \dfrac{\partial v_z}{\partial r} \\[2mm]
+ & 2\left(\dfrac{v_r}{r} + \dfrac{1}{r}\dfrac{\partial v_\theta}{\partial \theta}\right) & \dfrac{\partial v_\theta}{\partial z} + \dfrac{1}{r}\dfrac{\partial v_z}{\partial \theta} \\[2mm]
+ & + & 2\dfrac{\partial v_z}{\partial z}
\end{bmatrix}. \tag{A9a}
$$

Using $2\mathbf{w} = \mathbf{L} - \mathbf{L}^T$, one has the vorticity tensor:

$$\mathbf{w} = \frac{1}{2} \begin{bmatrix} 0 & \dfrac{\partial u}{\partial y} - \dfrac{\partial v}{\partial x} & \dfrac{\partial u}{\partial z} - \dfrac{\partial w}{\partial x} \\ - & 0 & \dfrac{\partial v}{\partial z} - \dfrac{\partial w}{\partial y} \\ - & - & 0 \end{bmatrix} \tag{A10}$$

where $-$ indicates that there is anti-symmetry about the diagonal. Clearly, $\mathbf{L} = \mathbf{d} + \mathbf{w}$.

A5 Operations on tensors—the UCM equation

Consider a second-order tensor $\mathbf{A}$, not necessarily symmetrical. The value of $\boldsymbol{\nabla} \cdot \mathbf{A}$ ($\equiv \partial A_{kj}/\partial x_k$ in component form) is, written out in Cartesian coordinates:

$$\mathbf{i}\left(\frac{\partial A_{xx}}{\partial x} + \frac{\partial A_{yx}}{\partial y} + \frac{\partial A_{zx}}{\partial z}\right) + \mathbf{j}\left(\frac{\partial A_{xy}}{\partial x} + \frac{\partial A_{yy}}{\partial y} + \frac{\partial A_{zy}}{\partial z}\right)$$
$$+ \mathbf{k}\left(\frac{\partial A_{xz}}{\partial x} + \frac{\partial A_{yz}}{\partial y} + \frac{\partial A_{zz}}{\partial z}\right). \tag{A11}$$

The corresponding cylindrical form is:

$$\boldsymbol{\nabla} \cdot \mathbf{A} = \mathbf{i}_r \left[\frac{1}{r}\frac{\partial}{\partial r}(rA_{rr}) + \frac{1}{r}\frac{\partial A_{\theta r}}{\partial \theta} - \frac{A_{\theta\theta}}{r} + \frac{\partial A_{zr}}{\partial z}\right]$$
$$+ \mathbf{i}_\theta \left[\frac{\partial A_{r\theta}}{\partial r} + \frac{A_{r\theta} + A_{\theta r}}{r} + \frac{1}{r}\frac{\partial A_{\theta\theta}}{\partial \theta} + \frac{\partial A_{z\theta}}{\partial z}\right]$$
$$+ \mathbf{k} \left[\frac{1}{r}\frac{\partial}{\partial r}(rA_{rz}) + \frac{1}{r}\frac{\partial A_{\theta z}}{\partial \theta} + \frac{\partial A_{zz}}{\partial z}\right]. \tag{A12}$$

We can use (A.12) to write the equations of motion (2.60) in cylindrical coordinates if we set $A = \boldsymbol{\sigma}$, the (symmetric) stress tensor:

$$\frac{1}{r}\frac{\partial}{\partial r}(r\sigma_{rr}) + \frac{1}{r}\frac{\partial}{\partial \theta}\sigma_{r\theta} + \frac{\partial}{\partial z}\sigma_{rz} - \frac{\sigma_{\theta\theta}}{r} + \rho f_r = \rho a_r$$

$$\frac{1}{r^2}\frac{\partial}{\partial r}(r^2\sigma_{\theta r}) + \frac{1}{r}\frac{\partial \sigma_{\theta\theta}}{\partial \theta} + \frac{\partial}{\partial z}\sigma_{\theta z} + \rho f_\theta = \rho a_\theta \tag{2.60a}$$

$$\frac{1}{r}\frac{\partial}{\partial r}(r\sigma_{zr}) + \frac{1}{r}\frac{\partial}{\partial \theta}\sigma_{z\theta} + \frac{\partial}{\partial z}\sigma_{zz} + \rho f_z = \rho a_z.$$

The steady-state components of the acceleration vector **a** are given in (A7) and the components of the body-force vector **f** are usually known. Now let **A** be a tensor, not necessarily symmetric, and **v** be a vector (v_r, v_θ, v_z). Define a scalar operator

$$Q \equiv v_r \frac{\partial}{\partial r} + \frac{v_\theta}{r} \frac{\partial}{\partial \theta} + v_z \frac{\partial}{\partial z}.$$

Then $\mathbf{v} \cdot \nabla \mathbf{A}$ in cylindrical polar coordinates (r, θ, z) can be written as

$$\mathbf{v} \cdot \nabla \mathbf{A} = Q\mathbf{A} + \frac{v_\theta}{r} \mathbf{M}, \tag{A.13}$$

where **M** is represented by

$$\mathbf{M} = \begin{bmatrix} -(A_{r\theta} + A_{\theta r}) & A_{rr} - A_{\theta\theta} & -A_{\theta z} \\ A_{rr} - A_{\theta\theta} & A_{r\theta} + A_{\theta r} & A_{rz} \\ -A_{z\theta} & A_{zr} & 0 \end{bmatrix} \tag{A.13a}$$

where $A_{r\theta}$, etc. are the components of **A**.

The $\nabla \cdot (\nabla \mathbf{v})$ ($\equiv \partial^2 v_j / \partial x_i \partial x_i$ in Cartesian components) becomes quite complex in cylindrical coordinates. We find

$$(\nabla \cdot \nabla \mathbf{v})_r = \frac{\partial}{\partial r} \left(\frac{1}{r} \frac{\partial}{\partial r} (rv_r) \right) + \frac{1}{r^2} \frac{\partial^2 v_r}{\partial \theta^2} + \frac{\partial^2 v_r}{\partial z^2} - \frac{2}{r^2} \frac{\partial v_\theta}{\partial \theta}$$

$$(\nabla \cdot \nabla v)_\theta = \frac{\partial}{\partial r} \left(\frac{1}{r} \frac{\partial}{\partial r} (rv_\theta) \right) + \frac{1}{r^2} \frac{\partial^2 v_\theta}{\partial \theta^2} + \frac{\partial^2 v_\theta}{\partial z^2} + \frac{2}{r^2} \frac{\partial v_r}{\partial \theta} \tag{A.14}$$

$$(\nabla \cdot \nabla \mathbf{v})_z = \frac{1}{r} \frac{\partial}{\partial r} \left(r \frac{\partial v_z}{\partial r} \right) + \frac{1}{r^2} \frac{\partial^2 v_z}{\partial \theta^2} + \frac{\partial^2 v_z}{\partial z^2}.$$

These operations arise in the Navier–Stokes viscous components (Problem A2).

By using formulas (A.13) and (A.13a) one can construct the second-order fluid operator, A_2, [Eqn (4.30c)]. Similarly, the Maxwell model (4.25) can be constructed in cylindrical coordinates from eqns (A.13) and (A.13a). We have, for this model, after dividing by λ and setting $\eta/\lambda = g$,

$$\frac{\partial \boldsymbol{\tau}}{\partial t} + (\mathbf{v} \cdot \nabla)\boldsymbol{\tau} - \boldsymbol{\tau}\mathbf{L}^T - \mathbf{L}\boldsymbol{\tau} + \boldsymbol{\tau}/\lambda = 2g\,\mathbf{d}, \tag{4.25}$$

and so the (symmetric) components of (4.25) in cylindrical coordinates become, since $\boldsymbol{\tau}$ is symmetric, those set out in Table A.1.

Table A.1 Upper convected Maxwell model [eqn (4.25)] in cylindrical coordinates

rr-component

$$\frac{\partial \tau_{rr}}{\partial t} + v_r \frac{\partial \tau_{rr}}{\partial r} + \frac{v_\theta}{r}\frac{\partial \tau_{rr}}{\partial \theta} + v_z \frac{\partial \tau_{rr}}{\partial z} - 2\left(\tau_{rr}\frac{\partial v_r}{\partial r} + \frac{\tau_{r\theta}}{r}\frac{\partial v_r}{\partial \theta} + \tau_{rz}\frac{\partial v_r}{\partial z}\right)$$
$$+ \tau_{rr}/\lambda = 2g\frac{\partial v_r}{\partial r}. \tag{A.15}$$

rθ-component

$$\frac{\partial \tau_{r\theta}}{\partial t} + v_r \frac{\partial \tau_{r\theta}}{\partial r} + \frac{v_\theta}{r}\frac{\partial \tau_{r\theta}}{\partial \theta} + v_z \frac{\partial \tau_{r\theta}}{\partial z} - \left(\frac{\tau_{\theta\theta}}{r}\frac{\partial v_r}{\partial \theta} + \tau_{\theta z}\frac{\partial v_r}{\partial z} + \tau_{rr}\frac{\partial v_\theta}{\partial r}\right.$$
$$\left. + \tau_{rz}\frac{\partial v_\theta}{\partial z} - \tau_{r\theta}\frac{\partial v_z}{\partial z} - \frac{v_\theta}{r}\tau_{rr}\right) + \tau_{r\theta}/\lambda = g\left(\frac{\partial v_\theta}{\partial r} + \frac{1}{r}\frac{\partial v_r}{\partial \theta} - \frac{v_\theta}{r}\right). \tag{A.16}$$

rz-component

$$\frac{\partial \tau_{rz}}{\partial t} + v_r \frac{\partial \tau_{rz}}{\partial r} + \frac{v_\theta}{r}\frac{\partial \tau_{rz}}{\partial \theta} + v_z \frac{\partial \tau_{rz}}{\partial z} - \left(\frac{\tau_{\theta z}}{r}\frac{\partial v_r}{\partial \theta} + \tau_{zz}\frac{\partial v_r}{\partial z} + \tau_{rr}\frac{\partial v_z}{\partial r} + \frac{\tau_{r\theta}}{r}\frac{\partial v_z}{\partial \theta}\right.$$
$$\left. - \left(\frac{v_r}{r} + \frac{1}{r}\frac{\partial v_\theta}{\partial \theta}\right)\tau_{rz}\right) + \tau_{rz}/\lambda = g\left(\frac{\partial v_r}{\partial z} + \frac{\partial v_z}{\partial r}\right). \tag{A.17}$$

θθ-component

$$\frac{\partial \tau_{\theta\theta}}{\partial t} + v_r \frac{\partial \tau_{\theta\theta}}{\partial r} + \frac{v_\theta}{r}\frac{\partial \tau_{\theta\theta}}{\partial \theta} + v_z \frac{\partial \tau_{\theta\theta}}{\partial z} - 2\left(\tau_{r\theta}\frac{\partial v_\theta}{\partial r} - \tau_{r\theta}\frac{v_\theta}{r} + \tau_{\theta\theta}\left(\frac{v_r}{r} + \frac{1}{r}\frac{\partial v_\theta}{\partial \theta}\right)\right.$$
$$\left. + \tau_{\theta z}\frac{\partial v_\theta}{\partial z}\right) + \tau_{\theta\theta}/\lambda = 2g\left(\frac{v_r}{r} + \frac{1}{r}\frac{\partial v_\theta}{\partial \theta}\right). \tag{A.18}$$

θz-component

$$\frac{\partial \tau_{\theta z}}{\partial t} + v_r \frac{\partial \tau_{\theta z}}{\partial r} + \frac{v_\theta}{r}\frac{\partial \tau_{\theta z}}{\partial \theta} + v_z \frac{\partial \tau_{\theta z}}{\partial z} - \left(\tau_{rz}\frac{\partial v_\theta}{\partial r} - \tau_{rz}\frac{v_\theta}{r} - \tau_{\theta z}\frac{\partial v_r}{\partial r} + \tau_{zz}\frac{\partial v_\theta}{\partial z}\right.$$
$$\left. + \tau_{r\theta}\frac{\partial v_z}{\partial r} + \frac{\tau_{\theta\theta}}{r}\frac{\partial v_z}{\partial \theta}\right) + \tau_{\theta z}/\lambda = g\left(\frac{\partial v_\theta}{\partial z} + \frac{1}{r}\frac{\partial v_z}{\partial \theta}\right). \tag{A.19}$$

zz-component

$$\frac{\partial \tau_{zz}}{\partial t} + v_r \frac{\partial \tau_{zz}}{\partial r} + \frac{v_\theta}{r}\frac{\partial \tau_{zz}}{\partial \theta} + v_z \frac{\partial \tau_{zz}}{\partial z} - 2\left(\tau_{rz}\frac{\partial v_z}{\partial r} + \frac{\tau_{\theta z}}{r}\frac{\partial v_z}{\partial \theta} + \tau_{zz}\frac{\partial v_z}{\partial z}\right)$$
$$+ \tau_{zz}/\lambda = 2g\frac{\partial v_z}{\partial z}. \tag{A.20}$$

A6 Strain measures

To find the deformation gradient $\mathbf{F}$ (2.27) in cylindrical co-ordinates we can express $d\mathbf{r}$ in the form

$$d\mathbf{r} = dR\,\mathbf{i}_r + R\,d\Theta\,\mathbf{i}_\theta + dZ\,\mathbf{k}, \tag{A.21}$$

where R, Θ, and Z are the components of the coordinate $\mathbf{r}$ in the cylindrical system; R, Θ, and Z are all to be regarded as functions of r, θ, z (the components of $\mathbf{x}$ in the cylindrical system). Hence,

$$dR = \frac{\partial R}{\partial r}dr + \frac{\partial R}{\partial \theta}d\theta + \frac{\partial R}{\partial z}dz \tag{A.22}$$

and similar expressions for $d\Theta$, dZ. So we can write

$$d\mathbf{r} = \begin{bmatrix} dR \\ R\,d\Theta \\ dZ \end{bmatrix} = \begin{bmatrix} \dfrac{\partial R}{\partial r} & \dfrac{1}{r}\dfrac{\partial R}{\partial \theta} & \dfrac{\partial R}{\partial z} \\[2mm] R\dfrac{\partial \Theta}{\partial r} & \dfrac{R}{r}\dfrac{\partial \Theta}{\partial \theta} & R\dfrac{\partial \Theta}{\partial z} \\[2mm] \dfrac{\partial Z}{\partial r} & \dfrac{1}{r}\dfrac{\partial Z}{\partial \theta} & \dfrac{\partial Z}{\partial z} \end{bmatrix} \begin{bmatrix} dr \\ r\,d\theta \\ dz \end{bmatrix} = \mathbf{F}\,d\mathbf{x}. \tag{A.23}$$

By computing $\mathbf{F}^T\mathbf{F}$, we can find the strain tensor components as shown in Table A.2.

Table A.2 Components of $\mathbf{C}$ in cylindrical polar coordinates

$$C_{11} = \left(\frac{\partial R}{\partial r}\right)^2 + R^2\left(\frac{\partial \Theta}{\partial r}\right)^2 + \left(\frac{\partial Z}{\partial r}\right)^2$$

$$C_{12} = C_{21} = \frac{1}{r}\frac{\partial R}{\partial r}\frac{\partial R}{\partial \theta} + \frac{R^2}{r}\frac{\partial \Theta}{\partial r}\frac{\partial \Theta}{\partial \theta} + \frac{1}{r}\frac{\partial Z}{\partial \theta}\frac{\partial Z}{\partial r}$$

$$C_{13} = C_{31} = \frac{\partial R}{\partial r}\frac{\partial R}{\partial z} + R^2\frac{\partial \Theta}{\partial r}\frac{\partial \Theta}{\partial z} + \frac{\partial Z}{\partial r}\frac{\partial Z}{\partial z}$$

$$C_{22} = \frac{1}{r^2}\left(\frac{\partial R}{\partial \theta}\right)^2 + \frac{R^2}{r^2}\left(\frac{\partial \Theta}{\partial \theta}\right)^2 + \frac{1}{r^2}\left(\frac{\partial Z}{\partial \theta}\right)^2$$

$$C_{23} = C_{32} = \frac{1}{r}\frac{\partial R}{\partial \theta}\frac{\partial R}{\partial z} + \frac{R^2}{r}\frac{\partial \Theta}{\partial \theta}\frac{\partial \Theta}{\partial z} + \frac{1}{r}\frac{\partial Z}{\partial \theta}\frac{\partial Z}{\partial z}$$

$$C_{33} = \left(\frac{\partial R}{\partial z}\right)^2 + R^2\left(\frac{\partial \Theta}{\partial z}\right)^2 + \left(\frac{\partial Z}{\partial z}\right)^2$$

From these results the inverse strain tensor $\mathbf{C}^{-1}$ can be found, if required, by matrix inversion.

A7   Spherical coordinates

For the spherical polar system (r, θ, ϕ) $(\theta = 0\text{–}\pi; \quad \phi = 0\text{–}2\pi)$ defined by $x = r \sin \theta \cos \phi$, $y = r \sin \theta \sin \phi$, $z = r \cos \theta$, the $\mathbf{\nabla}$-operator is

$$\mathbf{i}_r \frac{\partial}{\partial r} + \frac{\mathbf{i}_\theta}{r} \frac{\partial}{\partial \theta} + \frac{\mathbf{i}_\phi}{r \sin \theta} \frac{\partial}{\partial \phi},$$

and one has

$$\frac{\partial}{\partial r} (\mathbf{i}_r, \mathbf{i}_\theta, \mathbf{i}_\phi) \text{ equals } (\mathbf{0}, \mathbf{0}, \mathbf{0}),$$

$$\frac{\partial}{\partial \theta} (\mathbf{i}_r, \mathbf{i}_\theta, \mathbf{i}_\phi) = (\mathbf{i}_\theta, -\mathbf{i}_r, \mathbf{0}),$$

and

$$\frac{\partial}{\partial \phi} (\mathbf{i}_r, \mathbf{i}_\theta, \mathbf{i}_\phi) = (\mathbf{i}_\phi \sin \theta, \mathbf{i}_\phi \cos \theta, -\mathbf{i}_r \sin \theta - \mathbf{i}_\theta \cos \theta).$$

For the spherically symmetric case where $\partial/\partial \theta = \partial/\partial \phi = 0$ for all quantities, we get, for a symmetric tensor $\mathbf{B}$ and a vector $\mathbf{u}$

$$\mathbf{\nabla} \cdot \mathbf{u} = \frac{1}{r^2} \frac{\mathrm{d}}{\mathrm{d}r} (r^2 u_r) + \frac{\cot \theta}{r} u_\theta \tag{A.24}$$

$$\begin{aligned} \mathbf{\nabla} \cdot \mathbf{B} = {} & \frac{\mathbf{i}_r}{r^2} \frac{\mathrm{d}}{\mathrm{d}r} (r^2 B_{rr}) - \frac{(B_{\theta\theta} + B_{\phi\phi} - B_{r\theta} \cot \theta)}{r} \mathbf{i}_r \\ & + \frac{\mathbf{i}_\theta}{r^2} \frac{\mathrm{d}}{\mathrm{d}r} (r^3 B_{r\theta}) + \frac{(B_{\theta\theta} - B_{\phi\phi}) \cot \theta}{r} \mathbf{i}_\theta \\ & + \frac{\mathbf{i}_\phi}{r^3} \frac{\mathrm{d}}{\mathrm{d}r} (r^3 B_{r\phi}) + \frac{2 B_{\phi\theta} \cot \theta}{r} \mathbf{i}_\phi. \end{aligned} \tag{A.25}$$

In the case where the velocity vector $\mathbf{v}$ is

$$\mathbf{v} = \mathbf{i}_r v_r(r)$$

one has that the acceleration vector $\mathbf{a}$ is

$$\mathbf{a} = (\mathbf{v} \cdot \mathbf{\nabla})\mathbf{v} = \mathbf{i}_r v_r \frac{\mathrm{d} v_r}{\mathrm{d}r},$$

and the velocity gradient matrix is

$$\mathbf{L} = \mathbf{L}^T = \text{diag}\left[\frac{\mathrm{d}v_r}{\mathrm{d}r}, \frac{v_r}{r}, \frac{v_r}{r}\right]. \tag{A.26}$$

The operation $(\mathbf{v} \cdot \boldsymbol{\nabla}\mathbf{A})$ under these conditions is $v_r\mathrm{d}\mathbf{A}/\mathrm{d}r$ where $\mathbf{A} = \mathbf{A}(r)$ only. The strain tensor for a purely radial displacement which brings a particle from R at time t' to a place r at time t is

$$\mathbf{C} = \text{diag}\left[\left(\frac{\mathrm{d}R}{\mathrm{d}r}\right)^2, \left(\frac{R}{r}\right)^2, \left(\frac{R}{r}\right)^2\right]. \tag{A.27}$$

References

Bird, R. B., Armstrong, R. C., and Hassager, O. (1987). *Dynamics of complex fluids*, Vol. 1, *Fluid mechanics*. Wiley, New York.

Jeffreys, H. and Jeffreys, B. S. (1956). *Methods of mathematical physics*. Cambridge University Press.

Problems

1. Write the energy equation (2.62) in cylindrical coordinates, including the dissipation term.

2. Write out the Navier–Stokes equations in cylindrical coordinates.

3. Compute $\boldsymbol{\nabla} \times \mathbf{v}$ in cylindrical coordinates.

4. For the spherical polar system, develop the conservation of mass equations in the general case.

INDEX

Note: Figures and Tables are indicated by *italic* numbers